Advances in Experimental Medicine and Biology

Volume 1383

Advances in Experimental Medicine and Biology provides a platform for scientific contributions in the main disciplines of the biomedicine and the life sciences. This series publishes thematic volumes on contemporary research in the areas of microbiology, immunology, neurosciences, biochemistry, biomedical engineering, genetics, physiology, and cancer research. Covering emerging topics and techniques in basic and clinical science, it brings together clinicians and researchers from various fields.

Advances in Experimental Medicine and Biology has been publishing exceptional works in the field for over 40 years, and is indexed in SCOPUS, Medline (PubMed), EMBASE, BIOSIS, Reaxys, EMBiology, the Chemical Abstracts Service (CAS), and Pathway Studio.

2021 Impact Factor: 3.650 (no longer indexed in SCIE as of 2022)

Nick J. Spencer • Marcello Costa
Stuart M. Brierley
Editors

The Enteric Nervous System II

Springer

Editors
Nick J. Spencer
College of Medicine & Public Health
Flinders Medical Centre
Adelaide, SA, Australia

Marcello Costa
College of Medicine & Public Health
Flinders Medical Centre
Adelaide, SA, Australia

Stuart M. Brierley
South Australian Health & Medical Research Institute
Adelaide, SA, Australia

ISSN 0065-2598 ISSN 2214-8019 (electronic)
Advances in Experimental Medicine and Biology
ISBN 978-3-031-05842-4 ISBN 978-3-031-05843-1 (eBook)
https://doi.org/10.1007/978-3-031-05843-1

This Springer imprint is published by the registered company Springer Nature Switzerland AG
The registered company address is: Gewerbestrasse 11, 6330 Cham, Switzerland

Preface

In 1983, a number of international scientists working on the enteric nervous system (ENS) gathered at Flinders University in Adelaide, Australia, to discuss the advances and future of their research, marking informally the beginning of the discipline of neurogastreonterology. A second meeting nearly 30 years later was held in 2014, also in Adelaide, which consolidated the tremendous advances made in the field, and the proceedings of the meeting were published in a very successful book entitled: *The Enteric Nervous System 30 Years Later* (Springer), edited by Stuart M. Brierley and Marcello Costa. Springer.

A third one day meeting was organised on April 18, 2021, also in Adelaide, called "The Enteric Nervous System", which was a formal satellite meeting of the Federation of Neurogatroenterology and Motility (FNM 2020). The meeting was held at the same venue as the ENS meeting in 2014, the National Wine Centre of Australia. Both meetings were postponed by 1 year because of the COVID-19 pandemic. The third ENS meeting (held in 2021) consisted of an exciting and dynamic gathering of major players of this rapidly developing field of enteric neuroscience. Most Australian presenters and participants attended personally (see photo) while all overseas speakers and participants could access all pre-recorded talks.

Despite the challenges imposed by COVID-19, we were delighted with the calibre of the presentations and the collegiality and collaborative energy that arose during lunch time and evening discussion sessions. This international meeting represented an excellent cross section of research on the enteric nervous system, with major advances made at an increasing rate since the previous meeting. The meeting was opened by our guest of honour Professor Robin Warren, recipient of the 2005 Nobel prize in Medicine, who was born and educated in Adelaide. The ENS II meeting held in 2014 aimed to identify how far in the field of enteric neuroscience had progressed over 30 years, and where the future was heading with new technological advances. The 2021 meeting addressed many major unresolved mechanistic questions, shaping research for years to come. These proceedings cover a very broad spectrum of research from evolution of the ENS to the most sophisticated methodologies combining genetic engineering, spatio-temporal maps and clinical issues involving microbiota, all well integrated with advances of the fundamental organisation of the enteric circuits.

We are indebted to the professional help of Mr. Danny Brookes for the pre-recordings of presentations, the live presentations and the postproduction that allows participants to access to all presentations.

We are deeply grateful to the Flinders Foundation, the *British Journal of Pharmacology*, South Australian Neuroscience Institute (SANI) and Biosensis, who generously sponsored the meeting.

The organising committee of the meeting were made up of Nick J. Spencer, Stuart M. Brierley and Marcello Costa.

The editors of the proceedings (left to right): Stuart M. Brierley, Nick J. Spencer and Marcello Costa

Adelaide, SA, Australia

Nick J. Spencer
Marcello Costa
Stuart M. Brierley

Contents

Contribution of the Enteric Nervous System to Autoimmune Diseases and Irritable Bowel Syndrome

1

Anita Annaházi and Michael Schemann

Abstract

Anti-neuronal autoantibodies can lead to subacute gastrointestinal dysmotility, presenting with various symptoms typical of intestinal pseudoobstruction, achalasia, gastroparesis, or slow intestinal transit, among others. Such autoantibodies may be produced in response to a remote tumor and accelerate the diagnosis of malignancy, but in other cases they appear without an identifiable underlying cause. One example is the type I anti-neuronal nuclear antibody (ANNA-1 otherwise known as anti-Hu), which is usually linked to small cell–lung carcinoma. Anti-Hu can directly activate enteric neurons and visceral sensory nerve fibers and has a cytotoxic effect. Various other anti-neuronal antibodies have been described, targeting different ion channels or receptors on nerve cells of the central or the enteric nervous system. Autoimmune processes targeting enteric neurons may also play a role in more common disorders such as esophageal achalasia, celiac disease, or multiple sclerosis. Furthermore, anti-enteric neuronal antibodies have been found more abundant in the common functional gastrointestinal disorder, irritable bowel syndrome (IBS), than in controls. The pathogenesis of IBS is very complex, involving the release of various mediators from immune cells in the gut wall. Products of mast cells, such as histamine and tryptase, excite visceral afferents and enteric neurons, which may contribute to symptoms like abdominal pain and disturbed motility. Elevated serine- and cysteine-protease activity in stool of IBS-D and IBS-C patients, respectively, can be a factor leading to leaky gut and visceral hypersensitivity. More knowledge on these mediators in IBS may facilitate the development of novel diagnostic methods or therapies.

Keywords

Autoimmune gastrointestinal dysmotility · Anti-neuronal antibodies · Multiple sclerosis · Irritable bowel syndrome · Proteases · Mast cells

We describe the role of neurons in the enteric nervous system in autoimmune diseases and irritable bowel syndrome (as a graphical summary, see Fig. 1.1).

A. Annaházi · M. Schemann (✉)
Human Biology, Technical University of Munich, Freising, Germany
e-mail: schemann@wzw.tum.de

N. J. Spencer et al. (eds.), *The Enteric Nervous System II*, Advances in Experimental Medicine and Biology 1383, https://doi.org/10.1007/978-3-031-05843-1_1

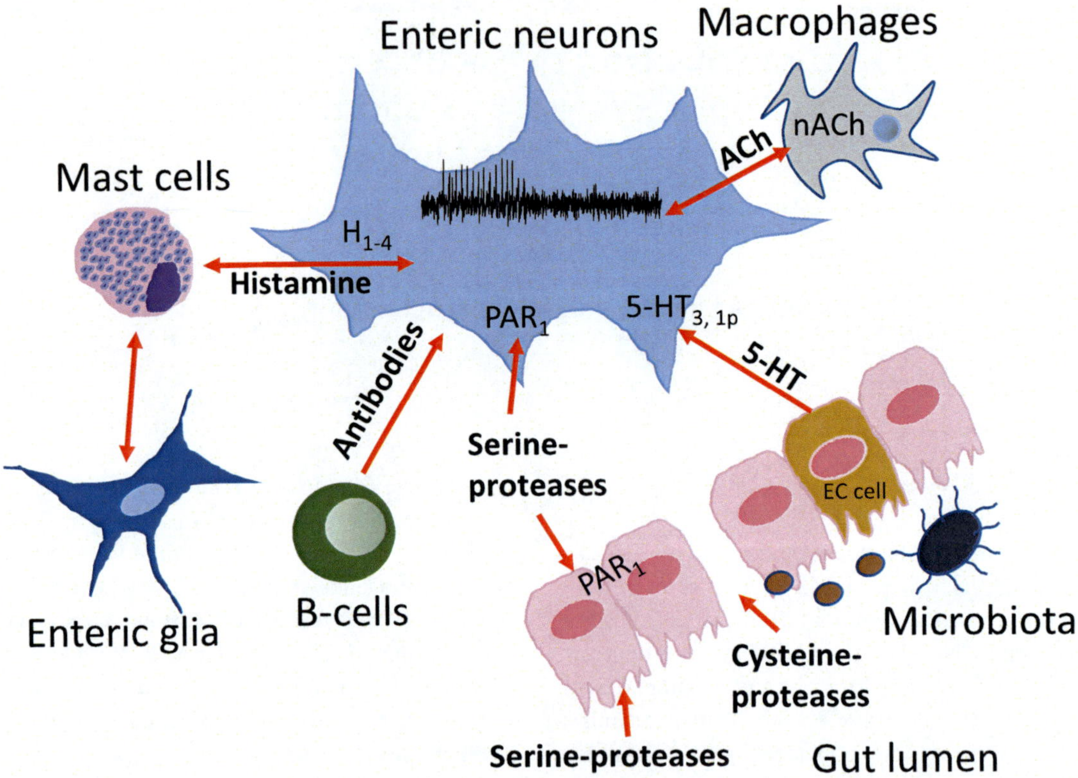

Fig. 1.1 The ENS integrates numerous inputs; the pharmacology applies to human gut. Activation of mast cells excites enteric neurons by release of histamine and proteases, the latter may also be released from nonimmune cells or increased in the gut lumen. Enteric interneurons or EC cells release serotonin, which also activates enteric neurons. Cysteine-proteases can directly disrupt tight junctions, enabling invasion of antigens. Acetylcholine released from enteric neurons modulate macrophages

1.1 Enteric Nervous System and Autoimmune Disease

In the last decades, considerable evidence has been accumulated that autoantibodies targeting neuronal structures play a role not only in diseases of the central nervous system (CNS) but also in several gastrointestinal (GI) disorders. Autoimmune gastrointestinal dysmotility (AGID) is a new clinical entity, which can appear with or without the presence of a remote tumor, such as lung, ovarian, breast, or endometrial cancer [15]. The disease represents a type of autoimmune dysautonomia characterized by subacute gastrointestinal dysmotility, such as intestinal pseudoobstruction, achalasia, gastroparesis, pyloric stenosis, slow intestinal transit, or anal spasm. One of the earliest descriptions of such autoantibodies included the presence of type I antineuronal nuclear antibodies (ANNA-1, also called anti-Hu) in small cell lung carcinoma (SCLC) patients with intestinal pseudoobstruction, which reacted with myenteric plexus neurons [21]. At this point it was not clear if these autoantibodies have a pathogenic role in gastrointestinal motility disorders, but they were already suggested to facilitate the diagnosis of eventual SCLC in the case of GI dysmotility. Later studies have demonstrated a direct cytotoxic effect of sera and IgG from lung cancer patients with paraneoplastic syndrome [31], and anti-HuD, which is produced in the presence of the tumor, on myenteric neurons [14]. To explore a possible link between the effect of ANNA-1/

anti-HuD autoantibodies on enteric neurons and gastrointestinal symptoms, we have tested ANNA-1-positive sera from patients with chronic intestinal pseudoobstruction (CIPO) on human and guinea pig enteric and sensory neurons [23]. Whole serum or purified IgG from patient sera, but not sera from healthy controls, labelled all nerve cell bodies of the guinea pig myenteric and submucous plexus, and the staining was due to anti-HuD antibodies. In neuroimaging experiments, purified IgG from patient sera as well as purified anti-HuD acted postsynaptically to evoke prompt action potential discharge lasting for several seconds in around 12% of myenteric neurons. In contrast, sera from healthy controls had no effect. We have demonstrated a similar, but even more prominent excitatory effect of ANNA-1-positive sera on human submucous neurons, where 50% of neurons responded, whereas glia were insensitive. We could also show that nicotinic acetylcholine receptor and P2X receptors, but not 5-HT3 receptor, were involved in spike discharge. Purified anti-HuD also evoked an excitation in visceral sensory nerve fibers. Our findings shed light to a possible role of anti-HuD in symptom generation by direct excitation of enteric and visceral sensory neurons. ANNA-1 antibodies are usually linked to SCLC and are rarely observed in tumor-negative cases. A recent report, however, described a patient with ANNA-1- and antiglial nuclear SOX-1 autoimmune antibody-positive limbic encephalitis with severe motility disturbances [17]. Although the limbic encephalitis was successfully treated with immunosuppressive therapy, the level of autoantibodies was still very high and GI symptoms persisted. One explanation could be the excitatory effect of the remaining autoantibodies on enteric neurons or a permanent damage already before the initiation of therapy. The patient successfully profited from a symptomatic treatment of the achalasia with pneumatic cardia dilation and the gastrointestinal dysmotility with prucalopride.

Besides ANNA-1, several other anti-neuronal antibodies have been described in AGID, some of them targeting, for example, voltage-gated calcium or potassium channels or acetylcholine receptors. Another well-characterized antibody, targeted against the dipeptidyl-peptidase-like protein 6 (DPPX), an auxiliary subunit of Kv4.2 potassium channels, is responsible for the new clinical entity called anti-DPPX encephalitis [3]. Apart from central nervous system symptoms, gastrointestinal complaints such as diarrhea or constipation, abdominal pain, nausea or vomiting, and severe weight loss are common in these patients. Serum from an anti-DPPX encephalitis patient bounded to enteric neurons of the submucous and myenteric plexus and increased the activity of guinea pig myenteric and human submucous neurons, supporting the pathogenic role of anti-DPPX autoantibodies in the observed gastrointestinal symptoms [28]. Gastrointestinal symptoms occur first, offering a chance of an early treatment with immunotherapy before the appearance of neurological symptoms.

In a study on 107 hospitalized adult patients with various GI symptoms, autoantibodies against enteric nervous system (ENS) and CNS neurons were over four times more common than in controls [24]. Established anti-enteric antibodies, like anti-Hu or anti-DPPX, were all excluded by screening on cell-based assays. Antibody-positive sera did not influence intestinal motility in organ bath experiments, but they enhanced tetrodotoxin (TTX)-insensitive epithelial secretion compared to antibody-negative sera. This study concluded that due to their low prevalence, the routine testing for well-established anti-neuronal antibodies in GI patients does not seem to be necessary, although autoantibodies may play a pathogenic role in GI dysfunction in a part of the patients.

The presence of anti-neuronal autoantibodies has been suggested in an increasing number of more common GI disorders. In esophageal achalasia, a significantly higher prevalence of anti-neuronal autoantibodies has been found compared to controls [20]. In a study testing the effect of serum from achalasia patients, alterations of neurochemical coding of cultured myenteric neurons and decreased relaxation of fundic muscle strips were found, mimicking the changes seen in the disease [4]. However, these effects were evoked by all serum samples but only 12% stained enteric neurons, suggesting that the anti-neuronal autoantibodies are rather an epiphenomenon. Motility disturbances are also common in inflammatory bowel disease (IBD), suggesting the involvement of the ENS. The infiltration of ganglia of the ENS with immune cells is a well-described phenomenon in IBD called plexitis or ganglionitis, often preceding severe inflammation of the bowel [33]. In a recent study on mice, a crucial role of autoimmune CD8-positive T cells was shown in enteric ganglionitis, leading to a rapid and dramatic destruction of enteric neurons via a cytotoxic effect, accompanied by decreased neuronally induced contractile force [30]. Neurological manifestations were observed in around 10% of patients with celiac disease, conventionally explained by vitamin deficiencies due to malabsorption. However, anti-neuronal antibodies against the central nervous system were found in 61% of celiac disease patients with neurological involvement but only in 5% of patients without neurological symptoms [35]. There was no significant difference in ENS-specific autoantibodies, which were shown in 10% of patients with neurological involvement and in 5% without. In another study autoantibodies to ENS were shown in 24% of celiac disease patients, and almost all of those with a high antibody titer suffered from severe constipation [10]. These data support that autoantibodies against the CNS and ENS in celiac disease have a pathogenic role in the accompanying neurological and motility disorders, respectively.

Recently, the involvement of the ENS as an autoimmune target in multiple sclerosis (MS) has been reported [38]. GI symptoms can be observed in almost two-third of MS patients, the most common being constipation, fecal incontinence, dyspeptic complaints, and dysphagia [22]. Motility disturbances are traditionally explained by the underlying musculoskeletal dysfunction or by side effects of medical therapy. However, by using validated diagnostic algorithms, in the same cohort, a surprisingly high rate of patients met criteria for functional GI disorders, such as functional dysphagia, functional dyspepsia, functional constipation, or irritable bowel syndrome (IBS). Furthermore, delayed colonic transit has been also observed in the whole colon in some MS patients, which suggests that ENS may also be involved in constipation besides the known rectoanal dysfunction in these cases [13]. These findings led to a new concept, which proposes that the autoimmune process attacking the CNS may directly target the ENS [38]. Experimental autoimmune encephalomyelitis (EAE) by immunization with various myelin antigens in mice mimics MS. In EAE mice, T cells, B cells, and macrophages invaded the myenteric plexus of the gut already in the preclinical state, and the number of T cells and macrophages increased with disease progression. Strikingly, degeneration and loss of myenteric neurons preceded the degeneration of the spinal cord and progressed with disease severity. In the small intestine, muscle atrophy and compensatory fibrosis were shown in EAE mice, parallel to a significantly decreased GI motility, reduced muscular cholinergic signaling, and NO release. The majority of mice with EAE developed antibodies against antigens of enteric neurons and glial cells. The authors also translated this data to humans by detecting one of the autoantibodies against the ENS identified in mice in 10 of 33 MS patients but in none of the controls. These results were corroborated by another study on EAE mice, demonstrating accelerated gastric emptying and reduced colonic motility in vivo, and significantly decreased motility ex vivo mimicking constipation seen in the majority of MS patients [32]. Sera of both MS patients and EAE mice stained neuronal and/or glial components of the ENS. Interestingly, no neuronal loss could be detected in myenteric ganglia in ENS mice, but the expression of glial fibrillary acidic protein (GFAP) was reduced. B-cell deficiency prevented motility disturbances in EAE mice supporting the role of autoantibodies in the pathogenesis.

1.2 Enteric Nervous System and Irritable Bowel Syndrome

IBS is considered a disease of the gut-brain axis. Observations that CIPO in its early phase is similar to IBS led to the idea of anti-enteric neuronal antibodies in IBS patients [36]. Such antibodies have been found in ~87% of IBS patients vs. 59% of controls. The study identified three antigens targeted by the autoantibodies : a nondescript ribonucleoprotein (RNP-complex), small nuclear ribonuclear polypeptide A, and Ro-5200 kDa. Similar to Hu, they are also involved in mRNA binding, processing, and transport. Selective depletion of anti-enteric neuronal antibodies may be a possible treatment of GI symptoms in autoimmune diseases. However, it needs to be considered that the percentage of controls with anti-enteric neuronal autoantibodies is surprisingly high. A more recent study has found a significantly higher rate of anti-enteric neuronal antibodies in IBS patients (76.8%) than in controls (33%) and in patients with slow transit constipation, but not significantly different from patients with IBD or CIPO [16]. The sensitivity and specificity of anti-enteric neuronal antibody to diagnose IBS were 76.8% and 42.3%, respectively. Moderate and strong positive IBS sera induced apoptosis in cultured guinea pig myenteric neurons and in the human neuroblastoma cell clone SH-Sy5Y. These data suggested a pathogenic role of anti-enteric neuronal antibodies in at least a subset of IBS patients.

Apart from being more numerous, the activation of mast cells was also observed in IBS-D patients, resulting in increased mediator release, such as histamine or mast cell tryptase, which are capable of neuronal activation [2]. The number of mast cells in close proximity to nerves significantly correlated with clinical symptoms, namely, severity and frequency of abdominal pain and/or discomfort in IBS patients. The increased number or activation of mast cells in the gut can be a consequence of dietary factors, pathogens, or stress events in IBS, which leads to the activation of enteric neurons [12]. The communication between mast cells and neurons is bidirectional [8]. In female IBS patients, activation of enteric glial cells and their disturbed communication with mast cells was observed in the colon, which may be a pathogenic factor in the increased epithelial permeability observed in many IBS patients [25]. In accordance with this, two- to threefold increased release of proteases (trypsin and tryptase) was measured in biopsy samples from the colon or rectum of IBS patients compared to controls [11]. Although the number of mast cells was not increased in the biopsies, the intense release of tryptase relates to their activation. Application of biopsy supernatants from IBS patients with elevated protease activity into the colon of mice evoked hyperalgesia and allodynia in response to colorectal distension. The hypersensitivity was due to protease-activated receptor (PAR)-2. In the following years, we repeatedly demonstrated that apart from activating visceral afferent neurons, biopsy supernatants of IBS patients can also excite enteric neurons [5–7, 9]. By pharmacological approaches, proteases, histamine, and serotonin were identified as components responsible for the neuronal activation, with the effect of proteases being most dominant [5]. In this study, a significantly elevated mast cell count was present in the lamina propria of biopsies from IBS patients, which correlated with the spike frequency evoked in human submucous enteric neurons. The excitatory effect was independent of IBS subtype (IBS-D or C), suggesting a general feature in IBS. Although the concentration of nerve-activating components in the biopsy supernatants is relatively small and individually would not cause a spike discharge in enteric neurons, we observed synergistic potentiation between histamine, serotonin, and tryptase explaining the effect of the supernatants [27]. Further, the nerve-activating effect is more prominent in the submucous than in myenteric plexus [6]. The effect of a cocktail of compounds mimicking IBS biopsy supernatants (a mixture of serotonin, histamine, tryptase, and tumor necrosis factor alpha (TNF-α)) was also tested on submucous neurons in biopsies from healthy controls and IBS patients [26]. Interestingly, the evoked responses of neurons in biopsies obtained from IBS patients were significantly lower than those from healthy controls, which may be caused by the presence of desensitized neurons in IBS biopsies due to the constant release of the mediators in the gut wall. In another study, histamine receptor-1 (HRH1)-mediated sensitization of the transient receptor potential channel V1 (TRPV1) was shown in submucous neurons in rectal biopsies from IBS patients compared to healthy controls [37]. Additionally, mucosal biopsy supernatants have been suggested as a biomarker for IBS with increased sensitivity to rectal distension, as biopsy supernatants of hypersensitive patients activated submucous neurons and dorsal root ganglion neurons significantly stronger [7]. Furthermore, recently 17 proteases were identified by proteome analysis in IBS biopsy supernatants as differentially expressed compared to ulcerative colitis and healthy controls. The combination of three proteases, elastase 3a, cathepsin L, and proteasome alpha subunit-4, showed the highest prediction accuracy of 98% to discriminate between IBS and healthy controls [9]. This study also identified PAR-1 receptors as most relevant for activation of human enteric neurons.

Elevated protease activity has been shown in the stool of IBS-D patients [29]. Importantly, in fecal supernatants from IBS-D patients, serine-protease activity was elevated, whereas in fecal supernatants from IBS-C patients, cysteine-protease activity was increased [29] [1]. The high serine-protease activity in the fecal supernatants of IBS-D patients was responsible for increased intestinal permeability and visceral hypersensitivity via the activation of PAR-2 when infused intracolonically to mice [18]. The increased protease activity is mostly of human origin and can be at least partially explained by decreased bacterial degradation due to the accelerated transit in IBS-D [34]. Similarly, elevated cysteine-protease activity in fecal supernatants from IBS-C patients caused disruption of the intestinal barrier and visceral hypersensitivity in the same model but only after prolonged and repeated applications [1]. The difference can be attributed to distinct mechanisms, as PAR-2 does not play a role in the effect of cysteine-proteases, which directly degrade tight junctions. Therefore, visceral hypersensitivity in IBS-C may be a consequence of the increased permeability rather than direct receptor activation on nerve cells.

The disturbed crosstalk between mast cells and enteric neurons triggered clinical studies. For example, in a double-blind placebo-controlled trial, 8 weeks of treatment with the mast cell stabilizer ketotifen increased the threshold of discomfort in patients with IBS with visceral hypersensitivity and relieved IBS symptoms such as bloating, flatulence, diarrhea, and incomplete evacuation, parallel to an improvement in health-related quality of life [19]. Likewise, the more selective HRH1 antagonist ebastine reduced visceral hypersensitivity in IBS patients, provided symptom relief, and reduced abdominal pain scores [37].

Improved understanding of the role of the ENS in autoimmune disorders and in IBS will help to develop novel diagnostic methods and therapies.

References

1. Annaházi A, Ferrier L, Bézirard V, Lévêque M, Eutamène H, Ait-Belgnaoui A, Coëffier M, Ducrotté P, Róka R, Inczefi O, Gecse K, Rosztóczy A, Molnár T, Ringel-Kulka T, Ringel Y, Piche T, Theodorou V, Wittmann T, Bueno L (2013) Luminal cysteine-proteases degrade colonic tight junction structure and are responsible for abdominal pain in constipation-predominant IBS. Am J Gastroenterol 108:1322–1331. https://doi.org/10.1038/ajg.2013.152
2. Barbara G, Stanghellini V, De Giorgio R, Cremon C, Cottrell GS, Santini D, Pasquinelli G, Morselli-Labate AM, Grady EF, Bunnett NW, Collins SM, Corinaldesi R (2004) Activated mast cells in proximity to colonic nerves correlate with abdominal pain in irritable bowel syndrome. Gastroenterology 126:693–702. https://doi.org/10.1053/j.gastro.2003.11.055
3. Boronat A, Gelfand JM, Gresa-Arribas N, Jeong H-Y, Walsh M, Roberts K, Martinez-Hernandez E, Rosenfeld MR, Balice-Gordon R, Graus F, Rudy B, Dalmau J (2013) Encephalitis and antibodies to dipeptidyl-peptidase-like protein-6, a subunit of Kv4.2 potassium channels. Ann Neurol 73:120–128. https://doi.org/10.1002/ana.23756
4. Bruley des Varannes S, Chevalier J, Pimont S, Le Neel J-C, Klotz M, Schafer K-H, Galmiche J-P, Neunlist M (2006) Serum from achalasia patients alters neurochemical coding in the myenteric plexus and nitric oxide mediated motor response in normal human fundus. Gut 55:319–326. https://doi.org/10.1136/gut.2005.070011
5. Buhner S, Li Q, Vignali S, Barbara G, De Giorgio R, Stanghellini V, Cremon C, Zeller F, Langer R, Daniel H, Michel K, Schemann M (2009) Activation of human enteric neurons by supernatants of colonic biopsy specimens from patients with irritable bowel syndrome. Gastroenterology 137:1425–1434. https://doi.org/10.1053/j.gastro.2009.07.005

6. Buhner S, Li Q, Berger T, Vignali S, Barbara G, De Giorgio R, Stanghellini V, Schemann M (2012) Submucous rather than myenteric neurons are activated by mucosal biopsy supernatants from irritable bowel syndrome patients. Neurogastroenterol Motil 24:1134–e572. https://doi.org/10.1111/nmo.12011
7. Buhner S, Braak B, Li Q, Kugler EM, Klooker T, Wouters M, Donovan J, Vignali S, Mazzuoli-Weber G, Grundy D, Boeckxstaens G, Schemann M (2014) Neuronal activation by mucosal biopsy supernatants from irritable bowel syndrome patients is linked to visceral sensitivity. Exp Physiol 99:1299–1311. https://doi.org/10.1113/expphysiol.2014.080036
8. Buhner S, Barki N, Greiter W, Giesbertz P, Demir IE, Ceyhan GO, Zeller F, Daniel H, Schemann M (2017) Calcium imaging of nerve-mast cell signaling in the human intestine. Front Physiol 8:971. https://doi.org/10.3389/fphys.2017.00971
9. Buhner S, Hahne H, Hartwig K, Li Q, Vignali S, Ostertag D, Meng C, Hörmannsperger G, Braak B, Pehl C, Frieling T, Barbara G, De Giorgio R, Demir IE, Ceyhan GO, Zeller F, Boeckxstaens G, Haller D, Kuster B, Schemann M (2018) Protease signaling through protease activated receptor 1 mediate nerve activation by mucosal supernatants from irritable bowel syndrome but not from ulcerative colitis patients. PLoS One 13:e0193943. https://doi.org/10.1371/journal.pone.0193943
10. Caio G, De Giorgio R, Venturi A, Giancola F, Latorre R, Boschetti E, Serra M, Ruggeri E, Volta U (2015) Clinical and immunological relevance of anti-neuronal antibodies in celiac disease with neurological manifestations. Gastroenterol Hepatol Bed Bench 8:146–152
11. Cenac N, Andrews CN, Holzhausen M, Chapman K, Cottrell G, Andrade-Gordon P, Steinhoff M, Barbara G, Beck P, Bunnett NW, Sharkey KA, Ferraz JGP, Shaffer E, Vergnolle N (2007) Role for protease activity in visceral pain in irritable bowel syndrome. J Clin Invest 117:636–647. https://doi.org/10.1172/JCI29255
12. Cheng L, Luo QQ, Chen SL (2021) The role of intestinal mast cell infiltration in irritable bowel syndrome. J Dig Dis 22:143–151. https://doi.org/10.1111/1751-2980.12971
13. Chia YW, Gill KP, Jameson JS, Forti AD, Henry MM, Swash M, Shorvon PJ (1996) Paradoxical puborectalis contraction is a feature of constipation in patients with multiple sclerosis. J Neurol Neurosurg Psychiatry 60:31–35. https://doi.org/10.1136/jnnp.60.1.31
14. De Giorgio R, Bovara M, Barbara G, Canossa M, Sarnelli G, De Ponti F, Stanghellini V, Tonini M, Cappello S, Pagnotta E, Nobile-Orazio E, Corinaldesi R (2003) Anti-HuD-induced neuronal apoptosis underlying paraneoplastic gut dysmotility. Gastroenterology 125:70–79. https://doi.org/10.1016/s0016-5085(03)00664-4
15. Dhamija R, Tan KM, Pittock SJ, Foxx–Orenstein A, Benarroch E, Lennon VA (2008) Serologic profiles aiding the diagnosis of autoimmune gastrointestinal dysmotility. Clin Gastroenterol Hepatol 6:988–992. https://doi.org/10.1016/j.cgh.2008.04.009
16. Fan W, Fei G, Li X, Wang X, Hu C, Xin H, Sun X, Li Y, Wood JD, Fang X (2018) Sera with anti-enteric neuronal antibodies from patients with irritable bowel syndrome promote apoptosis in myenteric neurons of guinea pigs and human SH-Sy5Y cells. Neurogastroenterol Motil 30:e13457. https://doi.org/10.1111/nmo.13457
17. Frieling T, Kreysel C, Blank M, Müller D, Melchior I, Euler P, Kuhlbusch-Zicklam R, Haarmeier T, Schemann M (2020) Autoimmune encephalitis and gastrointestinal dysmotility: achalasia, gastroparesis, and slow transit constipation. Z Gastroenterol 58:975–981. https://doi.org/10.1055/a-1233-2190
18. Gecse K, Róka R, Ferrier L, Leveque M, Eutamene H, Cartier C, Ait-Belgnaoui A, Rosztóczy A, Izbéki F, Fioramonti J, Wittmann T, Bueno L (2008) Increased faecal serine protease activity in diarrhoeic IBS patients: a colonic lumenal factor impairing colonic permeability and sensitivity. Gut 57:591–599. https://doi.org/10.1136/gut.2007.140210
19. Klooker TK, Braak B, Koopman KE, Welting O, Wouters MM, van der Heide S, Schemann M, Bischoff SC, van den Wijngaard RM, Boeckxstaens GE (2010) The mast cell stabiliser ketotifen decreases visceral hypersensitivity and improves intestinal symptoms in patients with irritable bowel syndrome. Gut 59:1213–1221. https://doi.org/10.1136/gut.2010.213108
20. Kraichely RE, Farrugia G, Pittock SJ, Castell DO, Lennon VA (2010) Neural autoantibody profile of primary achalasia. Dig Dis Sci 55:307–311. https://doi.org/10.1007/s10620-009-0838-9
21. Lennon VA, Sas DF, Busk MF, Scheithauer B, Malagelada JR, Camilleri M, Miller LJ (1991) Enteric neuronal autoantibodies in pseudoobstruction with small-cell lung carcinoma. Gastroenterology 100:137–142. https://doi.org/10.1016/0016-5085(91)90593-a
22. Levinthal DJ, Rahman A, Nusrat S, O'Leary M, Heyman R, Bielefeldt K (2013) Adding to the burden: gastrointestinal symptoms and syndromes in multiple sclerosis. Mult Scler Int 2013:319201. https://doi.org/10.1155/2013/319201
23. Li Q, Michel K, Annahazi A, Demir IE, Ceyhan GO, Zeller F, Komorowski L, Stöcker W, Beyak MJ, Grundy D, Farrugia G, De Giorgio R, Schemann M (2016) Anti-Hu antibodies activate enteric and sensory neurons. Sci Rep 6:38216. https://doi.org/10.1038/srep38216
24. Lütt A, Michel K, Krüger D, Volz MS, Nassir M, Schulz E, Poralla L, Tangermann P, Bojarski C, Höltje M, Teegen B, Stöcker W, Schemann M, Siegmund B, Prüss H (2018) High prevalence and functional effects of serum antineuronal antibodies in patients with gastrointestinal disorders. Neurogastroenterol Motil 30:e13292. https://doi.org/10.1111/nmo.13292
25. Meira de-Faria F, Casado-Bedmar M, Mårten Lindqvist C, Jones MP, Walter SA, Keita ÅV (2021) Altered interaction between enteric glial cells and mast cells in the colon of women with irritable bowel

syndrome. Neurogastroenterol Motil. https://doi.org/10.1111/nmo.14130
26. Ostertag D, Buhner S, Michel K, Pehl C, Kurjak M, Götzberger M, Schulte-Frohlinde E, Frieling T, Enck P, Phillip J, Schemann M (2015) Reduced responses of submucous neurons from irritable bowel syndrome patients to a cocktail containing histamine, serotonin, TNFα, and tryptase (IBS-cocktail). Front Neurosci 9:465. https://doi.org/10.3389/fnins.2015.00465
27. Ostertag D, Annahazi A, Krueger D, Michel K, Demir IE, Ceyhan GO, Zeller F, Schemann M (2017) Tryptase potentiates enteric nerve activation by histamine and serotonin: relevance for the effects of mucosal biopsy supernatants from irritable bowel syndrome patients. Neurogastroenterol Motil 29. https://doi.org/10.1111/nmo.13070
28. Piepgras J, Höltje M, Michel K, Li Q, Otto C, Drenckhahn C, Probst C, Schemann M, Jarius S, Stöcker W, Balint B, Meinck H-M, Buchert R, Dalmau J, Ahnert-Hilger G, Ruprecht K (2015) Anti-DPPX encephalitis: pathogenic effects of antibodies on gut and brain neurons. Neurology 85:890–897. https://doi.org/10.1212/WNL.0000000000001907
29. Róka R, Rosztóczy A, Leveque M, Izbéki F, Nagy F, Molnár T, Lonovics J, Garcia-Villar R, Fioramonti J, Wittmann T, Bueno L (2007) A pilot study of fecal serine-protease activity: a pathophysiologic factor in diarrhea-predominant irritable bowel syndrome. Clin Gastroenterol Hepatol 5:550–555. https://doi.org/10.1016/j.cgh.2006.12.004
30. Sanchez-Ruiz M, Brunn A, Montesinos-Rongen M, Rudroff C, Hartmann M, Schlüter D, Pfitzer G, Deckert M (2019) Enteric murine ganglionitis induced by autoimmune CD8 T cells mimics human gastrointestinal dysmotility. Am J Pathol 189:540–551. https://doi.org/10.1016/j.ajpath.2018.11.016
31. Schäfer KH, Klotz M, Mergner D, Mestres P, Schimrigk K, Blaes F (2000) IgG-mediated cytotoxicity to myenteric plexus cultures in patients with paraneoplastic neurological syndromes. J Autoimmun 15:479–484. https://doi.org/10.1006/jaut.2000.0454
32. Spear ET, Holt EA, Joyce EJ, Haag MM, Mawe SM, Hennig GW, Lavoie B, Applebee AM, Teuscher C, Mawe GM (2018) Altered gastrointestinal motility involving autoantibodies in the experimental autoimmune encephalomyelitis model of multiple sclerosis. Neurogastroenterol Motil 30:e13349. https://doi.org/10.1111/nmo.13349
33. Stavely R, Abalo R, Nurgali K (2020) Targeting enteric neurons and plexitis for the management of inflammatory bowel disease. Curr Drug Targets 21:1428–1439. https://doi.org/10.2174/1389450121666200516173242
34. Tooth D, Garsed K, Singh G, Marciani L, Lam C, Fordham I, Fields A, Banwait R, Lingaya M, Layfield R, Hastings M, Whorwell P, Spiller R (2014) Characterisation of faecal protease activity in irritable bowel syndrome with diarrhoea: origin and effect of gut transit. Gut 63:753–760. https://doi.org/10.1136/gutjnl-2012-304042
35. Volta U, Giorgio RD, Petrolini N, Stanghellini V, Barbara G, Granito A, Ponti FD, Corinaldesi R, Bianchi FB (2002) Clinical findings and anti-neuronal antibodies in coeliac disease with neurological disorders. Scand J Gastroenterol 37:1276–1281. https://doi.org/10.1080/003655202761020542
36. Wood JD, Liu S, Drossman DA, Ringel Y, Whitehead WE (2012) Anti-enteric neuronal antibodies and the irritable bowel syndrome. J Neurogastroenterol Motil 18:78–85. https://doi.org/10.5056/jnm.2012.18.1.78
37. Wouters MM, Balemans D, Van Wanrooy S, Dooley J, Cibert-Goton V, Alpizar YA, Valdez-Morales EE, Nasser Y, Van Veldhoven PP, Vanbrabant W, Van der Merwe S, Mols R, Ghesquière B, Cirillo C, Kortekaas I, Carmeliet P, Peetermans WE, Vermeire S, Rutgeerts P, Augustijns P, Hellings PW, Belmans A, Vanner S, Bulmer DC, Talavera K, Vanden Berghe P, Liston A, Boeckxstaens GE (2016) Histamine receptor H1-mediated sensitization of TRPV1 mediates visceral hypersensitivity and symptoms in patients with irritable bowel syndrome. Gastroenterology 150:875–887.e9. https://doi.org/10.1053/j.gastro.2015.12.034
38. Wunsch M, Jabari S, Voussen B, Enders M, Srinivasan S, Cossais F, Wedel T, Boettner M, Schwarz A, Weyer L, Göcer O, Schroeter M, Maeurer M, Woenckhaus M, Pollok K, Radbruch H, Klotz L, Scholz C-J, Nickel J, Friebe A, Addicks K, Ergün S, Lehmann PV, Kuerten S (2017) The enteric nervous system is a potential autoimmune target in multiple sclerosis. Acta Neuropathol 134:281–295. https://doi.org/10.1007/s00401-017-1742-6

Clinical and Pathological Features of Severe Gut Dysmotility

2

Francesca Bianco, Elena Bonora, Giulia Lattanzio, Paolo Clavenzani, Matteo Guarino, Maurizio Mazzoni, Vito Antonio Baldassarro, Luca Lorenzini, Giacomo Caio, Vincenzo Stanghellini, Catia Sternini, Gianrico Farrugia, Luciana Giardino, Laura Calzà, and Roberto De Giorgio

Abstract

Severe gut motility disorders are characterized by ineffective propulsion of intestinal contents. As a result, patients often develop extremely uncomfortable symptoms, ranging from nausea and vomiting along with alterations of bowel habits, up to radiologically confirmed subobstructive episodes. Chronic intestinal pseudo-obstruction (CIPO) is a typical clinical phenotype of severe gut dysmotility due to morphological and functional alterations of the intrinsic (enteric) innervation and extrinsic nerve supply (hence neuropathy), interstitial cells of Cajal (ICCs) (mesenchymopathy), and smooth muscle cells (myopathy). In this chapter, we highlight some molecular mechanisms of CIPO and review the clinical phenotypes and the genetics of the different types of CIPO. Specifically, we will detail the role of some of the most representative genetic mutations involving

F. Bianco
Department of Medical and Surgical Sciences, University of Bologna, Bologna, Italy

Department of Veterinary Sciences University of Bologna, Bologna, Italy

E. Bonora
Department of Medical and Surgical Sciences, University of Bologna, Bologna, Italy

G. Lattanzio · P. Clavenzani · M. Mazzoni
Department of Veterinary Sciences University of Bologna, Bologna, Italy

M. Guarino · G. Caio
Department of Translational Medicine, University of Ferrara, Ferrara, Italy

V. A. Baldassarro · L. Lorenzini · L. Giardino
L. Calzà
IRET Foundation, Ozzano Emilia, Italy

V. Stanghellini
Department of Medical and Surgical Sciences, University of Bologna, Bologna, Italy

Division of Internal Medicine, IRCCS Azienda Ospedaliero-Universitaria di Bologna, Bologna, Italy

C. Sternini
UCLA/DDRC, Division of Digestive Diseases, Departments Medicine and Neurobiology, David Geffen School of Medicine, Los Angeles, CA, USA

G. Farrugia
Enteric Neuroscience Program (ENSP), Mayo Clinic, Rochester, MN, USA

R. De Giorgio (✉)
Department of Translational Medicine, University of Ferrara, Ferrara, Italy

Unit of Internal Medicine, St. Anna Hospital, Ferrara, Italy
e-mail: dgrrrt@unife.it

N. J. Spencer et al. (eds.), *The Enteric Nervous System II*, Advances in Experimental Medicine and Biology 1383, https://doi.org/10.1007/978-3-031-05843-1_2

RAD21, *LIG3*, and *ACTG2* to provide a better understanding of CIPO and related underlying neuropathic or myopathic histopathological abnormalities. This knowledge may unveil targeted strategies to better manage patients with such severe disease.

Keywords

Chronic intestinal pseudo-obstruction · Enteric neurons · Enteric neuropathy · Neurogenetics · Severe gut dysfunction

2.1 Introduction

Severe gut dysmotility is a clinical condition characterized by a major impairment of intestinal propulsion due to changes of various key cells, such as enteric neurons and/or glial cells, interstitial cells of Cajal (ICCs), and smooth muscle cells of GI tract [6]. In very few cases, the functional impairment of digestive neuro-ICC-muscular integrated systems is so pronounced to hinder oral feeding leading to severe symptoms, which markedly compromise patients' quality of life. CIPO is a clinical phenotype of severe gut dysmotility usually with poor prognosis.

Patients with CIPO typically manifest with recurrent intestinal subocclusive episodes and clinical /radiological findings mimicking mechanical obstruction [5, 24]. In addition to the acute subobstructive phases, during the intercrisis period, patients complain of a variety of symptoms (e.g., chronic nausea, vomiting, abdominal distension, and constipation/diarrhea) requiring nutritional support to avoid weight loss and malnutrition [7]. Treatment options for CIPO are essentially aimed at preventing malnutrition (by providing adequate caloric intake), restore motility (via prokinetic/enterokinetic pharmacological agents), and control symptoms, for example, abdominal pain, nausea/vomiting, and diarrhea (often ascribable to bacterial overgrowth). Taken together, these options might allow for stabilization of the clinical picture at best, but certainly do not target the mechanisms underlying CIPO. The lack of established animal models, which could provide a basis for a better understanding of the molecular pathways contributing to neuro-ICC-myopathy underlying CIPO, is hindering preventative approaches and effective treatments. Highlighting molecular mechanisms of CIPO can be a strategy to decipher its clinical complexity, simplify the wide heterogeneity inherent to this condition, and ultimately pave the way to new treatment options. Herein, we review the clinical phenotypes and the genetics of the different types of CIPO with the goal of providing answers for a better understanding of severe dysmotility and related histopathological abnormalities with a special emphasis on neuro-myopathy.

2.2 Genetics of CIPO

In recent years, we and other groups have provided evidence for a genetic basis of the enteric neuronal and/or smooth muscle degeneration and loss observed in specific forms of CIPO. The discovery of novel genes mutated in different patients represents the first step in identifying the downstream molecular impairment in CIPO. The genes so far identified in CIPO patients are summarized in Table 2.1. An in-depth study of these gene variants in vitro and in vivo is required to understand the role of the different genes identified in the ENS and muscular alterations that can lead to a severe enteric dysmotility, such as CIPO. In the next paragraphs, we will highlight the role of genetic mutations involving *RAD21*, *LIG3*, and *ACTG2*.

2.2.1 RAD21

We identified the homozygous causative variant in a large consanguineous family segregating an autosomal-recessive form of CIPO. The affected family members also presented other clinical features including megaduodenum, long-segment Barrett's esophagus (up to 18 cm from the squamocolumnar junction or "Z-line"), and cardiac abnormalities of variable severity (OMIM 611376; Mungan syndrome). We performed

Table 2.1 Gene defects reported in patients with CIPO

Gene	Causative variants	References
FLNA	c.65-66delAC	Gargiulo A. et al., Am J Hum Genet. 2007.
ACTG2	c.443C>T (p.Arg148Leu)	Ravenscroft G. et al. Neurogastroenterol Motil. 2018;30.
RAD21	c. 1864 G>A (p.622 Ala>Thr)	Bonora E. et al., Gastroenterology. 2015; 148:771–782.
SGOL1	c.67A>G (p.Lys23Glu)	Chetaille P. et al., Nat Genet. 2014; 46:1245–1249.
TYMP	G1419A, G1443A, and A3371C	Nishino I. et al., science. 1999; 283:689–692.
POLG	c.679C>T and c.2542G>A	Giordano C. et al., Neurology. 2009; 72:1103–1105.
MYLK	c.3985+5C>A	Halim D. et al., Am J Hum Genet. 2017; 101: 123-9.
MYH11	c.5819delC (p.Pro1940HisfsTer91)	Dong W. et al., Clin Genet. 2019;96:473-477.
LMOD1	c.1108C>T; p.Arg370*	Halim D. et al., Proc. Natl. Acad. Sci. USA. 2017;114:E2739–E2747.

whole-exome sequencing analysis on the genomic DNA from two affected individuals and found the novel homozygous change c.1864 G>A in *RAD21* (NM_006265.2) producing the damaging missense variant p.Ala622Thr [3]. Any derangement to RAD21 molecular structure and function can result in significant changes to many tissues, including the gut neuro-muscular layer. In fact, RAD21 is part of the cohesin complex, involved in pairing and unpairing of sister chromatids during cell replication and division, and also regulates gene expression directly and independently of cell division [13]. The RAD21 subunit of the cohesin complex plays important structural and functional roles, as it serves as physical link between the SMC1/SMC3 (Structural Maintenance of Chromosome 1 and 3) heterodimer and the STAG subunit. RAD21 integrity regulates the association or disassociation of functional cohesin with chromatin and also plays a key role in double-strand breaks DNA repair [1]. Cohesin binds to many sites throughout the genome, in combination with the CCCTC-binding factor (CTCF) insulator protein, which is known to mediate chromatin loop formation. Cohesin colocalizes with CTCF along chromosome arms, cooperating with this protein in the regulation of gene expression and chromatin structure [23]. RAD21 is associated with other transcriptional regulators, such as estrogen receptor-α [8], and the RNA-polymerase-binding complex Mediator [14] in a cell-type-specific manner. In combination with other factors, cohesin selectively binds genes with paused RNA polymerase II and can regulate transcription by determining the amount of elongating RNA polymerase II on genes [18]. Cohesin-mediated chromatin organization plays an important role in the formation/stabilization of chromosome architecture and gene transcription and repression. RAD21 interacts with CTCF and other cohesin-associated proteins to maintain and stabilize multidimensional organizations of topologically associating domains and chromatin loops [12]. In addition, human RAD21 is linked to the apoptotic pathways and RAD21 cleavage can be induced by a broad spectrum of apoptotic stimuli at residue p.Asp279, which is different from the mitotic cleavage sites required for chromosomal segregation [4, 22].

In our study, in vitro and in vivo data (zebrafish model) showed that the mutant RAD21 p.Ala622Thr protein did not activate one of its well-known target genes, that is, *RUNX1*, which exerts a crucial role for RAD21 function [3]. Moreover, using the zebrafish as a model, we recapitulated the CIPO phenotype observed in patients with the homozygous *RAD21* causative variant, in terms of severe impairment of motility and marked reduction of neurons (hence a histopathology pattern reminiscent of an "oligo-neuronal hypoganglionosis"). After injecting a morpholino (mo) specific for the functional ablation of *rad21*, zebrafish embryos were allowed to develop to 5 days post-fertilization (dpf), when the digestive system has already developed. Control and mo-treated embryos were fed fluorescent beads through microgavage, a technique that allows to determine the rate of intestinal motility as a function of time. After 8 hours post microgavage of fluorescent beads, the GI tract of embryos was divided into four zones based on anatomical landmarks and the presence of fluorescent bead was scored in each segment. Compared to controls, *rad21* morphants showed delayed food transit along the gut and a significant depletion of enteric neurons as shown by staining the enteric neurons with antibodies against the neuronal marker HuC/D. The marked reduction of HuC/D-immunolabeled enteric neurons was highly suggestive of a neuropathic impairment responsible for the observed motility impairment (no evidence of any smooth muscle damage in the zebrafish embryos) [3]. Furthermore, Rad21 immunoreactivity was detected in about 43% of HuC/D-immunoreactive myenteric neurons/field of both adult and young mouse small intestine. A subset of ChAT-immunolabeled myenteric neurons was also immunoreactive for Rad21. Rad21 silencing confirmed specificity of the immunolabeled neurons [2]. Following this first evaluation of the normal Rad21 localization in the mouse ENS, and based on the evidence that *rad21*$^{-/-}$ mice were embryonic lethal [27], we designed and developed a conditional knock-in (KI) mouse with the p.Ala626Thr missense variant (equivalent to the p.Ala622Thr in humans) in order to generate a model reminiscent of the clinical phenotype and histopathological changes in the originally described affected family members. The design of the construct allowed us to selectively drive the p.Ala626Thr variant into the mouse ENS in response to Cre-driven recombination of the floxed alleles by crossing with mice carrying the expression of Cre under the *Wnt1* promoter. Compared to wild-type animals, preliminary data on the homozygous mice revealed that there was about 30% reduction of HuC/D myenteric neurons/ganglion in the Rad21KI mice [15], a finding indicative of the CIPO phenotype observed in patients. Also, subsets of HuC/D-IR myenteric neurons of wild-type mouse colon displayed either ChAT-immunoreactivity (43.71 ± 3.45) or nNOS-immunoreactivity (30.84 ± 5.40). However, in the Rad21KI mice, the HuC/D/ChAT-immunoreactive neurons/ganglions were 45.13 ± 4.27, whereas the HuC/D/nNOS-immunoreactive neurons/ganglions were 14.13 ± 1.40 (Bianco, unpublished data). These preliminary data point to an overall reduction of the myenteric neurons with a selective depletion of inhibitory motor neurons in Rad21KI mice, suggesting a role of this gene alteration in gut motility dysfunction. Further studies are currently ongoing to determine the full spectrum of abnormalities caused by the *RAD21* mutation.

2.2.2 LIG3

We recently characterized another mutant gene underlying CIPO and neurological manifestations (reminiscent of mitochondrial neurogastrointestinal encephalomyopathy, MNGIE) in seven patients from three unrelated families (Bonora 2021 in press). In addition to CIPO, the most prominent and consistent clinical signs were neurological abnormalities, including leukoencephalopathy, epilepsy, migraine, stroke-like episodes, and neurogenic bladder. DNA from these patients was subjected to whole-exome sequencing. Compound heterozygous variants were identified in the gene *LIG3* in all patients. All variants were predicted to have a damaging effect on the protein. The *LIG3* gene encodes the unique mitochondrial DNA (mtDNA) ligase that binds POLG and plays a pivotal role in mtDNA repair and replication. The study of the consequences of *LIG3* mutations was conducted in primary skin fibroblasts derived from patients and in transiently transfected cells expressing the different mutant versus wild-type proteins. All assays showed a severely reduced quantity of LIG3 protein in the mutant cells. In concordance, we demonstrated the lack of ligase activity in the mitochondrial extracts derived from patients' cells compared to control fibroblasts. The *LIG3* gene defects altered the mitochondrial network (Fig. 2.1a), affected mtDNA maintenance, leading to mtDNA depletion without the accumulation of multiple deletions observed in other mitochondrial disorders (e.g., MNGIE) and induced a severe imbalance in cell metabolism, with impaired ATP production and increased mitochondrial reactive oxygen species generation. The resultant mitochondrial dysfunction was key in the causative effects leading to the clinical phenotype observed in these patients. In the gut, the histopathological analysis and neuronal HuC/D immunoreactivity evaluation demonstrated a significant loss of the number of myenteric neurons in the colon (Fig. 2.1b). Likewise, in Rad21, the disruption of *lig3* in the zebrafish model reproduced leukoencephalopathy in the brain and impaired gut transit, with an alteration of mitochondrial markers in the mutant models. Biallelic heterozygous loss-of-function variants in the *LIG3* gene result therefore in a novel mitochondrial disease characterized by predominant gut dysmotility, encephalopathy, and neuronal abnormalities.

2.3 Smooth Muscle Actin–Related Diseases: Visceral Myopathy Driven by *ACTG2* Mutations

Visceral myopathy (MIM# 155310) causes gut dysmotility due to smooth muscle dysfunction with phenotypes ranging from functional intestinal dysmotility along with massively distended

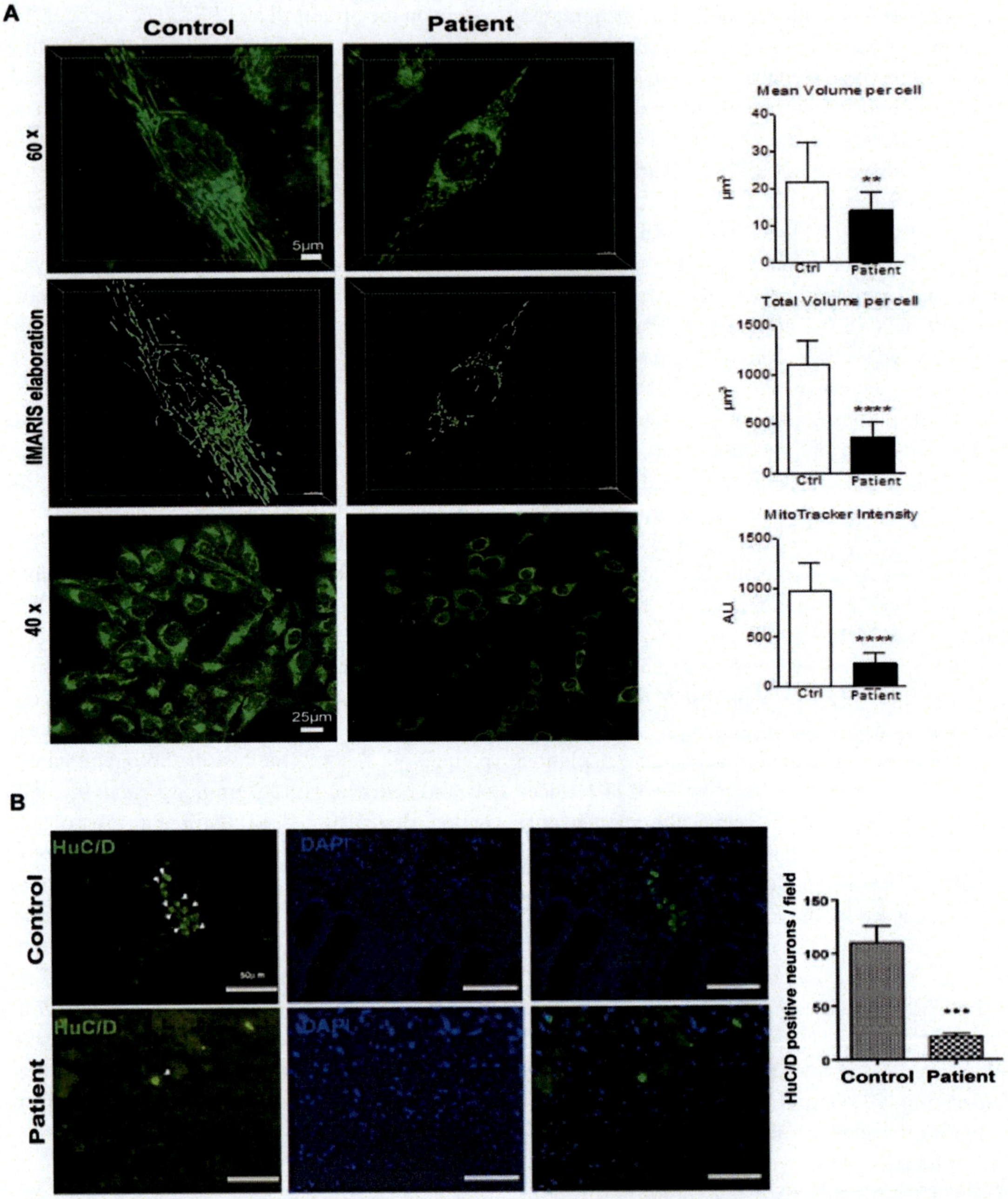

Fig. 2.1 Mitochondrial defects in vitro and ex vivo in presence of *LIG3* mutations underlie a severe reduction in ENS neurons. (**a**) Representative images of 3D z stacks and relative Imaris software reconstruction of the mitochondrial network using MitoTracker green staining. Quantification of mitochondrial network volume represented as the mean isosurface volume per cell ($n = 5$ pictures/group). A significant decrease in the mitochondrial isosurface volume and green intensity was observed in a patient with biallelic *LIG3* mutations ($P \leq 0.001$). Scale bars: 5 μm and 25 μm. (**b**) Histological features of gut (colon) biopsies. HuC/D immunofluorescence (green) in gut tissue biopsies (colon) of controls and patient carrying *LIG3* mutations; nuclear staining was obtained with DAPI (blue). Arrows point to labelled myenteric neurons. Myenteric neurons are significantly reduced in the LIG3 mutated patient (right panel, $P = 0.0005$; Student's t-test). Scale bar: 50 μm

bladder requiring catheterization, a condition also referred to as megacystis-microcolon intestinal hypoperistalsis syndrome (MMIHS), to cases characterized by predominant involvement of the GI tract with typical CIPO features. Causative variants in several genes have been identified in these phenotypes, but the majority of molecularly diagnosed cases are caused by heterozygous variants in *ACTG2*, resulting in dominant alleles running in families or arising de novo in the affected subjects [17, 25]. In vitro studies of the identified variants have shown an impairment of ACTG2 polymerization and a reduction of smooth muscle cell contractility [10, 11]. In a recent study, the molecular diagnostic rate of visceral myopathy was 64% (34/53), of which 97% (33/34) could be attributed to *ACTG2* variants. In particular, missense changes in five conserved arginine residues contributed for 49% (26/53) of disease in the cohort. The *ACTG2*-negative cases had a more favorable clinical outcome and more restricted disease. In the *ACTG2*-positive group, the poor outcome (i.e., total parenteral nutrition dependence, need for transplantation, and death) was always due to one of the arginine missense alleles. The analysis of the effect of the specific residues suggests a severity degree of the missense changes, with p.Arg178>p.Arg257>p.Arg40, along with other less frequent variant alleles at p.Arg63 and p.Arg211. Four novel missense variants were also reported, including one that is apparently transmitted according to a recessive mode of inheritance [19], indicating that the overall genetic architecture of visceral myopathy is still to be fully characterized.

Additional genes have been found to play a role in visceral myopathy pathogenesis. *ACTA2*, encoding for a smooth muscle actin gene, is mutated in the multisystemic smooth muscle dysfunction syndrome (MIM #613834). The clinical features include bladder hypotonicity and abnormal intestinal peristalsis as well as a significant involvement of vascular and ciliary smooth muscle, leading to vascular aneurysms and mydriasis [20]. Autosomal-recessive forms of MMIHS are caused by biallelic loss-of-function variants in genes encoding for proteins involved in actin–myosin interactions, such as *MYH11* (myosin heavy chain; [9]), *MYLK* (myosin-light chain kinase; [10, 11]), *LMOD1* (leiomodin 1, an actin-binding protein expressed primarily in vascular and visceral smooth muscles [10, 11]), and *MYL9* (regulatory myosin-light chain; [21]).

Several studies have highlighted the alterations in smooth muscle structural proteins and pathways related to smooth muscle function, providing mechanistic insights in the disease etiology. As an example, loss of LMOD1 in vitro and in vivo results in a reduction of filamentous actin, with elongated cytoskeletal dense bodies and impaired intestinal smooth muscle contractility [10, 11].

2.4 Conclusion and Future Perspectives

In this chapter, we have highlighted a possible strategy based on accurate clinical phenotyping followed by histopathology and in-depth molecular (genetic) analysis, which can be used to reconstruct a model to better understand neuro-ICC-muscular changes underlying severe gut dysmotility such as CIPO. Next-generation sequencing (NGS) has now made possible the analysis of multiple genomic regions simultaneously, shortening the time and cost of gene tests. Thus, several studies have unveiled the presence of many independent genes for severe gut dysmotility by adopting this technique (Fig. 2.2). The discovery of additional genes is crucial to identify altered pathways and is paramount for a better understanding of the disease and development of novel therapeutic targets. Moreover, identifying novel genetic factors is crucial to predict who is at risk of developing the disease, allowing for early detection. Regardless of the rates of clinical manifestations within each genetic group, patients with severe gut dysmotility represent difficult medical challenges, with a tremendous impact of the disease on the quality of life, severe clinical complications, and costly medical care, with minimally effective treatments.

Nonetheless, the future is full of exciting promises relying upon the differentiation of human-

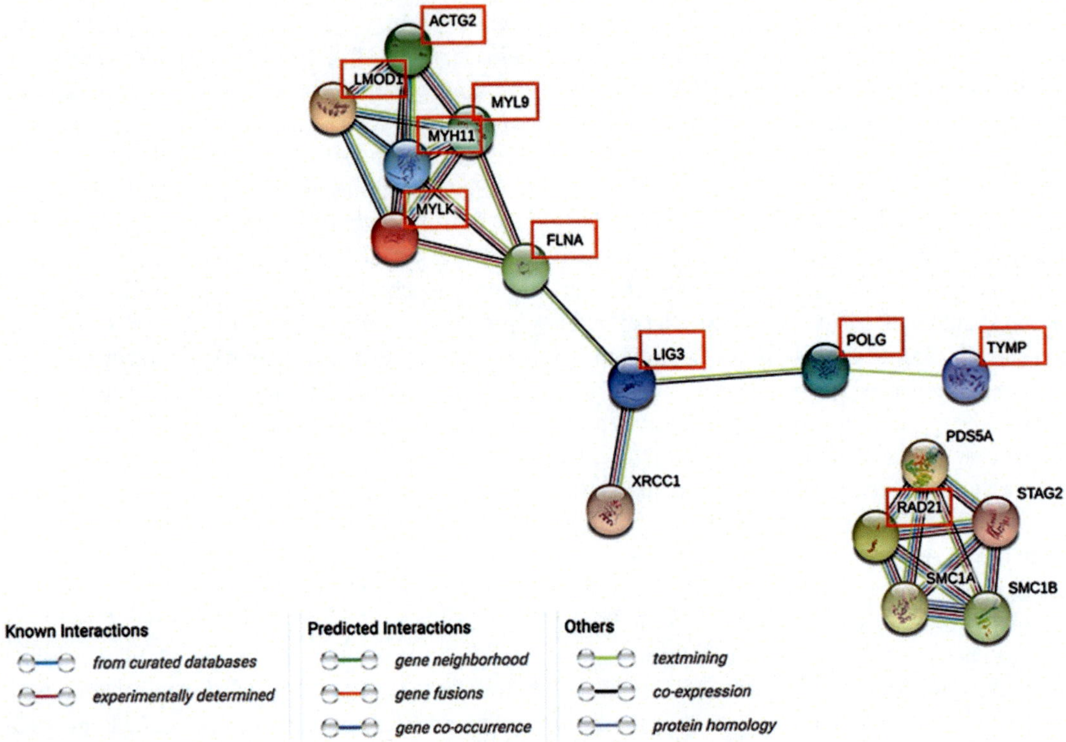

Fig. 2.2 STRING analysis of the proteins involved in severe gut dysmotility. The proteins mutated in severe gut dysmotility disorders (red boxes) cluster into two different groups: the myopathic (i.e., *ACTG2*-related) and/or mitochondrial-related neuro-myopathic forms vs. prevalent enteric neuropathies (i.e., *RAD21* related)

induced pluripotent stem cells (PSCs) directed to the formation of organoids, physiological three-dimensional human organ cultures [16]. A recent study applied a tissue-engineering approach with embryonic and PSC to generate a human intestinal tissue containing functional ENS. The normal intestinal ENS development was recapitulated in vitro by combining human-PSC-derived neural crest cells and developing human intestinal organoids (HIOs), with migration into the mesenchyme, differentiation into neurons and glial cells, and generation of neuronal activity exhibiting rhythmic waves of calcium transients. The ENS-containing HIOs in vivo formed neuroglial structures in a close fashion to a myenteric and submucosal plexus, contained functional interstitial cells of Cajal, and regulated waves of propagating contraction. Equivalent approach was applied to repopulate with a functional ENS a *PHOX2B* mutated Hirschsprung disease mouse model [26]. We see this as the dawn of a new era highlighting pluripotent stem cells as possible treatment options for patients with CIPO and broadly with severe gut dysmotility.

References

1. Bauerschmidt C (2010) Cohesin promotes the repair of ionizing radiation-induced DNA double-strand breaks in replicated chromatin. Nucleic Acids Res 38(2):477–487. https://doi.org/10.1093/nar/gkp976
2. Bianco F (2018) SJ. Expression of RAD21 immunoreactivity in myenteric neurons of the human and mouse small intestine. Neurogastroenterol Motil 30(9):e13429. https://doi.org/10.1111/nmo
3. Bonora E (2015) Mutations in RAD21 disrupt regulation of APOB in patients with chronic intestinal pseudo-obstruction. Gastroenterology 148(4): 771–782.e11. https://doi.org/10.1053/j.gastro.2014.12.034
4. Chen F (2002) Caspase proteolysis of the cohesin component RAD21 promotes apoptosis. J Biol Chem 277(19):16775–16781. https://doi.org/10.1074/jbc.M201322200

5. Cogliandro RF (2011) Patient-reported outcomes and gut dysmotility in functional gastrointestinal disorders. Neurogastroenterol Motil 23:1084–1091. https://doi.org/10.1111/j.1365-2982.2011.01783
6. De Giorgio R (2004) Advances in our understanding of the pathology of chronic intestinal pseudo-obstruction. Gut 53(11):1549–1552. https://doi.org/10.1136/gut.2004.043968
7. Di Nardo G (2017) Chronic intestinal pseudo-obstruction in children and adults: diagnosis and therapeutic options. Neurogastroenterol Motil 29:e12945. https://doi.org/10.1111/nmo.12945
8. Faure AJ (2012) Cohesin regulates tissue-specific expression by stabilizing highly occupied cis-regulatory modules. Genome Res 22(11):2163–2175. https://doi.org/10.1101/gr.136507.111
9. Gauthier J (2015) A homozygous loss-of-function variant in MYH11 in a case with megacystis-microcolon-intestinal hypoperistalsis syndrome. Eur J Hum Genet 23:1266–1268. https://doi.org/10.1038/ejhg.2014.256
10. Halim D (2017a) Loss of LMOD1 impairs smooth muscle cytocontractility and causes megacystis microcolon intestinal hypoperistalsis syndrome in humans and mice. Proc Natl Acad Sci U S A 114(13). https://doi.org/10.1073/pnas.1620507114
11. Halim D (2017b) Loss-of-function variants in MYLK cause recessive megacystis microcolon intestinal hypoperistalsis syndrome. Am J Hum Genet 101:123–129. https://doi.org/10.1016/j.ajhg.2017.05.011
12. Hansen AS (2017) CTCF and cohesin regulate chromatin loop stability with distinct dynamics. elife 6:e25776. https://doi.org/10.7554/eLife.25776
13. Horsfield JA (2012) Diverse developmental disorders from the one ring: distinct molecular pathways underlie the cohesinopathies. Front Genet 3:171. https://doi.org/10.3389/fgene.2012.00171
14. Kagey MH (2010) Mediator and cohesin connect gene expression and chromatin architecture. Nature 467:430–435. https://doi.org/10.1038/nature09380
15. Lattanzio G (2020) Modeling enteric neuropathy: the rad21 knock-in mouse. Eur J Histochem J Funct Cytol. ISSN 1121-760X volume 64/ supplement 3
16. Lehmann R (2019) Human organoids: a new dimension in cell biology. Mol Biol Cell 30:1129–1137. https://doi.org/10.1091/mbc.E19-03-0135
17. Lehtonen HJ (2012) Segregation of a missense variant in enteric smooth muscle actin γ-2 with autosomal dominant familial visceral myopathy. Gastroenterology 143(6):1482–1491.e3. https://doi.org/10.1053/j.gastro.2012.08.045
18. Mannini L (2015) Mutant cohesin affects RNA polymerase II regulation in Cornelia de Lange syndrome. Sci Rep 19(5):16803. https://doi.org/10.1038/srep16803
19. Matera I (2020) Novel ACTG2 variants disclose allelic heterogeneity and bi-allelic inheritance in pediatric chronic intestinal pseudo-obstruction. Clin Genet 99(3):430–436. https://doi.org/10.1111/cge.13895
20. Milewicz DM (2010) De novo ACTA2 mutation causes a novel syndrome of multisystemic smooth muscle dysfunction. Am J Med Genet 152A:2437–2443. https://doi.org/10.1002/ajmg.a.33657
21. Moreno CA (2018) Homozygous deletion in MYL9 expands the molecular basis of megacystis-microcolon-intestinal hypoperistalsis syndrome. Eur J Hum Genet 26(5):669–675. https://doi.org/10.1038/s41431-017-0055-5
22. Panigrahi AK (2012) A cohesin-RAD21 interactome. Biochem J 442(3):661–670. https://doi.org/10.1042/BJ20111745
23. Parelho V (2008) Cohesins functionally associate with CTCF on mammalian chromosome. Cell 132(3):422–433. https://doi.org/10.1016/j.cell.2008.01.011
24. Stanghellini V (2007) Chronic intestinal pseudo-obstruction: manifestations, natural history and management. Neurogastroenterol Motil 19(6):440–452. https://doi.org/10.1111/j.1365-2982.2007.00902
25. Wangler MF (2014) A heterozygous de novo and inherited mutations in the smooth muscle actin (ACTG2) gene underlie megacystis-microcolon-intestinal hypoperistalsis syndrome. PLoS Genet 10(3):e1004258
26. Workman MJ (2017) Human pluripotent-stem-cell-derived intestinal tissues with a functional enteric nervous system. Nat Med 23(1):49–59. https://doi.org/10.1038/nm.4233
27. Xu H (2010) Rad21-cohesin haploinsufficiency impedes DNA repair and enhances gastrointestinal radiosensitivity in mice. PLoS One 5(8):e12112. https://doi.org/10.1371/journal.pone.0012112

Check for updates

3 Luminal Chemoreceptors and Intrinsic Nerves: Key Modulators of Digestive Motor Function

John Dent and Phil G. Dinning

Abstract

This chapter reviews data on the pathways by which luminal, mainly duodenal, chemoreceptors modulate gastro-pyloro-duodenal motor function to control emptying of nutrients into the small intestine. The vagus mediates proximal gastric relaxation caused by nutrient stimulation of duodenal/jejunal mucosal chemoreceptors. Modulation of the spatial patterning and inhibition of antral contractions during duodenal chemoreceptor activation are somewhat conflicting: both vagal control and ascending intramural nerves appear to play a role. Intraduodenal nutrients stimulate the localized pyloric contractions that prevent transpyloric flow via ascending duodenal intramural nerve pathways. Though not yet formally investigated, patterns of activation of the duodenal brake motor mechanism suggest that duodenal loop mucosal chemoreceptors signal to a brake mechanism at the most aborad region of the duodenum via descending intramural duodenal nerves.

Intrinsic intramural pathways are important in the control of the first stages of digestion.

Keywords

Chemoreceptor · Jejunum · Duodenum · Vagus · Motility

Dedication
JD dedicates this chapter to Edwin ("Ed") E Daniel, the most scientifically omnivorous and inquisitive person he has ever had the good fortune to work with.

J. Dent
Royal Adelaide Hospital, Adelaide, SA, Australia

P. G. Dinning (✉)
Gastroenterology Unit, Flinders Medical Centre & College of Medicine and Public Health, Flinders University, Adelaide, SA, Australia
e-mail: phil.dinning@flinders.edu.au

3.1 Introduction

Understanding of the mechanisms of mucosal chemoreception [1] and how chemosensory receptors activate primary afferent nerves [2] have advanced substantially in recent years. By contrast, there is only limited information about the pathways via which mucosal chemoreceptors, mainly duodenal, signal to upper gastrointestinal motor components to vary the rate of movement of nutrients from the stomach and duodenum into the jejunum.

Demonstrations of gut-brain-gut vagal connections [3, 4] have led to the widely held view that the vagus is the primary modulator of normal gastric emptying. This chapter proposes that intrinsic duodenal intramural nerves are the dominant signaling pathways from duodenal chemo-

N. J. Spencer et al. (eds.), *The Enteric Nervous System II*, Advances in Experimental Medicine and Biology 1383, https://doi.org/10.1007/978-3-031-05843-1_3

receptors that cause variations of pyloric, duodenal, and upper jejunal motor function. These variations match the rate of gastric emptying to duodenal digestive function, an essential aspect of normal upper gastrointestinal nutrient digestion.

3.2 Spectrum of Duodenal Chemosensing

The duodenal mucosa is a very complex sensory organ [1, 2]. Chemoreceptor modulations of upper gastrointestinal motor function have been studied mainly by activating this system with "provocative" intraduodenal infusions of components of the postprandial duodenal luminal content. Components that have been studied include fatty emulsions and solutions of acid, isotonic and hypertonic dextrose, bile acids, and hypertonic saline.

3.3 Effects of Truncal Vagotomy on Duodenal Chemoreceptor Control of Gastric Emptying

In chronic studies in five awake, trained dogs, emptying of 600 ml of water from the stomach was measured repeatedly during intraduodenal delivery of either water or three different provocative infusates (250 mM HCl, 10% dextrose and 5% "fat") through an indwelling duodenal cannula (rate of delivery not stated), before and after bilateral transthoracic truncal vagotomy (75 and 95 experiments, respectively) [5]. Figure 3.1 shows that the hierarchy of increasing slowing of gastric emptying by the different intraduodenal infusions persisted after vagotomy, indicating that a major component of the signals from the duodenal chemoreceptors that caused the gradation of slowing of emptying with the provocative infusates, was not vagally transmitted. Compared to the control studies, vagotomy was however associated with slower gastric emptying for all of the 4 duodenal infusates (Fig. 3.1) (see below in Sect. 3.6, third paragraph). This study did not seek to assess gastric motor patterns.

3.4 Chemosensor Control of Proximal Gastric Motor Function

Tonic contraction of the proximal half of the stomach maintains a variable, nonpulsatile compression of the gastric content, which aids non-nutrient and nutrient emptying [4]. In six awake, trained dogs, a proximal jejunal infusion of a liquid nutrient supplement (osmolite) was shown to cause proximal gastric relaxation, which was prevented by reversible cervical or subdiaphragmatic vagal blockade by cooling [6]. This indicates that the vagus is the key control pathway for proximal jejunal chemoreceptor–induced upper gastric relaxation.

3.5 Duodenal Chemoreceptor Control of Antro-Pyloric Motor Function

The literature needs to be interpreted in the light of the significant methodological challenges associated with measurements of antral and pyloric motor functions.

3.5.1 Measurement of Antral and Pyloric Motor Functions

A manometric system that is suitably dampened to exclude rapidly changing contact pressures with the antral wall only detects an antral contraction when this causes active lumen-occlusion, which may only be in the very distal antrum [7]. Thus, though manometry underestimates the extent and number of antral contractions, it detects the lumen-occlusive antral contractions that usually lead to aborad pulsatile transpyloric flows, the major component of gastric emptying [8]. Therefore, manometry is informative about modulations of antral motor function relevant to changes of gastric emptying, especially when pyloric motor function is also reliably recorded. Fluoroscopy, especially when combined with manometry, is a sensitive method for evaluation of antro-pyloric motility but is subject to several

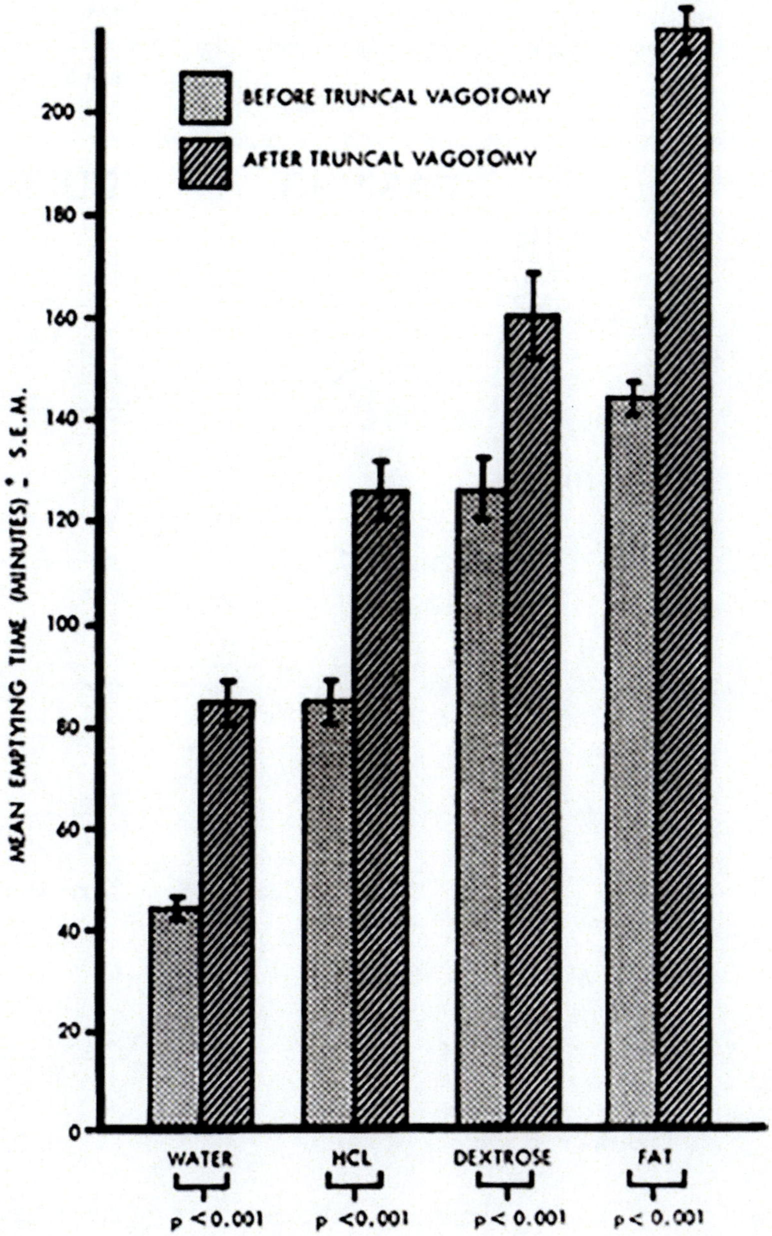

Fig. 3.1 Gastric emptying in five trained dogs before and after bilateral thoracic vagotomy during four types of intraduodenal infusion. The increasing level of stimulation of duodenal chemoreceptors by the 250 mM hydrochloric acid, 10% dextrose, and 5% fat infusates still slowed emptying progressively after vagotomy. The global slowing of emptying after vagotomy is discussed in Sect. 3.6, second paragraph. (Reproduced and redrawn with permission from Ref. [5])

major practical constraints, especially radiation exposure in humans [9].

The challenge of manometric recording of pyloric motor function was demonstrated in humans [10] during intraduodenal infusion of a high-calorie triglyceride emulsion; this stimulated highly localized or "isolated" phasic and tonic contractions of the pyloric ring (Fig. 3.2). These localized pressures were detected reliably by the sleeve sensor, as its 3.6 cm pressure-sensing length remained in contact with the less than 9-mm-wide region of isolated pyloric tonic and phasic contractions [10], despite unavoidable movements of the pylorus relative to the manometric assembly. Such movements confound recordings with single point pressure sensors, including high-resolution manometric catheters with their sensor spacings of no less than 10 mm (Fig. 3.2).

The sleeve, coupled with transmucosal potential difference monitoring to validate its position astride the pylorus [10], remains the only valid

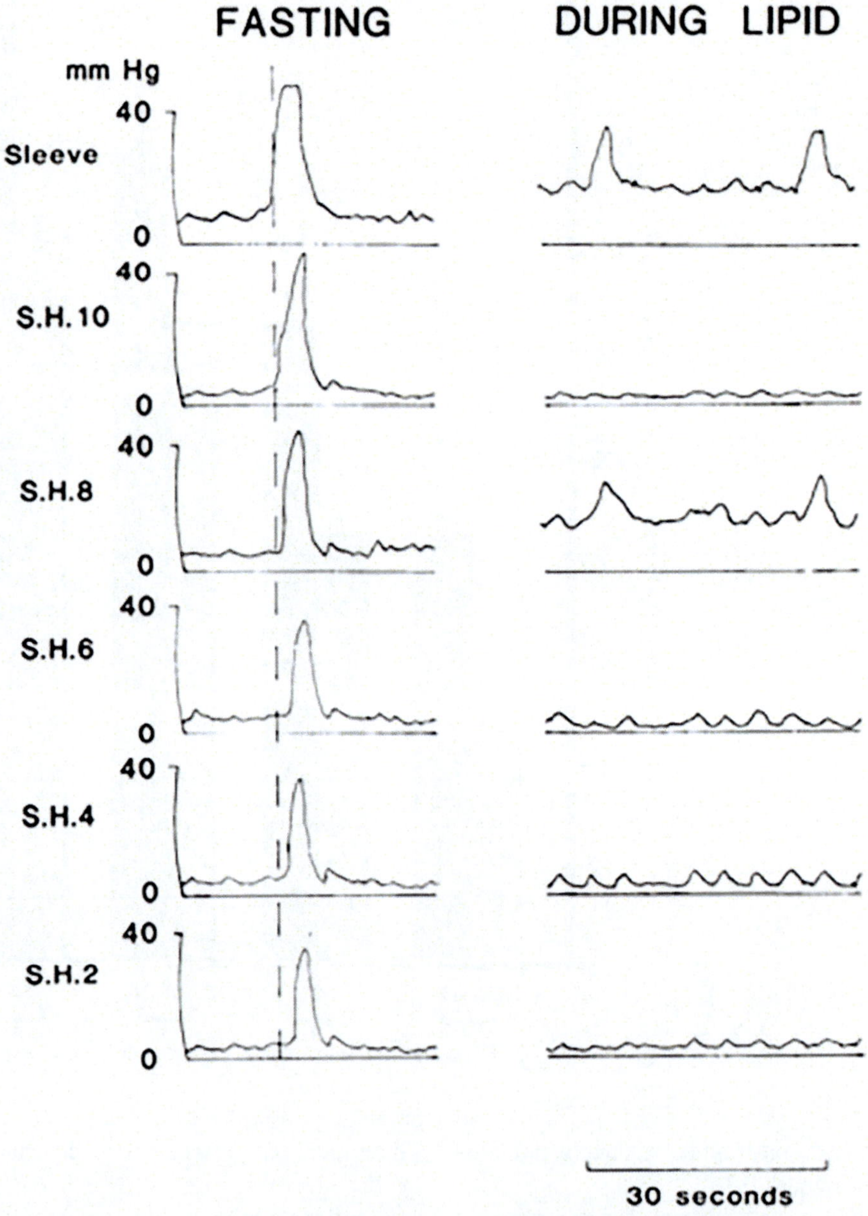

Fig. 3.2 Manometric tracings in a healthy volunteer from the sleeve, positioned astride the pylorus (uppermost trace), and an array of side-holes 6 mm apart (SH 10–2) along a 24 mm segment of the sleeve. The left-hand panel shows a fasting phase II antro-pyloro-duodenal pressure event sweeping across the pylorus. The right-hand panel recording was made during inhibition of antral pressure events and stimulation of localized contractions of the pyloric ring with intraduodenal fat infusion. The isolated pyloric tonic and phasic contractions are only present over side-hole 8, documenting the very narrow zone of pyloric activity. The sleeve tracing also records the phasic and tonic pyloric motor activities captured by side-hole 8. (Reproduced with permission from Ref. [10])

method for monitoring of pyloric motor function. This technique was used in all of the studies of pyloric motility cited below.

3.5.2 Chemoreceptor Control of Antral and Pyloric Motor Functions in Humans

In humans, emptying of a small radiolabeled nutrient meal into the duodenum or direct intraduodenal infusion of substances that stimulate duodenal chemoreceptors causes inhibition of antral motility and stimulation of localized tonic and phasic pyloric contractions [10–13]. This pattern of response is maintained for at least 90 minutes by continuous provocative intraduodenal infusions [10, 14–17] and is blocked by atropine [15, 17].

Concurrent fluoroscopy and manometry in 11 healthy men has directly confirmed that a provocative duodenal infusion causes active closure of the pylorus during localized pyloric tonic and phasic contractions [9] (Fig. 3.3). This finding emphasizes the capability of the pylorus to seal off the exit from the stomach, while inhibition of the proximal and distal stomach contractions can only suspend gastric pumping and grinding. The "plug in the bottle" function of the pylorus might be a mechanism that becomes progressively more important, as the larger the size of the meal, the greater is the passive force that favors emptying across a relaxed pylorus. However, there is a dearth of data on gastro-pyloro-duodenal motor function after consumption of a normal-sized "main" meal.

3.5.3 Studies in Pigs and Dogs of Pathways via Which Duodenal Chemoreceptors Alter Antral and Pyloric Motor Function

In chronic studies in awake pigs with intact vagi, pyloric excision abolished the inhibition of antral pressure events normally induced by duodenal infusion of an isotonic dextrose/saline solution [18]. This suggests that the antral inhibition is signaled by intramural nerves (divided by the pyloric excision) that travel orad from the duodenum to the antrum, a conclusion that is at odds with studies of duodenal transection discussed in the next paragraph.

Duodenal transection with immediate reanastomosis (and no resection) restores normal gross anatomy but results in persisting interruption of intramural nervous pathways. This intervention has only been used in pigs to investigate pathways activated by duodenal distension that slow gastric emptying. In six unsedated, trained pigs, midduodenal balloon distension caused a dose-related stimulation of localized pyloric contractions, inhibition of antral pressure events, and retardation of gastric emptying [19]. In another six pigs whose duodenum had been transected just aborad of the pylorus, duodenal balloon distension failed to stimulate pyloric motor function, consistent with this being caused by ascending intramural nerves. Confusingly though, given the findings discussed in the paragraph immediately above [18], the duodenal balloon distension-induced antral motor inhibition persisted [19], suggesting these effects were vagally induced.

In a study of antral transection in five awake pigs, control measurements before transection confirmed that intraduodenal infusion of 25% dextrose stimulated localized pyloric contractions and inhibited antral pressure events [20]. The antrum was then transected 2 cm orad of the pylorus: as expected, the 25% dextrose-induced stimulation of localized pyloric contractions persisted, but the inhibition of antral motility was unaffected by the antral transection, consistent with it being mediated via the vagus.

An approach developed by E.E Daniel has been used to investigate pathways of duodenal control of pyloric motor function in acute studies on laparotomized, anesthetized dogs [21, 22] (Fig. 3.4). Bilateral cervical vagotomy had no consistent effect on sleeve-recorded localized basal pyloric motility. Electrical field stimulation of the proximal duodenum stimulated localized

Fig. 3.3 Fluoroscopic appearances of the antrum (right side) pylorus (center) and proximal duodenum (left side) in a healthy human volunteer during stimulation of localized pyloric contractions by an intraduodenal fat infusion. The concurrent pyloric sleeve pressure recording displayed in the lower frame synchronized fluoroscopy with manometry. The narrow pyloric ring is closed over the sleeve assembly (the radio-opacity across the pyloric ring is due to a spring-wire sleeve stiffener), and during this motor pattern, no barium traversed the pylorus. (Reproduced with permission from Ref. [9])

pyloric contractions, an effect blocked by the targeted intra-arterial delivery of atropine, hexamethonium, and tetrodotoxin to the proximal duodenal/pyloric region. Duodenal transection orad of the duodenal field stimulation electrodes also abolished their effect on pyloric motility, indicating that the pylorus was stimulated by an ascending intramural pathway from the duodenum to the pyloric ring [21].

A further study in anesthetized dogs [22] documented stimulation of phasic and tonic localized pyloric and proximal duodenal pressure events within several seconds of the start of an intraduodenal infusion (0.92 ml/minute for 2 minute) of 0.1 normal hydrochloric acid. This response was mimicked by intraduodenal infusion of the sensory nerve-end stimulant, phenylbiguanide. In contrast, these acid-induced pyloric phasic and tonic pressure events were blocked by an intraduodenal infusion of 2% xylocaine and intravenous or close intra-arterial injection of atropine or hexamethonium. Bilateral cervical vagotomy had no effect on these pyloric responses.

3.6 Synthesis: Paths of Duodenal Chemoreceptor Control of Antral and Pyloric Motor Function During Normal Nutrient Processing

Though the data summarized above on the pathways of duodenal chemoreceptor control of antral motor function are conflicting, the weight of the evidence suggests that the vagi are the main signaling pathway through which duodenal chemoreceptors modulate antral motility via a vago-vagal pathway.

Data on the pylorus point overwhelmingly to duodenal chemoreceptors stimulating the pylorus via ascending duodenal intramural nerves. Pharmacological data, mainly from anesthetized dogs, are consistent with this pathway being a chain of orally projecting cholinergic nerves.

The existence of a rich extrinsic innervation in the stomach and pylorus [4] is a reminder that the vagus can influence gastric motor function as well as secretion. The retardation of emptying by

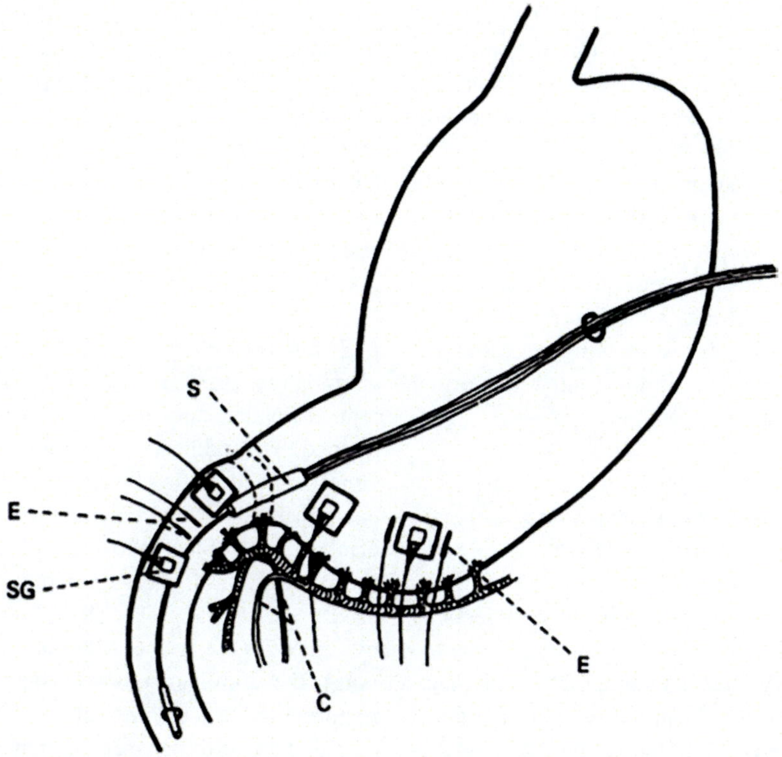

Fig. 3.4 Schematic of the technique developed by EE Daniel, which was used by Allescher et al. [22] to investigate pyloric motor function in the anesthetized dog. A multilumen sleeve/side-hole catheter was passed via a gastrotomy so that the 3 cm sleeve (S) was astride the pylorus and side-holes (SH) were in the antrum and duodenum at each end of the sleeve. Strain gauges (SG) were sutured to the antrum and proximal duodenum. Silver wire electrical field stimulation electrodes (E) were embedded in the proximal duodenum. Two cannulae enabled close intra-arterial injections of drugs to the pylorus and antrum. The perfusion territory covered by these cannulae was confirmed initially by observing blanching following an injection of a bolus of Krebs solution and at the end of each study by injection of black ink. (Reproduced with permission from Ref. [21])

truncal vagotomy [5] mentioned in Sect. 3.3 (Fig. 3.1) was probably due to interruption of descending vagal nerves that modulate the spatial patterning of gastric motor function. These extrinsic nerves enter the wall of the stomach and then project intramurally along the antrum. Studies in pigs have shown that division of these nerves by antral transection with reanastomosis retards gastric emptying by causing earlier-than-normal closure of the pyloric ring during antropyloric contractions, with reduction of the volume of flow pulses delivered into the duodenum [20].

The modulation of motor mechanisms by duodenal chemoreceptors that occurs within seconds, and the data from physical and pharmacological interventions on duodenal intramural nervous pathways indicate that neural pathways are of paramount importance for driving duodenal chemoreceptor–induced modulations of antropyloric motor function. This does not exclude the possibility that hormone release can, in certain circumstances, also play a modulatory role.

3.7 Duodenal Chemoreceptors and the Duodenal Brake Mechanism

The term "duodenal brake" appears to have first been used by Shadidullah et al. in 1975 [5], but this term was then used to refer to the ability of duodenal luminal receptors to slow gastric emptying, through undefined motor mechanisms. Rao et al. have revived and extended the concept of a "duodenal brake" with limited fluoroscopic data that indicate that there is a motor mechanism, which can prevent outflow of duodenal content into the proximal jejunum [23].

In a recent manometric study summarized below, we have described a specialized zone of motor function, which extends from the very distal duodenum into the proximal jejunum; we have proposed that this is the brake mechanism. Furthermore, we suggest that this mechanism plays a significant role in the control of normal nutrient emptying.

3.7.1 Fluoroscopic Demonstration of the Duodenal Brake

Fluoroscopic observations in healthy subjects by Rao et al. (1996) have revealed a motor mechanism at the very distal duodenum of healthy subjects, which causes closure of the extreme distal duodenal lumen [23]. This occlusion was triggered by an intraduodenal infusion of the lipid-rich sodium oleate solution made radio-opaque with a barium suspension. The duodenal closure developed within about 15 seconds of starting the 1 minute 0.33 ml/second infusion (Fig. 3.5). Control normal saline/barium duodenal infusions passed freely into the jejunum.

The above duodenal infusates were probably supraphysiological and so, activation of the duodenal brake could have been either due to a mechanism normally active during gastric emptying of nutrient or to a duodenal "alarm" response to a pathologically intense stimulus.

3.7.2 Manometric Definition of the Duodenal Brake

The study of Rao et al. [23] included measurements from several duodenal manometric sensors, but all of these were positioned well orad of the duodenal brake zone defined on fluoroscopy. In a study of healthy subjects, we have used a high-resolution, fiber-optic manometric catheter incorporating 72 sensors, each spaced at 1 cm intervals [24]. This had been modified especially for upper gastrointestinal intubation [25] and enabled sensors to be placed into the region of the duodenal brake, allowing for the first detailed manometric recordings of fed-state pressure patterns within the region. After a period of fasting recordings, fed-state pressure patterns were recorded for at least an hour after ingestion of a 200 ml, 480 kcal nutrient drink.

Fed-state recordings In all 15 subjects, the most aborad pressure sensor was beyond the duodeno-jejunal flexure, and in 12, it was at least 15 cm beyond this landmark. Thus, the zone of the duodenal brake identified by Rao et al. [24] was well encompassed. After ingestion of the nutrient drink, each subject showed development of a sharp transition of pressure patterns at a mean of 18.8 ± 3.7 cm (range 13–28 cm) aborad from the pylorus [25] (Fig. 3.6). Beyond this transition point, the dominant motor pattern was regular, nonpropagating pressure events at 11.5 ± 0.5 cycles per minute, which extended to the end of the recording sensors in all subjects, in some cases for more than 20 cm aborad from the transition point. We named this motor pattern the duodeno-jejunal complex (DJC).

Orad of the transition point, in the duodenal loop, pressure events were less frequent and dominated by aborad-propagated events at a rate of 4–6/minute, more than half of which travelled more than 10 cm along the duodenal loop region (Figs. 3.6 and 3.7). These data show that the duodenum and upper jejunum have two major zones from the perspective of fed-state motor function: the zone between the pylorus and transition point has been named the duodenal loop (DL) region and that aborad from the transition the duodeno-jejunal (DJ) region [25].

Typically, vigorous DJC activity started within 90 seconds after complete ingestion of the nutrient drink. In some subjects, DJC activity occurred in clusters, the onsets of which were usually associated with pressure events that propagated from the antrum across the pylorus deep into the DL region (Fig. 3.7), a pattern previously shown to propel a pulse of gastric content into the duo-

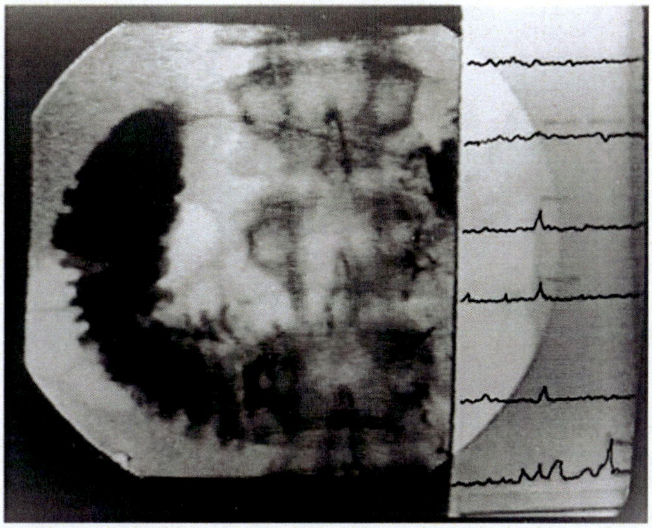

Fig. 3.5 This shows simultaneous fluoroscopic imaging and manometry of the duodenum. The brake was stimulated with sodium oleate mixed with a 20% barium suspension. Fluoroscopy showed occlusion of the most distal part of the duodenum for the entire 52 seconds of fluoroscopic observation: the infusate remained within the duodenal loop. The right-hand panel displays concurrently recorded duodenal loop pressures, but the most aborad sensor was not within the occluded distal duodenum/proximal jejunum. (Reproduced with permission from Ref. [23])

denum [8]. In all subjects, DJC activity remained prominent for the hour of fed-state recordings.

Fasting recordings Review of the period of fasting recording immediately before ingestion of the nutrient drink also revealed a less obvious transition point of motor function in the same position as in the fed state (Fig. 3.8). Aborad of the transition, there was short-extent, low-amplitude DJC activity, which usually occurred in brief clusters, associated with just-prior groups of vigorous, propagated phase II antro-duodenal pressure events, most of which extended more than 10 cm into the duodenal loop region (Fig. 3.8). These events are known to empty pulses of acid into and along the duodenum [26], suggesting that fasting DJC activity is driven at least in part by episodic entry of presumably small volumes of acid into the duodenum during the highly expulsive Phase II interdigestive motor activity. The right-hand panel of Fig. 3.8 shows a period of fed-state recording in the same subject for comparison.

3.8 Synthesis: Interpretation of Duodeno-Jejunal Complex Activity

3.8.1 Spatial Correlation of DJC Activity

An abdominal X-ray at the end of recordings described above defined the positions of the radio-opaque fiber-optic sensors, so the position of the transition point of motor function at the orad margin of DJC activity could be accurately located in each subject. This position was consistent with the localization of the duodenal brake by Rao et al. [23]

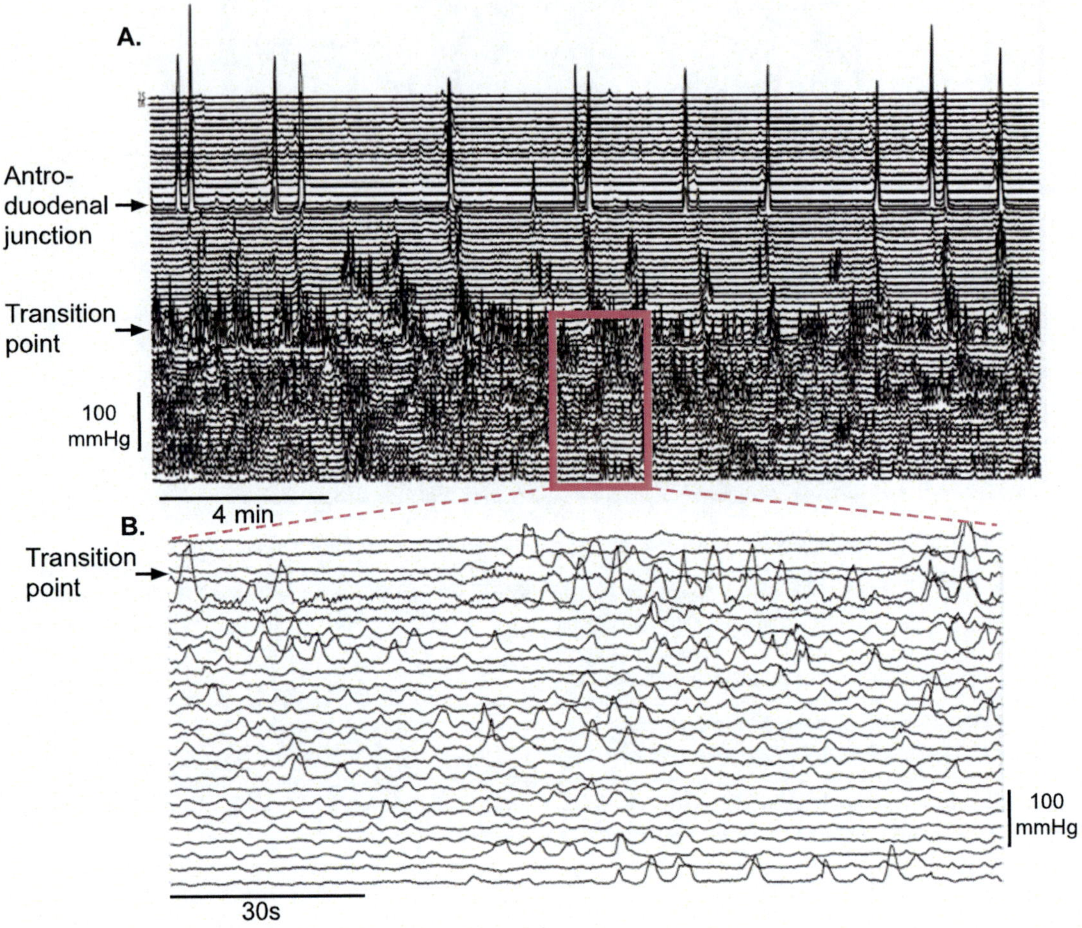

Fig. 3.6 The upper panel (**a**) shows a 22-minute period of very intense fed-state duodeno-jejunal complex (DJC) activity, which emphasizes the existence of the transition point between the differing motor patterns in the duodenal loop and duodeno-jejunal regions. The lower panel (**b**) shows a very time-expanded small section of panel A. This shows the nonpropagated obstructive pattern of DJC activity, which extends well into the proximal jejunum. (Reproduced with permission from Ref. [25])

3.8.2 The Mechanical Outcome of DJC Activity

The repetitive, nonperistaltic, and high-frequency pattern of DJC pressure events is consistent with it impeding luminal flow (Fig. 3.6). The pattern of DJC activity resembles phase III of the migrating motor complex, except it is somewhat "ragged," originates close to the duodeno-jejunal flexure, and does not propagate aborad along the duodenum and jejunum in a stereotyped pattern. Correlation of fluoroscopy with manometry by Schemann and Ehrlein (1986) has shown that the segment of small intestine involved in phase III activity of the migrating motor complex is occluded [27].

3.8.3 Entry of Chyme into the Duodenum Stimulates DJC Activity

The intensity, prevalence, and distal extent of DJC activity were strikingly greater in the fed state compared to fasting, consistent with

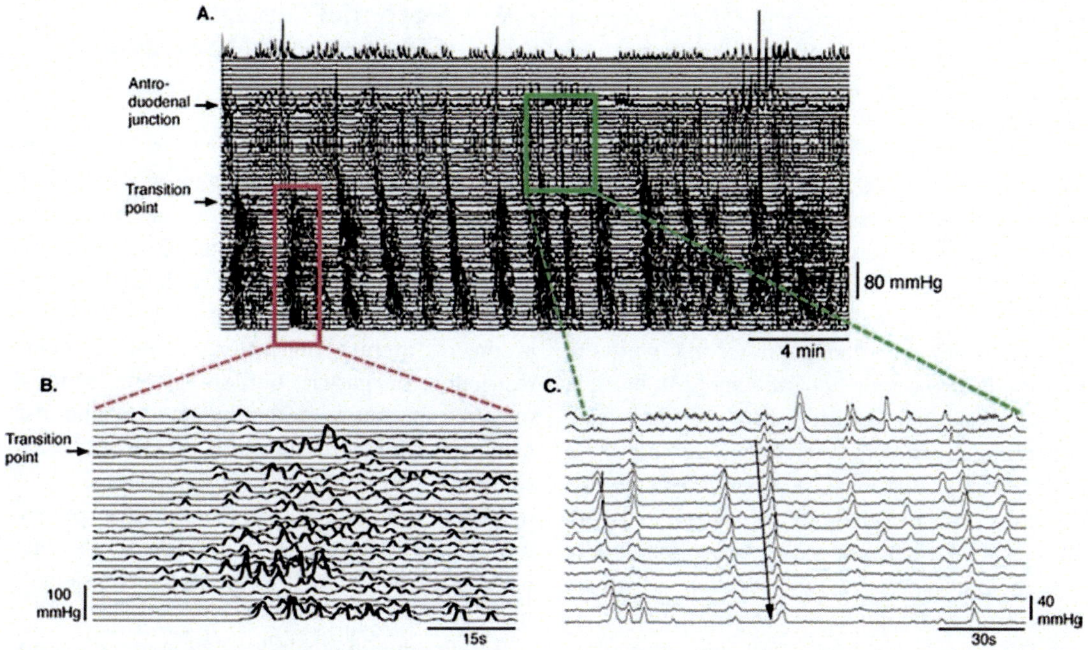

Fig. 3.7 The top panel (**a**) shows a 26-minute period of fed-state clustered DJC activity. The two panels below are time-expanded tracings (see time-base at the bottom of each panel). The left panel (**b**) mainly shows pressures aborad of the transition point. The right panel (**c**) is entirely from the duodenal loop region. Most of the very long segment, relatively slowly propagated antro-duodenal pressure events travel to the transition point. (Reproduced with permission from Ref. [25])

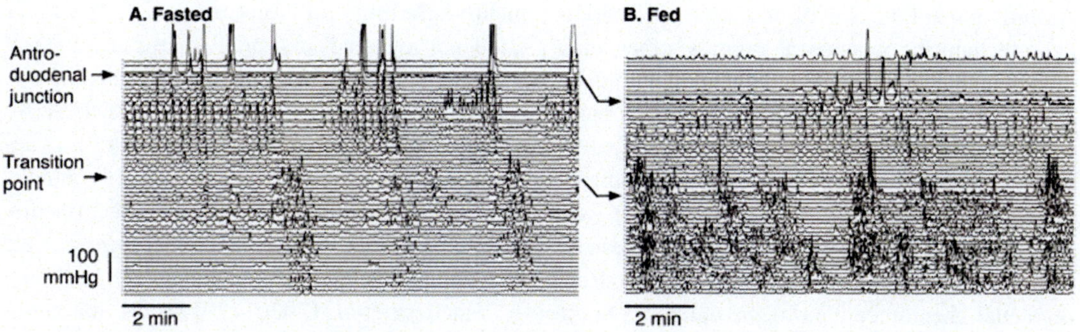

Fig. 3.8 The two panels show fasting and fed-state recordings from the same subject. The left-hand panel shows three main clusters of fasting DJC activity, following a series of phase II antro-duodenal pressure events. Following the nutrient drink (right-hand panel), DJC activity is much more sustained and vigorous. (Reproduced with permission from Ref. [25])

their stimulation by chyme. As described above, clusters of DJC activity were usually closely preceded by expulsive antro-duodenal pressure events both during fasting and after the nutrient drink was consumed (Figs. 3.7 and 3.8).

3.8.4 Physiological Significance of DJC Activity

The considerable DJC activity that occurred during emptying of a very modest volume of highly nutrient liquid indicates that DJC activity is a physiological mechanism, active during normal gastric emptying. An obvious next step in researching DJC activity would be to determine directly whether and when this motor activity prevents duodenal outflow and whether this is stimulated in a dose-related fashion by graded rates of delivery of provocative solutions into the duodenum. Such studies could be done in animals.

We have proposed [25] that the retention of content within the duodenum, along with the mixing duodenal loop contractions [23], ensures that the normal gastric, duodenal mucosal, hepatic, and pancreatic secretions that all come together in the duodenal loop (the most complex digestive region of the gut) are well mixed with chyme. We further propose that once the duodenal chemoreceptors detect adequate chemical processing of the duodenal content, the stimulus for DJC activity subsides, allowing emptying into the jejunum.

3.8.5 Pathways of Stimulation of DJC Activity

Assuming that DJC activity is a basic physiological mechanism present in animals other than humans, the pathways that control it could be studied directly in acute experiments with techniques similar to those used by Allescher et al. in dogs [21] (Fig. 3.4). The clustered patterning of DJC activity in close temporal association with propagated expulsive antro-duodenal pressure events is consistent with aborad signaling from duodenal chemoreceptors to the duodenal brake region on a second-by-second basis. Neural signaling is the only feasible mechanism.

Given the data that support signaling from duodenal chemoreceptors via ascending intramural nerve pathways to the pylorus discussed in Sect. 3.6, by far the most likely neural signaling pathway would seem to be descending intramural nerves to the region of the duodenal brake.

3.9 Potential Clinical Significance of the Duodenal Brake

This can only be a "blue sky" commentary, given the limited data available. An abnormally overactive duodenal brake has the potential to hold the normal upstream motor mechanisms that control gastric emptying to ransom. Duodenal dilatation activates localized pyloric contractions and inhibition of antral contractions, with attendant restriction of gastric outflow [19]. Abnormal retention of content in the duodenum could also cause persistent stimulation of duodenal chemoreceptors. The other side of the coin, a "lazy" duodenal brake mechanism, could allow premature release of the duodenal content into the jejunum, with attendant symptoms due to jejunal maldigestion.

By determining the rate of emptying of carbohydrates into the duodenum, duodenal and probably jejunal chemoreceptors also influence glucose tolerance. A high sugar and fat diet causes hypertrophy of the duodenal mucosa in rats with an unusually high density of enterochromaffin cells, in association with development/worsening of diabetes [28]. Similar duodenal mucosal and small intestinal mucosal hypertrophy has also been reported in diabetic patients compared to healthy controls [28]. Furthermore, when the duodenum (and its chemosensors) is bypassed surgically in patients with type 2 diabetes and nonalcoholic liver disease, blood glucose control and measures of liver disease activity have improved [28].

Van Baar et al. (2020) [29] report that endoscopic hydrothermal ablation of all of the duodenal mucosa aborad of the sphincter of Oddi in 37 poorly-controlled type 2 diabetic patients was associated with clinically valuable improvements of measures of blood glucose control and blood levels of markers of nonalcoholic fatty liver disease ($p < 0.001$). These effects persisted at 12 months. Unfortunately, this study did not include any measurements of motor function or detailed reporting of duodenal mucosal histology. These remarkable results emphasize the importance of working toward better understanding of duodenal chemoreceptor signaling pathways.

References

1. Iwasaki M, Akiba Y, Kaunitz JD (2018) Duodenal chemosensing. Curr Opin Gastroenterol 34:422–427
2. Bellono NW, Bayrer JR, Leitch DB et al (2017) Enterochromaffin cells are gut Chemosensors that couple to sensory neural pathways. Cell 170(185–198):e16
3. Iggo A (1957) Gastric mucosal chemoreceptors with vagal afferent fibres in the cat. Q J Exp Physiol Cogn Med Sci 42:398–409
4. Goyal RK, Guo Y, Mashimo H (2019) Advances in the physiology of gastric emptying. Neurogastroenterol Motil 31:e13546
5. Shahidullah M, Kennedy TL, Parks TG (1975) The vagus, the duodenal brake, and gastric emptying. Gut 16:331–336
6. Azpiroz F, Malagelada JR (1986) Vagally mediated gastric relaxation induced by intestinal nutrients in the dog. Am J Phys 251:G727–G735
7. Fone DR, Akkermans LM, Dent J et al (1990) Evaluation of patterns of human antral and pyloric motility with an antral wall motion detector. Am J Phys 258:G616–G623
8. Anvari M, Dent J, Malbert C et al (1995) Mechanics of pulsatile transpyloric flow in the pig. J Physiol 488(Pt 1):193–202
9. Tougas G, Anvari M, Dent J et al (1992) Relation of pyloric motility to pyloric opening and closure in healthy subjects. Gut 33:466–471
10. Heddle R, Dent J, Toouli J et al (1988) Topography and measurement of pyloric pressure waves and tone in humans. Am J Phys 255:G490–G497
11. Houghton LA, Read NW, Heddle R et al (1988) Motor activity of the gastric antrum, pylorus, and duodenum under fasted conditions and after a liquid meal. Gastroenterology 94:1276–1284
12. Houghton LA, Read NW, Heddle R et al (1988) Relationship of the motor activity of the antrum, pylorus, and duodenum to gastric emptying of a solid-liquid mixed meal. Gastroenterology 94: 1285–1291
13. Heddle R, Collins PJ, Dent J et al (1989) Motor mechanisms associated with slowing of the gastric emptying of a solid meal by an intraduodenal lipid infusion. J Gastroenterol Hepatol 4:437–447
14. Heddle R, Dent J, Read NW et al (1988) Antropyloroduodenal motor responses to intraduodenal lipid infusion in healthy volunteers. Am J Phys 254:G671–G679
15. Fraser R, Fone D, Heddle R et al (1992) Stimulation of pyloric contractions by intraduodenal triglyceride is persistent and sensitive to atropine. J Gastroenterol Hepatol 7:563–568
16. Heddle R, Fone D, Dent J et al (1988) Stimulation of pyloric motility by intraduodenal dextrose in normal subjects. Gut 29:1349–1357
17. Fone DR, Horowitz M, Dent J et al (1989) Pyloric motor response to intraduodenal dextrose involves muscarinic mechanisms. Gastroenterology 97:83–90
18. Treacy PJ, Jamieson GG, Dent J (1994) Pyloric motility and liquid gastric emptying during barostatic control of gastric pressure in pigs. J Physiol 474:361–366
19. Treacy PJ, Jamieson GG, Dent J (1996) The effect of duodenal distention upon antro-pyloric motility and liquid gastric emptying in pigs. ANZ J Surg 66:37–40
20. Anvari M, Yu P, Dent J et al (1995) Role of antral intramural neural pathways in control of gastric emptying in the pig. J Physiol 488(Pt 1):203–209
21. Allescher HD, Daniel EE, Dent J et al (1988) Extrinsic and intrinsic neural control of pyloric sphincter pressure in the dog. J Physiol 401:17–38
22. Allescher HD, Daniel EE, Dent J et al (1989) Neural reflex of the canine pylorus to intraduodenal acid infusion. Gastroenterology 96:18–28
23. Rao SS, Lu C, Schulze-Delrieu K (1996) Duodenum as a immediate brake to gastric outflow: a videofluoroscopic and manometric assessment. Gastroenterology 110:740–747
24. Dinning PG, Wiklendt L, Maslen L et al (2014) Quantification of in vivo colonic motor patterns in healthy humans before and after a meal revealed by high-resolution fiber-optic manometry. Neurogastroenterol Motil 26:1443–1457
25. Dent J, Deloose E, Dinning P et al (2020) Manometric demonstration of duodenal/jejunal motor function consistent with the duodenal brake mechanism. Neurogastroenterol Motil 32:e13835
26. Dent J, Chir B (1976) A new technique for continuous sphincter pressure measurement. Gastroenterology 71:263–267
27. Schemann M, Ehrlein H (1986) Mechanical characteristics of phase II and phase III of the interdigestive migrating complex in dogs. Gastroenterology 91:117–123
28. Cherrington AD, Rajagopalan H, Maggs D et al (2017) Hydrothermal duodenal mucosal resurfacing: role in the treatment of metabolic disease. Gastrointest Endosc Clin N Am 27:299–311
29. van Baar ACG, Holleman F, Crenier L et al (2020) Endoscopic duodenal mucosal resurfacing for the treatment of type 2 diabetes mellitus: one year results from the first international, open-label, prospective, multicentre study. Gut 69:295–303

4 Nitrergic and Purinergic Nerves in the Small Intestinal Myenteric Plexus and Circular Muscle of Mice and Guinea Pigs

Alberto Perez-Medina and James J. Galligan

Abstract

ATP is an excitatory and inhibitory neurotransmitter, while nitric oxide (NO) is an inhibitory neurotransmitter in the enteric nervous system (ENS). We used a vesicular nucleotide transporter (SLC17A9, VNUT) antibody and a nitric oxide synthase (NOS) antibody to identify purinergic and nitrergic nerves in mouse and guinea ileum. *Mouse*: VNUT-immunoreactivity (ir) was detected in nerve fibers in myenteric ganglia and circular muscle. VNUT-ir fibers surrounded choline acetyltransferase (ChAT), nitric oxide synthase (nNOS), and calretinin-ir neurons. VNUT-ir nerve cell bodies were not detected. Tyrosine hydroxylase (TH)-ir nerves were detected in myenteric ganglia and the tertiary plexus. *Guinea pig*: VNUT-ir was detected in neurons and nerves fibers and did not overlap with NOS-ir nerve fibers. VNUT-ir was detected in nerve fibers in ganglia but not nerve cell bodies. VNUT-ir nerve fibers surrounded NOS-ir and NOS$^-$ neurons. NOS-ir and VNUT-ir nerve fibers did not overlap in myenteric ganglia or circular muscle. VNUT-ir nerves surrounded some ChAT-ir neurons. VNUT-ir and ChAT-ir were detected in separate nerves in the CM. VNUT-ir nerve fibers surrounded calretinin-ir neurons.

Conclusions: VNUT-ir neurons likely mediate purinergic signaling in small intestinal myenteric ganglia and circular muscle. ATP and NO are likely released from different inhibitory motorneurons. VNUT-ir and ChAT-ir interneurons mediate cholinergic and purinergic synaptic transmission in the myenteric plexus.

Keywords

Purinergic nerve · Inhibitory Neurotransmission · ATP · Enteric nervous system

A. Perez-Medina
Department of Pharmacology and Toxicology, Michigan State University, East Lansing, MI, USA

J. J. Galligan (✉)
Department of Pharmacology and Toxicology, Michigan State University, East Lansing, MI, USA

The Neuroscience Program, Michigan State University, East Lansing, MI, USA
e-mail: galliga1@msu.edu

Highlights

- ATP and nitric oxide are important synaptic and inhibitory neuromuscular transmitters in the myenteric plexus.
- Nitric oxide synthase (NOS) is a marker for inhibitory motorneurons, while the vesicular

N. J. Spencer et al. (eds.), *The Enteric Nervous System II*, Advances in Experimental Medicine and Biology 1383, https://doi.org/10.1007/978-3-031-05843-1_4

nucleotide transmitter (VNUT) is a purinergic nerve marker.
- Nitric oxide (NO) and ATP may be cotransmitters, which cause muscle relaxation that is needed to establish intestinal pressure gradients allowing for efficient propulsion of gastrointestinal content.

4.1 Introduction

ATP (or related purine) and nitric oxide (NO) are enteric neurotransmitters. Electrophysiological studies show that purinergic signaling produces fast and slow synaptic excitation of myenteric neurons [3, 21, 26] in addition to fast inhibitory junction potentials (fIJPs) recorded from circular smooth muscle [14, 15, 17, 23], and possibly excitatory junction potentials (EJPs) in the longitudinal smooth muscle layer [25, 51]. P2X ionotropic receptors and G-protein-coupled P2Y purinergic receptors mediate neuronal excitatory and smooth muscle inhibitory and excitatory responses, respectively, and these mechanisms participate in control of gastrointestinal (GI) motility [1]. Although there is substantial functional data that indicate there are purinergic neurons in the ENS, attempts to identify these neurons using neurochemical tools have been unsuccessful due primarily to lack of suitable purinergic markers. Identification of purinergic nerves has been attempted using quinacrine-induced fluorescence of ATP stores [48], or by identification of tritiated ATP in vesicles [8]. These methods are not ideal because they target ATP indiscriminately, which is ubiquitously expressed in all cells.

Identification and cloning of the vesicular nucleotide transporter (VNUT, SLC17A9) in pheochromacytoma-12 (PC-12) cells facilitated studies of purinergic neurons using VNUT antibodies [45]. VNUT is a 430 amino acid, 12 transmembrane spanning nucleotide transporter from the SLC17 family that transports ATP into synaptic vesicles [45]. ATP signaling contributes to neuropathic hypersensitivity after peripheral nerve injury (PNI), and decreased expression of VNUT in SC17A9-/- mice is associated with diminished neuropathic pain after spinal nerve injury. These data highlight the role of VNUT-dependent ATP exocytosis in the nervous system. [35]. In this study, we used a VNUT antibody to identify the subsets of myenteric neurons and nerve fibers that may coexpress VNUT in the mouse and guinea pig ileum.

4.2 Materials and Methods

4.2.1 Animals

Male guinea pigs (200–400 g weight) obtained from Charles River Laboratories (Matawan, MI USA) and male C57/Bl6J mice (2–3 months old) were obtained from Jackson Laboratories (Bar Harbor, ME). Mice and guinea pigs were deeply anesthetized via isoflurane inhalation and were killed by bilateral pneumothorax. All procedures were approved by the Institutional Animal Use and Care Committee at Michigan State University (Animal use protocol #PROTO201900058).

4.2.2 Tissue Preparation

The proximal small intestine was harvested following euthanasia and fixed for immunohistochemistry. Tissues were cleaned and pinned with the mucosa facing up in a Sylgard-lined petri dish and stored overnight at 4 °C in Zamboni's fixative (4% formaldehyde with 5% picric acid in 0.1 M sodium phosphate buffer, pH 7.2). After overnight fixation, tissues were washed with 0.1 M phosphate buffer (PB) solution, (pH 7.2), 3 times at 10-minute intervals, or until the wash buffer was free of yellow coloring due to the presence of picric acid. Whole-mount longitudinal muscle myenteric plexus (LMMP) preparations were made following microdissection of the mucosa, submucosa, and circular muscle layers. Circular muscle strips were kept for immunohistochemical labeling. LMMP preps and circular muscle strips were then incubated overnight at 4 °C with primary antibodies. The tissue preps were then washed 3 X at 10-minute intervals and incubated with the secondary antibody (Table 4.1) for 1 hour at room

Table 4.1 Secondary antibodies used in this study

Antibody	Source	Cat. #	Lot #	Dilution	Supplier
Anti-rabbit-Cy^3	Donkey	711-166-152	131,954	1:200	Jackson Labs
Anti-rabbit-Cy^3	Donkey	705-166-147	131,306	1:200	Jackson Labs
Anti-rabbit-Cy^3	Donkey	711-096-152	125,792	1:200	Jackson Labs
Anti-sheep-FITC	Donkey	713-096-147	123,217	1:200	Jackson Labs

temperature. Double-staining was accomplished by pairing the VNUT antibody with other primary antibodies followed by labeling with appropriate secondary antibodies conjugated to red and green fluorophores (Table 4.1). Secondary antibodies were removed by 3 × 10 min washes in PB and then mounted on a glass slide with ready-to-use mounting medium (ProlongTM Gold Antifade, Thermofisher, Cat No: P36961).

4.2.3 Fluorescence Microscopy

LMMP and CM preps were examined by conventional microscopy using a Nikon TE2000-U Inverted Microscope (Nikon TE2000-U series, Nikon Corporation, Tokyo, Japan) with MetaMorph image acquisition and analysis software. All images were obtained with a CFI Plan Fluor 20X NA: 0.50 (air) and CFI Plan Fluor 40X NA: 0.75 (air) objectives. Epitopes tagged with Cyanine 3 (Cy3), or fluorescein isothiocyanate (FITC) fluorophores were visualized using Nikon Filter cubes Cy GFP and B-2E/C. Colocalization of VNUT-ir with markers for subsets of myenteric neurons was determined by merging images taken from both fluorophores at the same focal plane.

4.3 Results

4.3.1 Nitric Oxide Synthase (NOS)

Myenteric plexus. NOS-immunoreactive (ir) nerve cell bodies and nerve fibers were detected in the small intestinal myenteric plexus of guinea pigs and mice (Fig. 4.1). In guinea pigs, VNUT-ir nerve fibers were detected in the ileal myenteric plexus, but these did not overlap with NOS-ir nerve fibers or nerve cell bodies (Fig. 4.1a). VNUT-ir nerve fibers formed pericellular baskets that surrounded some NOS-ir neurons and VNUT-ir nerve fibers formed pericellular baskets around NOS-positive and NOS-negative neurons in guinea pig myenteric ganglia. NOS-ir did not overlap with VNUT-ir nerve fibers in guinea pig myenteric ganglia (Fig. 4.1a). VNUT-ir and NOS-ir nerve fibers were detected in the circular smooth muscle with colocalization of some nerve fibers (Fig. 4.1b). Data obtained with mouse myenteric ganglia (Fig. 4.1c) showed VNUT-ir nerve fibers in myenteric ganglia and in the circular muscle layer with occasional overlap between VNUT-ir and NOS-ir nerve fibers in myenteric ganglia and in the circular muscle layer (Fig. 4.1d).

4.3.2 Choline Acetyltransferase (ChAT)

ChAT-ir was expressed in myenteric neurons in the ileum and in nerve fibers supplying the circular muscle layer (Fig. 4.2a, b). In addition, VNUT-ir varicose nerve fibers surrounded ChAT-ir nerve cell bodies in the guinea pig and mouse myenteric plexus. ChAT-ir neurons were also detected in the mouse ileum myenteric plexus with varicose VNUT-ir nerve fibers surrounding ChAT-ir neurons and ChAT-negative neurons (Fig. 4.2c). There was little or no overlap between ChAT-ir and VNUT-ir nerve fibers in the circular muscle layers (Fig. 4.2b, d).

4.3.3 Calbindin (CalB)

VNUT-ir and CalB-ir were expressed in separate groups of myenteric nerve fibers and cell bodies (Fig. 4.3a, c). VNUT-ir nerve cell bodies were not

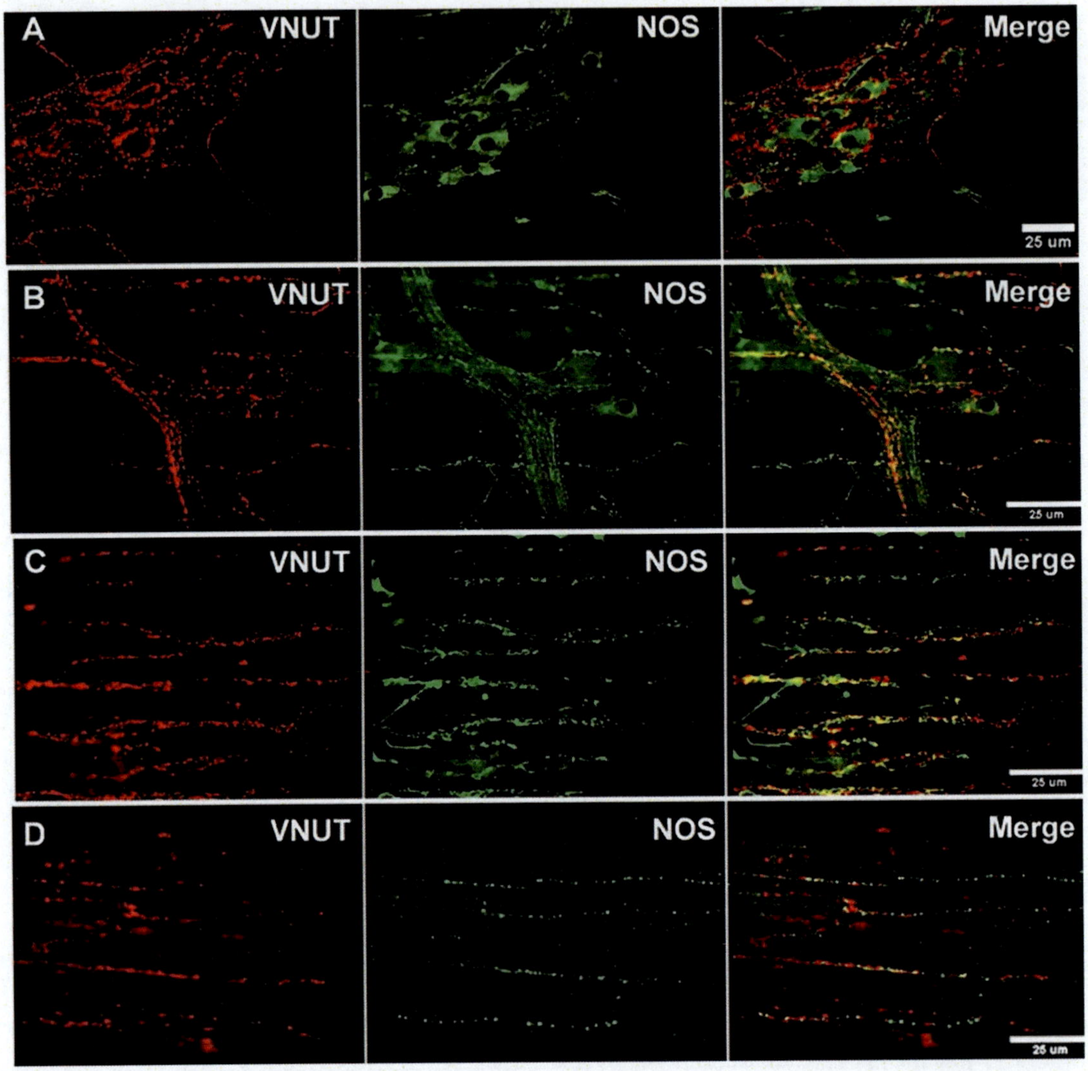

Fig. 4.1 Localization of nNOS-ir and VNUT-ir in the myenteric plexus and circular muscle of the guinea pig (**a**, **c**) and mouse ileum (**b**, **d**). VNUT-ir is present in nerve fibers in myenteric ganglia of guinea pigs (**a**) and mice (**b**), but nerve cell bodies are not detected in ganglia. nNOS-ir is expressed in myenteric neurons of guinea pigs (**a**) and mice (**b**). The merged images for nNOS-ir and VNUT-ir show that there is no overlap between VNUT-ir and NOS-ir nerve fibers in myenteric ganglia of guinea pigs (**a**) or mice (**b**) or in the circular muscle layer of guinea pigs (**c**) and mice (**d**)

detected, but there was dense innervation of myenteric ganglia by VNUT-ir nerve fibers. CalB-ir nerve cell bodies were restricted to the myenteric plexus in guinea pig and mouse intestine (Fig. 4.3a, c), and CalB-ir was not detected in the circular muscle layer of guinea pigs or mice (Fig. 4.3b, d).

4.3.4 Calretinin (CalR)

CalR-ir is expressed in neurons in the myenteric plexus of the guinea pig (A) and mouse (C) small intestine. VNUT-ir nerve varicosities surrounded many CalR-ir nerve cell bodies in guinea pig and mouse myenteric ganglia. CalR-ir and VNUT-ir

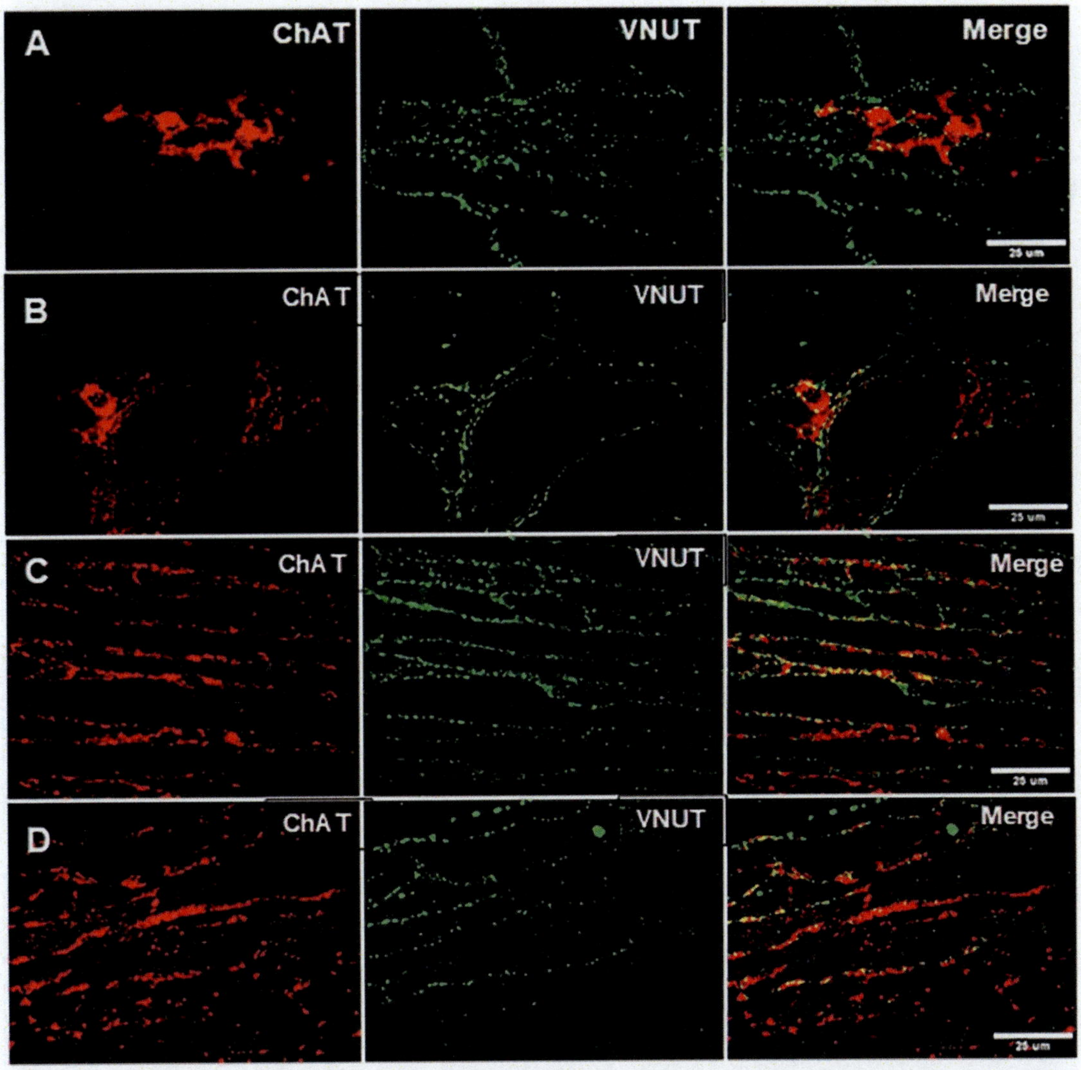

Fig. 4.2 Localization of ChAT-ir and VNUT-ir in myenteric ganglia and circular muscle of guinea pig (**a**, **b**) and mouse ileum (**c**, **d**). VNUT-ir is present in myenteric nerve fibers of guinea pigs (**a**) and mice (**c**), but neurons are not detected in myenteric ganglia. nNOS-ir is in guinea pig (**a**) and mouse (**b**) myenteric neurons. Merged images for nNOS-ir and VNUT-ir show little overlap between VNUT-ir and NOS-ir nerve fibers in myenteric ganglia of guinea pigs (**a**) or mice (**c**) or in the circular muscle of guinea pigs (**b**) and mice (**d**)

nerve fibers were detected in myenteric ganglia and circular muscle, but there was little overlap between CalR-ir and VNUT-ir neurons or nerve fibers in myenteric ganglia (Fig. 4.4a, c). VNUT-ir and CalR-ir nerve fibers were localized to the circular muscle layer of guinea pigs and mice, but there was little overlap between VNUT-ir and CalR-ir nerve fibers in the circular muscle (Fig. 4.4b, d).

4.4 Discussion

4.4.1 VNUT-ir and NOS-ir

NOS-ir is found in nerve cell bodies and nerve fibers in the myenteric plexus of the guinea pig and mouse ileum [5, 12, 33]. NOS synthesizes NO as an inhibitory neuromuscular transmitter in the ENS [6, 15]. In ileal LMMP preparations,

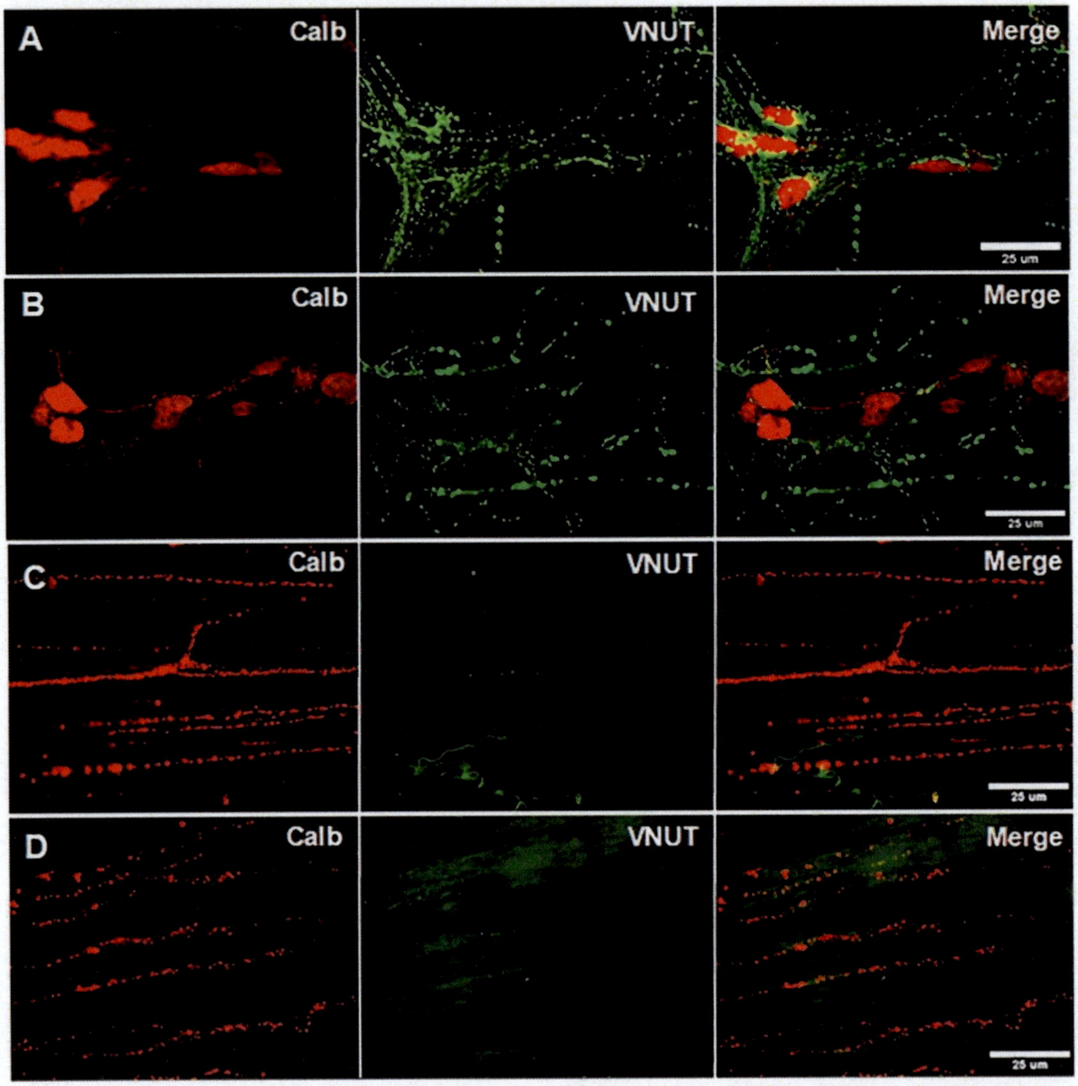

Fig. 4.3 Localization of calbindin-ir (Calb) and VNUT-ir in the myenteric plexus and circular muscle of guinea pig (**a**, **c**) and mouse duodenum (**b**, **d**). VNUT-ir is in nerve fibers in myenteric ganglia of guinea pigs (**a**) and mice (**c**), but nerve cell bodies are not detected in myenteric ganglia. Calb-ir is expressed in myenteric neurons of guinea pigs (**a**) and mice (**c**). Merged images show no overlap between VNUT-ir and NOS-ir nerve fibers in myenteric ganglia of guinea pigs (**a**) or mice (**c**) or in circular muscle layer of guinea pigs (**b**) and mice (**d**)

VNUT-ir varicose nerve fibers form pericellular baskets that surround most cell bodies labeled with the NOS antibody. Pericellular baskets are indicative of synaptic sites of neurotransmitter release to the cells they surround [34, 40]. This finding is not surprising as NOS-ir neurons express purinergic receptors and require purinergic innervation in order to generate fEPSPs during receptor activation [4, 22]. In the small intestine, NOS neurons comprise 21% of all neurons, they all have Dogiel Type I morphology, and they are interneurons and excitatory and inhibitory motorneurons (IMNs) [13]. NOS-ir IMNs are immunoreactive for vasoactive intestinal peptide (VIP) [5], and about 25% of IMNs do not contain NOS-ir [44].

In addition to ACh and 5-HT, descending interneurons can release ATP to stimulate post-

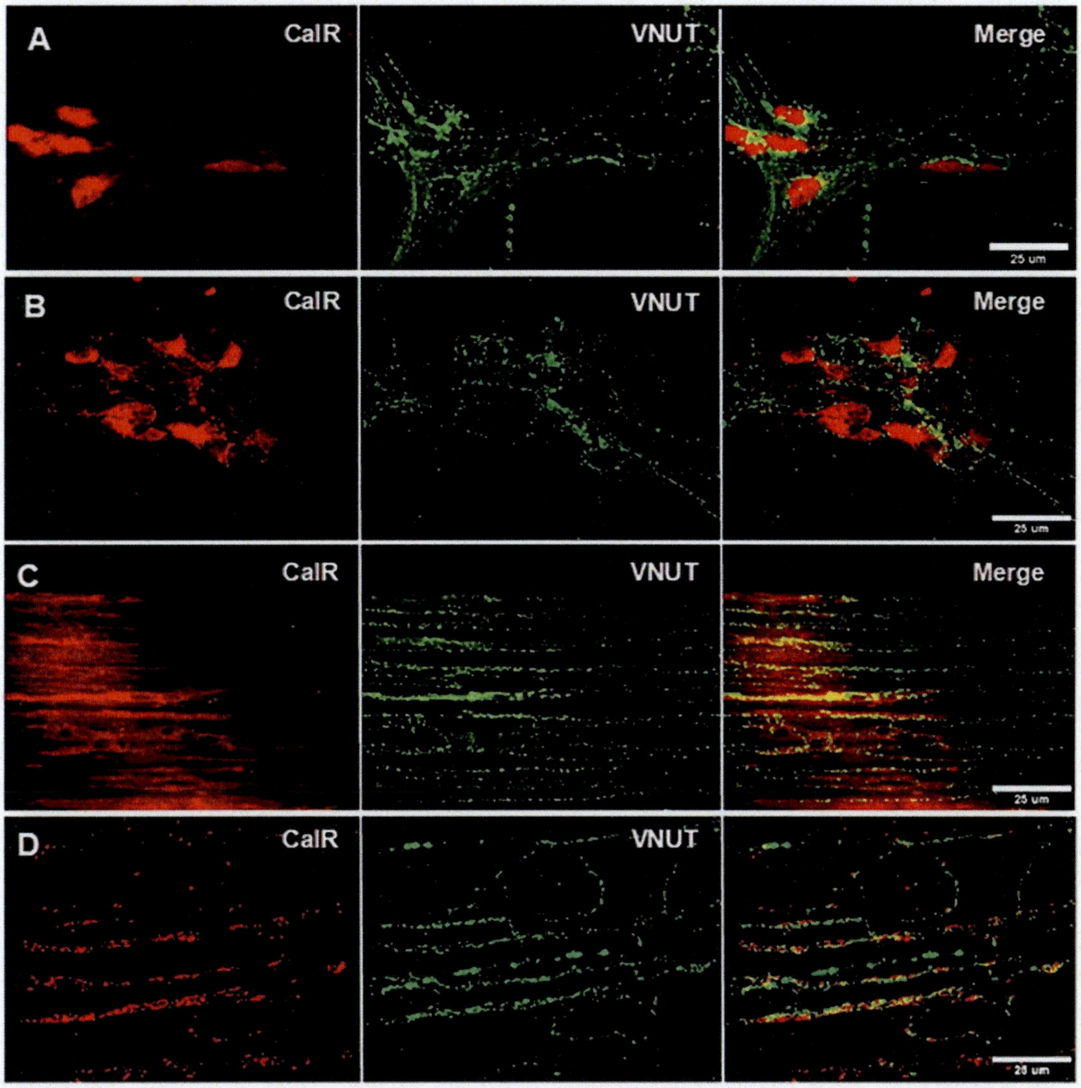

Fig. 4.4 Myenteric neurons and nerve fibers in the circular muscle of guinea pig (**a**, **c**) and ileum (**b**, **d**). Calretinin-ir is expressed in cell bodies and nerve fibers in ganglia of guinea pigs (**a**) and mice (**c**). VNUT-ir neurons are not detected in ganglia. Merged images for nNOS-ir and VNUT-ir show little or no overlap between VNUT-ir and NOS-ir nerve fibers in ganglia of guinea pigs (**a**) or mice (**c**) or in the circular muscle of guinea pigs (**b**) and mice (**d**)

synaptic P2X receptor activation of IMNs and other descending interneurons [3, 27, 30, 31]. Alternatively, it has been shown in guinea pigs [47], mice [18], pigs [20], rats [24], and humans [19] that IMNs innervating the muscle layers release a purine, arguably ATP, to trigger fast IJPs causing muscle relaxation via P2Y1 receptor activation. Consistent with previous studies, we found that VNUT-ir varicose nerve fibers project from the myenteric plexus to the tertiary plexus and circular muscle. Additionally, a significant portion of these purinergic nerves also contain NOS-ir. These findings align with the current model that one population of IMNs coreleases ATP and NO to cause smooth muscle relaxation. However, VNUT and NOS are localized in separate neurons projecting to the tertiary plexus. Several studies reveal that ATP is an excitatory neurotransmitter in the longitudinal muscle of guinea pig ileum [25, 43, 51], suggesting that purinergic nerves innervating the longitudinal muscle may cause contractions and relaxations needed for propulsive reflexes.

4.4.2 VNUT-ir and ChAT-ir

VNUT-ir nerve fibers were found to innervate cholinergic neurons of the guinea-pig ileum myenteric plexus. ACh is the prominent excitatory neurotransmitter in the ENS, and ACh stimulates contraction [39, 46]. ChAT is the enzyme that synthesizes Ach and, therefore, it is an ideal marker to study enteric cholinergic neurons [5]. VNUT-ir nerve fibers innervate Dogiel Type I and II neurons, because ChAT-ir is expressed by both classes of myenteric neurons [41], plus both types of neurons are immunoreactive for purinergic P2X receptors [16, 32]. However, Dogiel Type I cholinergic neurons comprise about half of the total myenteric neuronal population, which include interneurons and motorneurons. Conversely, Dogiel Type II cholinergic neurons are intrinsic primary afferent neurons (IPANs) that have immunoreactivity for the calcium-binding protein calbindin-D28K. However, in mice, some IPANS are immunoreactive for calretinin, another calcium-binding protein [5, 41].

In contrast to these findings, myenteric VNUT-ir nerve fibers that project from one ganglion to another, or to the tertiary plexus, only contain VNUT-ir. A subpopulation of descending interneurons may mediate mixed cholinergic/purinergic fast excitatory postsynaptic potentials (fEPSPs) [30, 31], yet our finding indicates a purine-only population of descending interneurons may exist as well. ACh is the primary excitatory neurotransmitter controlling GI motility [9, 29], while purines have multiple roles in the GI tract, they contribute significantly to smooth muscle relaxation [10]. However, there does appear to be fiber bundles in the circular smooth muscle that express VNUT and ChAT, yet they rarely colocalize. GI motility requires synchronized muscle contraction and relaxation to allow mixing and propulsion of gut content. Therefore, having both markers present in the same nerve bundles is conceivable.

4.4.3 VNUT-ir and CalB-ir

Calbindin (CalB)-ir is visible on cell bodies and nerve fibers in the myenteric plexus, but rarely visible in the circular muscle or the tertiary plexus. These observations are consistent with past descriptions of calbindin-ir in guinea pig tissues [13, 38, 41, 42, 44]. Also, VNUT-ir nerve fibers were not immunoreactive for CalB. CalB-ir is localized to Dogiel type II neurons, which are intrinsic primary afferent neurons (IPANs) [15]. IPANs express $P2X_2$ purinergic ion channel receptors [11, 14, 32], but our data suggest that VNUT-ir nerve fibers are less likely to synapse with IPANS. One explanation is that control of gut motility requires fast purine release, and thus, only myenteric neurons involved with this mechanism, such as motorneurons, maintain direct synaptic innervation by VNUT-ir varicose fibers. Conversely, many IPANs have long processes that extend into mucosal villi and activation of these fibers is due to changes in luminal chemistry and mechanical distortion of mucosal villi. IPANs contribute to nerve circuits in the myenteric plexus by connecting to interneurons that in turn trigger motor and secretory reflexes [2, 16, 28]. Therefore, purinergic receptor activation of IPANs may play a modulatory role and does not require direct purinergic innervation.

4.4.4 VNUT-ir and CalR-ir

The calcium-binding protein calretinin (CalR) is used as an immunological marker for cholinergic excitatory motor neurons (EMNs) in the guinea pig, [34]. In the guinea pig ileum, CalR is expressed in nerve fibers located in the myenteric plexus, tertiary plexus, but not in the circular smooth muscle [7]. In contrast to this, CalR-ir is present in the circular smooth muscle of the guinea pig ileum [36, 37], which is consistent with our findings. Double immunolabeling of CalR with VNUT, however, revealed no colocalization of the markers in nerve fibers within the myenteric plexus, suggesting that EMNs are less likely to be purinergic [41]. On the other hand, purinergic varicose fibers formed perivascular baskets that surrounded CalR-ir cell bodies. These baskets are less prominent in the ileum. High numbers of CalR neurons are immunoreactive for purinergic receptors, including the ionotropic P2X2 receptors [16, 49]. Therefore, it likely that VNUT-ir nerve fibers innervate EMNs to generate fEPSPs via P2X receptors, resulting in contraction of GI muscles [50]. In contrast, CalR-ir is not expressed by Dogiel type II neurons in guinea pigs [5], which means that EMNs are innervated by purinergic neurons. VNUT-ir and calretinin-ir do not colocalize in nerve fibers in the tertiary plexus and circular muscle layer. In addition, VNUT-ir and CalR-ir coexpression in some nerve bundles innervating the ileum, duodenum, and stomach circular muscle layers was observed. These findings are not surprising as the primary function of purines in the circular muscle is to cause GI smooth muscle relaxation along with NO [20, 23]. Moreover, CalR-ir and NOS-ir are expressed in different motorneuron subtypes [16] and our data show VNUT-ir and NOS-ir colocalization in nerve fibers at the muscle layers. Therefore, VNUT is likely expressed in a subclass of IMNs that innervate the circular muscle layer.

4.4.5 Summary and conclusions

VNUT-ir neurons are a subset of myenteric neurons that likely mediate purinergic signaling in myenteric ganglia and circular muscle. ATP and NO may be released from separate groups of inhibitory motorneurons to cause circular muscle relaxation. There are also separate groups of VNUT-ir and ChAT-ir myenteric nerves, suggesting that cholinergic and purinergic synaptic transmission is mediated by separate groups of interneurons. ATP (or related purine) is an important excitatory synaptic and inhibitory neuromuscular transmitter in the ENS of mice and guinea pigs. NOS is a marker for neurons that use NO as a transmitter, while VNUT is a marker for purinergic nerves. NO is an important inhibitory neuromuscular transmitter that collaborates with ATP (or related purine) to cause muscle relaxation. Muscle relaxation is essential for efficient propulsion of gastrointestinal content.FundingThis project was funded by the National Institutes of Digestive Diseases and Kidney (NIDDK) Grant #R01 DK121272.

References

1. Antonioli L, Colucci R, Pellegrini C, Giustarini G, Tuccori M, Blandizzi C, Fornai M (2013) The role of purinergic pathways in the pathophysiology of gut diseases: pharmacological modulation and potential therapeutic applications. Pharmacol Ther 139:157–188
2. Bertrand PP, Kunze WA, Bornstein JC, Furness JB, Smith ML (1997) Analysis of the responses of myenteric neurons in the small intestine to chemical stimulation of the mucosa. Am J Phys 273:G422–G435
3. Bian XC, Bertrand PP, Bornstein JC (2000) Descending inhibitory reflexes involve P2X receptor-mediated transmission from interneurons to motor neurons in guinea-pig ileum. J Physiol 528:551–560
4. Bornstein JC (2008) Purinergic mechanisms in the control of gastrointestinal motility. Purinergic Signal 4:197–212
5. Brookes SJ (2001) Classes of enteric nerve cells in the guinea-pig small intestine. Anat Rec 262:58–70
6. Brookes SJH (1993) Neuronal nitric-oxide in the gut. J Gastroenterol Hepatol 8:590–603
7. Brookes SJH, Steele PA, Costa M (1991) Calretinin immunoreactivity in cholinergic motor neurons, interneurons and vasomotor neurons in the guinea-pig small-intestine. Cell Tissue Res 263:471–481
8. Burnstock G (1972) Purinergic nerves. Pharmacol Rev 24:509-+
9. Burnstock G (1981) Neurotransmitters and trophic factors in the autonomic nervous-system. J Physiol 313:1
10. Burnstock G (2014) Purinergic signalling in the gastrointestinal tract and related organs in health and disease. Purinergic Signal 10:3–50
11. Castelucci P, Robbins HL, Poole DP, Furness JB (2002) The distribution of purine P2X(2) receptors in the guinea-pig enteric nervous system. Histochem Cell Biol 117:415–422
12. Chen BN, Sharrad DF, Hibberd TJ, Zagorodnyuk VP, Costa M, Brookes SJH (2015) Neurochemical characterization of extrinsic nerves in myenteric ganglia of the guinea pig distal colon. J Comp Neurol 523:742–756

13. Costa M, Brookes SJ, Steele PA, Gibbins I, Burcher E, Kandiah CJ (1996) Neurochemical classification of myenteric neurons in the guinea-pig ileum. Neuroscience 75:949–967
14. Crist JR, He XD, Goyal RK (1992) Both ATP and the peptide VIP are inhibitory neurotransmitters in guinea-pig ileum circular muscle. J Physiol 447:119–131
15. Furness JB (2000) Types of neurons in the enteric nervous system. J Auton Nerv Syst 81:87–96
16. Furness JB, Jones C, Nurgali K, Clerc N (2004a) Intrinsic primary afferent neurons and nerve circuits within the intestine. Prog Neurobiol 72:143–164
17. Gallego D, Gil V, Aleu J, Auli M, Clave P, Jimenez M (2008a) Purinergic and nitrergic junction potential in the human colon. Am J Physiol Gastrointest Liver Physiol 295:G522–G533
18. Gallego D, Gil V, Martinez-Cutillas M, Mane N, Martin MT, Jimenez M (2012) Purinergic neuromuscular transmission is absent in the colon of P2Y(1) knocked out mice. J Physiol 590:1943–1956
19. Gallego D, Hernandez P, Clave P, Jimenez M (2006) P2Y1 receptors mediate inhibitory purinergic neuromuscular transmission in the human colon. Am J Physiol Gastrointest Liver Physiol 291:G584–G594
20. Gallego D, Vanden Berghe P, Farre R, Tack J, Jimenez M (2008b) P2Y1 receptors mediate inhibitory neuromuscular transmission and enteric neuronal activation in small intestine. Neurogastroenterol Motil 20:159–168
21. Galligan JJ, Bertrand PP (1994) ATP mediates fast synaptic potentials in enteric neurons. J Neurosci 14:7563–7571
22. Giaroni C, Knight GE, Ruan HZ, Glass R, Bardini M, Lecchini S, Frigo G, Burnstock G (2002) P2 receptors in the murine gastrointestinal tract. Neuropharmacology 43:1313–1323
23. Gil V, Gallego D, Grasa L, Martin MT, Jimenez M (2010) Purinergic and nitrergic neuromuscular transmission mediates spontaneous neuronal activity in the rat colon. Am J Physiol Gastrointest Liver Physiol 299:G158–G169
24. Grasa L, Gil V, Gallego D, Martin MT, Jimenez M (2009) P2Y(1) receptors mediate inhibitory neuromuscular transmission in the rat colon. Br J Pharmacol 158:1641–1652
25. Ivancheva C, Itzev D, Radomirov R (2000) Functional antagonism between nitric oxide and ATP in the motor responses of guinea-pig ileum. J Auton Pharmacol 20:147–156
26. Johnson PJ, Shum OR, Thornton PD, Bornstein JC (1999a) Evidence that inhibitory motor neurons of the guinea-pig small intestine exhibit fast excitatory synaptic potentials mediated via P2X receptors. Neurosci Lett 266:169–172
27. Johnson PJ, Shum OR, Thornton PDJ, Bornstein JC (1999b) Evidence that inhibitory motor neurons of the guinea-pig small intestine exhibit fast excitatory synaptic potentials mediated via P-2X receptors. Neurosci Lett 266:169–172
28. Kirchgessner AL, Tamir H, Gershon MD (1992) Identification and stimulation by serotonin of intrinsic sensory neurons of the submucosal plexus of the guinea pig gut: activity-induced expression of Fos immunoreactivity. J Neurosci 12:235–248
29. Komori K, Suzuki H (1986) Distribution and properties of excitatory and inhibitory junction potentials in circular muscle of the guinea-pig stomach. J Physiol 370:339–355
30. Lepard KJ, Galligan JJ (1999) Analysis of fast synaptic pathways in myenteric plexus of guinea pig ileum. Am J Phys 276:G529–G538
31. Lepard KJ, Messori E, Galligan JJ (1997) Purinergic fast excitatory postsynaptic potentials in myenteric neurons of guinea pig: distribution and pharmacology. Gastroenterology 113:1522–1534
32. Li ZS, Furness JB (1998) Immunohistochemical localisation of cholinergic markers in putative intrinsic primary afferent neurons of the guinea-pig small intestine. Cell Tissue Res 294:35–43
33. Lomax AE, Furness JB (2000) Neurochemical classification of enteric neurons in the guinea-pig distal colon. Cell Tissue Res 302:59–72
34. Mann PT, Southwell BR, Young HM, Furness JB (1997) Appositions made by axons of descending interneurons in the guinea-pig small intestine, investigated by confocal microscopy. J Chem Neuroanat 12:151–164
35. Masuda T, Ozono Y, Mikuriya S, Kohro Y, Tozaki-Saitoh H, Iwatsuki K, Uneyama H, Ichikawa R, Salter MW, Tsuda M, Inoue K (2016) Dorsal horn neurons release extracellular ATP in a VNUT-dependent manner that underlies neuropathic pain. Nat Commun 7:12529
36. Mcconalogue K, Furness JB (1996) Calretinin immunoreactivity of motor neurons in the guinea-pig distal colon and Taenia coli. Cell Tissue Res 284:367–372
37. Mcconalogue K, Low AM, Williamson S, Bornstein JC, Furness JB (1994) Calretinin-immunoreactive neurons and their projections in the guinea-pig colon. Cell Tissue Res 276:359–365
38. Messenger JP, Furness JB (1990) Projections of chemically-specified neurons in the guinea-pig colon. Arch Histol Cytol 53:467–495
39. Okasora T, Bywater RA, Taylor GS (1986) Projections of enteric motor neurons in the mouse distal colon. Gastroenterology 90:1964–1971
40. Pompolo S, Furness JB (1995) Sources of inputs to longitudinal muscle motor neurons and ascending interneurons in the guinea-pig small intestine. Cell Tissue Res 280:549–560
41. Qu ZD, Thacker M, Castelucci P, Bagyanszki M, Epstein ML, Furness JB (2008) Immunohistochemical analysis of neuron types in the mouse small intestine. Cell Tissue Res 334:147–161
42. Reiche D, Pfannkuche H, Michel K, Hoppe S, Schemann M (1999) Immunohistochemical evidence for the presence of calbindin containing neurons in the myenteric plexus of the guinea-pig stomach. Neurosci Lett 270:71–74

43. Rodriguez-Tapia ES, Naidoo V, Devries M, Perez-Medina A, Galligan JJ (2017) R-type Ca(2+) channels couple to inhibitory neurotransmission to the longitudinal muscle in the guinea-pig ileum. Exp Physiol 102:299–313
44. Sang Q, Young HM (1996) Chemical coding of neurons in the myenteric plexus and external muscle of the small and large intestine of the mouse. Cell Tissue Res 284:39–53
45. Sawada K, Echigo N, Juge N, Miyaji T, Otsuka M, Omote H, Yamamoto A, Moriyama Y (2008) Identification of a vesicular nucleotide transporter. Proc Natl Acad Sci U S A 105:5683–5686
46. Unekwe PC, Savage AO (1991) The effects of electrical-stimulation, adenosine and adenosine-5′-triphosphate (ATP) on mouse rectal muscle. Pharmacol Res 23:389–398
47. Wang GD, Wang XY, Hu HZ, Liu S, Gao N, Fang X, Xia Y, Wood JD (2007) Inhibitory neuromuscular transmission mediated by the P2Y(1) purinergic receptor in guinea pig small intestine. Am J Physiol Gastrointest Liver Physiol 292:G1483–G1489
48. White PN, Thorne PR, Housley GD, Mockett B, Billett TE, Burnstock G (1995) Quinacrine staining of marginal cells in the stria vascularis of the guinea-pig cochlea - a possible source of extracellular ATP. Hear Res 90:97–105
49. Xiang Z, Burnstock G (2005) Distribution of P2Y2 receptors in the guinea pig enteric nervous system and its coexistence with P2X2 and P2X3 receptors, neuropeptide Y, nitric oxide synthase and calretinin. Histochem Cell Biol 124:379–390
50. Xiang Z, Burnstock G (2006) Distribution of P2Y6 and P2Y12 receptor: their colocalization with calbindin, calretinin and nitric oxide synthase in the guinea pig enteric nervous system. Histochem Cell Biol 125:327–336
51. Zizzo MG, Mule F, Serio R (2007) Evidence that ATP or a related purine is an excitatory neurotransmitter in the longitudinal muscle of mouse distal colon. Br J Pharmacol 151:73–81

Mechanosensitive Enteric Neurons (MEN) at Work

5

Gemma Mazzuoli-Weber

Abstract

In the last decade, we characterized an enteric neuronal subpopulation of multifunctional mechanosensitive enteric neurons (MEN) while studying the gastrointestinal peristalsis. MEN have been described in a variety of gastrointestinal regions and species. This chapter summarizes existing data on MEN, describing their proportions, firing behaviors, adaptation musters, and chemical phenotypes. We also discuss MEN sensitivity to different mechanical stimulus qualities such as compression and tension along with pharmacology of their responses.

Keywords

Mechanosensitivity · Myenteric neurons · Multifunctionality · Compression/distention

G. Mazzuoli-Weber (✉)
Institute for Physiology and Cell Biology, University of Veterinary Medicine Hannover, Foundation, Hannover, Germany
e-mail: gemma.mazzuoli-weber@tiho-hannover.de

5.1 Enteric Nervous System and Mechanosensitive Enteric Neurons

The enteric nervous system (ENS), as third division of the autonomous nervous system, is located directly inside the gastrointestinal wall and it is able to control all gut functions autonomously. These functions are, among others, regulation of absorption, secretion, blood flow, immune response and motility, during digestive and interdigestive states. In order to control and regulate gastrointestinal functions, the ENS needs sensors capable of perceiving biochemical and mechanical stimuli and effectors to adequately respond to them. Enteric neurons, differently from central neurons, due to their particular localization embedded into the gastrointestinal wall, are continuously subjected to deformations of their cell body and neurites. This makes them perfect candidates to sense mechanical stimuli and react to them. The idea of an enteric neuronal subpopulation with a specific "sensory" function, as for instance the intrinsic primary afferent neurons [7, 10], has been challenged during the past decades by new data describing multifunctional enteric neurons [4, 14, 21], which are able to act not only as sensory neurons but also as interneurons or motoneurons. We described these neurons as mechanosensitive enteric neurons (MEN) and we could identify them in a variety of regions and species (Table 5.1). After identifying them, we

N. J. Spencer et al. (eds.), *The Enteric Nervous System II*, Advances in Experimental Medicine and Biology 1383, https://doi.org/10.1007/978-3-031-05843-1_5

Table 5.1 Proportion of compression and tension sensitive MEN per ganglion in different animals and regions [5, 11–15, 17]

Species	Region	Plexus	Percent of compression-sensitive MEN pro ganglion/cluster	Percent of tension-sensitive MEN pro ganglion/cluster
Mouse	Ileum	Myenteric	22	
Mouse	Distal colon	Myenteric	14	
Guinea pig	Stomach (corpus)	Myenteric	27	27
Guinea pig (primary culture)	Stomach	Myenteric	44	
Guinea pig	Ileum	Myenteric	25	
Guinea pig (primary culture)	Ileum	Myenteric	45 neurites (14 soma)	14
Guinea pig	Distal colon	Myenteric	12	
Guinea pig	Distal colon	Submucosal	27	14
Pig	Proximal colon	Submucosal	8	8
Pig	Distal colon	Submucosal	10	6
Human (primary culture)	Small intestine	Myenteric	27	
Human (primary culture)	Large intestine	Myenteric	14	
Human	Colon	Submucosal	14	24

studied their properties, we analyzed their responses, and we characterized their functions. In this chapter, we will summarize existing data about MEN and additionally, we will provide unpublished data that would reveal new insights on their functionality.

5.2 Methods Used to Identify MEN

In our studies, we used wholemount preparations of different gastrointestinal plexus, regions, and species. Part of the experiments was also performed in cultured myenteric neurons, in order to study the behavior of isolated neurons avoiding the influences of neighboring cells. The ultrafast neuroimaging technique with the voltage-sensitive dye Di-8-ANEPPS was used to detect neuronal activity in response to various mechanical stimuli [12–14, 17]. Mechanical stimuli were distinguished between compression, tension and shear stress. Compression was applied classically by von Frey hair probe and/or with the technique of intraganglionic injection of a small volume of the same Krebs solution perfusing the tissue [5, 14, 15, 17]. Tension was applied by strain of the matrix where neurons were grown and ganglionic stretch was applied with a self-developed stretching tool [5, 13, 17]. For application of shear stress, the neurons were grown in a microchannel and then they were exposed to different flow rates of Krebs solution [13]. Neuronal responses were recorded, as mentioned, with an ultrafast neuroimaging technique. This technique combines the advantage of recording simultaneously from all the neurons in the field of view and of being noninvasive.

5.3 Nature of the Mechanosensitive Stimuli Activating MEN: Sensitivity to Compression, Tension and Shear Stress

Traditionally, mechanosensitivity has been tested with compression stimuli (e.g., von Frey hair) in the skin and with different forms of distention of the gastrointestinal wall. However, distention is not the only stimulus that enteric neurons undergo. Enteric neurons are subject to different kinds of mechanical stimuli during gastrointestinal circular and longitudinal muscle contractions/relaxations. Indeed, enteric ganglia, especially from the myenteric plexus, which is embedded between the two gut main muscle layers, are elongated, compressed and deformed due to muscle activity (see Movie 1 in [14]). With experiments on isolated primary cultured myenteric neurons, we showed that MEN are sensitive to different mechanical stimulus qualities: in particular, they react to compression and tension and to a lesser extent to shear stress [13]. Similarly, enteric neurons within intact tissue are sensitive to both compression and tension [5, 17]. Data from different regions and species showed that three different subpopulations of MEN exist: (1) a subpopulation of compression sensitive MEN, (2) a subpopulation of tension sensitive MEN and (3) a subpopulation of MEN, which is both compression sensitive and tension sensitive. The nature of receptor/receptors involved is still under debate. The question is whether the same mechanoreceptor is able to sense both stimuli, or if more mechanoreceptors are located on the same neuron.

5.4 Properties of Isolated Myenteric MEN

Using neuroimaging in combination with an independent component analysis, we were able to reconstruct neuronal morphology and follow neuronal signals in order to describe innovative features of cultured guinea pig MEN [12]. With this particular setting, we recorded responses to target stimulations of the soma and/or of the neurites of isolated enteric neurons. Of the cultured neurons, 45% pro cluster responded to neurite probing (compression sensitive MEN). Action potentials generated at the stimulation site invaded the soma and other neurites. Mechanosensitive sites were expressed across large areas of the neurites. Many mechanosensitive neurites appeared to have afferent and efferent functions as those that responded to deformation also conducted spikes coming from the soma. Mechanosensitive neurites were also activated by nicotine application. This supported the concept of multifunctionality of MEN. 14% of the neurons pro cluster responded to soma deformation. The majority of the cultured MEN showed slowly adapting spiking behavior, although significantly more MEN showed a rapidly adapting behavior after neurite probing. In addition, each mechanosensitive neuron appears to have multiple mechanosensitive neurites, which, when stimulated, responded with a similar spike discharge pattern [12].

Cultured myenteric neurons of human small and large intestines had similar properties. This is another important evidence of the transferability of our concept between regions and species. To compare MEN with another type of neurons with mechanosensitive properties, we cultivated dorsal root ganglion (DRG) neurons of the guinea pig and we performed the same kind of experiments on them. Compared to MEN, DRG neurons were activated by neurite but not by soma deformation. All mechanosensitive DRG neurons showed a slow adaptation of the firing rate [12].

In order to answer the question if enteric neurons could be activated by different mechanical stimuli, we recorded the activity of isolated myenteric neurons from 25 guinea pigs in response to normal stress, shear stress and strain/tension [13]. All the stimuli induced excitatory responses. Tension activated 14% of the stimulated myenteric neurons per cluster. They responded with a spike frequency of 1.9 (0.7/3.2) Hz (median with 25% and 75% quartiles). Shear stress excited only a few neurons (5.6%) with a very low spike frequency of 0 (0/0.6) Hz [13]. Paired experiments with tension and compression allowed to divide the mechanosensitive enteric neuronal in the three subpopulations: (1) tension sensitive (37%), (2) compression sensitive (17%) and (3) tension and compression sensitive (46%) MEN. MEN responded to shear stress with both rapid and slow adapting behavior, whereas they respond to tension only with a slow adapting behavior [13]. We concluded that shear stress is not an adequate stimulus for MEN, while compression and tension can activate enteric neurons. These data revealed that shear stress does not play a role in the neuronal control of motility but compression and tension.

5.5 Regional- and Species-Specific Differences in the Properties of MEN

Not all enteric neurons are mechanosensitive but just a subpopulation of them. The proportion of MENs in the ENS varies between species and localizations. Figure 5.1 displays the proportion of compression sensitive MEN per ganglion that we identified in the different plexuses, gut regions, and species.

Responses to compression and tension in all regions and species were reproducible when the same stimulus was applied twice or thrice, as expected, in physiology, for mechanosensors. Moreover, the responses became stronger with increased stimulus strength.

MEN responsiveness was positively correlated with the degree of ganglionic deformation [5, 14, 15, 17]. This could be an indication of the existence of low and high threshold MEN.

MEN showed rapid or slow adaptation behavior. Interestingly, even after adaptation, a second deformation evoked spike discharge again, suggesting a resetting mode of the receptors. The adaptation behavior appeared to be region and stimulus specific [5, 14, 16, 17]. In the stomach and in the small intestine, compression sensitive MEN showed mostly a rapid adaptation. In the large intestine, MEN responded to compression with both rapid and slow adaptation muster. When stimulated with tension, MEN responded in all regions mostly with a slow adaptation muster [5, 14, 17].

All compression and tension sensitive MEN received fast excitatory postsynaptic potentials (EPSPs).

Based on data coming from our own studies, we can conclude that the basic properties of MEN, such as firing pattern, adaptation behavior, and proportion on the total neuronal population, are highly conserved between gut regions with different functional specialization and even between species (Table 5.1).

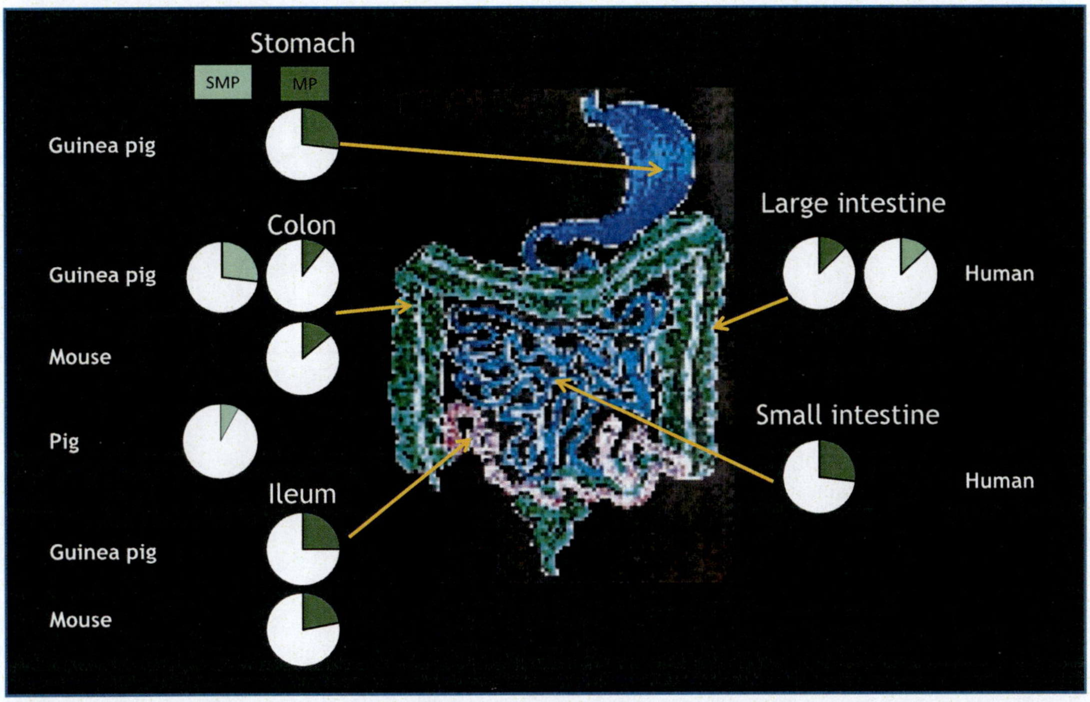

Fig. 5.1 Compression sensitive MEN. Proportions in different plexus, regions and species

5.6 Deformation Rate

In 1917, Ulrich Trendelenburg, a German scientist, meticulously described recordings of controlled distension of the intestinal segment under physiological conditions [22]. His method to record intestinal motility is well known and recognized. Curiously, he showed that the initiation of the classical "peristaltic reflex" varied with the speed of intestinal distension: a rapid distension initiated the reflex, while with slow distension, the threshold was much higher or eventually, the reflex was not even evoked [22]. Thus, we had the idea that one parameter, which needs to be considered in order to study MEN activation, was the velocity of the deformation. In this perspective, we performed parallel experiments changing the speed of the deformation of our compression and tension stimuli. Experiments performed in the guinea pig ileum ($n = 5$) changing the velocity of deformation revealed neither significant changes in the neuronal responsiveness nor in the response patterns of MEN (Fig. 5.2).

In order to address the question whether compression and tension stimuli evoking neuronal responses had similar deformation velocity, we performed parallel experiments with both stimuli ($n = 7$). There was no significant difference between compression and tension stimuli ($p = 0.836$; Mann-Whitney Rank Sum Test; unpublished data). We also did not observe any correlation between the deformation velocity and neuronal responsiveness ($n = 7$).

5.7 Myenteric and Submucosal MEN

Neurons residing in the myenteric plexus are the ones mainly involved in the control of gastrointestinal motility. Thus, we first concentrated our studies on this plexus. Our findings revealed, indeed, the presence of MEN in the myenteric plexus of various gastrointestinal regions (Table 5.1). However, we hypothesized that MEN are also present in the submucosal plexus. This idea was based on two facts. First fact is a study, which demonstrated in the submucosal plexus of the guinea pig intestine an increased early gene marker Fos immunoreactivity in response to a

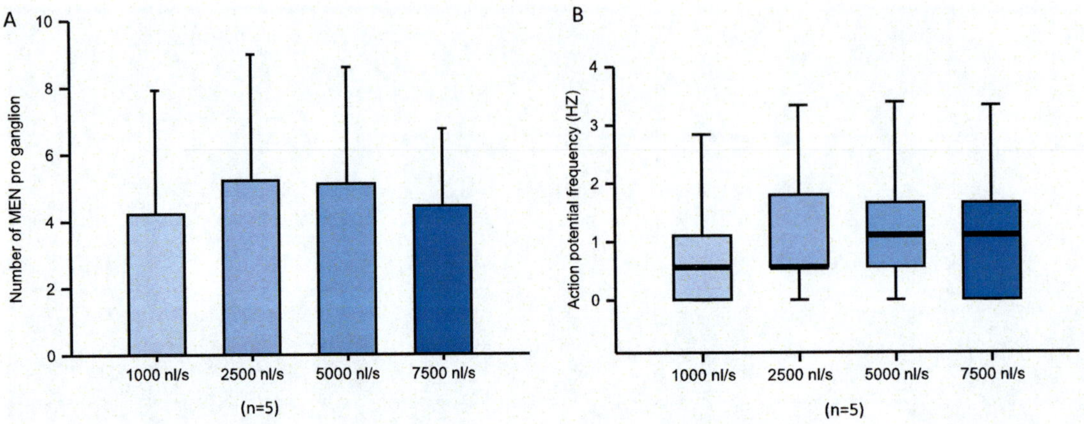

Fig. 5.2 The effect of increasing stimulation velocities was tested in paired experiments in 9 ganglia from the ileum of 5 guinea pigs. Neither the number of MEN pro ganglion nor their action potential frequency showed a significant difference between the randomly applied different stimulus velocities (one-way ANOVA with repeated measures test, not significant)

puff of nitrogen directly on the mucosa [10]. Second fact is the observation that in Ussing chamber voltage clamp experiments, distension of submucosa/mucosa intestinal preparations evoked chloride secretion, which was significantly reduced in tetrodotoxin (TTX), indicating that submucosal sensory neurons involved in secretion must respond to deformation ([3, 23]; personal observations). In order to prove or disprove our hypothesis, we applied the same type of mechanical stimuli in the submucosal plexus. Such stimuli have been successfully used in the myenteric plexus to activate MEN. We performed these experiments in the guinea pig colon (n = 10), porcine proximal (n = 25) and distal colon (n = 21) and human colon (n = 27). Results indicated the presence of compression sensitive, tension sensitive, and compression and tension sensitive MEN in the colonic submucosal plexus [5].

5.8 MEN Neurochemical Phenotype

In myenteric plexus of the ileum, 72% of compression sensitive MEN were cholinergic and 22% were nitric oxide synthase (NOS) immunoreactive [14]. In the gastric corpus, 55% of compression sensitive MEN showed cholinergic phenotype and 45% were NOS antibody immunoreactive [17]. The proportion of nitrergic neurons was even higher (64%) for tension sensitive MEN: the functional relevance of this finding is probably to be linked to the nerve-dependent, nitric oxide-mediated gastric volume accommodation.

In the guinea pig submucosal plexus of the distal colon, ~65% of the compression sensitive MEN were cholinergic. In the same region, we identified ~60% cholinergic tension sensitive MEN (unpublished data).

Data from the porcine large intestine revealed that in the proximal colon, ~70% of the compression or tension sensitive neurons were cholinergic, with ~65% of them coexpressing substance P (SP). NOS-expressing compression or tension sensitive neurons were ~3%. In the distal colon of the pig, ~64% of the neurons were cholinergic, with ~60% of them also coexpressing SP. NOS-expressing compression or tension sensitive neurons were here 12% and 19%, respectively [5].

We speculate that these submucosal MENs function as multifunctional sensory-secretomotor neurons.

5.9 Pharmacology of MEN

Experiments with pharmacology are critical to understand the mechanotransduction machinery of MEN. Since these experiments have been performed mostly with compression stimuli, we can draw our conclusions only for compression sensitive MEN. In the guinea pig ileum, compression stimuli evoking spike discharge were not influenced by synaptic blockade with hexamethonium, ω-conotoxin or low Ca^{2+}/high Mg^{2+}, defunctionalization of extrinsic afferents with capsaicin or muscle paralysis with nifedipine, suggesting a direct activation of compression sensitive MEN. In the stomach, corpus MEN responses were significantly decreased after perfusion of low Ca^{2+}/high Mg^{2+}, capsaicin, and capsazepine, indicating a modulatory role in MEN responses of transient potential receptor vanilloid 1–expressing extrinsic nerves.

Perfusion with the fast Na^{+} channel blocker TTX, as expected, lowered the number of MEN both in the ileum and in the stomach. However, there was still a MEN subpopulation, which was responding in TTX ([17] and unpublished data). We hypothesized that these neurons fired calcium-driven action potentials or that our mechanical stimulus activated TTX-insensitive sodium channels, which must be voltage independent.

Recently, searching plausible candidates for the putative channels responsible for MEN mechanotransduction, we bumped into Piezo proteins. These proteins have been identified as mechano-activated cation channels in different eukaryotic cells [1, 2, 9, 24]. Results of pharmacological experiments using the toxin GsMTx4 to block Piezo channels did not confirm the hypothesis of an involvement of these channels in enteric neurons mechanosensitivity [18].

We are keen performing other pharmacological experiments to reveal receptors or channels involved in MEN mechanotransduction. All in all, the results of the pharmacological experiments led us to the hypothesis that more than one channel/mechanism is involved in mechanosensitivity of MEN. It still remains to be cleared if tension sensitive MEN present the same pharmacology.

5.10 Multifunctionality

We performed experiments stimulating enteric neurons mechanically in preparations where we previously identified interneurons and motor neurons with the DiI retrograde tracing technique [14]. These experiments proved that MEN are able not only to sense the mechanical stimulus, but also to function as interneurons or even motoneurons; hence, they can be defined as multifunctional neurons. Experiments on isolated enteric neurons reinforced this hypothesis, showing that MEN possess mechanosensitive processes (afferent) and, at the same time, they have processes with motor (effector) functions [12]. Differently from the central nervous system (CNS), multifunctionality is not a new concept in the ENS. Indeed, examples of multifunctional mechanosensitive neurons have been described in the myenteric plexus of the esophagus, stomach, and colon [4, 20, 21].

From an evolutionary point of view, neurons with more than a single function have been described in different invertebrate organisms, for instance, in the gastropod mollusc well recognized as "model organisms" in neuroscience Aplysia [8, 19]. Neuronal multifunctionality appears to be a specific property of enteric neurons, and we like to speculate that it could be derived from an independent evolution of the ENS from the CNS. The fascinating idea that the ENS evolved independently from the CNS and could be defined as the "first brain", has been recently discussed in light of different studies on central and peripheral nervous systems of species with different developmental degrees [6].

5.11 Outlook

All collected data resumed in this chapter about MEN reinforced our novel concept of sensory processing in the ENS. This idea highlights the cooperative activity of multifunctional neurons acting as part of a sensory network.

We have learnt many things about MEN. Their multifunctionality, in other words, their ability to function not only as sensors but also as interneurons, motoneurons, and probably secretomotoneurons is fascinating. Unfortunately, these are only some pieces of a big puzzle and this means that there are still many questions, which remains open; for instance, the role of MEN in the regulation of the peristalsis or in the nerve-mediated secretory response. We hope to add other pieces to the puzzle in the next few years. For instance, pharmacological experiments, which we are now running, will reveal us the molecular targets transducing the mechanical stimulus. This will represent an important step and it will be economically valuable: understanding physiology of sensory signaling can help us to develop targeted therapeutic treatment in response to gastrointestinal disorders.

References

1. Coste B, Mathur J, Schmidt M, Earley TJ, Ranade S, Petrus MJ, Dubin AE, Patapoutian A (2010) Piezo1 and Piezo2 are essential components of distinct mechanically activated cation channels. Science 330:55–60
2. Coste B, Xiao B, Santos JS, Syeda R, Grandl J, Spencer KS, Kim SE, Schmidt M, Mathur J, Dubin AE et al (2012) Piezo proteins are pore-forming subunits of mechanically activated channels. Nature 483:176–181
3. Diener M, Rummel W (1990) Distension-induced secretion in the rat colon: mediation by prostaglandins and submucosal neurons. Eur J Pharmacol 178:47–57
4. Dong H, Jiang Y, Dong J, Mittal RK (2015) Inhibitory motor neurons of the esophageal myenteric plexus are mechanosensitive. Am J Physiol Cell Physiol 308:C405–C413
5. Filzmayer AK, Elfers K, Michel K, Buhner S, Zeller F, Demir IE, Theisen J, Schemann M, Mazzuoli-Weber G (2020) Compression and stretch sensitive submucosal neurons of the porcine and human colon. Sci Rep 10:13791
6. Furness JB, Stebbing MJ (2018) The first brain: species comparisons and evolutionary implications for the enteric and central nervous systems. Neurogastroenterol Motil 30:e13234
7. Furness JB, Kunze WA, Bertrand PP, Clerc N, Bornstein JC (1998) Intrinsic primary afferent neurons of the intestine. Prog Neurobiol 54:1–18
8. Gardner D, Kandel ER (1977) Physiological and kinetic properties of cholinergic receptors activated by multiaction interneurons in buccal ganglia of Aplysia. J Neurophysiol 40:333–348
9. Gottlieb PA, Sachs F (2012) Piezo1: properties of a cation selective mechanical channel. Channels Austin Tex 6:214–219
10. Kirchgessner AL, Tamir H, Gershon MD (1992) Identification and stimulation by serotonin of intrinsic sensory neurons of the submucosal plexus of the guinea pig gut: activity-induced expression of Fos immunoreactivity. J Neurosci 12:235–248
11. Kugler EM (2014) Mechanosensitivity in isolated myenteric neuronal networks. PhD thesis. Fakultät für Ernährung, Landnutzung und Umwelt; Lehrstuhl für Humanbiologie; TU München
12. Kugler EM, Michel K, Zeller F, Demir IE, Ceyhan GO, Schemann M, Mazzuoli-Weber G (2015) Mechanical stress activates neurites and somata of myenteric neurons. Front Cell Neurosci 9:342
13. Kugler EM, Michel K, Kirchenbüchler D, Dreissen G, Csiszár A, Merkel R, Schemann M, Mazzuoli-Weber G (2018) Sensitivity to strain and shear stress of isolated mechanosensitive enteric neurons. Neuroscience 372:213–224
14. Mazzuoli G, Schemann M (2009) Multifunctional rapidly adapting mechanosensitive enteric neurons (RAMEN) in the myenteric plexus of the guinea pig ileum. J Physiol 587:4681–4693
15. Mazzuoli G, Schemann M (2012) Mechanosensitive enteric neurons in the myenteric plexus of the mouse intestine. PLoS One 7:e39887
16. Mazzuoli-Weber G, Schemann M (2015a) Mechanosensitivity in the enteric nervous system. Front Cell Neurosci 9:408
17. Mazzuoli-Weber G, Schemann M (2015b) Mechanosensitive enteric neurons in the guinea pig gastric corpus. Front Cell Neurosci 9:430
18. Mazzuoli-Weber G, Kugler EM, Bühler CI, Kreutz F, Demir IE, Ceyhan OG, Zeller F, Schemann M (2019) Piezo proteins: incidence and abundance in the enteric nervous system. Is there a link with mechanosensitivity? Cell Tissue Res 375:605–618
19. Rosen SC, Teyke T, Miller MW, Weiss KR, Kupfermann I (1991) Identification and characterization of cerebral-to-buccal interneurons implicated in the control of motor programs associated with feeding in Aplysia. J Neurosci 11:3630–3655
20. Schemann M, Schaaf C (1995) Differential projection of cholinergic and nitroxidergic neurons in the myenteric plexus of guinea pig stomach. Am J Phys 269:G186–G195
21. Spencer NJ, Smith TK (2004) Mechanosensory S-neurons rather than AH-neurons appear to generate a rhythmic motor pattern in guinea-pig distal colon. J Physiol 558:577–596

22. Trendelenburg P (1917) Physiologische und pharmakologische Versuche uber die Dunndarmperistaltik. Naunyn Schmiedeberg's Arch Pharmacol 8:55–129
23. Weber E, Neunlist M, Schemann M, Frieling T (2001) Neural components of distension-evoked secretory responses in the guinea-pig distal colon. J Physiol 536:741–751
24. Wang F, Knutson K, Alcaino C, Linden DR, Gibbons SJ, Kashyap P, Grover M, Oeckler R, Gottlieb PA, Li HJ, Leiter AB, Farrugia G, Beyder A (2017) Mechanosensitive ion channel Piezo2 is important for enterochromaffin cell response to mechanical forces. J Physiol 595(1):79-91. https://doi.org/10.1113/JP272718. Epub 2016 Aug 13. PMID: 27392819; PMCID: PMC5199733.].

New Concepts of the Interplay Between the Gut Microbiota and the Enteric Nervous System in the Control of Motility

6

Fernando A. Vicentini, Tanner Fahlman, Stephanie G. Raptis, Laurie E. Wallace, Simon A. Hirota, and Keith A. Sharkey

Abstract

Propulsive gastrointestinal (GI) motility is critical for digestive physiology and host defense. GI motility is finely regulated by the intramural reflex pathways of the enteric nervous system (ENS). The ENS is in turn regulated by luminal factors: diet and the gut microbiota. The gut microbiota is a vast ecosystem of commensal bacteria, fungi, viruses, and other microbes. The gut microbiota not only regulates the motor programs of the ENS but also is critical for the normal structure and function of the ENS. In this chapter, we highlight recent research that has shed light on the microbial mechanisms of interaction with the ENS involved in the control of motility. Toll-like receptor signaling mechanisms have been shown to maintain the structural integrity of the ENS and the neurochemical phenotypes of enteric neurons, in part through the production of trophic factors including glia-derived neurotrophic factor. Microbiota-derived short-chain fatty acids and/or single-stranded RNA regulates the synthesis of serotonin in enterochromaffin cells, which are involved in the initiation of enteric reflexes, among other functions. Further evidence suggests a crucial role for microbial modulation of serotonin in maintaining the integrity of the ENS through enteric neurogenesis. Understanding the

Fernando A. Vicentini, Tanner Fahlman and Stephanie G. Raptis contributed equally.

F. A. Vicentini
Hotchkiss Brain Institute, Calgary, AB, Canada

Snyder Institute for Chronic Diseases, Calgary, AB, Canada

Inflammation Research Network, Calgary, AB, Canada

Department of Physiology and Pharmacology, Cumming School of Medicine, University of Calgary, Calgary, AB, Canada

T. Fahlman · S. G. Raptis · L. E. Wallace
K. A. Sharkey (✉)
Hotchkiss Brain Institute, Calgary, AB, Canada

Snyder Institute for Chronic Diseases, Calgary, AB, Canada

Department of Physiology and Pharmacology, Cumming School of Medicine, University of Calgary, Calgary, AB, Canada
e-mail: ksharkey@ucalgary.ca

S. A. Hirota
Snyder Institute for Chronic Diseases, Calgary, AB, Canada

Inflammation Research Network, Calgary, AB, Canada

Alberta Children's Hospital Research Institute, Calgary, AB, Canada

Department of Physiology and Pharmacology, Cumming School of Medicine, University of Calgary, Calgary, AB, Canada

N. J. Spencer et al. (eds.), *The Enteric Nervous System II*, Advances in Experimental Medicine and Biology 1383, https://doi.org/10.1007/978-3-031-05843-1_6

microbial pathways of enteric neural control sheds new light on digestive health and provides novel treatment strategies for GI motility disorders.

Keywords

5-HT · GDNF · Neurogenesis · Serotonin · Toll-like receptors · Myenteric plexus

6.1 Introduction

The gastrointestinal (GI) tract hosts a vast and diverse microbial ecosystem, consisting of bacteria, archaea, fungi, protozoa, and viruses [31, 71]. The absolute number of microbes in the GI tract has been approximated to be in the order of 10^{14} [55]. By mutualistic coevolution, mammals have developed an interdependency on the gut microbiota to provide metabolic, nutritional, and immunological support crucial to maintaining health [5, 44]. Gut microbes are involved in the metabolism of nutrients, particularly harvesting energy from indigestible complex carbohydrates such as dietary fibers, as well as the processing of proteins and lipids and the synthesis of vitamins [60, 71, 72]. Nowadays, we appreciate that microbes play a vital role in many other functions in the GI tract, including educating the host immune system, providing trophic support for the intestinal epithelium, contributing to the maintenance of the epithelial barrier, and occupying a niche in the intestinal lumen that allows for protection against pathogens [44, 77].

With such large, complex, and metabolically active microbial communities residing along the entire length of the GI tract, it is not surprising that the gut microbiota is also involved in the regulation of physiological functions. The gut microbiota has been shown to be essential for the normal regulation of secretion, motility, barrier, and local immune functions [4, 57, 77, 93, 100, 102]. Since these critical gut functions are controlled by the enteric nervous system (ENS) [34, 84], it has been suggested that there is a functional interaction between the gut microbiota and the intrinsic nerves that control the gut [45, 69]; in fact, there is an intricate and sophisticated relationship that we will discuss in this chapter.

Enteric neurons reside in two interconnected ganglionated plexuses in the gut wall: the myenteric and submucosal plexuses [34, 36]. The myenteric plexus is located between the longitudinal and circular muscle layers of the gut wall [36, 84]. Together with interstitial cells of Cajal and platelet-derived growth factor receptor α cells, myenteric neurons regulate basal, interdigestive, and digestive motility patterns [43, 53, 78]. The myenteric plexus also has neurons that project to the mucosa, which play a critical role in mucosal functions, including growth, barrier, secretomotor, and sensory functions [36]. The submucosal plexus, on the other hand, is located between the mucosa and circular muscle layers, and primarily innervates mucosal and submucosal tissues and controls secretion, mucosal barrier function, and blood flow [34, 36]. In addition to neurons, the ENS contains equal or greater numbers of enteric glia [41, 79]. These specialized peripheral glial cells are actively involved in signaling between neurons that modulate GI reflexes, regulate immune processes in the gut wall, and contribute to neurogenesis in pathophysiological states [41, 79].

Enteric neurons are characterized as sensory neurons, referred to as intrinsic primary afferent neurons (IPANs), interneurons, and motor neurons; the latter include cholinergic excitatory and nitrergic inhibitory motor neurons [34, 36, 84]. Recent single cell RNA sequencing analyses of the ENS have revealed a far greater diversity of neuronal subtypes within these broad categories than previously recognized [28, 63, 66, 99, 104]. Assigning molecular and genetic signatures to the functional classes of enteric neurons remains an exciting future development for the field.

The regulation of gut motility occurs through modulation of the synaptic circuitry of the ENS. Mechanical, chemical, and/or microbial signals from the lumen initiate action potentials in IPANs that are integrated by interneuronal circuits, leading to appropriate coordinated motor outputs [36, 78, 84]. Enteric nerves do not penetrate beyond the epithelium, and therefore receive information from the luminal environment via specialized epithelial cells, including enteroendocrine and enterochromaffin cells, that transduce the luminal signals [9, 48, 62, 96]. Our understanding of how microbial signals modulate the ENS circuitry and influence gut motility has been gradually increasing in recent years [45, 69]. In this chapter, we will focus on current findings discussing some of the proposed mechanisms of microbial interactions with the ENS circuits that regulate GI motility. Most studies to date have focused on gut bacteria, but other microbial constituents such as fungi have been implicated in the control of aspects of GI physiology in health and disease [11, 93].

6.2 Microbial Regulation of Enteric Neurons and Enteric Glia

Early studies of germ-free rats revealed that there is a remarkable plasticity of neurons in the myenteric plexus of the markedly enlarged cecum, a characteristic of germ-free animals [29, 30]. Myenteric neurons demonstrate substantial hypertrophy in the enlarged cecum, whereas those in the ileum remain unchanged, despite the absence of a microbiota, suggesting that this neuroplasticity is linked to the expanded tissue volume that is innervated and not the complete absence of gut microbes per se [29]. When the germ-free rats were allowed to regain a conventional microbiota, neuronal diameters were similar to conventionally raised rats, illustrating the reversibility of this phenomenon [29]. Further studies that examine the mechanisms of these interesting observations would shed new light on the factors involved in the regulation of enteric neuronal growth.

In more recent studies of germ-free mice, Collins et al. reported a reduction in the number of myenteric neurons in the jejunum (~40%) and ileum (~50%), but not the duodenum 3 days after birth [23]. Despite the overall reduction in total neurons, the proportion of nitrergic neurons was increased in this study [23]. However, in adult mice, there appears to be no marked changes in the number of myenteric neurons in the jejunum [64], but a small reduction in the ileum [67]. The number of myenteric neurons in the colon has been reported to be unchanged [25, 67], or reduced [4, 101], while the density of the innervation of the colonic myenteric plexus in germ-free mice was comparable to conventionally raised animals [25]. Whether there are alterations in the number of myenteric neurons, the absence of a gut microbiota changes the proportion of neurons expressing various neuropeptides, transmitters, or other molecular markers of enteric neurons. McVey Neufeld et al. found that the proportion of calbindin (Calb)$^+$ neurons was markedly reduced in the jejunum from adult germ-free mice, and this reduction correlated with changes in the electrophysiological properties of the Calb$^+$ IPAN neurons [64]. Muller et al. described reductions in somatostatin and cocaine and amphetamine-related transcript (CART)$^+$ neurons in the ileum and colon [67], while many authors observed marked reductions in colonic nitrergic neurons [3, 4, 101] and an increased density of cholinergic neurons [3]. Most importantly, there are functional correlations with these neurochemical changes, including changes in gut motility linked to the loss of nitrergic neurons [3, 4] and altered glucose homeostasis linked to the loss of intestinofugal CART$^+$ neurons of the ileum [67]. Interestingly, Obata et al. reported similar molecular and neurochemical characteristics of colonic myenteric neurons from germ-free and conventional mice, despite showing considerable transcriptomic differences between these two groups [68]. One of the key transcriptional changes was an upregulation of the gene encoding the aryl hydrocarbon receptor in mice that were colonized with a normal microbiota. In elegant and detailed studies, they then went on to show that this transcription factor controlled the expression of an inwardly rectifying potassium channel that controls the firing of colonic myenteric neurons and that aryl hydrocarbon receptor signaling regulates propulsive motility in the colon [68].

Enteric glial density as determined by S100B labeling has been reported to be reduced [25] or unaffected [3] in the colonic myenteric plexus of germ-free mice, though the total number of

Sox10⁺ glial cells remains unaltered [25]. The enteric glial network in the ileal and colonic mucosa is also markedly reduced in the absence of gut microbiota [3, 47]. The functional significance of these findings remains to be fully understood. Together, these studies illustrate the need for further detailed systematic examination of the neural and glial elements of the ENS along the length of the GI tract in germ-free mice.

Since germ-free mice lack microbiota during development, various laboratories, including our own, have treated adult mice with oral antibiotics to examine the impact of the gut microbiota when developmental factors are not involved. When antibiotics are used to deplete bacteria in the gut, there is a consistent reduction in the number of myenteric neurons in both the ileum [67, 94] and colon [4, 67, 94, 101], but not in the duodenum [67]. It has been argued these observations might be due in part to neurotoxic effects of the antibiotics [19], which are often administered in high doses. However, the completely reversible nature of the reductions after withdrawal of antibiotics [67, 94], and the fact that antibiotics administered directly to enteric neurons [101], systemically to adult mice [67], or orally to germ free-mice [101], did not change the numbers of myenteric neurons, strongly suggests that these changes are related to the depletion of gut bacteria. It remains to be shown exactly why there are differences between antibiotic-treated and germ-free mice. However, in studies that evaluated both models, the neurochemical changes observed in germ-free mice are similar to those seen after antibiotic administration. For example, Muller et al. reported similar reductions in $CART^+$ neurons in the ileal and colonic, but not duodenal, myenteric plexus, as they observed in germ-free mice [67]. Likewise, nitrergic neurons are reduced in the myenteric plexus of the ileum and colon in both antibiotic-treated and germ-free mice [4, 94, 101], and these changes appear to underlie the reduced motility seen in these models.

In antibiotic-treated mice, there are also changes associated to enteric glia. Interestingly, Vicentini et al. found a reduced number of $S100B^+$ glial cells in the myenteric plexus of the ileum, but not the colon of antibiotic-treated mice as might be expected from the findings of De Vadder et al. [25]. Contrary to previous reports [47], no changes in the mucosal glial network were observed in this study [94]. Further studies of enteric glia are needed to verify these disparate findings.

An important feature of the ENS is that it maintains a considerable neurogenic potential in the adult animal [8, 46, 51, 54, 56]. Neurogenesis in the ENS was initially demonstrated by Liu et al. [56], and is still an area of intense debate in the field. Emerging evidence suggests that the gut microbiota is also involved in the modulation of enteric neurogenesis, with reports showing microbiota-modulated neurogenesis both in germ-free and antibiotic-treated mice [25, 94, 101]. This microbiota-mediated phenomenon is novel, and the molecular mechanisms that regulate it are largely unknown. Understanding enteric neurogenesis brings huge implications for the development of tools to treat enteric neuropathies, such as Hirschsprung's disease.

Together, these findings illustrate the remarkable role the gut microbiota plays in regulating enteric neural homeostasis. In the subsequent sections of this chapter, we will focus on specific mechanisms that gut bacteria use to signal to the ENS and control motility.

6.3 Gut Bacteria Interact with Toll-Like Receptors and Enterochromaffin Cells to Regulate the Integrity of the ENS and Control GI Motility

It has been demonstrated that luminal bacteria are capable of modulating enteric neuronal activity [52], and as we have already discussed, they regulate the structure and neurochemistry of the ENS. Due to the anatomical separation provided by the epithelium, it is hypothesized that interaction with the ENS occurs indirectly through bacterial molecules, such as polysaccharide A, lipoteichoic acid (LTA), lipopolysaccharide, formyl peptides, and short chain fatty acids (SCFAs) [4, 20, 61, 82, 101].

Some structural components of enteric bacteria are ligands for toll-like receptors (TLRs) [1]. Recent progress has been made in understanding how the TLR system impacts the ENS and controls motility [4, 12, 16, 17, 101]. In addition to structural components, the bacterial metabolites SCFAs are well known to be involved in mediating host-microbe interactions, such as regulating mucosal serotonin (5-hydroxytryptamine, or 5-HT) [33, 59, 75, 100]. 5-HT is synthesized and secreted by enterochromaffin cells and initiates propulsive motility by activating IPANs [62, 89]. An increasing number of studies have demonstrated that 5-HT levels are tightly regulated by the enteric microbiota [25, 75, 80, 98, 100], suggesting a promising mechanism of microbiota-ENS interaction that affects gut motility.

The next sections of this chapter will focus on the evidence for TLR and 5-HT mechanisms by which bacteria and their metabolic products are able to influence the structure and function of the ENS. Understanding this link is of great importance to GI pathophysiology. Frequently, diseases affecting the GI tract, including the inflammatory bowel diseases (IBD) and irritable bowel syndrome, show microbial dysbiosis and patterns of GI dysmotility leading to some of the clinical symptoms of these diseases [72]. A better understanding of how the control of GI motility is influenced by the gut microbiota may provide new avenues for therapeutic approaches to treat GI dysmotility disorders.

6.4 Toll-Like Receptors as Regulators of Intestinal Motility

Host cells recognize microorganisms through a family of proteins known as pattern recognition receptors [2]. TLRs belong to the pattern recognition receptor family and play a crucial role in host innate immunity [1, 2]. TLRs are classically known for recognizing microbe-associated molecular patterns (MAMPs). When activated, TLRs recruit adaptor molecules, such as myeloid differentiation primary response 88 (MyD88), initiating a signaling cascade. Induction of specific target genes ultimately activates the innate immune system [1].

Beyond their roles in innate immunity, TLRs have been associated with physiological regulation of the GI tract, acting to maintain gut-microbial homeostasis [14, 70]. Among the many types of TLRs (TLR 1–10 in humans and 1–9, 11–13 in mice) [49], some are expressed on key cells involved in the regulation of gut motility, such as enteric neurons, enteroendocrine cells, and smooth muscle cells [6, 10, 13]. To date, most research has focused on TLR2 and TLR4, both of which have been strongly linked to the regulation of gut motility and ENS integrity [4, 12].

6.4.1 Deletion of TLR2 and TLR4 Impacts the Integrity of the ENS and Alters Motility

TLR2 and TLR4 are expressed on enteric neurons (soma and nerve fibers) and enteric glial cells of both plexuses [6, 10, 12, 15, 70, 101]. Interestingly, higher TLR expression is observed in more distal regions of the GI tract, which have a higher microbial load [6, 70]. To better understand the role of TLRs on gut motility, different types of approaches have been applied, including evaluating the effects of deleting the receptor. Deletion of either TLR2 or TLR4 has consistently been shown to cause dysmotility in the GI tract of mice [12, 16, 18, 32, 101]. In the seminal work of Anitha et al., TLR4-deficient mice were found to have delayed intestinal transit time [4]. This delayed intestinal transit is accompanied by a reduction in the nitrergic relaxation response to electrical field stimulation and a reduction in the number of nitrergic (neuronal nitric oxide synthase [$nNOS^+$]) neurons. Moreover, they reported similar results abolishing TLR signaling specifically in enteric neurons using a $Wnt1Cre^{+/-}$/$Myd88^{fl/fl}$ mouse, suggesting that alterations observed in ENS integrity and gut motility caused by TLR deficiency are directly dependent on ENS mechanisms [4]. Other reports corroborate these results describing slower GI motility, reduced amplitude and frequency of spontaneous contractile events, and reduced nNOS+ neurons in TLR4 KO mice [16, 32].

Like TLR4, TLR2 has been directly implicated in the control of ENS integrity and intestinal motility. Brun et al. demonstrated that TLR2-deficient mice exhibited faster gastric emptying and intestinal transit, a higher frequency and amplitude of spontaneous contractions, and enhanced response to contractile electrical field stimulation in the ileum [12]. The enhanced contractile responses were reproduced by others [32]. Furthermore, Brun et al. reported a reduction in the total number of ileal myenteric neurons, along with reduction in nNOS+ neurons and a reduced density of cholinergic nerve fibers [12]. Interestingly, Yarandi et al. recently showed a reduced GI transit time when TLR2 signaling was blocked using a TLR2-blocking antibody [101]. These apparently contradictory results might be explained by the different assays of motility or that there are differences in the role of TLR2 along the length of the GI tract [12, 101]. A consistent observation among studies blocking TLR2 is the reduction in the number of enteric neurons, specifically nNOS+ nitrergic neurons in both the ileum and colon [12, 101], strongly implicating TLR2 in neuronal survival. TLR signaling pathways are integral to maintaining structural integrity of the ENS and normal GI motility (Fig. 6.1). However, the mechanisms by which this occurs need to be further investigated.

6.4.2 TLRs Are Modulated by Gut Microbiota Affecting Gut Transit, Neurogenesis, and Glial-Derived Neurotrophic Factor (GDNF)

The absence or depletion of the microbiota has been widely reported to induce gut dysmotility [3, 4, 16, 40, 101], a phenotype similar to the one observed in mice with impaired or absent TLR2 and TLR4 signaling [4, 12, 16, 101]. Yarandi et al. recently showed that dysmotility induced by antibiotic treatment is prevented by supplementation of LTA, a structural molecule found in Gram-positive bacteria and a potent TLR2 agonist [101]. Furthermore, they showed that the effect of microbiota depletion on enteric neurons is reversed by LTA likely due to TLR2-mediated neurogenesis [101]. TLR modulation of neurogenesis in the brain has been suggested before, with TLR2 playing a major role in inducing proliferation and differentiation of neural progenitor cells [76]. In the gut, in vivo neurogenesis in adult animals has been observed in several contexts [7, 51, 56, 88], but is an area of continued debate. With the high levels of mechanical and biochemical stresses that the ENS faces, and largely having constant number of enteric neurons throughout the life, it has been proposed that the ENS is in fact maintained through natural replacement of old enteric neurons in favor of new ones [51]. While TLR4 has not been implicated in enteric neurogenesis [101], TLR4 signaling is important for neuronal survival in vitro [4], and after depletion of bacteria in vivo [94]. However, the balance regarding activation of TLR4 signaling is very important, as, for instance, changes in diet (e.g., Western diet) lead to increased activation of TLR4, resulting in loss of nitrergic neurons and impaired motility [74]. Thus, as the reduction in the number of enteric neurons is correlated with dysmotility [43], this host-microbe pathway, in which microbiota modulate enteric neurogenesis via TLRs, is potentially important in GI diseases that present with both dysbiosis and dysmotility (e.g., IBD).

Another important feature related to neurogenesis, which is also modulated by TLR signaling, is the production of neurotrophic factors [12, 39]. Ret signaling, via its ligand GDNF, has been well described as an important modulator of ENS integrity [39, 65]. GDNF-Ret signaling plays a crucial role in enteric neurogenesis, as it controls ENS precursor proliferation, therefore determining the number of mature enteric neurons [39, 91, 92]. Interestingly, upregulation of GDNF expression correlates with activation of the NF-kB and p38 MAPK signaling pathways [12], both regulated by TLR [1]. Brun et al. demonstrated that TLR2-deficient mice have reduced GDNF expression, and that administration of GDNF restored motility and ENS defects seen in these mice [12]. Further, the same group showed that

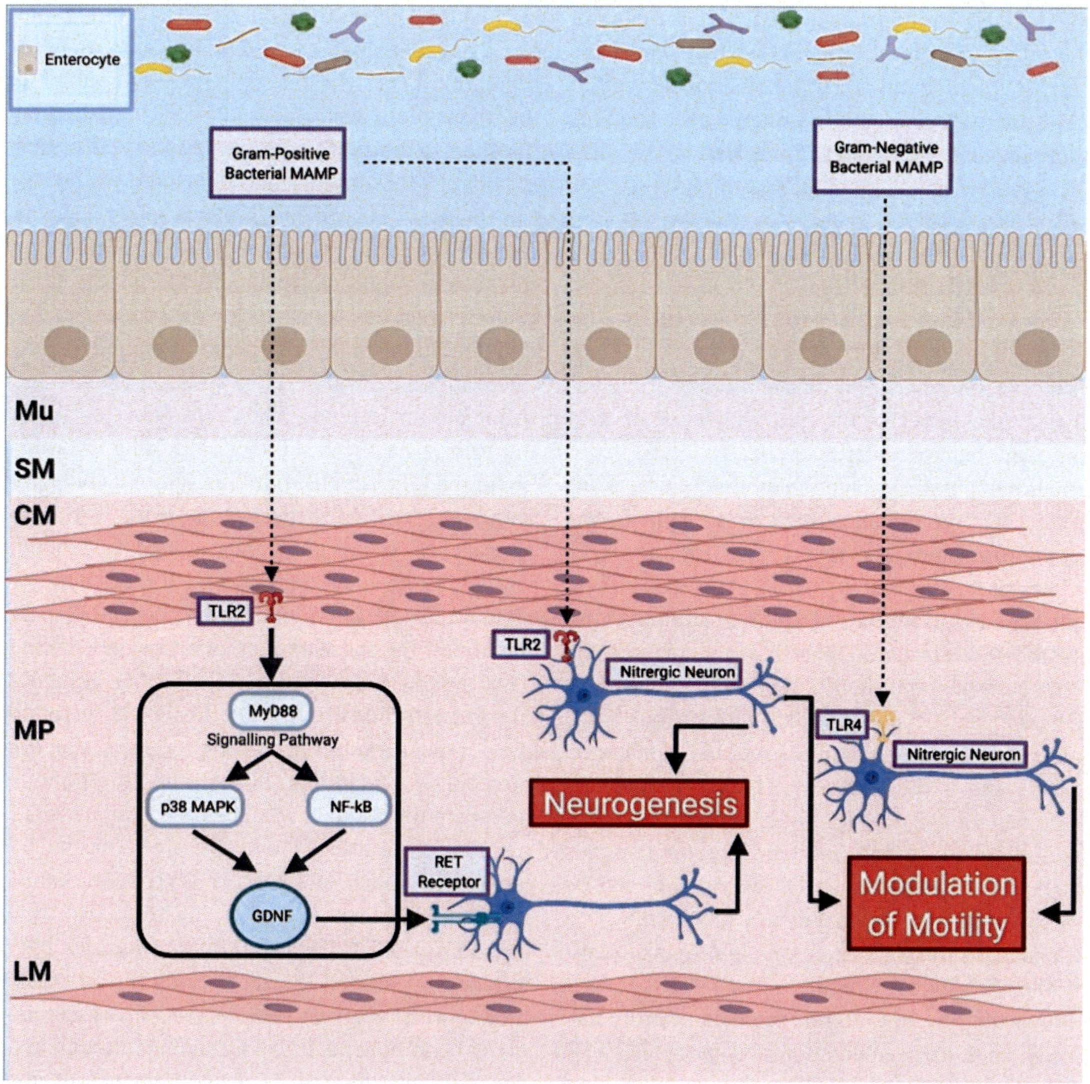

Fig. 6.1 Microbiota modulates intestinal motility and enteric neurogenesis acting through toll-like receptor (TLR) signaling. Anatomical overview of the gut wall with the mechanisms by which microbiota modulates the enteric nervous system via TLR. Microbe-associated molecular patterns (MAMPs) derived from Gram-positive and Gram-negative bacteria activates TLR2 and TLR4, respectively, in enteric neurons and smooth muscle cells. Activation of TLR2 and TLR4 in neurons results in modulation of gut motility, with TLR2 signaling associated with enteric neurogenesis. TLR2 activation in smooth muscle cells induces glial-derived neurotrophic factor (GDNF) production. GDNF, via the Ret receptor, is also linked to enteric neurogenesis. Mu mucosa, SM submucosa, CM circular muscle layer, MP myenteric plexus, LM longitudinal muscle layer. (Figure created with BioRender.com)

when activated, TLR2 stimulates intestinal smooth muscle cells to produce GDNF, contributing to the maintenance signaling cascade [13]. These results illustrate the importance of this TLR signaling pathway in maintaining the integrity of the GI tract through the production of trophic factors, such as GDNF. Recently, Soret et al. showed that rectal administration of GDNF induces neurogenesis in mouse models of Hirschsprung's disease, improving GI motility [83]. Together, these findings illustrate a mechanism by which the microbiota could modulate local GDNF (or other neurotrophins) to maintain the integrity of the ENS (Fig. 6.1). If this is the case, it could be utilized as a potential strategy to treat enteric neuropathies, prompting the need for more mechanistic studies in this area.

6.5 Microbial Regulation of Serotonin in Enterochromaffin Cells

It is widely understood that ~90% of 5-HT synthesis takes place in the gut, with most of it in mucosal enteroendocrine cells, followed by a small percentage in enteric serotonergic neurons [38, 62]. Tryptophan hydroxylase (Tph) is the rate-limiting enzyme responsible for the 5-HT synthesis, expressed in two isoforms, Tph1 in enteroendocrine cells and Tph2 in neurons. After synthesis and secretion, 5-HT has pleiotropic effects throughout the GI tract and beyond [86]. Once released, 5-HT interacts with various 5-HT receptors to modulate GI motility, including 5-HT_3 and 5-HT_4 receptors, both of which are involved in the control of propulsive motility, including the colonic migrating motor complex [103]. After secretion, 5-HT is metabolized by monoamine oxidase enzymes or taken up by epithelial cells through the serotonin reuptake transporter. This intricate regulation of 5-HT in the gut allows the system to tightly regulate motility patterns, such as segmentation and peristalsis [38, 62]. The effects of 5-HT on gut motility have been debated over the last decade, with earlier research suggesting that 5-HT was necessary for the initiation of peristalsis [27, 81], and more recent investigations suggesting that 5-HT plays more of modulatory role regulating the frequency of motility patterns, rather than being responsible for initiating these processes [50, 85, 95]. Though not crucial for initiation, the role of 5-HT in the modulation of contractile frequency and duration within the smooth muscle layers of the GI tract is of great importance to the normal physiological functioning of the muscle movements, which allow the ability of efficient macronutrient absorption and waste expulsion.

Due to the crucial role of 5-HT as a modulator of GI motility, understanding how mucosal 5-HT regulation occurs in the context of alterations to the microbiota has emerged as a fundamental question to be answered. Initial reports described the effects of probiotics in rodents in the modulation of the 5-HT precursor (tryptophan) and its metabolites, suggesting a potential role for the microbiota to influence the 5-HT system [26]. Later, it was shown that germ-free rodents exhibited alterations in 5-HT concentration and metabolism in the central nervous system [22, 42] and in the serum [80, 98]. Therefore, Yano et al. set out to explore the 5-HT modulation by microbiota in the GI tract [100]. In this seminal study, by using a series of experiments using different approaches (e.g., germ-free, conventionalized, and antibiotic-treated mice), the authors demonstrated that the microbiota influenced 5-HT production in the gut via modulation of the enzyme Tph1 in enterochromaffin cells, further influencing gut motility [100]. Specifically, they found that spore-forming bacteria produced specific metabolites, including SCFA, that through an unknown mechanism was able to modulate 5-HT metabolism [100]. These results are supported by the work of other labs [25, 37, 75]. Remarkably, it has been shown that the opposite is also true: alterations in 5-HT levels influence luminal bacterial composition [35], reinforcing the complex, bidirectional relation between these two fundamental components of the GI tract. Although evidence points toward a clear influence of the microbiota on 5-HT metabolism, the exact mechanisms by which microbes act on host cells remain an area of active investigation.

6.5.1 Potential Mechanisms for Microbiota Modulation of 5-HT Metabolism and GI Motility

Acetate, propionate, and butyrate are the main SCFAs found in the gut. Being products of fermentation by anaerobic microbes, they are thought to be a major link between the microbiota, the ENS, and gut motility [24]. Administration of SCFAs has long been associated with increased colonic motility [21, 33, 82, 95]. Yano et al. have pointed out that SCFAs may be one of the metabolites missing in germ-free mice that would be responsible for regulation of control levels of 5-HT [100]. In vitro data corroborates this hypothesis, as SCFA treatment induced an increased expression of Tph1 in BON cells (human 5-HT-producing cell line) [75]. Evaluating colonic motility ex vivo, Vincent et al. showed no effect of SCFA in samples from Tph1 KO mice, suggesting a 5-HT-dependent mechanism mediating the promotility responses [95]. However, this process was observed only with low concentrations of SCFA, where higher concentrations induced alterations in colonic motility through a 5-HT-independent mechanism [95]. Together these data support SCFA as an important microbial mediator that modulates 5-HT, and thereby alters the regulation of enteric reflex-modulation of motility.

Although SCFA may have direct effects on enterochromaffin cells, in the small intestine, evidence suggests that SCFA may modulate 5-HT release indirectly, through the action of the glucagon-like peptide (GLP)-1 receptor [59]. Within the small intestine, it was shown that SCFAs increase the secretion of 5-HT from enterohromaffin cells [100]. However, Lund et al. were unable to detect SCFA receptors on the enterochromaffin cells in this region of the GI tract [59]. Instead, adjacent GLP-1-secreting L-cells have receptors that can sense and respond to SCFA. A mechanism was suggested wherein SCFA acts on L-cells, inducing the release of GLP-1 [90], which can then diffuse to the neighboring enterochromaffin cells, where it binds to the GLP-1 receptor, triggering 5-HT release [59]. In contrast, colonic enterochromaffin cells were found to have SCFA receptors [59], illustrating interesting regional differences in the gut, that might be linked to the gradient of increasing microbial density from the small bowel to the large bowel (Fig. 6.2).

Serotonin has also been linked to enteric neurogenesis [8, 25, 56, 88]. One of the first to publish on the idea of adult neurogenesis in the ENS was Gershon and colleagues [56]. Working on the precedent of neurogenesis involving the 5-HT_4 receptor in the hippocampus [58], they showed that 5-HT_4 KO mice were born with an equal number of enteric neurons compared to wild-type mice. However, the number of enteric neurons increased after birth in wild-type mice, but not in 5-HT_4-deficient mice, suggesting a link between neurogenesis and this receptor [56]. Further, they demonstrated that mice treated with a 5-HT_4 agonist had an increased incorporation of the nucleotide analogue bromodeoxyuridine in myenteric neurons and Nestin$^+$ neural precursor cells, suggesting the potential for cell proliferation to maintain neuronal numbers regardless of cell death of enteric neurons through the life of the animal [56]. Furthermore, 5-HT_4-deficient animals had slowed GI transit, which brought forth the idea that in the absence of enteric neurogenesis, normal GI motility was compromised [56]. Interestingly, the gut microbiota is involved in the modulation of 5-HT_4 receptor expression in the myenteric plexus [25]. As described above, germ-free mice have reduced numbers of enteric neurons, which are recovered by microbial recolonization [101]. De Vadder et al. showed that an increase in Nestin$^+$ neuronal progenitor cells by microbial recolonization is blunted using a 5-HT_4 receptor antagonist [25]. Moreover, they showed a recovery of myenteric neurons and gut motility with 5-HT_4 receptor agonist in germ-free mice, linking this receptor with ENS integrity and GI physiology [25]. In our lab, we observed that SCFA supplementation is capable of inducing neuronal recovery in antibiotic-treated mice [94], suggesting a mechanism by which microbiota may modulate neurogenesis through SCFA-dependent release of 5-HT (Fig. 6.2).

The expression levels of Tph1 in enterochromaffin cells have been shown to be highly regulated by the gut microbiota, determining the levels of 5-HT [37, 75, 100]. A recent development in this area has come from Sugisawa et al., who discovered another mechanism for the regulation of Tph1 in enterochromaffin cells through the Piezo1 receptor [87]. The Piezo1 receptor is a mechanosensitive ion channel with a variety of

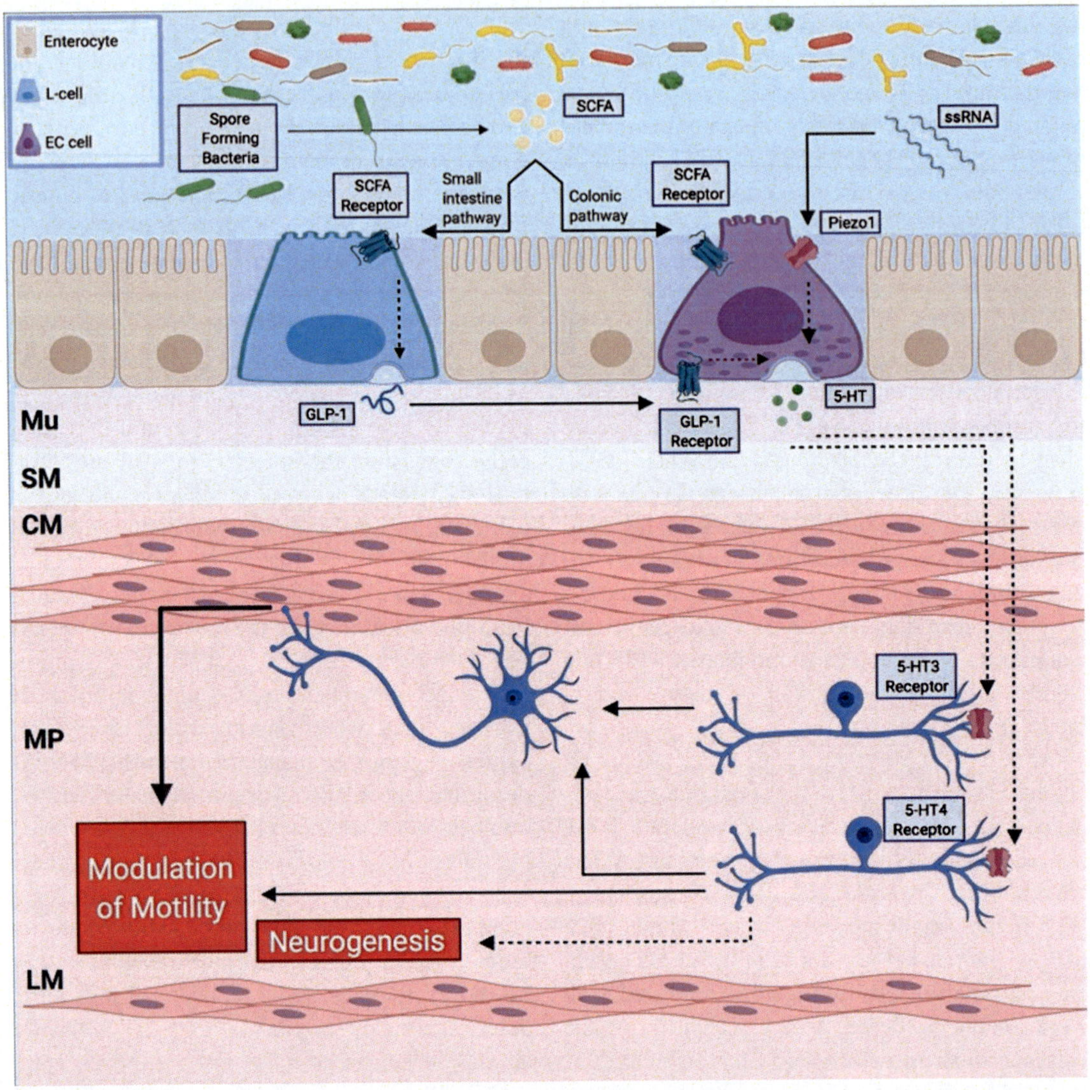

Fig. 6.2 Microbiota modulates serotonin (5-HT) metabolism to regulate intestinal motility and enteric neurogenesis. Spore-forming bacteria, likely via the production of short-chain fatty acids (SCFA), regulate 5-HT synthesis in enterochromaffin (EC) cells. In the small intestine, SCFA stimulates glucagon-like peptide (GLP)-1 secretion from L-cells, which in turn leads to increased 5-HT levels in EC cells. In the colon, SCFA may act directly on EC cells, regulating 5-HT synthesis. Single-strand RNA (ssRNA) derived from the microbiota has been also associated with the regulation of 5-HT in EC cells via the Piezo1 receptor. 5-HT secretion influences neuronal function, and motility, via 5-HT_3 and 5-HT_4 receptors. 5-HT_4 receptors have also been strongly linked to enteric neurogenesis. Mu mucosa, SM submucosa, CM circular muscle layer, MP myenteric plexus, LM longitudinal muscle layer. (Figure created with BioRender.com)

functions [73]. Piezo1 modulates Tph1 expression by responding to extracellular single-stranded RNA (ssRNA) originating from microbes in the gut. This channel regulates gut motility, as observed with a decreased peristalsis in mice with a targeted deletion of Piezo1 in the intestinal epithelium [87]. These findings suggest that microbial-derived ssRNA is an important determinant of 5-HT availability, providing a new mechanism by which the microbiota may modulate gut function. It would be interesting to know the nature of the interactions between mechanical and microbial mechanisms in enterochromaffin cells of the intestinal epithelium in the regulation of enteric reflexes controlling propulsive motility.

6.6 Conclusions and Future Directions

The regulation of the spatiotemporal coordination of GI function relies mainly on the ENS, augmented by autonomic input from the central nervous system. The ENS senses the luminal environment indirectly, through enteroendocrine/enterochromaffin cell mediators, and responds accordingly. GI motility is finely regulated by the intramural reflex pathways of the ENS that are modulated through interactions with the gut microbiota. Microbial signaling is critical for the normal structure and function of the ENS, as we have described in this chapter. There is a sophisticated relationship between the ENS, and the local immune and enteroendocrine systems of the gut that is shaped by the presence of the microbiota. Together they form a network from which the ENS can function efficiently to gather information on the state of the gut and respond by modulating various gut functions including motility patterns such as peristalsis.

Recent research has shed light on the microbial mechanisms of interaction with the ENS for the control of motility. Toll-like receptor signaling pathways activated by MAMPs maintain the structural integrity of the ENS and the neurochemical phenotypes of enteric neurons, in part through the production of trophic factors including GDNF [4, 12, 101]. Microbial SCFA and/or ssRNA regulate the synthesis of 5-HT, which is involved in the initiation of enteric reflexes, among other functions, with evidence suggesting a crucial role for microbiota modulation of 5-HT in maintaining the integrity of the ENS through enteric neurogenesis [25, 59, 87, 100]. Interestingly, 5-HT and TLRs may also work together to modulate gut motility and normal ENS function, as TLR2 signaling in nonhematopoietic cells has been shown to stimulate 5-HT production in the gut [97], and deletion of TLR2 and TLR4 modifies the expression of specific 5-HT receptors [32].

The examples outlined in this chapter are unlikely to be the only ways in which the microbiota can regulate host physiology but serve to illustrate the complexity and sophistication of the control of gut function. Exactly how these mechanisms are altered in disease remains an area for future study. Work focused on the interplay of the serotonergic and TLR-mediated responses to the microbiota will provide useful insights into how motility is controlled. Understanding these microbial signaling pathways and their targets in the ENS could provide novel approaches for treating GI motility disorders.

Conflict of Interest The authors declare no conflicts of interest.

Author Contributions All authors reviewed the literature and drafted the manuscript, critically reviewed the content, revised the manuscript, and approved it for submission.

Funding This work was supported by grants from the Canadian Institutes of Health Research (SAH and KAS). FAV was supported by the National Council for Scientific and Technological Development (CNPq), Brazil. TF and SGR were supported by Alberta Innovates Summer Research Studentships.

References

1. Akira S, Takeda K (2004) Toll-like receptor signalling. Nat Rev Immunol 4:499–511
2. Akira S, Uematsu S, Takeuchi O (2006) Pathogen recognition and innate immunity. Cell 124:783–801
3. Aktar R, Parkar N, Stentz R, Baumard L, Parker A, Goldson A, Brion A, Carding S, Blackshaw A, Peiris M (2020) Human resident gut microbe Bacteroides thetaiotaomicron regulates colonic neuronal innervation and neurogenic function. Gut Microbes 11:1745–1757

4. Anitha M, Vijay-Kumar M, Sitaraman SV, Gewirtz AT, Srinivasan S (2012) Gut microbial products regulate murine gastrointestinal motility via Toll-like receptor 4 signaling. Gastroenterology 143:1006–1016.e1004
5. Bäckhed F, Ley RE, Sonnenburg JL, Peterson DA, Gordon JI (2005) Host-bacterial mutualism in the human intestine. Science 307:1915–1920
6. Barajon I, Serrao G, Arnaboldi F, Opizzi E, Ripamonti G, Balsari A, Rumio C (2009) Toll-like receptors 3, 4, and 7 are expressed in the enteric nervous system and dorsal root ganglia. J Histochem Cytochem 57:1013–1023
7. Belkind-Gerson J, Graham HK, Reynolds J, Hotta R, Nagy N, Cheng L, Kamionek M, Shi HN, Aherne CM, Goldstein AM (2017) Colitis promotes neuronal differentiation of Sox2+ and PLP1+ enteric cells. Sci Rep 7:2525
8. Belkind-Gerson J, Hotta R, Nagy N, Thomas AR, Graham H, Cheng L, Solorzano J, Nguyen D, Kamionek M, Dietrich J et al (2015) Colitis induces enteric neurogenesis through a 5-HT4-dependent mechanism. Inflamm Bowel Dis 21:870–878
9. Bellono NW, Bayrer JR, Leitch DB, Castro J, Zhang C, O'Donnell TA, Brierley SM, Ingraham HA, Julius D (2017) Enterochromaffin cells are gut chemosensors that couple to sensory neural pathways. Cell 170:185–198.e116
10. Bogunovic M, Davé SH, Tilstra JS, Chang DT, Harpaz N, Xiong H, Mayer LF, Plevy SE (2007) Enteroendocrine cells express functional Toll-like receptors. Am J Physiol GI Liver Physiol 292:G1770–G1783
11. Botschuijver S, Roeselers G, Levin E, Jonkers DM, Welting O, Heinsbroek SEM, de Weerd HH, Boekhout T, Fornai M, Masclee AA et al (2017) Intestinal fungal dysbiosis is associated with visceral hypersensitivity in patients with irritable bowel syndrome and rats. Gastroenterology 153:1026–1039
12. Brun P, Giron MC, Qesari M, Porzionato A, Caputi V, Zoppellaro C, Banzato S, Grillo AR, Spagnol L, De Caro R et al (2013) Toll-like receptor 2 regulates intestinal inflammation by controlling integrity of the enteric nervous system. Gastroenterology 145:1323–1333
13. Brun P, Gobbo S, Caputi V, Spagnol L, Schirato G, Pasqualin M, Levorato E, Palù G, Giron MC, Castagliuolo I (2015) Toll like receptor-2 regulates production of glial-derived neurotrophic factors in murine intestinal smooth muscle cells. Mol Cell Neurosci 68:24–35
14. Burgueño JF, Abreu MT (2020) Epithelial Toll-like receptors and their role in gut homeostasis and disease. Nat Rev Gastroenterol Hepatol 17:263–278
15. Burgueño JF, Barba A, Eyre E, Romero C, Neunlist M, Fernández E (2016) TLR2 and TLR9 modulate enteric nervous system inflammatory responses to lipopolysaccharide. J Neuroinflammation 13:187
16. Caputi V, Marsilio I, Cerantola S, Roozfarakh M, Lante I, Galuppini F, Rugge M, Napoli E, Giulivi C, Orso G et al (2017) Toll-like receptor 4 modulates small intestine neuromuscular function through nitrergic and purinergic pathways. Front Pharmacol 8:350
17. Caputi V, Marsilio I, Filpa V, Cerantola S, Orso G, Bistoletti M, Paccagnella N, De Martin S, Montopoli M, Dall'Acqua S et al (2017) Antibiotic-induced dysbiosis of the microbiota impairs gut neuromuscular function in juvenile mice. Br J Pharmacol 174:3623–3639
18. Cerantola S, Caputi V, Marsilio I, Ridolfi M, Faggin S, Bistoletti M, Giaroni C, Giron MC (2020) Involvement of enteric glia in small intestine neuromuscular dysfunction of Toll-like receptor 4-deficient mice. Cell 9:838
19. Champagne-Jorgensen K, Kunze WA, Forsythe P, Bienenstock J, McVey Neufeld KA (2019) Antibiotics and the nervous system: more than just the microbes? Brain Behav Immun 77:7–15
20. Chandrasekharan B, Saeedi BJ, Alam A, Houser M, Srinivasan S, Tansey M, Jones R, Nusrat A, Neish AS (2019) Interactions between commensal bacteria and enteric neurons, via FPR1 induction of ROS, increase gastrointestinal motility in mice. Gastroenterology 157:179–192.e172
21. Cherbut C, Ferrier L, Rozé C, Anini Y, Blottière H, Lecannu G, Galmiche JP (1998) Short-chain fatty acids modify colonic motility through nerves and polypeptide YY release in the rat. Am J Physiol GI Liver Physiol 275:G1415–G1422
22. Clarke G, Grenham S, Scully P, Fitzgerald P, Moloney RD, Shanahan F, Dinan TG, Cryan JF (2013) The microbiome-gut-brain axis during early life regulates the hippocampal serotonergic system in a sex-dependent manner. Mol Psychiatry 18:666–673
23. Collins J, Borojevic R, Verdu EF, Huizinga JD, Ratcliffe EM (2014) Intestinal microbiota influence the early postnatal development of the enteric nervous system. Neurogastroenterol Motil 26:98–107
24. Dalile B, Van Oudenhove L, Vervliet B, Verbeke K (2019) The role of short-chain fatty acids in microbiota-gut-brain communication. Nat Rev Gastroenterol Hepatol 16:461–478
25. De Vadder F, Grasset E, Mannerås Holm L, Karsenty G, Macpherson AJ, Olofsson LE, Bäckhed F (2018) Gut microbiota regulates maturation of the adult enteric nervous system via enteric serotonin networks. Proc Natl Acad Sci U S A 115:6458
26. Desbonnet L, Garrett L, Clarke G, Bienenstock J, Dinan TG (2008) The probiotic Bifidobacteria infantis: an assessment of potential antidepressant properties in the rat. J Psychiatr Res 43:164–174
27. Dickson EJ, Heredia DJ, Smith TK (2010) Critical role of 5-HT1A, 5-HT3, and 5-HT7 receptor subtypes in the initiation, generation, and propagation of the murine colonic migrating motor complex. Am J Physiol GI Liver Physiol 299:G144–G157
28. Drokhlyansky E, Smillie CS, Van Wittenberghe N, Ericsson M, Griffin GK, Eraslan G, Dionne D, Cuoco

MS, Goder-Reiser MN, Sharova T et al (2020) The human and mouse enteric nervous system at single-cell resolution. Cell 182:1606–1622.e1623
29. Dupont JR, Jervis H, Sprinz H (1965) A kinetic study of the myenteric neurones: size variations in the neurovegetative periphery with body weight and organ size. Am J Anat 116:335–339
30. Dupont JR, Jervis HR, Sprinz H (1965) Auerbach's plexus of the rat cecum in relation to the germfree state. J Comp Neurol 125:11–18
31. Eckburg PB, Bik EM, Bernstein CN, Purdom E, Dethlefsen L, Sargent M, Gill SR, Nelson KE, Relman DA (2005) Diversity of the human intestinal microbial flora. Science 308:1635–1638
32. Forcén R, Latorre E, Pardo J, Alcalde AI, Murillo MD, Grasa L (2016) Toll-like receptors 2 and 4 exert opposite effects on the contractile response induced by serotonin in mouse colon: role of serotonin receptors. Exp Physiol 101:1064–1074
33. Fukumoto S, Tatewaki M, Yamada T, Fujimiya M, Mantyh C, Voss M, Eubanks S, Harris M, Pappas TN, Takahashi T (2003) Short-chain fatty acids stimulate colonic transit via intraluminal 5-HT release in rats. Am J Physiol Regul Integr Comp Physiol 284:R1269–R1276
34. Fung C, Vanden Berghe P (2020) Functional circuits and signal processing in the enteric nervous system. Cell Mol Life Sci 77:4505–4522
35. Fung TC, Vuong HE, Luna CDG, Pronovost GN, Aleksandrova AA, Riley NG, Vavilina A, McGinn J, Rendon T, Forrest LR et al (2019) Intestinal serotonin and fluoxetine exposure modulate bacterial colonization in the gut. Nat Microbiol 4:2064–2073
36. Furness JB (2012) The enteric nervous system and neurogastroenterology. Nat Rev Gastroenterol Hepatol 9:286–294
37. Ge X, Ding C, Zhao W, Xu L, Tian H, Gong J, Zhu M, Li J, Li N (2017) Antibiotics-induced depletion of mice microbiota induces changes in host serotonin biosynthesis and intestinal motility. J Transl Med 15:13
38. Gershon MD, Tack J (2007) The serotonin signaling system: from basic understanding to drug development for functional GI disorders. Gastroenterology 132:397–414
39. Gianino S, Grider JR, Cresswell J, Enomoto H, Heuckeroth RO (2003) GDNF availability determines enteric neuron number by controlling precursor proliferation. Development 130:2187–2198
40. Grasa L, Abecia L, Forcén R, Castro M, de Jalón JAG, Latorre E, Alcalde AI, Murillo MD (2015) Antibiotic-induced depletion of murine microbiota induces mild inflammation and changes in Toll-like receptor patterns and intestinal motility. Microb Ecol 70:835–848
41. Gulbransen BD, Sharkey KA (2012) Novel functional roles for enteric glia in the gastrointestinal tract. Nat Rev Gastroenterol Hepatol 9:625–632
42. Heijtz RD, Wang S, Anuar F, Qian Y, Björkholm B, Samuelsson A, Hibberd ML, Forssberg H, Pettersson S (2011) Normal gut microbiota modulates brain development and behavior. Proc Natl Acad Sci U S A 108:3047–3052
43. Heuckeroth RO (2018) Hirschsprung disease - integrating basic science and clinical medicine to improve outcomes. Nat Rev Gastroenterol Hepatol 15:152–167
44. Hooper LV, Gordon JI (2001) Commensal host-bacterial relationships in the gut. Science 292:1115–1118
45. Hyland NP, Cryan JF (2016) Microbe-host interactions: influence of the gut microbiota on the enteric nervous system. Dev Biol 417:182–187
46. Joseph NM, He S, Quintana E, Kim Y-G, Núñez G, Morrison SJ (2011) Enteric glia are multipotent in culture but primarily form glia in the adult rodent gut. J Clin Invest 121:3398–3411
47. Kabouridis PS, Lasrado R, McCallum S, Chng SH, Snippert HJ, Clevers H, Pettersson S, Pachnis V (2015) Microbiota controls the homeostasis of glial cells in the gut lamina propria. Neuron 85:289–295
48. Kaelberer MM, Buchanan KL, Klein ME, Barth BB, Montoya MM, Shen X, Bohórquez DV (2018) A gut-brain neural circuit for nutrient sensory transduction. Science 361:eaat5236
49. Kawasaki T, Kawai T (2014) Toll-like receptor signaling pathways. Front Immunol 5:461
50. Keating DJ, Spencer NJ (2010) Release of 5-hydroxytryptamine from the mucosa is not required for the generation or propagation of colonic migrating motor complexes. Gastroenterology 138:659–670 670.e651–652
51. Kulkarni S, Micci MA, Leser J, Shin C, Tang SC, Fu YY, Liu L, Li Q, Saha M, Li C et al (2017) Adult enteric nervous system in health is maintained by a dynamic balance between neuronal apoptosis and neurogenesis. Proc Natl Acad Sci U S A 114:E3709–e3718
52. Kunze WA, Mao YK, Wang B, Huizinga JD, Ma X, Forsythe P, Bienenstock J (2009) Lactobacillus reuteri enhances excitability of colonic AH neurons by inhibiting calcium-dependent potassium channel opening. J Cell Mol Med 13:2261–2270
53. Kurahashi M, Zheng H, Dwyer L, Ward SM, Koh SD, Sanders KM (2011) A functional role for the 'fibroblast-like cells' in gastrointestinal smooth muscles. J Physiol 589:697–710
54. Laranjeira C, Sandgren K, Kessaris N, Richardson W, Potocnik A, Vanden Berghe P, Pachnis V (2011) Glial cells in the mouse enteric nervous system can undergo neurogenesis in response to injury. J Clin Invest 121:3412–3424
55. Ley RE, Peterson DA, Gordon JI (2006) Ecological and evolutionary forces shaping microbial diversity in the human intestine. Cell 124:837–848
56. Liu M-T, Kuan Y-H, Wang J, Hen R, Gershon MD (2009) 5-HT4 receptor-mediated neuroprotection and neurogenesis in the enteric nervous system of adult mice. J Neurosci 29:9683–9699

57. Lomasney KW, Houston A, Shanahan F, Dinan TG, Cryan JF, Hyland NP (2014) Selective influence of host microbiota on cAMP-mediated ion transport in mouse colon. Neurogastroenterol Motil 26:887–890
58. Lucas G, Rymar VV, Du J, Mnie-Filali O, Bisgaard C, Manta S, Lambas-Senas L, Wiborg O, Haddjeri N, Piñeyro G et al (2007) Serotonin 4 (5-HT4) receptor agonists are putative antidepressants with a rapid onset of action. Neuron 55:712–725
59. Lund ML, Egerod KL, Engelstoft MS, Dmytriyeva O, Theodorsson E, Patel BA, Schwartz TW (2018) Enterochromaffin 5-HT cells – a major target for GLP-1 and gut microbial metabolites. Mol Metab 11:70–83
60. Makki K, Deehan EC, Walter J, Bäckhed F (2018) The impact of dietary fiber on gut microbiota in host health and disease. Cell Host Microbe 23:705–715
61. Mao YK, Kasper DL, Wang B, Forsythe P, Bienenstock J, Kunze WA (2013) *Bacteroides fragilis* polysaccharide A is necessary and sufficient for acute activation of intestinal sensory neurons. Nat Commun 4:1465
62. Mawe GM, Hoffman JM (2013) Serotonin signalling in the gut—functions, dysfunctions and therapeutic targets. Nat Rev Gastroenterol Hepatol 10:473
63. May-Zhang AA, Tycksen E, Southard-Smith AN, Deal KK, Benthal JT, Buehler DP, Adam M, Simmons AJ, Monaghan JR, Matlock BK et al (2021) Combinatorial transcriptional profiling of mouse and human enteric neurons identifies shared and disparate subtypes in situ. Gastroenterology 160:755–770.e726
64. McVey Neufeld KA, Perez-Burgos A, Mao YK, Bienenstock J, Kunze WA (2015) The gut microbiome restores intrinsic and extrinsic nerve function in germ-free mice accompanied by changes in calbindin. Neurogastroenterol Motil 27:627–636
65. Moore MW, Klein RD, Fariñas I, Sauer H, Armanini M, Phillips H, Reichardt LF, Ryan AM, Carver-Moore K, Rosenthal A (1996) Renal and neuronal abnormalities in mice lacking GDNF. Nature 382:76–79
66. Morarach K, Mikhailova A, Knoflach V, Memic F, Kumar R, Li W, Ernfors P, Marklund U (2021) Diversification of molecularly defined myenteric neuron classes revealed by single-cell RNA sequencing. Nat Neurosci 24:34–46
67. Muller PA, Matheis F, Schneeberger M, Kerner Z, Jové V, Mucida D (2020) Microbiota-modulated CART+ enteric neurons autonomously regulate blood glucose. Science 370:314–321
68. Obata Y, Castaño Á, Boeing S, Bon-Frauches AC, Fung C, Fallesen T, de Agüero MG, Yilmaz B, Lopes R, Huseynova A et al (2020) Neuronal programming by microbiota regulates intestinal physiology. Nature 578:284–289
69. Obata Y, Pachnis V (2016) The effect of microbiota and the immune system on the development and organization of the enteric nervous system. Gastroenterology 151:836–844
70. Price AE, Shamardani K, Lugo KA, Deguine J, Roberts AW, Lee BL, Barton GM (2018) A map of Toll-like receptor expression in the intestinal epithelium reveals distinct spatial, cell type-specific, and temporal patterns. Immunity 49:560–575.e566
71. Qin J, Li R, Raes J, Arumugam M, Burgdorf KS, Manichanh C, Nielsen T, Pons N, Levenez F, Yamada T et al (2010) A human gut microbial gene catalogue established by metagenomic sequencing. Nature 464:59–65
72. Quigley EM (2013) Gut bacteria in health and disease. Gastroenterol Hepatol 9:560–569
73. Ranade SS, Syeda R, Patapoutian A (2015) Mechanically activated ion channels. Neuron 87:1162–1179
74. Reichardt F, Chassaing B, Nezami BG, Li G, Tabatabavakili S, Mwangi S, Uppal K, Liang B, Vijay-Kumar M, Jones D et al (2017) Western diet induces colonic nitrergic myenteric neuropathy and dysmotility in mice via saturated fatty acid- and lipopolysaccharide-induced TLR4 signalling. J Physiol 595:1831–1846
75. Reigstad CS, Salmonson CE, Rainey JF, Szurszewski JH, Linden DR, Sonnenburg JL, Farrugia G, Kashyap PC (2015) Gut microbes promote colonic serotonin production through an effect of short-chain fatty acids on enterochromaffin cells. FASEB J 29:1395–1403
76. Rolls A, Shechter R, London A, Ziv Y, Ronen A, Levy R, Schwartz M (2007) Toll-like receptors modulate adult hippocampal neurogenesis. Nat Cell Biol 9:1081–1088
77. Round JL, Mazmanian SK (2009) The gut microbiota shapes intestinal immune responses during health and disease. Nat Rev Immunol 9:313–323
78. Sanders KM, Koh SD, Ro S, Ward SM (2012) Regulation of gastrointestinal motility--insights from smooth muscle biology. Nat Rev Gastroenterol Hepatol 9:633–645
79. Seguella L, Gulbransen BD (2021) Enteric glial biology, intercellular signalling and roles in gastrointestinal disease. Nat Rev Gastroenterol Hepatol 18:571-587
80. Sjögren K, Engdahl C, Henning P, Lerner UH, Tremaroli V, Lagerquist MK, Bäckhed F, Ohlsson C (2012) The gut microbiota regulates bone mass in mice. J Bone Miner Res 27:1357–1367
81. Smith TK, Koh SD (2017) A model of the enteric neural circuitry underlying the generation of rhythmic motor patterns in the colon: the role of serotonin. Am J Physiol GI Liver Physiol 312:G1–G14
82. Soret R, Chevalier J, De Coppet P, Poupeau G, Derkinderen P, Segain JP, Neunlist M (2010) Short-chain fatty acids regulate the enteric neurons and control gastrointestinal motility in rats. Gastroenterology 138:1772–1782
83. Soret R, Schneider S, Bernas G, Christophers B, Souchkova O, Charrier B, Righini-Grunder F, Aspirot A, Landry M, Kembel SW et al (2020) Glial cell derived neurotrophic factor induces enteric

neurogenesis and improves colon structure and function in mouse models of Hirschsprung disease. Gastroenterology 159:1824–1838.e1817
84. Spencer NJ, Hu H (2020) Enteric nervous system: sensory transduction, neural circuits and gastrointestinal motility. Nat Rev Gastroenterol Hepatol 17:338–351
85. Spencer NJ, Nicholas SJ, Robinson L, Kyloh M, Flack N, Brookes SJ, Zagorodnyuk VP, Keating DJ (2011) Mechanisms underlying distension-evoked peristalsis in guinea pig distal colon: is there a role for enterochromaffin cells? Am J Physiol GI Liver Physiol 301:G519–G527
86. Spohn SN, Mawe GM (2017) Non-conventional features of peripheral serotonin signalling - the gut and beyond. Nat Rev Gastroenterol Hepatol 14:412–420
87. Sugisawa E, Takayama Y, Takemura N, Kondo T, Hatakeyama S, Kumagai Y, Sunagawa M, Tominaga M, Maruyama K (2020) RNA sensing by gut Piezo1 is essential for systemic serotonin synthesis. Cell 182:609–624.e621
88. Takaki M, Goto K, Kawahara I (2014) The 5-hydroxytryptamine 4 receptor agonist-induced actions and enteric neurogenesis in the gut. J Neurogastroenterol Motil 20:17–30
89. Terry N, Margolis KG (2017) Serotonergic mechanisms regulating the GI tract: experimental evidence and therapeutic relevance. Handb Exp Pharmacol 239:319–342
90. Tolhurst G, Heffron H, Lam YS, Parker HE, Habib AM, Diakogiannaki E, Cameron J, Grosse J, Reimann F, Gribble FM (2012) Short-chain fatty acids stimulate glucagon-like peptide-1 secretion via the G-protein-coupled receptor FFAR2. Diabetes 61:364–371
91. Uesaka T, Nagashimada M, Enomoto H (2013) GDNF signaling levels control migration and neuronal differentiation of enteric ganglion precursors. J Neurosci 33:16372–16382
92. Uesaka T, Nagashimada M, Enomoto H (2015) Neuronal differentiation in Schwann cell lineage underlies postnatal neurogenesis in the enteric nervous system. J Neurosci 35:9879–9888
93. van Tilburg Bernardes E, Pettersen VK, Gutierrez MW, Laforest-Lapointe I, Jendzjowsky NG, Cavin J-B, Vicentini FA, Keenan CM, Ramay HR, Samara J et al (2020) Intestinal fungi are causally implicated in microbiome assembly and immune development in mice. Nat Commun 11:2577
94. Vicentini FA, Keenan CM, Wallace LM., Woods C, Cavin J-B, Flockton AR, Macklin WB, Belkind-Gerson J, Hirota SA, Sharkey KA (2021) Intestinal microbiota shapes gut physiology and regulates enteric neurons and glia. Microbiome 9:210
95. Vincent AD, Wang XY, Parsons SP, Khan WI, Huizinga JD (2018) Abnormal absorptive colonic motor activity in germ-free mice is rectified by butyrate, an effect possibly mediated by mucosal serotonin. Am J Physiol GI Liver Physiol 315:G896–G907
96. Walsh KT, Zemper AE (2019) The enteric nervous system for epithelial researchers: basic anatomy, techniques, and interactions with the epithelium. Cell Mol Gastroenterol Hepatol 8:369–378
97. Wang H, Kwon YH, Dewan V, Vahedi F, Syed S, Fontes ME, Ashkar AA, Surette MG, Khan WI (2019) TLR2 plays a pivotal role in mediating mucosal serotonin production in the gut. J Immunol 202:3041–3052
98. Wikoff WR, Anfora AT, Liu J, Schultz PG, Lesley SA, Peters EC, Siuzdak G (2009) Metabolomics analysis reveals large effects of gut microflora on mammalian blood metabolites. Proc Natl Acad Sci U S A 106:3698–3703
99. Wright CM, Schneider S, Smith-Edwards KM, Mafra F, Leembruggen AJL, Gonzalez MV, Kothakapa DR, Anderson JB, Maguire BA, Gao T et al (2021) scRNA-Seq reveals new enteric nervous system roles for GDNF, NRTN, and TBX3. Cell Mol Gastroenterol Hepatol 11:1548–1592.e1541
100. Yano JM, Yu K, Donaldson GP, Shastri GG, Ann P, Ma L, Nagler CR, Ismagilov RF, Mazmanian SK, Hsiao EY (2015) Indigenous bacteria from the gut microbiota regulate host serotonin biosynthesis. Cell 161:264–276
101. Yarandi SS, Kulkarni S, Saha M, Sylvia KE, Sears CL, Pasricha PJ (2020) Intestinal bacteria maintain adult enteric nervous system and nitrergic neurons via Toll-like receptor 2-induced neurogenesis in mice. Gastroenterology 159:200–213.e208
102. Yoon H, Schaubeck M, Lagkouvardos I, Blesl A, Heinzlmeir S, Hahne H, Clavel T, Panda S, Ludwig C, Kuster B et al (2018) Increased pancreatic protease activity in response to antibiotics impairs gut barrier and triggers colitis. Cell Mol Gastroenterol Hepatol 6:370–388.e373
103. Yu Y, Chen JH, Li H, Yang Z, Du X, Hong L, Liao H, Jiang L, Shi J, Zhao L et al (2015) Involvement of 5-HT3 and 5-HT4 receptors in colonic motor patterns in rats. Neurogastroenterol Motil 27:914–928
104. Zeisel A, Hochgerner H, Lönnerberg P, Johnsson A, Memic F, van der Zwan J, Häring M, Braun E, Borm LE, La Manno G et al (2018) Molecular architecture of the mouse nervous system. Cell 174:999–1014.e1022

7 Optical Approaches to Understanding Enteric Circuits Along the Radial Axis

Pieter Vanden Berghe and Candice Fung

Abstract

The gastrointestinal tract operates in a highly dynamic environment. The gut is typically exposed to continually changing and highly convoluted luminal compositions comprising not only ingested content but also a multitude of resident microbes and microbial factors. It is therefore critical that the gut is capable of distinguishing between nutritious components from noxious substances. This is facilitated by specialized cellular sensory machinery that are in place in the intestinal epithelium and the ENS. However, the specific chemosensory processes and enteric neuronal pathways that enable the gut to discern and respond appropriately to different chemicals remain unclear. A major hurdle in studying the neural processing of luminal information has been the complex spatial organization of the mucosal structures and their innervation along the radial axis. Much of our current knowledge of enteric neuronal responses to luminal stimuli stems from studies that used semi-dissected guinea pig small intestine preparations with the mucosa and submucosa removed in one-half in order to record electrical activity from exposed myenteric neurons or in the circular muscle. Building on this, we ultimately strive to work towards integrated systems with all the gut layers intact. With advanced microscopy techniques including multiphoton intravital imaging, together with transgenic technologies utilizing cell-type specific activity-dependent reporters, we stand in good stead for studying the ENS in more intact preparations and even in live animals. In this chapter, we highlight recent contributions to the knowledge of sensory gut innervation by the developing and mature ENS. We also revisit established work examining the functional connectivity between the myenteric and submucosal plexus, and discuss the methodologies that can help advance our understanding of the enteric circuitry and signaling along the mucosa-serosa axis.

Keywords

Enteric nervous system · Luminal sensing · Microbiota · Calcium imaging

7.1 Introduction

The gastrointestinal tract operates in a highly dynamic environment. The gut is typically exposed to continually changing and highly convoluted luminal compositions comprising not

P. Vanden Berghe (✉) · C. Fung
Laboratory for Enteric NeuroScience (LENS), Translational Research Center for Gastrointestinal Disorders (TARGID), University of Leuven, Leuven, Belgium
e-mail: pieter.vandenberghe@kuleuven.be

N. J. Spencer et al. (eds.), *The Enteric Nervous System II*, Advances in Experimental Medicine and Biology 1383, https://doi.org/10.1007/978-3-031-05843-1_7

only ingested content but also a multitude of resident gut microbes and microbial factors. While one of its key roles is to digest and take up nutritious components from ingested meals, its ability to detect and exclude invading pathogens or noxious agents is equally important. Accordingly, the gut possesses a variety of cellular sensory machinery within the intestinal epithelium and ENS, namely, enteroendocrine cells (EECs) and intrinsic primary afferent neurons (IPANS, intrinsic sensory neurons), respectively [1, 11]. Nonetheless, it remains unclear how these chemosensory processes occur, particularly regarding whether different luminal stimuli activate distinct enteric neuronal pathways, such that the gut can respond accordingly. In addition to physiological function, it is also pertinent to understand the development of this sensory system, its adaptations to dietary alterations or shifts in microbial diversity, and the contribution of these factors to shaping overall gut health. However, investigating the neural processing of luminal information in the intestine is especially challenging. A major hurdle has been the complex spatial organization of the mucosal structures and their innervation along the radial axis. Furthermore, their relation to the underlying interconnected nerve plexuses embedded between contractile muscle layers also remains elusive.

The foundation of our current knowledge on enteric neuronal responses to chemical and nutrient stimulation of the mucosa stems from an extensive body of work by Bertrand, Bornstein, and colleagues [2–5, 14]. These studies were performed on semidissected guinea pig small intestine preparations with the mucosa and submucosa removed in one-half to record electrical activity from exposed myenteric neurons or in the circular muscle. Various chemicals including 5-HT and acetate applied onto the mucosa were shown to activate AH neurons in a nonsynaptic manner, while S-type neurons were stimulated synaptically. Specific amino acids and putative mucosal mediators (5-HT, ATP, and CCK-8) puffed onto the mucosa could also elicit neurogenic inhibitory junction potentials in the smooth muscle [14]. These responses to amino acids were found to involve serotonergic and purinergic signaling pathways. To build on these findings and advance our understanding on how luminal contents are perceived by the ENS, we ultimately strive to work toward integrated systems with all the gut layers intact and connections between them preserved. Given the advantage of imaging techniques in enabling the survey of activity and spatiotemporal relation in large populations of neurons, as well as the availability of transgenic animals encoding cell-type-specific activity-dependent reporters, such technologies are increasingly adopted for ENS studies [6]. Together with the development of advanced microscopy techniques including multiphoton intravital imaging [35], we stand in good stead for studying the ENS in more intact ex vivo preparations and even in live animals via an optical approach.

Here, we highlight recent contributions to the knowledge of sensory mucosal innervation of the small and large bowel by the developing and mature ENS, as well as some of the methodologies that facilitate this process in parallel. We also revisit some established work examining the functional connectivity between the myenteric and submucosal plexus and their role in coordinating overall gut function. Finally, we conclude by discussing future perspectives on the study of the enteric circuitry, particularly along the mucosa-serosa axis.

7.2 The Development of the Intrinsic Sensory Innervation of Mucosa

Recent work in mice shows that developing neurites from the myenteric plexus (MP) project toward the mucosa from E13.5, and this occurs even earlier than the formation of the submucosal plexus (SMP) and villi [15]. Further, these neurites were shown to be functionally active by E15.5. While putative mucosal mediators were previously examined by localized applications to the mucosal surface, this study details a technique developed to mimic the basolateral release mediators from EECs whereby a micropipette is used to impale a single villus to inject different substances. In combination with transgenic mice encoding the fluorescent Ca^{2+} reporter GCaMP3, responses to mucosal stimulation in the underlying nerve plexus could be imaged in full-thickness preparations (Fig. 7.1). Using this approach, the injection of 5-HT into the villus tip and focal electrical stimulation of a villus were both shown to elicit responses in underlying myenteric neurons at E15.5. Hence, these neurites in the developing gut were already capable of transmitting mucosal signals to the MP at this early age [15]. Coinciding with the development of this mucosal innervation is the beginning of the differentiation of EECs in the epithelium, which occurs between ~E14-E15 [9]. Thus, the prenatal gut may already be able to detect luminal contents. It will be important to understand how the interaction between these sensory nerves and the intestinal epithelium develops postnatally, particularly relating to the introduction of a liquid diet from birth and later the shift to a solid diet during weaning, as well as the concomitant increase in microbial diversity. The influence of nutritional and microbial factors on how these connections establish potentially impacts the health of the gut

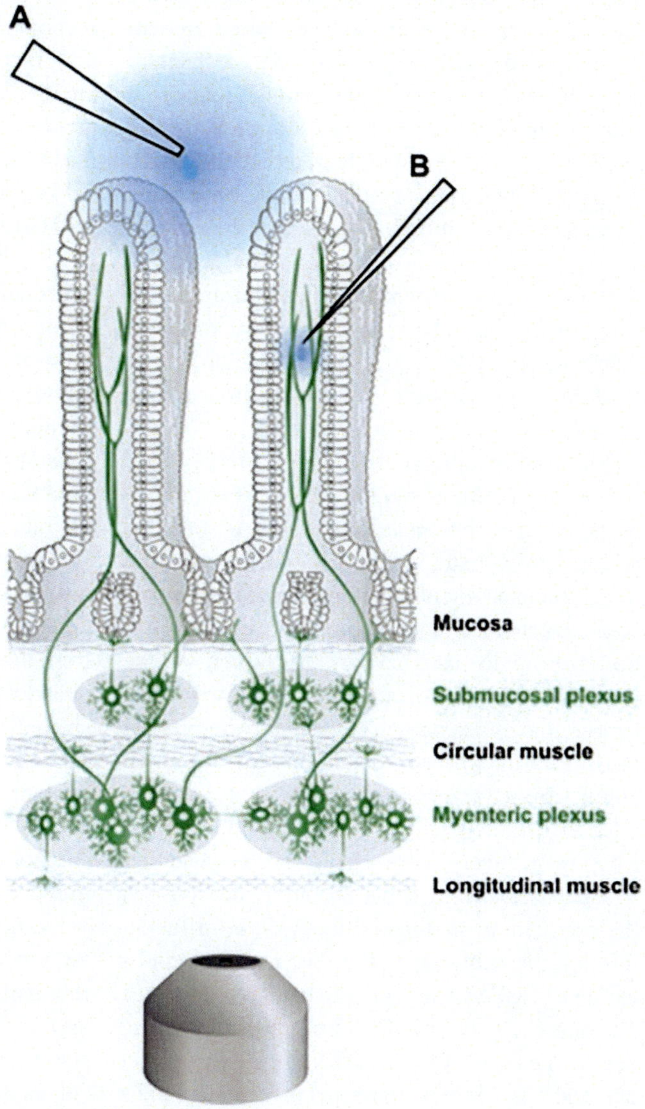

Fig. 7.1 A schematic depicting the experimental setup that enables imaging of responses to mucosal stimuli in the underlying nerve plexus. Using tissues from transgenic animals that express a fluorescent-activity-dependent reporter (e.g., GCaMP) specifically in the ENS [6], neural activity in the myenteric and submucosal plexus can be imaged through the layers of the intact gut. Chemical stimuli can be (**a**) perfused onto the mucosal surface or (**b**) injected into a single villus via micropipette pushed through the epithelial layer to mimic the basolateral release of mucosal mediators targeting the containing mucosal nerve endings

and its predisposition to disease in later life. For instance, neonatal antibiotic administration is associated with neurochemical changes in the gut and functional alterations in gut motility [18], and is also associated with an increased susceptibility to various diseases [10].

7.3 Sensing Microbial Metabolites

Despite the immense attention on the impact of gut microbiota in recent years given their implication in multifaceted aspects of health and disease, how microbes influence neurogenic function is only beginning to be explored. Recent work by Obata et al. [29] uncovered a mechanism whereby the gut microbiota can modulate the transcriptome of enteric neurons and, by acting through the aryl hydrocarbon receptor (AhR), can ultimately regulate colonic motility. Intestinal microbial composition can also affect the electrophysiological properties of enteric neurons [25]. Yet, taking into account the intestinal epithelial barrier separating the luminal contents from the ENS, the question remains of how microbes or microbial metabolites interact with enteric neurons.

Short-chain fatty acids (SCFAs) are key candidates for mediating signaling between microbiota and the ENS. Indeed, the SCFAs, namely, acetate, propionate, and butyrate, are among the most abundant metabolites produced by microbial fermentation and are present at concentrations of up to 100 mM in the lumen of the proximal colon [8]. In our recent work, we took a similar imaging approach as that described earlier [15]. Full-thickness preparations of proximal colon from transgenic mice that express GCaMP3 in their intestinal epithelium or in their ENS, respectively, were used to address whether luminal SCFAs may signal to enteric neurons via the epithelium or directly, that is, via diffusion through the epithelium [12]. The mucosal application of SCFAs was found to acutely elicit Ca^{2+} responses in the epithelium and underlying enteric neurons. On the other hand, the direct application of SCFAs to peeled myenteric plexus preparations did not evoke consistent neuronal responses. This suggests that SCFAs do not activate myenteric neuronal cell bodies directly and that the neuronal responses to luminal SCFAs observed require the mucosa.

Gut microbes may also modulate ENS activity via SCFAs through serotonergic signaling by increasing both enterochromaffin (EC) cell density and their expression of TPH1 (5-HT synthesizing enzyme) [34, 37]. Luminal 5-HT levels, which are modulated by the gut microbiota, can also reciprocally regulate microbial composition [16]. However, the effect of luminal 5-HT on the ENS is less clear. 5-HT acutely applied to mucosal surface of the colon elicits epithelial Ca^{2+} responses, but it was not an effective stimulus of Ca^{2+} responses in myenteric neurons, with only a small proportion of responding neurons within the field of view [12]. Myenteric responses to 5-HT injected into villi or applied onto the mucosal surface have also been compared in the study by Hao et al. [15]. Notably, 5-HT applied to the epithelial surface evoked detectable myenteric neuronal responses in only 30% of preparations, while responses to 5-HT injections were observed in over 80% of preparations examined. Thus, while 5-HT released from EC cells acts as a sensory mediator and acutely activates enteric neurons at the basolateral site, luminal 5-HT likely modulates ENS activity by broadly stimulating the mucosal epithelium.

The mouse proximal colon contains transverse mucosal folds with specific microbial populations residing beneath the folds that are distinct from those in the luminal digesta [28]. This raises questions of whether there are localized regional specializations in the innervation of these mucosal structures and in their detection of microbial content. With fluorescence and second harmonic generation (SHG) microscopy, we found that submucosal ganglia were organized such that they closely align with the striations of the transverse mucosal folds in the proximal colon [12]. By contrast, myenteric ganglia were distributed relatively evenly throughout the plexus (Fig. 7.2), and responses to SCFAs either applied to the ridge or to the mucosal surface between the ridges were also comparable. Conceivably, secre-

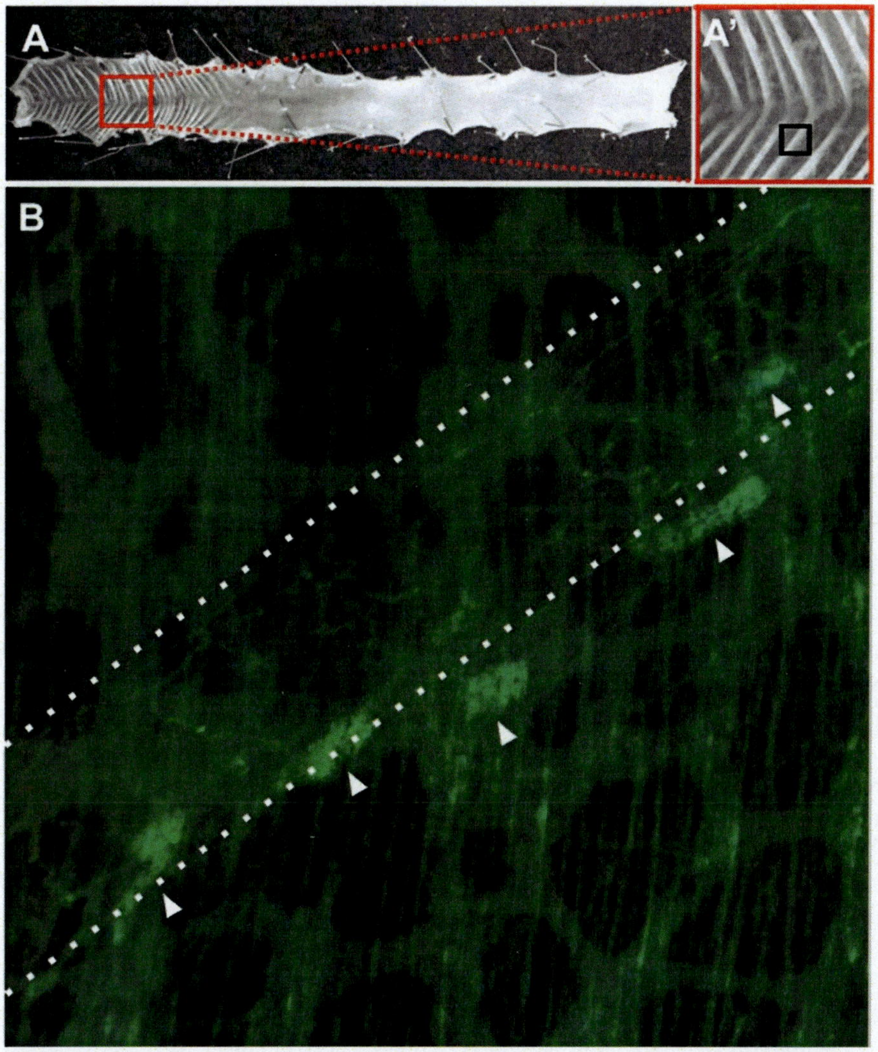

Fig. 7.2 (**a-a'**) A flat sheet preparation of mouse colon illustrating the striated mucosal folds in the proximal region. (**b**) A flat sheet preparation of peeled proximal colon (with the epithelium removed) from a Wnt1xGCaMP3 mouse, which expresses the fluorescent Ca^{2+} indicator in its ENS. The black square in (**a'**) represents the approximate location along a mucosal ridge at which this image was taken. The dotted lines outline the mucosal ridge running across the preparation. The arrowheads indicate underlying submucosal ganglia that are closely aligned along the ridge. The neural network that is out of focus in the background shows that myenteric ganglia by comparison are evenly distributed throughout the plexus

tomotor neurons more densely innervate the mucosal folds to regulate mucus and fluid secretions for the retention of bacterial content. Another possibility is that this SMP organization reflects an influence of microbial colonization on its development during early postnatal ages. Perhaps developing submucosal neurons and their projections are drawn toward the accumulation of microbes at the mucosal folds. Indeed, the developmental time window during which the postnatal maturation of submucosal ganglia in the colon occurs overlaps with the period during which the early gut microbiome is established [10, 24]. It remains to be determined whether there is specific innervation of the mucosal ridges from the underlying submucosal ganglia aligned along these structures and how the microbiota may influence their development.

7.4 Coordinating Activity in the Myenteric and Submucosal Plexus

The integrated activity of the myenteric and submucosal neurons is presumably critical in coordinating motor, secretomotor, and vasodilator output for normal gut function. For instance, organized changes in motility, secretion, and blood flow would be necessary for optimizing digestion and nutrient absorption in the fed state. It has been proposed that secretomotor and vasomotor reflexes running through the myenteric plexus may be involved in integrating responses to luminal stimuli [32, 33, 36]. The presence of such secretomotor reflexes was demonstrated in guinea pig ileum where the mucosa was removed in half of the preparation and electrical stimulation of the mucosa evoked fast EPSPs in 86% of submucosal neurons situated 0.7–1 cm anal to the site of stimulation [33]. The number of neurons displaying fast EPSPs was reduced to 11% in preparations where connections between the submucosa and myenteric plexus were severed. Furthermore, fast excitatory postsynaptic potentials (EPSPs) were absent after an incision was made in the myenteric plexus between the stimulus site and the recording site but were still observed when an incision was made in the submucosal plexus. Using the same tissue preparation, vasodilator reflexes elicited by mucosal stroking or balloon distension were similarly reduced by an incision made in the myenteric plexus between the stimulus and recording sites [32]. A lesion made in the submucosal plexus also had no effect. These studies highlight the functional connectivity between the mucosa, submucosa, and myenteric plexus. Examining activity in the enteric network at a larger scale (albeit at a lower resolution), Okamoto et al. [30] have also demonstrated that there is a degree of synchrony in myenteric and submucosal neuronal activities during colonic migrating motor complexes (CMMCs) using Ca^{2+} imaging of the murine large intestine. A key limitation here is motion artefacts due to contractile activity and therefore the continued development of analytical tools will be critical for addressing such issues [17, 20].

From a developmental perspective, Lasrado et al. [22] provided evidence of a lineage-dependent organization of enteric neuroglial units distributed in radial columns using a genetic approach to fate map individual Sox10-expressing enteric neural crest-derived cells (ENCCs) in mice. Lineally related myenteric neurons were found to be more likely to display synchronous Ca^{2+} activity in response to electrical stimulation. The authors propose that such neural units potentially underlie the coordinated signaling of the two nerve plexuses and may give rise to synchronous changes in gut motility and secretion. Nonetheless, how this synchronous activity is coordinated and its implications on integrated gut function need to be further explored.

Clearly, there is communication between the myenteric and submucosal plexus, but our understanding of this interaction is still limited. Current knowledge of the interconnectivity between the two plexus layers is primarily based on studies of the guinea pig small intestine. Early work by Bornstein et al. [7] demonstrated that submucosal neurons receive synaptic input from the myenteric neurons using intracellular recording on preparations from guinea pigs with a surgically induced lesion in the myenteric plexus. The number of submucosal neurons displaying excitatory synaptic inputs was significantly reduced following the ablation of the myenteric plexus and degradation of its nerve terminals. Moore and Vanner [27] further demonstrated a functional connection between the plexus layers using flat sheet preparations with the mucosa completely removed and the submucosal plexus removed in half of the preparation to expose the myenteric plexus. They performed intracellular recordings of submucosal neurons while applying focal electrical stimuli to the myenteric plexus at varying distances oral to the recording site. Both fast and slow excitatory, but not inhibitory, synaptic inputs were observed in 62% of submucosal S-type neurons anal to the stimulation site. Slow EPSPs were observed up to 10 mm from the point of stimulation, whereas fast EPSPs were detected up to 25 mm. Communication between the myenteric and submucosal plexus can also occur bidirectionally as Monro et al. [26] have shown that focal electrical stimulation of submucosal neurons can similarly evoke fast and slow EPSPs in some myenteric S-type neurons. Myenteric responses could be evoked by stimulating submucosal ganglia located orally, anally, or circumferentially at distances between 0.42 and 1.33 mm from the recording site. However, most myenteric neurons did not respond, with only 24% of neurons examined displaying synaptic potentials [26]. This suggests that the functional innervation of myenteric neurons by submucosal neurons is sparse compared to myenteric projections to submucosal neurons [27]. However, the specificity of such sparse connections is unclear [26]. That is, whether these connections enable precise control of a specific circuit or whether it simply has a less substantial modulatory role remains to be determined. One caveat of using such preparations is that it is technically unfeasible to record from myenteric neurons lying directly beneath the stimulated submucosal neurons or vice versa. Thus, connections running directly vertically along the radial axis could not be examined and the synaptic connectivity between the two plexuses is likely underrepresented in these studies. Using current technologies, we are now capable of imaging neuronal Ca^{2+} activity in both the myenteric and submucosal plexus of full-thickness tissues in response to the luminal application of different stimuli. Furthermore, together with *post-hoc* immunostaining, it is possible to identify specific neurochemical subtypes of enteric neurons that respond to nutrients such as glucose [13].

7.5 Future Perspectives

Our present understanding of the development and function of the enteric circuitry in the radial serosa-mucosa axis is less comprehensive compared to that along the longitudinal axis within the nerve plexus [19, 21, 23]. However, we now have the ability to generate recordings of ENS activity over time using Ca^{2+} imaging in full-thickness preparations [13, 15] or even in vivo [31, 35]. Thus, in employing and further refining such techniques, we expect to expand on this knowledge and gain insight into not only the coordination of signals in myenteric and submucosal neurons, but also their integrated responses to luminal stimuli in the intact gut.

Acknowledgments The authors acknowledge support from Methusalem (METH/014/05, KU Leuven) and FWO (Fonds voor wetenschappelijk onderzoek) G.093818N. Images were recorded on microscopy equipment in LENS / Cell and tissue Imaging Cluster (CIC), supported by Hercules AKUL/11/37; AKUL/09/50 and FWO G.0929.15.

References

1. Bertrand PP (2009) The cornucopia of intestinal chemosensory transduction. Front Neurosci 3:48
2. Bertrand PP, Bornstein JC (2002) ATP as a putative sensory mediator: activation of intrinsic sensory neurons of the myenteric plexus via P2X receptors. Neuroscience 22:4767–4775
3. Bertrand PP, Kunze WA, Bornstein JC, Furness JB (1998) Electrical mapping of the projections of intrinsic primary afferent neurones to the mucosa of the guinea-pig small intestine. Neurogastroenterol Motil 10:533–541
4. Bertrand PP, Bian X, Thornton PD, Bornstein JC (2000) Enteric micro-circuits activated upon stimulation of the mucosa. Gastroenterology 118:A667
5. Bertrand PP, Kunze WA, Bornstein JC, Furness JB, Smith ML (1997) Analysis of the responses of myenteric neurons in the small intestine to chemical stimulation of the mucosa. Am J Physiol 273:G422–G435
6. Boesmans W, Hao MM, Vanden Berghe P (2017) Optogenetic and chemogenetic techniques for neurogastroenterology. Nat Rev Gastroenterol Hepatol 15:21
7. Bornstein JC, Furness JB, Costa M (1987) Sources of excitatory synaptic inputs to neurochemically identified submucous neurons of guinea-pig small intestine. J Auton Nerv Syst 18:83–91
8. Cummings JH, Pomare EW, Branch WJ, Naylor CP, Macfarlane GT (1987) Short chain fatty acids in human large intestine, portal, hepatic and venous blood. Gut 28:1221–1227
9. Desai S, Loomis Z, Pugh-Bernard A, Schrunk J, Doyle MJ, Minic A, McCoy E, Sussel L (2008) Nkx2.2 regulates cell fate choice in the enteroendocrine cell lineages of the intestine. Dev Biol 313:58–66
10. Foong JPP, Hung LY, Poon S, Savidge TC, Bornstein JC (2020) Early life interaction between the microbiota and the enteric nervous system. Am J Physiol-Gastroint Liver Physiol 319:G541–G548

11. Fung C, Vanden Berghe P (2020) Functional circuits and signal processing in the enteric nervous system. Cell Mol Life Sci 77:4505–4522
12. Fung C, Cools B, Malagola S, Martens T, Tack J, Kazwiny Y, Vanden Berghe P (2021) Luminal short chain fatty acids and 5-HT acutely activate myenteric neurons in the mouse proximal colon. Neurogastroenterol Motil 33: e14186
13. Fung C, Hao MM, Obata Y, Tack J, Pachnis V, Boesmans W, Vanden Berghe P (2021) Luminal nutrients activate distinct patterns in submucosal and myenteric neurons in the mouse small intestine. bioRxiv:2021.2001.2019.427232
14. Gwynne RM, Bornstein JC (2007) Local inhibitory reflexes excited by mucosal application of nutrient amino acids in guinea pig jejunum. Am J Physiol Gastrointest Liver Physiol 292:G1660–G1670
15. Hao MM, Fung C, Boesmans W, Lowette K, Tack J, Vanden Berghe P (2020) Development of the intrinsic innervation of the small bowel mucosa and villi. Am J Physiol Gastroint Liver Physiol 318:G53–G65
16. Hata T, Asano Y, Yoshihara K, Kimura-Todani T, Miyata N, Zhang X-T, Takakura S, Aiba Y, Koga Y, Sudo N (2017) Regulation of gut luminal serotonin by commensal microbiota in mice. PLoS One 12:e0180745
17. Hennig GW, Gould TW, Koh SD, Corrigan RD, Heredia DJ, Shonnard MC, Smith TK (2015) Use of genetically encoded calcium indicators (GECIs) combined with advanced motion tracking techniques to examine the behavior of neurons and glia in the enteric nervous system of the intact murine colon. Front Cell Neurosci 9:436
18. Hung LY, Parathan P, Boonma P, Wu Q, Wang Y, Haag A, Luna RA, Bornstein JC (2020) Antibiotic exposure postweaning disrupts the neurochemistry and function of enteric neurons mediating colonic motor activity. Am J Physiol Gastrointest Liver Physiol 318:G1042–G1053
19. Kang YN, Fung C, Vanden Berghe P (2021) Gut innervation and enteric nervous system development: a spatial, temporal and molecular tour de force. Development 148(3):dev182543
20. Kazwiny Y, Pedrosa J, Zhang Z, Boesmans W, D'Hooge J, Vanden Berghe P (2021) Extracting neuronal activity signals from microscopy recordings of contractile tissue using B-spline Explicit Active Surfaces (BEAS) cell tracking. Sci Rep 11:10937
21. Lake JI, Heuckeroth RO (2013) Enteric nervous system development: migration, differentiation, and disease. Am J Physiol Gastrointest Liver Physiol 305:G1–24
22. Lasrado R, Boesmans W, Kleinjung J, Pin C, Bell D, Bhaw L, McCallum S, Zong H, Luo L, Clevers H, Vanden Berghe P, Pachnis V (2017) Lineage-dependent spatial and functional organization of the mammalian enteric nervous system. Science 356:722–726
23. Li Z, Hao MM, Van den Haute C, Baekelandt V, Boesmans W, Vanden Berghe P (2019) Regional complexity in enteric neuron wiring reflects diversity of motility patterns in the mouse large intestine. Elife 8:e42914
24. McKeown SJ, Chow CW, Young HM (2001) Development of the submucous plexus in the large intestine of the mouse. Cell Tissue Res 303:301–305
25. McVey Neufeld KA, Mao YK, Bienenstock J, Foster JA, Kunze WA (2013) The microbiome is essential for normal gut intrinsic primary afferent neuron excitability in the mouse. Neurogastroenterol Motil 25:183–e188
26. Monro RL, Bornstein JC, Bertrand PP (2008) Synaptic transmission from the submucosal plexus to the myenteric plexus in Guinea-pig ileum. Neurogastroenterol Motil 20:1165–1173
27. Moore BA, Vanner S (2000) Properties of synaptic inputs from myenteric neurons innervating submucosal S neurons in guinea pig ileum. Am J Physiol Gastrointest Liver Physiol 278:G273–G280
28. Nava GM, Friedrichsen HJ, Stappenbeck TS (2011) Spatial organization of intestinal microbiota in the mouse ascending colon. ISME J 5:627–638
29. Obata Y, Castano A, Boeing S, Bon-Frauches AC, Fung C, Fallesen T, de Aguero MG, Yilmaz B, Lopes R, Huseynova A, Horswell S, Maradana MR, Boesmans W, Vanden Berghe P, Murray AJ, Stockinger B, Macpherson AJ, Pachnis V (2020) Neuronal programming by microbiota regulates intestinal physiology. Nature 578:284–289
30. Okamoto T, Bayguinov PO, Broadhead MJ, Smith TK (2012) Ca2+ transients in submucous neurons during the colonic migrating motor complex in the isolated murine large intestine. Neurogastroenterol Motil 24:769–e354
31. Rakhilin N, Barth B, Choi J, Munoz NL, Kulkarni S, Jones JS, Small DM, Cheng YT, Cao Y, LaVinka C, Kan E, Dong X, Spencer M, Pasricha P, Nishimura N, Shen X (2016) Simultaneous optical and electrical in vivo analysis of the enteric nervous system. Nat Commun 7:11800
32. Reed DE, Vanner SJ (2003) Long vasodilator reflexes projecting through the myenteric plexus in guinea-pig ileum. J Physiol 553:911–924
33. Reed DE, Vanner S (2007) Mucosal stimulation activates secretomotor neurons via long myenteric pathways in guinea pig ileum. Am J Physiol Gastrointest Liver Physiol 292:G608–G614
34. Reigstad CS, Salmonson CE, Rainey JF 3rd, Szurszewski JH, Linden DR, Sonnenburg JL, Farrugia G, Kashyap PC (2015) Gut microbes promote colonic serotonin production through an effect of short-chain fatty acids on enterochromaffin cells. FASEB J 29:1395–1403
35. Ritsma L, Steller EJA, Ellenbroek SIJ, Kranenburg O, Borel Rinkes IHM, van Rheenen J (2013) Surgical

implantation of an abdominal imaging window for intravital microscopy. Nat Protoc 8:583
36. Vanner S (2000) Myenteric neurons activate submucosal vasodilator neurons in guinea pig ileum. Am J Physiol Gastrointest Liver Physiol 279:G380–G387
37. Yano JM, Yu K, Donaldson GP, Shastri GG, Ann P, Ma L, Nagler CR, Ismagilov RF, Mazmanian SK, Hsiao EY (2015) Indigenous bacteria from the gut microbiota regulate host serotonin biosynthesis. Cell 161:264–276

Serotonergic Paracrine Targets in the Intestinal Mucosa

8

Jackie D. Wood

Abstract

Serotonin functions as a neurotransmitter in the enteric nervous system. Aside from its neurotransmitter role, serotonin also is a paracrine mediatorial signal in the digestive tract. It is a major paracrine signaling molecule in the integrated physiology of several classes of cells in the intestinal mucosa. Paracrine action can be initiation or suppression of activity in populations of cells that make up divergent phenotypic classes. This underlies phenotypic plasticity in single classes and links single classes to other neighboring phenotypic classes, thereby forming a single and higher-order organization in which different categories of function are integrated to work in harmony as a single homeostatic entity at higher levels of physiological organization. Phenotypic classes of cells that are linked by serotonergic paracrine signaling at upper levels of functional organization in the small intestine are (1) enterochromaffin cells; (2) enteric mast cells; (3) spinal sensory afferents; (4) sympathetic postganglionic neurons; (5) enteric neurons.

J. D. Wood (✉)
Department of Physiology and Cell Biology, The Ohio State University College of Medicine, Columbus, OH, USA
e-mail: Jackie.Wood@osumc.edu

Keywords

Serotonin · Enterochromaffin cells · Mast cells · Spinal afferents · Serotonergic receptors

8.1 Overview

Serotonin functions as a neurotransmitter in the enteric nervous system [1]. Aside from its neurotransmitter role, serotonin also is a paracrine mediatorial signal in the digestive tract. It is a major paracrine signaling molecule in the integrated physiology of several classes of cells in the intestinal mucosa. Paracrine action can be initiation or suppression of activity in populations of cells that make up divergent phenotypic classes. This underlies phenotypic plasticity in single classes and links single classes to other neighboring phenotypic classes, thereby forming a single and higher-order organization in which different categories of function are integrated to work in harmony as a single homeostatic entity at higher levels of physiological organization. Phenotypic classes of cells that are linked by serotonergic paracrine signaling at upper levels of functional organization in the small intestine are (1) enterochromaffin cells; (2) enteric mast cells; (3) spinal sensory afferents; (4) sympathetic postganglionic neurons; (5) enteric neurons.

N. J. Spencer et al. (eds.), *The Enteric Nervous System II*, Advances in Experimental Medicine and Biology 1383, https://doi.org/10.1007/978-3-031-05843-1_8

8.2 Enterochromaffin Cells

Enterochromaffin cells are dispersed among columnar epithelial cells, M-cells, and goblet cells (Fig. 8.1). Columnar epithelial cells are the phenotypic class of cells that secrete sodium chloride, sodium bicarbonate, and water into the intestinal lumen. Goblet cells, as neighbors of columnar epithelial cells, are a class that synthesize mucus and secrete it as a luminal lubricant and protective barrier that "traps" and immobilizes hazards inside the lumen. *M-cells* are a specialized class of epithelial cells belonging to the immune system. They ingest bacteria, viruses, and multiple kinds of particulate antigens by endocytosis from the intestinal luminal milieu. Enteric mast cells are a class of inflammatory-immune cells that express detection and paracrine signaling functions and provide a first line of defense at a dangerous interface separating the "dirty" outside world from the body's interior.

8.3 Enterochromaffin Cells

Enterochromaffin cells (ECs) synthesize and store large amounts of serotonin, estimated to be as high as 95 percent of bodily content. ECs release serotonin in paracrine signaling manner to interact with serotonergic receptors expressed by ENS neurons, enteric mast cells, and terminals of spinal sensory afferent neurons in the mucosa.

Postprandial acid secretion is one of several stimuli for release of 5-HT from storage vesicles at the basolateral membrane of ECs, as is mechanical stimulation at Piezo 2 ionic channels, associated with movement of luminal contents and postprandial motility [2–4]. When released, serotonin diffuses into the lamina propria where it encounters secretomotor neurons of the ENS, enteric mast cells, and spinal afferent nerve terminals (Fig. 8.1) [5]. Evidence, emerging from intestinal organoid cultures, suggests that ECs express functional voltage–gated ionic channels and are electrically excitable. Voltage-gated calcium channels appear to be mediators of release of serotonin in response to luminal irritants, metabolites, and catecholamines [2].

8.4 Enteric Mast Cells

Enteric mast cells represent another phenotypic class of mucosal cells that release paracrine signaling mediators. Multiple kinds of preformed mediators are synthesized and stored in granules inside mast cells. Newly synthesized lipid mediators are derived from enzymatic cleavage of membrane lipids. Preformed mediators are

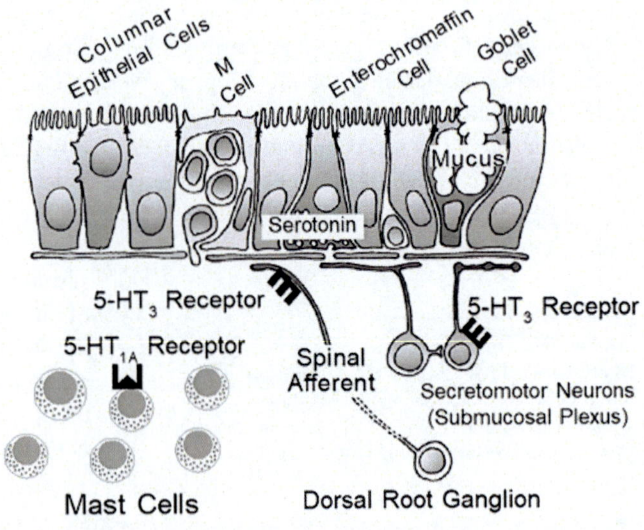

Fig. 8.1 Mucosal enterochromaffin cells release serotonin as a paracrine signaling molecule at receptors expressed by enteric mast cells, spinal afferent nerve terminals, and secretomotor neurons in the enteric nervous system

released into the extracellular environment by mast cell degranulation. Serotonin, histamine, and serine proteases are prominent among preformed mediators released as paracrine signals from mast cells. Prostaglandins and platelet-activating factor are examples of lipid mediators. Preformed and membrane-derived mediators are released when a mast cell is activated by antigen cross-linking to immunoglobulin E antibodies attached to high-affinity FcεRI receptors at the mast cell surface or by paracrine mediators, one of which is serotonin.

Selective recognition of antigens by antibodies attached to FcεRI receptors on sensitized mast cells triggers degranulation and release of the mast cell's paracrine mediators. Among the mediators are serotonin and the proteases, trypsin, and chymotrypsin. Once released, the mediators become paracrine signals to the ENS, which responds with programed defensive behavior that removes the antigen from the intestinal lumen. Copious glandular secretion, vasodilation, and increased blood flow followed by powerful orthograde propulsion of the luminal contents are operational stages of the behavior [8–11].

8.5 Paracrine Linkage of Enterochromaffin Cells to Mast Cells

Serotonin, released from EC cells onto mucosal mast cells, degranulates the mast cells and releases mediators that become paracrine signals to the ENS, spinal afferent nerve terminals, and epithelial secretory glands (Fig. 8.1). Mucosal mast cells express the 5-HT_{1A} serotonergic receptor subtype [12]. Activation of the receptor results in degranulation and release of preformed histamine and mast cell proteases. Exposure to the selective 5-HT_{1A} receptor agonist, 8-hydroxy-PIPAT, in vitro, evokes mast cell release of histamine followed by paracrine action of the released histamine to evoke a cyclical pattern of intestinal motility that is simultaneously matched with a cyclical pattern of glandular secretion of H_2O and NaCl [12, 13]. Mast cell–stabilizing drugs, doxantrazole and cromolyn sodium, suppress release of histamine and mast cell proteases when evoked by serotonin or 8-hydroxy-PIPAT in preparations from guinea pig and human intestinal preparations in vitro [14].

8.6 Paracrine Linkage of Enteric Mast Cells to Spinal Afferents

Degranulation of mucosal mast cells discharges paracrine mediators that sensitize "silent" spinal afferent nociceptors and lowers the threshold for painful responses to intestinal distension [15]. Painful responses of this nature are suppressed by treatment with mast cell–stabilizing drugs, as well as pharmacological blockade of the serotonergic 5-HT_3 receptor subtype expressed by afferent nerve terminals [15–17]. Effects of mast cell degranulation that sensitizes spinal afferents are reminiscent of the canonical hypersensitivity to intestinal distension in IBS [18]. Mucosal biopsies from IBS patients have elevated numbers of mast cells and this raises a question of whether hypersensitivity to distension reflects sensitization of intramural endings of mechanosensitive spinal afferents by mediators released from mast cells (Fig. 8.2) [7, 19].

Aside from mast cell mediators that sensitize intestinal afferents, degranulating mast cells release chemoattractant factors, such as TNFα and IL-1 cytokines that stimulate inflammatory leukocyte extravasation into the mucosa. This can be demonstrated by placement of purified *Clostridium difficile* toxin-A into loops of intestine in vivo. *Clostridium difficile* toxin-A stimulates extravasation coincident with profuse mucosal secretion linked with powerful orthograde propulsive motility and association with abdominal pain. Blockade of conduction in nerves of the ENS, blockade of receptors for the spinal afferent neurotransmitters NK-1 and CGRP, or application of mast cell–stabilizing drugs prevents the acute inflammatory response to *C. difficile* toxin, all of which reflects a mast cell to spinal afferent link [20–22]. Intestinal inflammatory events are not evoked by intraluminal application *C. difficile* toxin in mast cell–deficient mice [23].

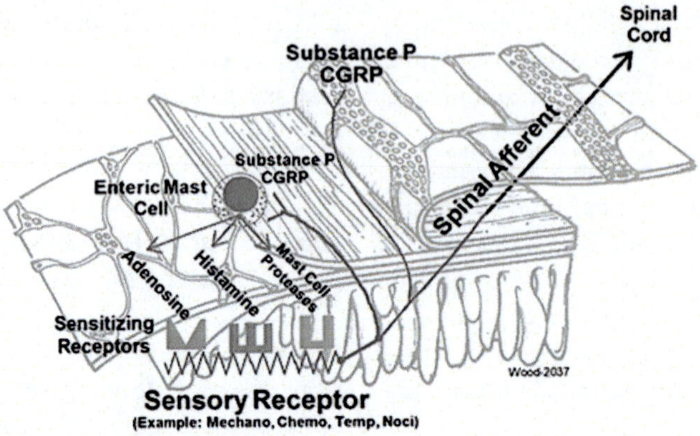

Fig. 8.2 Serotonergic paracrine signals link enteric mast cells and spinal afferents and thereby form a high-level interactive organization consisting of spinal afferents, enteric mast cells, and the enteric nervous system. Enteric mast cells are major intersections in the interactive induction of spinal afferent hypersensitivity. Afferent terminals discharge action potentials when receptors for paracrine mediators, such as adenosine, histamine, or mast cell proteases and/or channels for a sensory modality, such as chemosensitivity, mechano-sensitivity, or temperature response to the preferred stimulus. Interactive signaling in the form of action potential codes releases substance P and CGRP at afferent-mast cell junctions, as well as at synaptic connections with neurons in the enteric nervous system and second-order neurons in the spinal cord. Mast cells express receptors for substance P and CGRP, which degranulate the mast cells and release mediators, such as mast cell proteases and histamine. Histamine and mast cell proteases, and additional mediators, diffuse to their receptors on afferent terminals, in paracrine fashion, and act to enhance the sensitivity for a stimulus to trigger action potentials and augment firing frequency in the terminal. In this event, afferent input to mast cells, enteric neuronal microcircuits, and the spinal cord is elevated. Mast cells and afferents connected in this manner form a positive feed-forward loop that self-perpetuates in the sensitization of afferent terminals. Neurophysiological connectivity of an interactive enteric mast cell-afferent-enteric neuronal network, like this, would be expected to underlie hypersensitivity to stimuli in the gut

Substance P is found to be a principal mediator in the sequence of events underlying mast cell degranulation and release of chemoattractant factors for circulating leucocytes in *C. difficile* toxin A–induced mast cell degranulation [24, 25]. It is expressed and released by interneurons, secretomotor neurons, and musculomotor neurons in the ENS, as well as by intestinal spinal afferents [6].

Substance P is a putative neurotransmitter for local spinal afferent reflexes, in the gut, that are reminiscent of the cutaneous triple response of Lewis. The triple response of Lewis is due to the release of histamine. Moreover, substance P is a secretagogue for histamine and cytokine release from mast cells [6]. Known firing of ENS neurons by excitatory action of *C. difficile* neurotoxin releases neuronal substance P, which in turn acts to degranulate mast cells in the neighborhood where it is released.

C. difficile toxin-A depolarizes the membrane potential and elevates excitability of myenteric and submucosal neurons [26]. This occurs coincident with presynaptic suppression of nicotinic fast excitatory synaptic transmission in the integrative microcircuitry of both the myenteric and submucosal divisions of the ENS and with suppression of inhibitory noradrenergic neurotransmission to secretomotor neurons in the submucosal plexus [26]. Suppression of noradrenergic neurotransmission removes sympathetic braking action from ENS secretomotor neurons and thereby expedites mucosal glandular secretion (Fig. 8.1). When this occurs in combination with neurotoxin-evoked excitation of secretomotor neurons, it becomes an underlying factor in diarrhea associated with antibiotic-induced *C. difficile* overgrowth.

8.7 Spinal Afferents Degranulate Enteric Mast Cells

Spinal afferents innervate enteric mast cells (Fig. 8.2). Substance P and calcitonin gene–related peptide (CGRP) are released from afferent collaterals at their junctions with mucosal mast cells. Acting as neurotransmitters, substance P and CGRP degranulate the mast cells and release mediators that become paracrine signals to the ENS, epithelial secretory glands, and mucosal blood vessels. Antidromic electrical stimulation of spinal afferents, as they leave the intestine in perivascular nerve bundles in the mesentery, is a method for study of local reflexes involving spinal afferents inside the intestinal tract. Along with reflexes involving the muscularis and secretory glands, retrograde stimulation activates reflex degranulation of mast cells. Backfiring of the afferents has the added effect of evoking slow excitatory postsynaptic potentials (slow EPSPs) in neurons of the myenteric and submucosal divisions of the ENS [27]. These slow EPSPs reflect release of substance P and calcitonin gene–related peptide (CGRP) from the afferents [27].

Responses to application of the mast cell secretagogue, compound 48/80, to guinea pig ileum and human jejunal preparations in vitro mimic all aspects of slow EPSPs, evoked by stimulation of spinal afferents in the myenteric and submucosal divisions of the ENS [27]. Slow EPSP-like depolarizing responses to compound 48/80 are suppressed by mast cell–stabilizing drugs, cromolyn and doxantrasole, suggesting that slow EPSP-like depolarization in ENS neurons reflects release of mast cell products that reach ENS neurons in paracrine signaling fashion.

Expression of the transient receptor potential cation channel, TrpV1, also known as the capsaicin receptor, is a marker for spinal afferent fibers in the lamina propria of intestinal preparations in vitro [27]. Application of the VR1 receptor agonist, capsaicin, mimics the action of compound 48/80 to depolarize and elevate the excitability of neurons in the ENS. Responses to capsaicin are reduced when applied in the presence of PAR3888, a selective antagonist at PAR2 serine protease-activated receptors. This implies that capsaicin acts to fire intramural afferents that connect with enteric mast cells. Release of substance P and CGRP, at the afferent-mast cell junction, degranulates the mast cells and releases mast cell proteases as excitatory paracrine signals to neurons in the ENS. Mast cell proteases are known agonists at excitatory PAR receptors on ENS Neurons [28].

Data confirm that stimulation of intramural spinal afferents, either by antidromic electrical stimulation or by capsaicin, degranulates enteric mast cells and releases significant concentrations of histamine and mast cell proteases into the tissues and bathing medium in vitro [27] Histamine and mast cell proteases, when released, diffuse to the ENS in paracrine signaling fashion, where they evoke slow EPSP-like excitation. These responses mimic antigen-evoked mast cell degranulation in intestinal preparations from laboratory animals that have been immune-sensitized to food proteins (milk) or antigens isolated from intestinal parasites, such as *Trichinella spiralis* [9, 10].

In addition to being paracrine signals to ENS neurons, mast cell degranulation products sensitize and evoke action potential discharge in the terminals of intramural afferents, with the information transmitted via spinal dorsal roots into the spinal cord, as well as to ENS neurons by intramural spinal afferent collateral projections [15]. The afferent-mast cell connection in the intestine is a reflex reminiscent of the cutaneous "triple response of Lewis" reflex. The triple response of Lewis is evoked by noxious stroking of the skin, which results in an initial red line, followed by a flare around the line and then a reddened wheal. The triple response is due to afferent-evoked release of histamine by cutaneous mast cells in like manner to the afferent-enteric mast cell reflex in the bowel.

References

1. Wood JD, Mayer CJ (1978) Slow synaptic excitation mediated by serotonin in Auerbach's plexus. Nature 276:836–837
2. Bellono NW, Bayner JR, Leitch DB, Brierley SM, Ingraham HA, David J (2017) Enterochromaffin cells are gut chemosensors that couple to sensory neural pathways. Cell 170:185–198
3. Wang F, Knutson K, Alcaino C, Linden DR, Gibbons SJ, Keshap P, Grover M, Odeker R, Gottlieb PA, Li JH, Farrugia G, Beyder A (2017) Mechanosensitive ion channel Piezo2 is important for enterochromaffin cell response to mechanical forces. J Physiol 595:79–91
4. Kellum JM Jr, Jaffe BM (1976) Release of immunoreactive serotonin following acid perfusion of the duodenum. Ann Surg 184:633–636
5. Cooke HJ, Montakhab M, Wade PR, Wood JD (1983) Transmural fluxes of 5-hydroxytryptamine in guinea-pig ileum. Am J Physiol 244:G421–G425
6. Wood JD (2012) Cellular neurophysiology of enteric neurons. In: Johnson LR, Kaunitz JD, Ghishan FK, Merchant JL, Said HM, Wood JD (eds) Physiology of the gastrointestinal tract, 5th edn. Elsevier, San Diego, pp 629–669
7. O'Sullivan M, Clayton N, Breslin NP, Harman I, Bountra C, McLaren A, O'Morain CA (2000) Increased mast cells in the irritable bowel syndrome. Neurogastroenterol Motil 12:449–457
8. Wood JD (2018) Enteric nervous system: brain-in-the-gut. In: Said HM, Ghishan RK, Kaunitz JD, Merchant JL, Wood JD (eds) Physiology of the gastrointestinal tract, 6th edn. Academic Press, San Diego, pp 361–372
9. Frieling T, Palmer JM, Cooke HJ, Wood JD (1994) Neuroimmune communication in the submucous plexus of guinea pig colon after infection with *Trichinella spiralis*. Gastroenterology 107:1602–1609
10. Liu S, Hu HZ, Gao N, Gao C, Wang G, Wang X, Peck OC, Kim G, Gao X, Xia Y, Wood JD (2003) Neuroimmune interactions in guinea pig stomach and small intestine. Am J Physiol Gastrointest Liver Physiol 284:G154–G164
11. Wood JD (2012) Nonruminant nutrition symposium: neurogastroenterology and food allergies. J Anim Sci 90:1213–1223
12. Wang GD, Wang XY, Zou F, Qu M, Liu S, Fei G, Xia Y, Needleman BJ, Mikami DJ, Wood JD (2013) Mast cell expression of the serotonin1A receptor in guinea pig and human intestine. Am J Physiol Gastrointest Liver Physiol 304:G855-63
13. Bozarov A, Wang YZ, Yu JG et al (2009) Activation of adenosine low-affinity A3 receptors inhibits the enteric short interplexus neural circuit triggered by histamine. Am J Physiol Gastrointest Liver Physiol 297(6):G1147–G1162
14. Wang GD, Wang XY, Zou F, Qu M, Liu S, Fei G, Xia Y, Needleman BJ, Mikami DJ, Wood JD (2013) Mast cell expression of the serotonin1A receptor in guinea pig and human intestine. Am J Physiol Gastrointest Liver Physiol 304:G855–G863
15. Harrington AM, Castro J, Erickson A, Grundy L, Brierley SM (2018) Extrinsic sensory afferent innervating the gastrointestinal tract. In: Said HM, Ghishan RK, Kaunitz JD, Merchant JL, Wood JD (eds) Physiology of the gastrointestinal tract, 6th edn. Academic Press, San Diego, pp 387–418
16. Coelho AM, Fioramonti J, Bueno L (1998) Mast cell degranulation induces delayed rectal allodynia in rats: role of histamine and 5-HT. Dig Dis Sci 43:727–737
17. Hillsley K, Kirkup AJ, Grundy D (1998) Direct and indirect actions of 5-hydroxytryptamine on the discharge of mesenteric afferent fibres innervating the rat jejunum. J Physiol 506:551–561
18. Ritchie J (1973) Pain from distension of the pelvic colon by inflating a balloon in the irritable colon syndrome. Gut 14:125–132
19. Barbara G, Stanghellini V, De Giorgio R, Cremon C, Cottrell GS, Santini D, Pasquinelli G, Morselli-Labate AM, Grady EF, Bunnett NW, Collins SM, Corinaldesi R (2004) Activated mast cells in proximity to colonic nerves correlate with abdominal pain in irritable bowel syndrome. Gastroenterology 126:693–697
20. Castagliuolo I, Riegler M, Pasha A, Nikulasson S, Lu B, Gerard C, Gerard NP, Pothoulakis C (1998) Neurokinin-1 (NK-1) receptor is required in Clostridium difficile- induced enteritis. J Clin Invest 101:1547–1550
21. Pothoulakis C, Castagliuolo I, LaMont JT, Jaffer A, O'Keane JC, Snider RM, Leeman SE (1994) CP-96,345, a substance P antagonist, inhibits rat intestinal responses to Clostridium difficile toxin A but not cholera toxin. Proc Natl Acad Sci U S A 91:947–951
22. Pothoulakis C, Karmeli F, Kelly CP, Eliakim R, Joshi MA, O'Keane CJ, Castagliuolo I, LaMont JT, Rachmilewitz D (1993) Ketotifen inhibits Clostridium difficile toxin A-induced enteritis in rat ileum. Gastroenterology 105:701–707
23. Wershil BK, Castagliuolo I, Pothoulakis C (1998) Direct evidence of mast cell involvement in Clostridium difficile toxin A-induced enteritis in mice. Gastroenterology 114:956–964
24. Cocchiara R, Albeggiani G, Lampiasi N, Bongiovanni A, Azzolina A, Geraci D (1999) Histamine and tumor necrosis factor-alpha production from purified rat brain mast cells mediated by substance P. Neuroreport 10:575–578
25. Cocchiara R, Bongiovanni A, Albeggiani G, Azzolina A, Lampiasi N, Di Blasi F, Geraci D (1997) Inhibitory effect of neuraminidase on SP-induced histamine release and TNF-alpha mRNA in rat mast cells: evidence of a receptor-independent mechanism. J Neuroimmunol 75:9–118

26. Xia Y, Hu HZ, Liu S, Pothoulakis C, Wood JD (2000) Clostridium difficile toxin A excites enteric neurones and suppresses sympathetic neurotransmission in the guinea pig. Gut 46:481–486
27. Wang GD, Wang XY, Liu S, Qu M, Xia Y, Needleman BJ, Dean J, Mikami DJ, Wood JD (2014) Innervation of enteric mast cells by primary spinal afferents in guinea pig and human small intestine. Am J Physiol Gastrointest Liver Physiol 307:G719–G731
28. Gao C, Liu S, Hu H-Z, Gao N, Kim GY, Xia Y, Wood JD (2002) Serine proteases excite myenteric neurons through protease-activated receptors in guinea-pig small intestine. Gastroenterology 123:1554–1564

9 Enteric Control of the Sympathetic Nervous System

Tim Hibberd, Nick J. Spencer, Simon Brookes, Marcello Costa, and Wai Ping Yew

Abstract

The autonomic nervous system that regulates the gut is divided into sympathetic (SNS), parasympathetic (PNS), and enteric nervous systems (ENS). They inhibit, permit, and coordinate gastrointestinal motility, respectively. A fourth pathway, "extrinsic sensory neurons," connect gut to the central nervous system, mediating sensation. The ENS resides within the gut wall and its activities are critical for life; ENS failure to populate the gut in development is lethal without intervention.

"Viscerofugal neurons" are a distinctive class of enteric neurons, being the only type that escapes the gut wall. They form a unique circuit: their axons project out of the gut wall and activate sympathetic neurons, which then project back to the gut, and inhibit gut movements.

For 80 years viscerofugal/sympathetic circuits were thought to have a restricted role, mediating simple sensory-motor reflexes. New data shows viscerofugal and sympathetic neurons behaving unexpectedly, compelling a re-evaluation of these circuits: both viscerofugal and sympathetic neurons transmit higher order, synchronized firing patterns that originate within the ENS. This identifies them as driving long-range motility control between different gut regions.

There is need for gut motor control over distances beyond the range of ENS circuits, yet no mechanism has been identified to date. The entero-sympathetic circuits are ideally suited to meet this need. Here we provide an overview of the structure and functions of these peripheral sympathetic circuits, including new data showing the firing patterns generated by enteric networks can transmit through sympathetic neurons.

Keywords

Enteric nervous system · Sympathetic nervous system · Motility · Viscerofugal neuron · Intestinofugal neuron · Intestino-intestinal reflex · Colonic motor complex · Prevertebral ganglia

T. Hibberd (✉) · N. J. Spencer · S. Brookes · M. Costa · W. P. Yew
College of Medicine and Public Health, Flinders University, Adelaide, Australia
e-mail: tim.hibberd@flinders.edu.au

The gut's intrinsic enteric nervous system (ENS) contained within the gut wall confers a unique capacity for independent coordination of motor

N. J. Spencer et al. (eds.), *The Enteric Nervous System II*, Advances in Experimental Medicine and Biology 1383, https://doi.org/10.1007/978-3-031-05843-1_9

behaviors. The persistence of gut motor activities deprived of connections to the rest of the body was noted early [119], and an intrinsic nervous system to coordinate motor behaviors was proposed even prior to description of the myenteric plexus [2, 66] – later shown to be the essential network for neurogenic motility [13]. While neurogenic motor behaviors such as peristalsis [70, 102] and motor complexes [112, 122] can operate by coordination from the ENS alone, motility is "liable to be stimulated or checked by other parts of the nervous system" [66].

9.1 Central Sympathetic Circuits

Such other parts of the nervous system liable to exert control over the gut include sympathetic neurons in splanchnic pathways. Electrical stimulation of splanchnic nerves or the thoracolumbar spinal cord from which they emerged causes widespread intestinal motor inhibition [66, 72, 98]. Conversely, severing these pathways increased motility [72, 98] and enhanced intestinal motor responses to vagal stimulation [46]. Reflex activation was soon discovered: acting principally through the splanchnic spinal pathways, they are evoked by mechanical stimulation of peritoneum [71], or gut handling [6]. They are prevented or abolished by splanchnic nerve transection [6]. Thus, it was shown reflex inhibition of the gut can be evoked by mechanical or chemical stimulation from multiple sites, including bladder, skin, urogenital organs, peritoneum, mesentery, and gut [10, 50, 76]. Indeed, laparotomy alone is sufficient to produce inhibition [84] and, along with gut handling and other stimuli, they are key contributors to the phenomenon of ileus following surgery [5, 120]. In our hands, mouse colonic motility is enhanced by extrinsic denervation in preparations containing spinal cord and sympathetic ganglia, in vitro (Spencer, unpublished data; Fig. 9.1).

Among the locations from which gut inhibition can be evoked, the gut itself is particularly effective [6]. Mechanical or chemical stimulations of the intestine that caused inhibition of the same and distant gut segments were termed the intestino-intestinal reflex [91]. Here we refer also to those reflexes that affect the colon, rectum, and stomach by splanchnic pathways (e.g., Hukuhara et al. [44]). The long circuit intestino-intestinal reflexes depend on the spinal cord [44, 97, 124] and splanchnic nerves [7, 38, 50, 90, 97, 123, 124], but not the vagus [38, 50, 90, 123, 124]. Inhibition is mediated primarily through $\alpha 2$ adrenoreceptors [48, 90]. The reflex can also be triggered by electrical stimulation of the central end of a sectioned splanchnic nerve [91], or mesenteric nerves [45, 48]. Thus, the afferent arms of this reflex are thoracolumbar spinal sensory neurons and the efferent arm is formed by sympathetic neurons, both of which travel in the splanchnic nerves. A schematic diagram of the long central sympathetic circuit is shown in Fig. 9.2. Reflex activation typically requires abrupt, noxious levels of stimulation and are thus regarded a defense mechanism [11, 19]. Indeed, intense intestinal distensions showed inhibition of motility and decreased tension in the distended segment, but sympathectomy unmasked active contractions, leading to development of even greater intestinal pressures than distension alone [123].

9.2 Peripheral Sympathetic Circuits

Some studies reported residual motor inhibition after severing connectivity between prevertebral ganglia and the CNS [35, 63], suggesting independent tonic postganglionic sympathetic neuron activity or a peripheral circuit. However, others found no inhibitory role for the PVG after decentralization [32, 44]. The existence of a peripheral circuit was firmly established by demonstrations that distension-evoked inhibitory reflexes persisted between two colonic segments linked only by PVG, decentralized acutely [60], and chronically [61]. Similar results were obtained in dog small intestine [104]; also see Szurszewski and Miller [114] for review. These studies thus provided the basis for an additional circuit underlying intestino-intestinal reflexes, which feature enteric neurons in the afferent arm and postgan-

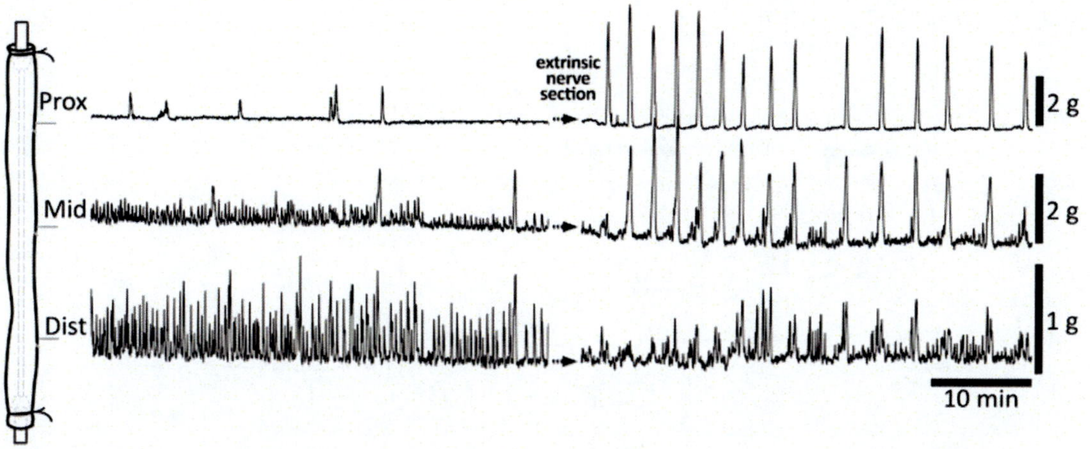

Fig. 9.1 Mouse colonic motility before and after extrinsic nerve section, in vitro. These unpublished traces show standard force recordings [31], from whole mouse colon with intact connections to spinal cord, in vitro (Spencer, unpublished data). Prior to cutting extrinsic nerves, motility was dominated by phasic myogenic patterns with few CMCs. Regular CMCs appeared promptly after complete extrinsic (mesenteric) nerve section. These data are consistent with the well-known central sympathetic reflex inhibition following noxious stimulation caused by cutting and dissecting, as was evoked by simple laparotomy in rabbits [84]

glionic sympathetic neurons in the efferent arm (see schematic diagram, Fig. 9.2). The population of enteric neurons that project to PVG are commonly referred to as "viscerofugal" or "intestinofugal." Here we use the more generic term "viscerofugal."

9.3 Enteric Viscerofugal Neurons

Initial structural evidence for viscerofugal neurons was provided by nerve degeneration studies showing axons persisted on the distal side of chronically sectioned mesenteric/colonic nerve trunks using silver staining [28, 59–61, 99, 101, 103, 118] and from the accumulation of immunohistochemically detected neuropeptides [14, 20, 22, 73]. Subsequently, neuroanatomical tracing techniques enabled direct visualization of viscerofugal nerve cell bodies [21, 29, 30], and immunohistochemical analysis of their nerve cell bodies ([37, 64], see [116] for review). Viscerofugal nerve cell bodies occur exclusively in the myenteric plexus in rat, mouse, guinea pig, and dog [62, 65, 69, 85, 88], while in cats and pigs, they are in the myenteric plexus with a minority in the outer submucous plexus [3, 30]. The number of PVG-projecting viscerofugal neurons increases distally along the gastrointestinal tract in guinea pig [85, 86] and pig [3, 4]. Aside their unique axonal projections out of the gut, viscerofugal neuron morphological characteristics resemble other enteric neurons. In guinea pigs, rats, and mice, most viscerofugal neurons have been described as uniaxonal, Dogiel type I cells, with or without short lamellar dendrites [12, 34, 39–41, 62, 67, 85, 86, 88, 94, 105, 117, 125, 126], although minor populations of multiaxonal viscerofugal neurons have been reported [25, 26, 41, 69]. Examples of viscerofugal nerve cell body morphology are shown in Fig. 9.3.

9.4 The ENS-SNS Interface in Prevertebral Ganglia

Intracellular recordings from single prevertebral sympathetic neurons demonstrate they receive synaptic inputs from multiple enteric viscerofugal neurons in guinea pig [18] and mouse [87]. They release acetylcholine [96], acting on postsynaptic nicotinic receptors, giving rise to fast excitatory postsynaptic potentials in sympathetic neurons [18]. Neuropeptides contained in vis-

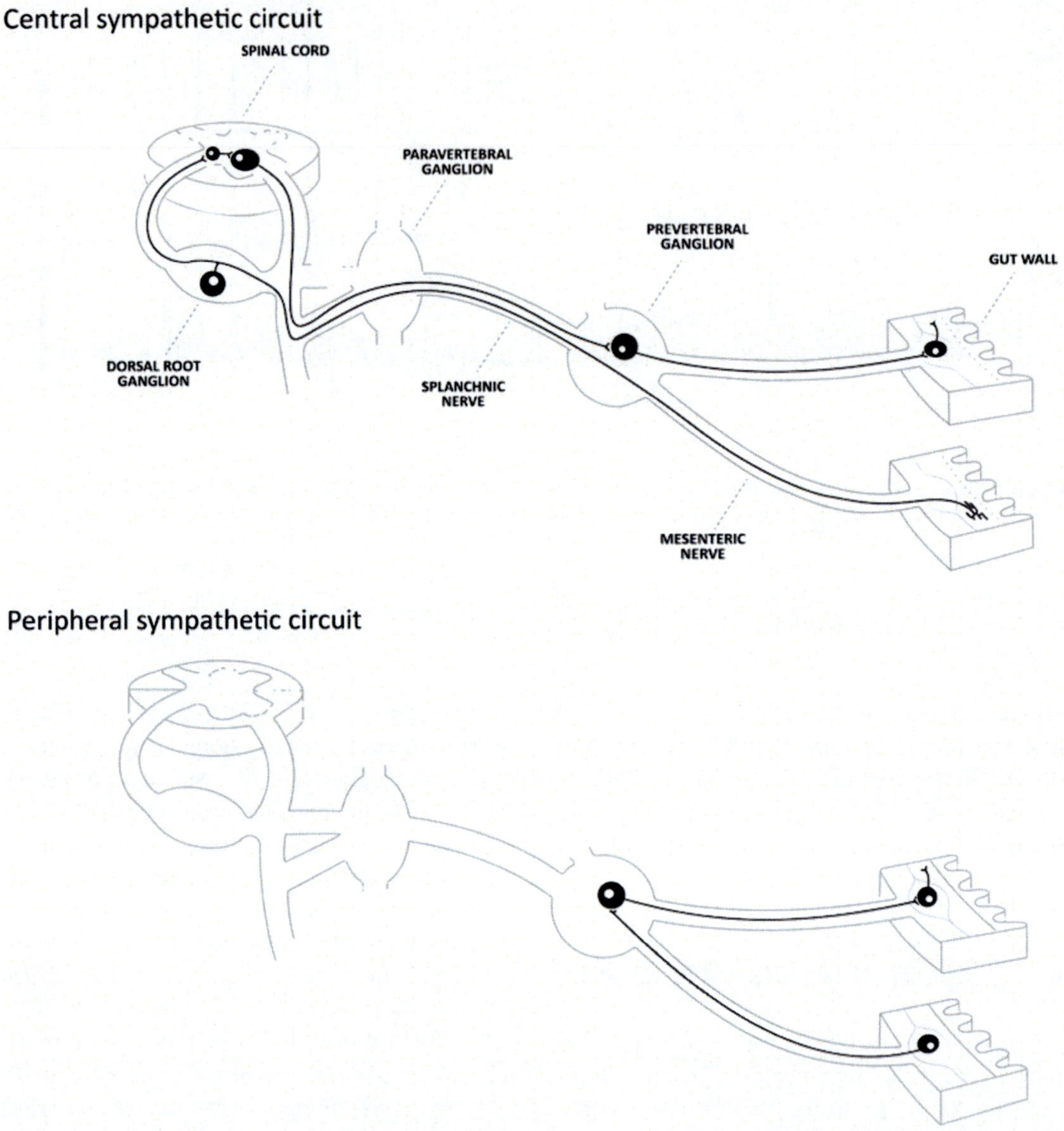

Fig. 9.2 Schematic diagrams of the two intestino-intestinal reflex circuits. The upper diagram shows the long central sympathetic circuit comprised of visceral spinal sensory afferent nerves, spinal interneurons, and the pre- and postganglionic sympathetic neurons to the gut. This circuit is implicated in paralysis of the gut following noxious stimuli. The lower diagram shows the shorter peripheral circuit comprising the viscerofugal afferent arm and postganglionic sympathetic efferent arm. This circuit is the major focus of this chapter and, in contrast to the long circuit, may underlie physiological regulatory functions, since viscerofugal neurons were recently observed transmitting firing patterns organized by enteric motor circuits [40, 43]. (Diagrams adapted from Furness and Costa [33])

cerofugal neurons can also be released, mediating slow membrane depolarizations, enhancing the efficacy of fast inputs [27, 68, 96]. Nicotinic EPSPs from viscerofugal neurons are numerous but subthreshold in guinea pig, requiring summation to evoke action potentials [81, 82]. Sympathetic neurons that receive viscerofugal inputs also have relatively low-amplitude EPSPs from central preganglionic inputs, which McLachlan [81] noted highlights the importance of signal integration. As we note below, our recent work suggests the ENS may be a synchro-

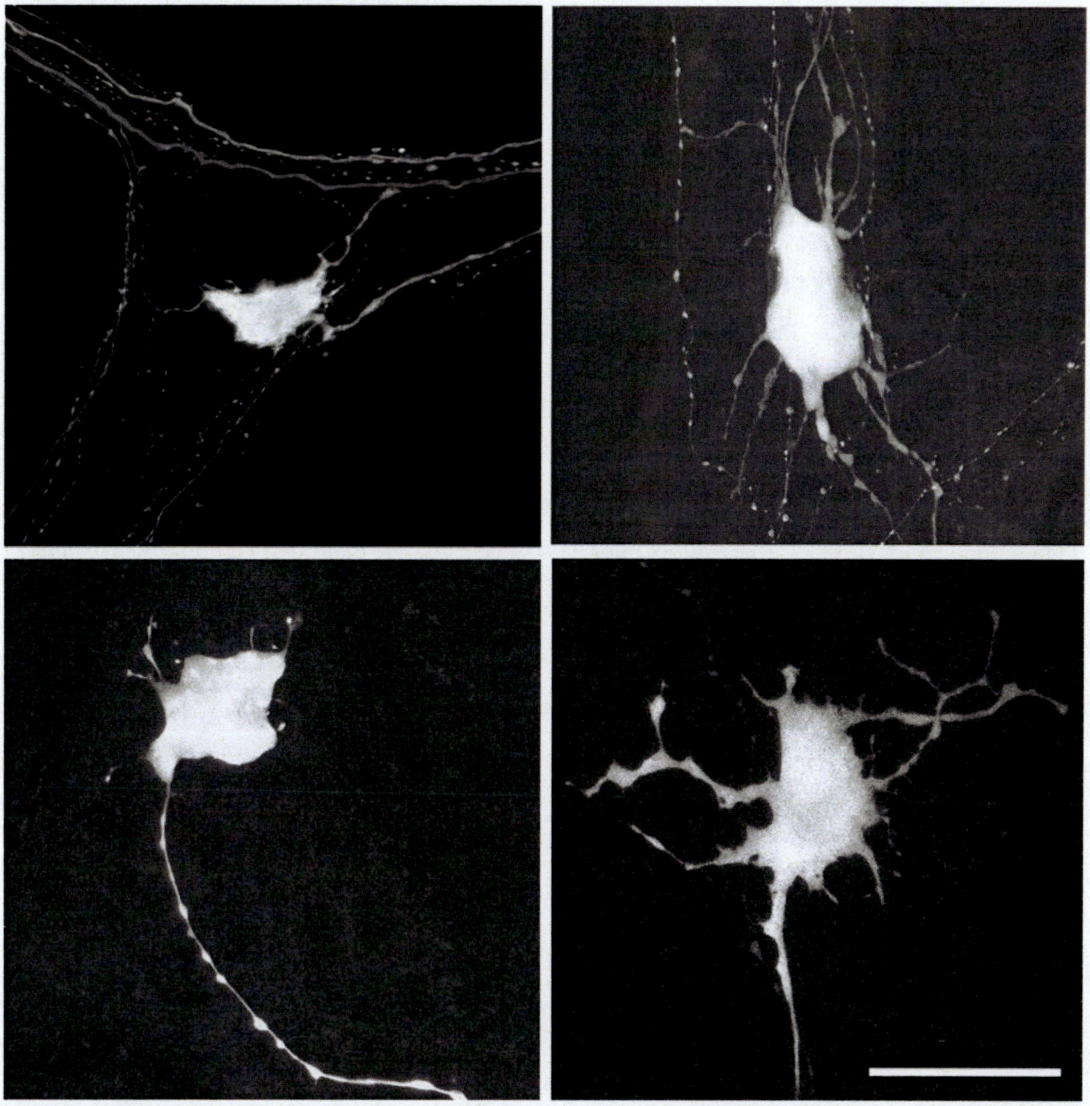

Fig. 9.3 Viscerofugal nerve cell body morphology. Examples of viscerofugal nerve cell bodies revealed by retrograde tracing from colonic nerve trunks to the guinea pig distal colon. Most viscerofugal nerve cell bodies in this species have a single axon and short flat dendrites. Calibration 50 μm

nizing mechanism that enhances signaling through these sympathetic neurons via subthreshold viscerofugal inputs.

In guinea pig celiac ganglion, viscerofugal nerve terminals target postganglionic neurons classified electrophysiologically as "tonic" neurons [36, 83], which contain immunohistochemically detectable tyrosine hydroxylase (TH), either alone or in combination with somatostatin [36]. These characteristics correspond to the visceromotor and secretomotor functional classes of sympathetic neurons [15, 75]. Vasomotor neurons contain TH and neuropeptide Y and lacked viscerofugal inputs [36, 75, 82]. On this basis, peripheral sympathetic reflexes are expected to affect secretion as well as motility in the guinea pig. Some evidence for secretomotor intestino-intestinal reflexes in rat has been presented [100]. Viscerofugal neurons also project to PVG in mice, but the neurochemistry and electrophysiology of mouse prevertebral neurons are different from that of guinea pigs [47, 49] and are currently less well characterized.

9.5 Viscerofugal Neurons as Enteric Mechanoreceptors

Like earlier studies of the intestino-intestinal reflex, physiological studies of viscerofugal neuron function make extensive use of distension as an activating stimulus. Colonic distension, which causes motor inhibition in adjacent gut segments [55], evokes nicotinic EPSPs and action potential firing in guinea pig inferior mesenteric ganglia [18, 56, 115, 121], guinea pig coeliac- and superior mesenteric ganglia [54, 113], and mouse superior mesenteric ganglia [87–89]. A component of distension-evoked viscerofugal firing is directly mechanosensitive, since nicotinic and synaptic blockade in the gut fails to completely abolish distension-evoked fEPSPs in guinea pig and mouse PVG [9, 87, 96, 113]. This was similarly the case during synaptic blockade in direct electrophysiological recordings of viscerofugal axons in guinea pig colon [39] and found to be related to total gut strain [95]. This finding parallels several studies that demonstrate widespread capacity for direct mechanotransduction among enteric neurons [24, 57, 58, 77–80, 110].

9.6 Viscerofugal Neurons as Interneurons

Most synaptic inputs to PVG provided by viscerofugal pathways are second order in nature, implying synaptic drive by other cells [18, 87]. Nerve cell bodies of viscerofugal neurons are surrounded by varicosities from other cholinergic enteric neurons, particularly those with descending projections [67]. Intracellular recordings from viscerofugal nerve cell bodies have shown they receive nicotinic fEPSPs [105] that arise from enteric neurons located orally, anally, and circumferentially [41]. Recordings from viscerofugal axons in guinea pig colon revealed multiple viscerofugal neurons fired together in synchronized bursts of action potentials, implying receipt of common interneuronal connections in the ENS [40].

9.7 Viscerofugal Neurons and Neurogenic Motor Behaviors

Viscerofugal neurons are hardwired into the neural network principally responsible for coordinating motor behavior so that they may be reasonably suspected to show synaptically driven behavior associated with neurogenic motor behaviors. Studies of viscerofugal inputs to PVG neurons during colonic motility recordings note examples of both poor [18] and good correlation [1] between ongoing motility patterns and fEPSP inputs to PVG. Specifically, nonpropulsive motor patterns under lower distension conditions poorly correlated with viscerofugal inputs [121], while increased distension induced cyclic propulsive motor patterns that associated with fluctuations in viscerofugal fEPSPs input to PVG [1, 121]. The pressure dependence is consistent with the dependence of neurally organized cyclic motor complexes and propulsion on ongoing distension in guinea pig colon [17]. Later it was observed that muscle paralysis in guinea pig colon also disrupts the neural events underlying cyclic neurogenic motor patterns [16, 17], precluding the opportunity to observe ENS-driven viscerofugal inputs to PVG without confounding muscular/mechanical activity. Under this constraint, the best correlations between viscerofugal inputs and the mechanical state of the gut were observed under isovolumetric conditions in both guinea pig [1] and mouse [88].

9.8 Enteric Control of the Sympathetic Nervous System

An association between viscerofugal neuron firing behavior and enteric motor circuits was raised in the first direct electrophysiological recordings from identified viscerofugal neuron axons [40]. Multiple viscerofugal neurons showed an ongoing discharge in brief synchronized bursts of action potentials at ~25 cpm that was dependent on nicotinic transmission. These characteristics are remarkably similar to transient neural events we have since characterized in the guinea pig colon [17], which is probably the same motor pattern as the rhythmic discharge of synchronized junction potentials recorded by Spencer et al. [109] electrophysiologically. In addition, large fluctuations in firing were associated with motor activity in a way that could not be explained solely by an enteric mechanoreceptor model of viscerofugal neurons. In that study of guinea pig colon, multiple viscerofugal neurons discharged large bursts of action potential with rapid firing rates (up to 50 Hz) *before* muscular contractions, prior to changes in gut length [40]. Figure 9.4 shows examples of this activity.

Recently, we reported the pattern of enteric neural activity underlying colonic motor complexes in mice [111]. This was characterized as a ~2 Hz synchronized firing pattern among myenteric neurons spanning many rows of ganglia, driving a similar pattern of voltage oscillation in smooth muscle at ~2 Hz. Recruitment of myenteric neurons into the firing pattern was virtually ubiquitous, strongly implicating viscerofugal neurons.

Colonic motor complexes in mice typically persist during muscle paralysis and remain readily detected electrophysiologically [8, 42, 74], making possible an isolation of underlying neural dynamics from changes in gut strain or stress. Indeed, axonal recordings of viscerofugal neurons in rectal nerve trunks revealed a 2 Hz firing pattern during motor complexes, synchronized across multiple single units, with or without muscle paralysis [43]. See the examples of this synchronized firing pattern shown in Fig. 9.5.

This not only confirmed a role for viscerofugal neurons as interneurons in enteric motor circuits, but also provided a higher-resolution window into the firing activities of neurons underlying the motor complex than is currently achieved with calcium imaging. The 2 Hz firing pattern of the motor complex is, at least for viscerofugal neurons, a burst firing pattern. Here, variable numbers of action potentials are discharged by individual neurons in any given cycle in the 2 Hz pattern (from 0 up to 12 were observed).

Motor complexes persisted after isolating the gut for 4 days in organ culture, enabling involvement of spinal afferents to be excluded. Finally, sympathetic fibers emerging from inferior mesenteric ganglia were recorded from the central side of sectioned lumbar colonic nerves. Remarkably, all colonic motor complexes led to parallel discharge of sympathetic neurons with the 2 Hz firing pattern coordinated by enteric motor circuits. See examples of sympathetic nerve firing in Fig. 9.6.

It may be speculated that synchronized viscerofugal firing enhances the likelihood their synaptic inputs reach threshold through temporal summation. In any case, these data show the ENS can drive sympathetic firing during colonic motor complexes, independent of preganglionic input. This suggests the ENS may intermittently use the SNS to feedback onto itself at other locations, presumably to modify motility and/or secretion.

9.8.1 Effector Function of ENS-Driven Sympathetic Firing

The physiological consequences of the enteric-driven mode of activation on the sympathetic effector side now present an open question. Since electrical stimulation of sympathetic efferents does not block motor complexes [108] and its effects on their characteristics are subtle [106], it might be speculated motor complexes should be little affected. However, neurotransmission associated with peristaltic reflexes along ascending and descending cholinergic interneuronal pathways is inhibited in guinea pig [107] and in ascending pathways in mouse proximal, but not distal colon [106]. Thus, it is possible intermittent sympathetic activity associated with

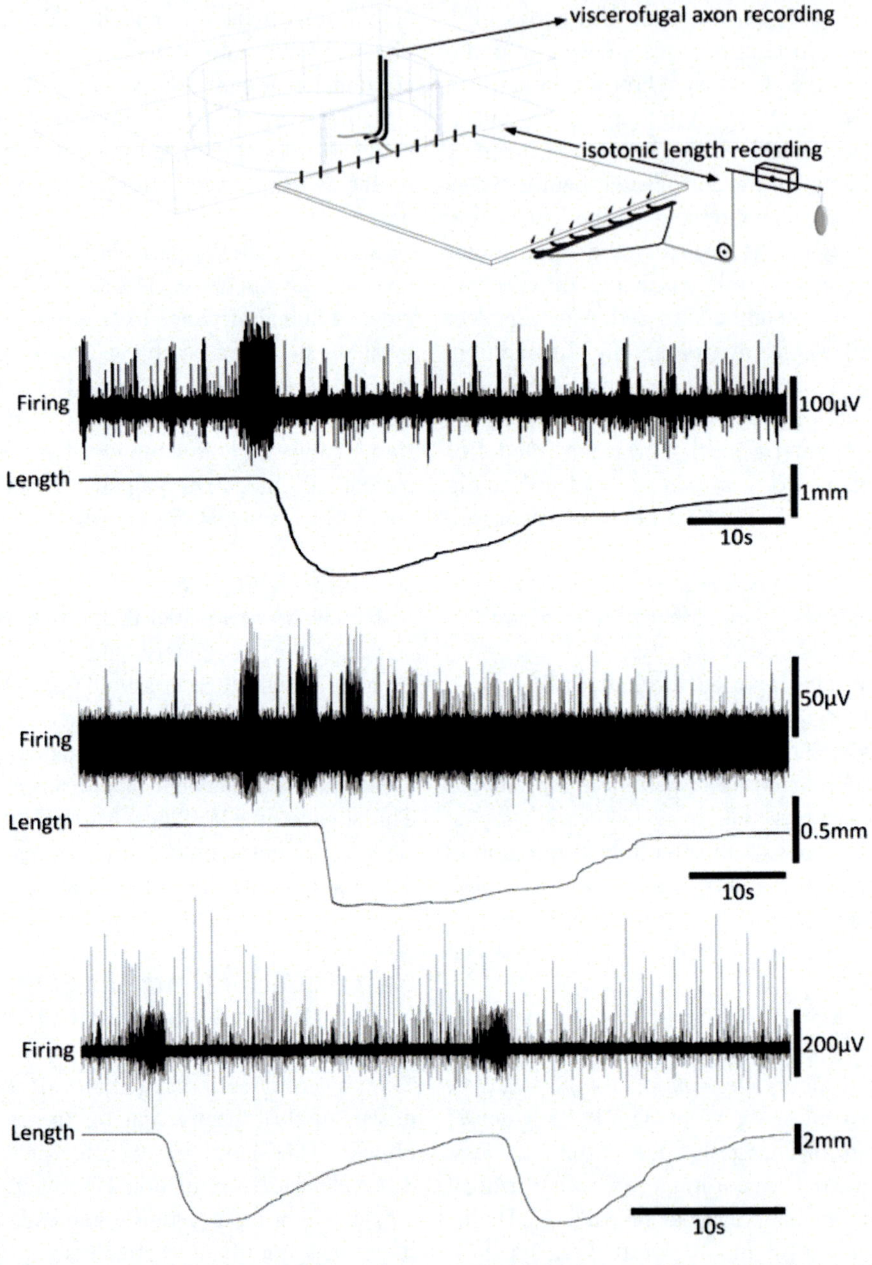

Fig. 9.4 Viscerofugal firing behavior before contractions. Three examples of viscerofugal discharge before "spontaneous" contractions in the guinea pig distal colon [40]. These data strongly suggested viscerofugal firing activity was driven by enteric motor circuits

motor complexes may inhibit peristalsis elsewhere. However, the effects of the ENS-coordinated pattern of sympathetic firing on either of these activities remain to be examined. Other interesting possibilities include sympathetic excitation of colonic ICC-MY and epithelia [106], and whether obscure viscerofugal populations like the rectospinal neurons [23, 93],

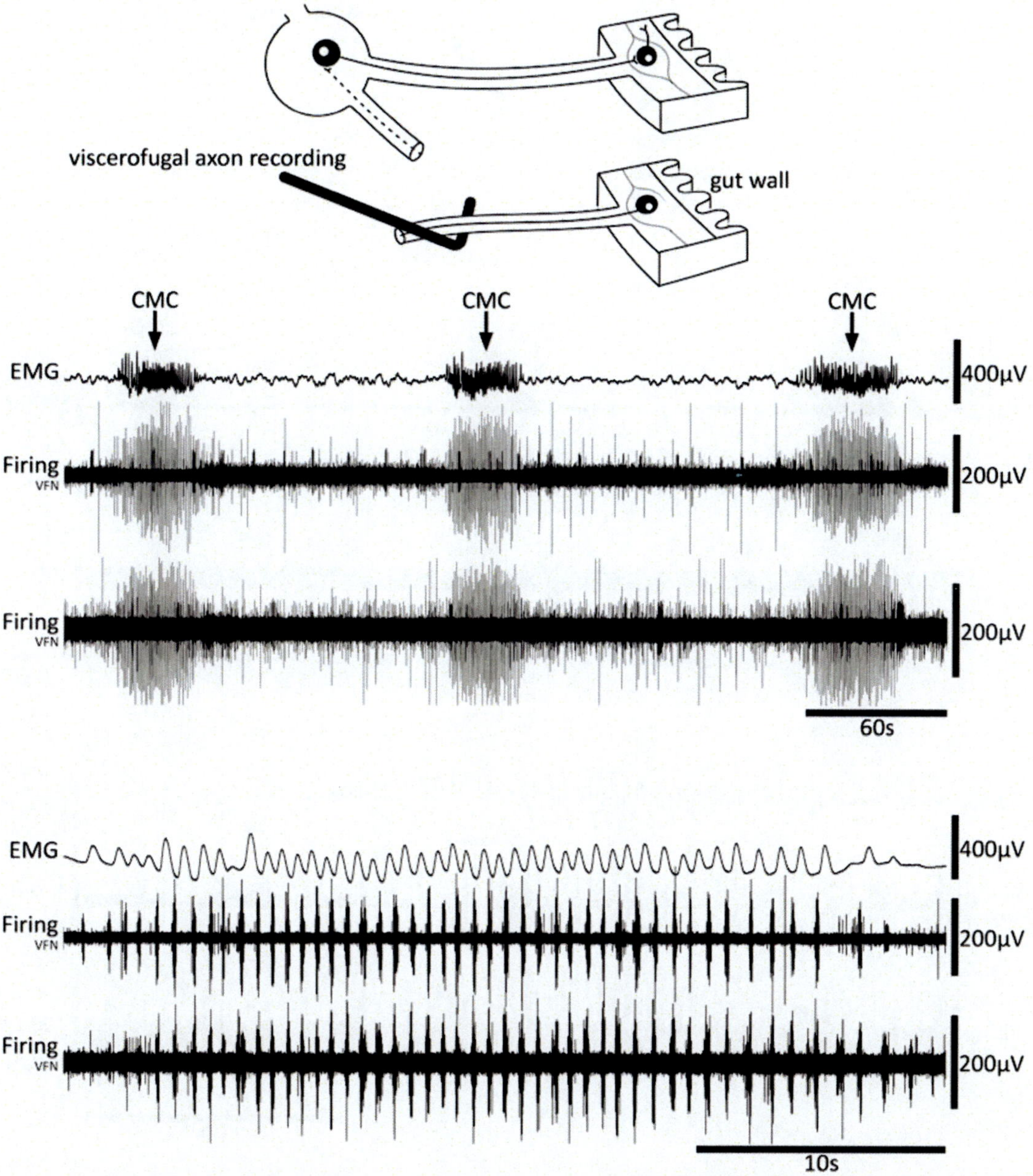

Fig. 9.5 Viscerofugal neurons transmit the ENS-organized 2 Hz firing pattern underlying the CMC. The upper traces show three CMCs, detected by suction electrodes (EMG). Each CMC was associated with a large increase in viscerofugal firing activity. The lower trace demonstrates the firing behavior was organized into bursts at a frequency of 2 Hz, which aligned with the 2 Hz voltage oscillations in smooth muscle EMG. Note the burst firing activity synchronized across two separate rectal nerve trunks. These activities occurred with or without muscle paralysis by nicardipine and thus are not explained by mechanoreception [43]

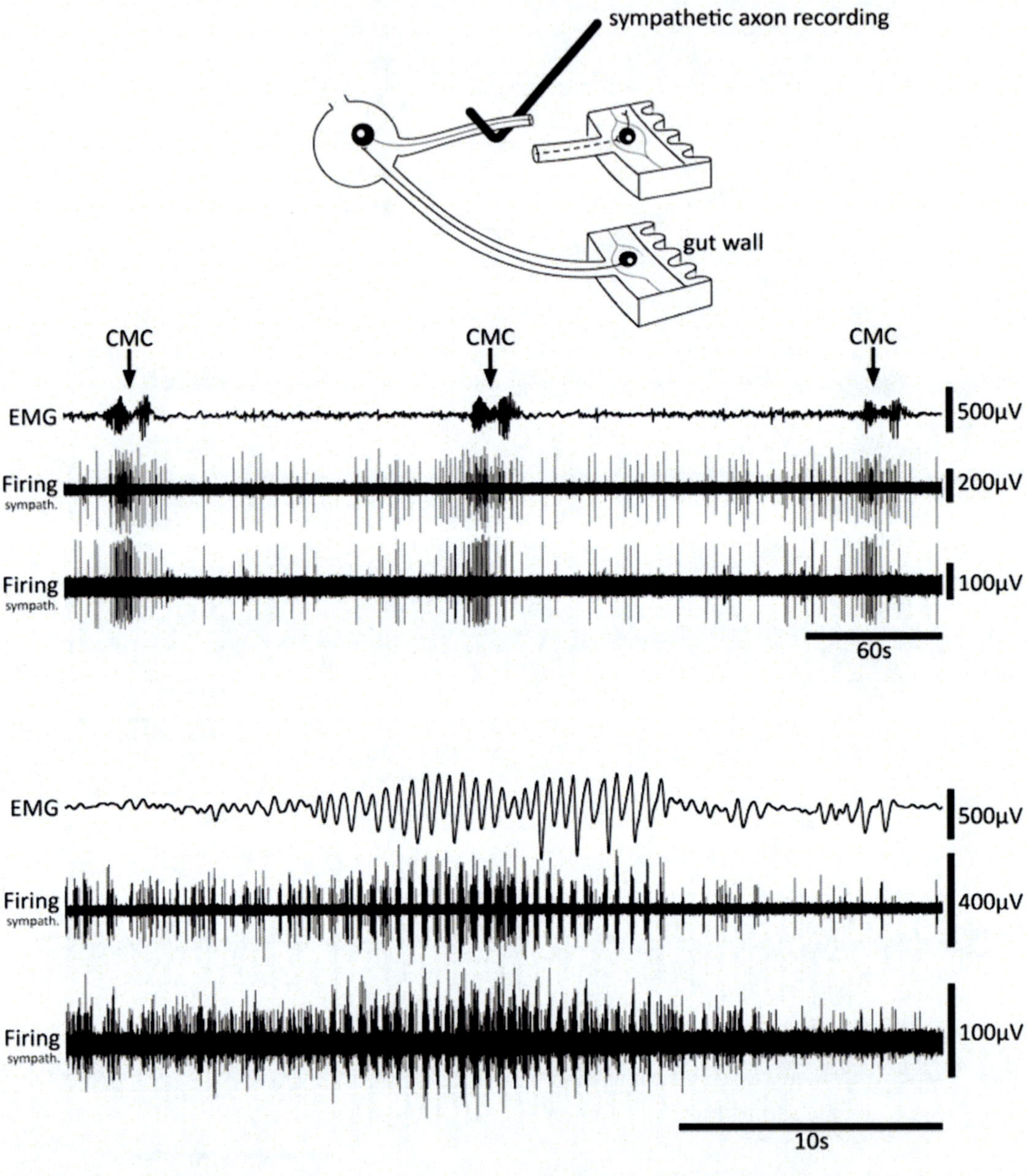

Fig. 9.6 Parallel sympathetic firing during CMCs. All CMCs were associated with increased firing in sympathetic neurons (upper traces). As with viscerofugal firing, sympathetic firing was organized into bursts at a frequency of 2 Hz [43]. Thus, the ENS is independently capable of driving sympathetic nerves

or enteropancreatic neurons [51–53] are also activated by motor circuits. Finally, evidence of gut-pancreatic and gut-liver cross organ circuits involving viscerofugal neurons and prevertebral sympathetic neurons [92] raises the question whether gut motor behaviors are functionally linked to glucoregulatory processes through peripheral sympathetic circuits.

References

1. Anthony TL, Kreulen DL (1990) Volume-sensitive synaptic input to neurons in guinea pig inferior mesenteric ganglion. Am J Physiol 259:G490–G497
2. Auerbach L (1862) Über einen plexus myentericus, einen bisher unbekannten ganglio-nervösen apparat im darmkanal der wirbelthiere. Verlag von E. Morgenstern, Breslau

3. Barbiers M, Timmermans JP, Adriaensen D, De Groodt-Lasseel MH, Scheuermann DW (1993) Topographical distribution and immunocytochemical features of colonic neurons that project to the cranial mesenteric ganglion in the pig. J Auton Nerv Syst 44:119–127
4. Barbiers M, Timmermans JP, Scheuermann DW, Adriaensen D, Mayer B, De Groodt-Lasseel MHA (1994) Nitric oxide synthase-containing neurons in the pig large intestine: topography, morphology, and viscerofugal projections. Microsc Res Tech 29:72–78
5. Bauer AJ, Boeckxstaens GE (2004) Mechanisms of postoperative ileus. Neurogastroenterol Motil 16(Suppl 2):54–60
6. Bayliss WM, Starling EH (1899) The movements and innervation of the small intestine. J Physiol 24:44
7. Brunemeier E, Carlson A (1915) Contributions to the physiology of the stomach: XIX. Reflexes from the intestinal mucosa to the stomach. Am J Physiol-Legacy Content 36:191–195
8. Bywater RA, Small RC, Taylor GS (1989) Neurogenic slow depolarizations and rapid oscillations in the membrane potential of circular muscle of mouse colon. J Physiol 413:505–519
9. Bywater RA (1993) Activity following colonic distension in enteric sensory fibres projecting to the inferior mesenteric ganglion in the guinea pig. J Auton Nerv Syst 46:19–26
10. Cannon WB, Murphy FT (1906) The movements of the stomach and intestines in some surgical conditions. Ann Surg 43:512
11. Chang P-Y, Hsu F-Y (1942) The localization of the intestinal inhibitory reflex arc. Q J Exp Physiol Cogn Med Sci 31:311–318
12. Chen BN, Sharrad DF, Hibberd TJ, Zagorodnyuk VP, Costa M, Brookes SHJ (2014) Neurochemical characterisation of extrinsic nerves in myenteric ganglia of the guinea-pig distal colon. J Comp Neurol 523:742–756
13. Costa M, Furness J (1976) The peristaltic reflex: an analysis of the nerve pathways and their pharmacology. Naunyn Schmiedebergs Arch Pharmacol 294:47–60
14. Costa M, Furness JB (1983) The origins, pathways and terminations of neurons with VIP-like immunoreactivity in the guinea-pig small intestine. Neuroscience 8:665–676
15. Costa M, Furness J (1984) Somatostatin is present in a subpopulation of noradrenergic nerve fibres supplying the intestine. Neuroscience 13:911–919
16. Costa M, Wiklendt L, Keightley L, Brookes SJH, Dinning PG, Spencer NJ (2017) New insights into neurogenic cyclic motor activity in the isolated guinea-pig colon. Neurogastroenterol Motil 29:1–13
17. Costa M, Keightley LJ, Wiklendt L, Hibberd TJ, Arkwright JW, Omari TI, Wattchow DA, Brookes SJH, Dinning PG, Spencer NJ (2019) Identification of multiple distinct neurogenic motor patterns that can occur simultaneously in the guinea-pig distal colon. Am J Physiol Gastrointest Liver Physiol 316:G32–G44
18. Crowcroft PJ, Holman ME, Szurszewski JH (1971) Excitatory input from the distal colon to the inferior mesenteric ganglion in the guinea-pig. J Physiol 219:443–461
19. Da Cunha MJ, Summers R, Thompson H, Wingate D, Yanda R (1981) Effects of intestinal secretagogues and distension on small bowel myoelectric activity in fasted and fed conscious dogs. J Physiol 321:483–494
20. Dalsgaard C, Vincent S, Hökfblt T, Christensson I, Terenius L (1983) Separate origins for the dynorphin and enkephalin immunoreactive fibers in the inferior mesenteric ganglion of the guinea pig. J Comp Neurol 221:482–489
21. Dalsgaard CJ, Elfvin LG (1982) Structural studies on the connectivity of the inferior mesenteric ganglion of the guinea pig. J Auton Nerv Syst 5:265–278
22. Dalsgaard CJ, Hokfelt T, Schultzberg M, Lundberg JM, Terenius L, Dockray GJ, Goldstein M (1983) Origin of peptide-containing fibers in the inferior mesenteric ganglion of the guinea-pig: immunohistochemical studies with antisera to substance P, enkephalin, vasoactive intestinal polypeptide, cholecystokinin and bombesin. Neuroscience 9:191–211
23. Doerffler-Melly J, Neuhuber WL (1988) Rectospinal neurons: evidence for a direct projection from the enteric to the central nervous system in the rat. Neurosci Lett 92:121–125
24. Dong H, Jiang Y, Dong J, Mittal R (2015) Inhibitory motor neurons of the esophageal myenteric plexus are mechanosensitive. Am J Physiol Gastrointest Liver Physiol 308:C405–C413
25. Ermilov LG, Hanani M, Schmalz PF, Miller SM, Szurszewski JH (1998) Distribution of nicotinic receptors in intestinofugal neurons of the guinea pig. Neurogastroenterol Motil 10:69: A50
26. Ermilov LG, Miller SM, Schmalz PF, Hanani M, Lennon VA, Szurszewski JH (2003) Morphological characteristics and immunohistochemical detection of nicotinic acetylcholine receptors on intestinofugal afferent neurones in Guinea-pig colon. Neurogastroenterol Motil 15:289–298
27. Ermilov LG, Schmalz PF, Miller SM, Szurszewski JH (2004) PACAP modulation of the colon-inferior mesenteric ganglion reflex in the Guinea pig. J Physiol 560:231–247
28. Feher E, Vajda J (1974) Degeneration analysis of the extrinsic nerve elements of the small intestine. Acta Anat 87:97–109
29. Feher E (1982) Electron microscopic study of retrograde axonal transport of horseradish peroxidase in the wall of the small intestine in the cat. Acta Anat 112:69–78
30. Feher E, Vajda J (1982) Evidence of sensory neurons in the wall of the small intestine revealed by horseradish peroxidase technique. Zeitschrift fur Mikroskopisch-Anatomische Forschung 96:2
31. Fida R, Lyster DJ, Bywater RA, Taylor GS (1997) Colonic migrating motor complexes (CMMCs) in

the isolated mouse colon. Neurogastroenterol Motil 9:99–107
32. Freund S, Sheehan D (1943) Experimental investigation of visceral afferent synapses in coeliac ganglia. J Neurophysiol 6:263–268
33. Furness JB, Costa M (1974) Adynamic ileus, its pathogenesis and treatment. Med Biol 52:82–89
34. Furness JB, Kuramoto H, Messenger JP (1990) Morphological and chemical identification of neurons that project from the colon to the inferior mesenteric ganglia in the Guinea-pig. J Auton Nerv Syst 31:203–210
35. Garry R (1933) The nervous control of the caudal region of the large bowel in the cat. J Physiol 77:9
36. Gibbins IL, Teo EH, Jobling P, Morris JL (2003) Synaptic density, convergence, and dendritic complexity of prevertebral sympathetic neurons. J Comp Neurol 455:285–298
37. Hamaji M, Kawai Y, Kawashima Y, Tohyama M (1987) Projections of bombesin-like immunoreactive fibers from the rat stomach to the celiac ganglion revealed by a double-labeling technique. Brain Res 416:192–194
38. Hermann H, Morin G (1934) Mise en évidence d'un réflexe inhibiteur intestino-intestinal. CR Soc Biol (Paris) 115:529–531
39. Hibberd TJ, Zagorodnyuk VP, Spencer NJ, Brookes SJH (2012) Identification and mechanosensitivity of viscerofugal neurons. Neuroscience 225:118–129
40. Hibberd TJ, Zagorodnyuk VP, Spencer NJ, Brookes SJH (2012) Viscerofugal neurons recorded from Guinea-pig colonic nerves after organ culture. Neurogastroenterol Motil 24:1041–e1548
41. Hibberd TJ, Spencer NJ, Zagorodnyuk VP, Chen BN, Brookes SJH (2014) Targeted electrophysiological analysis of viscerofugal neurons in the myenteric plexus of Guinea pig colon. Neuroscience 275:272–284
42. Hibberd TJ, Costa M, Travis L, Brookes SJH, Wattchow DA, Feng J, Hu H, Spencer NJ (2017) Neurogenic and myogenic patterns of electrical activity in isolated intact mouse colon. Neurogastroenterol Motil 29:e13089
43. Hibberd TJ, Yew WP, Chen BN, Costa M, Brookes SJ, Spencer NJ (2020) A novel mode of sympathetic reflex activation mediated by the enteric nervous system eNeuro 7
44. Hukuhara T, Nakayama S, Yamagami M, Miyake T (1959) On the intestinal extrinsic reflexes elicited from the small intestine. Acta Med Okayama 13:113–121
45. Hukuhara T, Nakayama S, Nanba R (1960) Locality of receptors concerned with the intestino-intestinal extrinsic and intestinal muscular intrinsic reflexes. Jpn J Physiol 10:414–419
46. Jacobj C (1891) Beiträge zur physiologischen und pharmakologischen Kenntniss der Darmbewegungen mit besonderer Berücksichtigung der Beziehung der Nebenniere zu denselben. Archiv für experimentelle Pathologie und Pharmakologie 29:171–211
47. Jobling P, Gibbins IL (1999) Electrophysiological and morphological diversity of mouse sympathetic neurons. J Neurophysiol 82:2747–2764
48. Jönson C, Fändriks L (1988) Afferent electrical stimulation of mesenteric nerves inhibits duodenal HCO3 secretion via a spinal reflex activation of the splanchnic nerves in the rat. Acta Physiol Scand 133:545–550
49. Kaestner CL, Smith EH, Peirce SG, Hoover DB (2019) Immunohistochemical analysis of the mouse celiac ganglion: an integrative relay station of the peripheral nervous system. J Comp Neurol 527:2742–2760
50. King CE (1924) Studies on intestinal inhibitory reflexes. Am J Physiol 70:183–193
51. Kirchgessner A, Gershon M (1990) Innervation of the pancreas by neurons in the gut. J Neurosci 10:1626–1642
52. Kirchgessner AL, Liu MT, Gershon MD (1994) NADPH diaphorase (nitric oxide synthase)-containing nerves in the enteropancreatic innervation: sources, co-stored neuropeptides, and pancreatic function. J Comp Neurol 342:115–130
53. Kirchgessner AL, Liu M-T, Gershon MD (1996) In situ identification and visualization of neurons that mediate enteric and enteropancreatic reflexes. J Comp Neurol 371:270–286
54. Kreulen DL, Szurszewski JH (1979) Nerve pathways in celiac plexus of the Guinea pig. Am J Physiol 237:E90–E97
55. Kreulen DL, Szurszewski JH (1979) Reflex pathways in the abdominal prevertebral ganglia: evidence for a Colo-colonic inhibitory reflex. J Physiol 295:21–32
56. Kreulen DL, Peters S (1986) Non-cholinergic transmission in a sympathetic ganglion of the Guinea-pig elicited by colon distension. J Physiol 374:315–334
57. Kugler EM, Michel K, Zeller F, Demir IE, Ceyhan GO, Schemann M, Mazzuoli-Weber G (2015) Mechanical stress activates neurites and somata of myenteric neurons. Front Cell Neurosci 9:342
58. Kugler EM, Michel K, Kirchenbüchler D, Dreissen G, Csiszár A, Merkel R, Schemann M, Mazzuoli-Weber G (2018) Sensitivity to strain and shear stress of isolated mechanosensitive enteric neurons. Neuroscience 372:213–224
59. Kuntz A (1938) The structural organization of the celiac ganglia. J Comp Neurol 69:12
60. Kuntz A (1940) The structural organization of the inferior mesenteric ganglia. J Comp Neurol 72:371–382
61. Kuntz A, Saccomanno G (1944) Reflex inhibition of intestinal motility mediated through decentralized prevertebral ganglia. J Neurophysiol 7:163–170
62. Kuramoto H, Furness JB (1989) Distribution of enteric nerve cells that project from the small intestine to the coeliac ganglion in the Guinea-pig. J Auton Nerv Syst 27:241–248
63. Lawson H, Holt J (1937) The control of the large intestine by the decentralized IMG. Am J Physiol 118:5

64. Lee Y, Shiosaka S, Hayashi N, Tohyama M (1986) The presence of vasoactive intestinal polypeptide-like immunoreactive structures projecting from the myenteric ganglion of the stomach to the celiac ganglion revealed by a double-labelling technique. Brain Res 382:392–394
65. Li MZ, Masuko S (1997) Neuronal circuitry between the inferior mesenteric ganglion and lower intestine of the dog. Arch Histol Cytol 60:391–404
66. Lister J (1859) III. Preliminary account of an inquiry into the functions of the visceral nerves, with special reference to the so-called'inhibitory system'. By Joseph Lister, Esq., FRCS Eng. & Edin., assistant surgeon to the Royal Infirmary of Edinburgh; in a letter to Dr. Sharpey, sec. RS communicated by Dr. Sharpey. Proceedings of the Royal Society of London 367–380
67. Lomax AE, Zhang JY, Furness JB (2000) Origins of cholinergic inputs to the cell bodies of intestinofugal neurons in the Guinea pig distal colon. J Comp Neurol 416:451–460
68. Love JA, Szurszewski JH (1987) The electrophysiological effects of vasoactive intestinal polypeptide in the Guinea-pig inferior mesenteric ganglion. J Physiol 394:67–84
69. Luckensmeyer GB, Keast JR (1995) Distribution and morphological characterization of viscerofugal projections from the large intestine to the inferior mesenteric and pelvic ganglia of the male rat. Neuroscience 66:663–671
70. Lüderitz C (1889) Experimentelle Untersuchungen über die Entstehung der Darmperistaltik. Archiv für pathologische Anatomie und Physiologie und für klinische Medicin 118:19–36
71. Lüderitz C (1890) Experimentelle untersuchungen über die entstehung der darmperistaltik. Archiv für pathologische Anatomie und Physiologie und für klinische Medicin 122:1–28
72. Ludwig C (1852) Lehrbuch der Physiologie des Menschen. C. F. Winter, Heidelberg
73. Lundberg JM, Hökfelt T, Schultzberg M, Uvnäs-Wallensten K, Köhler C, Said SI (1979) Occurrence of vasoactive intestinal polypeptide (VIP)-like immunoreactivity in certain cholinergic neurons of the cat: evidence from combined immunohistochemistry and acetylcholinesterase staining. Neuroscience 4:1539–1559
74. Lyster DJK, Bywater RAR, Taylor GS (1993) Regulation of the frequency of myoelectric complexes in the isolated mouse colon. Neurogastroenterol Motil 5:143–151
75. Macrae IM, Furness JB, Costa M (1986) Distribution of subgroups of noradrenaline neurons in the coeliac ganglion of the Guinea-pig. Cell Tissue Res 244:173–180
76. Markowitz J, Campbell WR (1927) The relief of experimental ileus by spinal anesthesia. Am J Physiol 81:101–106
77. Mazzuoli-Weber G, Schemann M (2015) Mechanosensitive enteric neurons in the Guinea pig gastric corpus. Front Cell Neurosci 9:430
78. Mazzuoli G (2009) Mechanosensitivity in the myenteric plexus of the Guinea pig ileum
79. Mazzuoli G, Schemann M (2009) Multifunctional rapidly adapting mechanosensitive enteric neurons (RAMEN) in the myenteric plexus of the Guinea pig ileum. J Physiol 587:4681–4694
80. Mazzuoli G, Schemann M (2012) Mechanosensitive enteric neurons in the myenteric plexus of the mouse intestine. PLoS One 7:e39887
81. McLachlan E (2003) Transmission of signals through sympathetic ganglia–modulation, integration or simply distribution? Acta Physiol Scand 177:227–235
82. McLachlan EM, Meckler RL (1989) Characteristics of synaptic input to three classes of sympathetic neurone in the coeliac ganglion of the Guinea-pig. J Physiol 415:109–129
83. Meckler RL, McLachlan EM (1988) Axons of peripheral origin preferentially synapse with tonic neurones in the Guinea pig coeliac ganglion. Neurosci Lett 86:189–194
84. Meltzer SJ, Auer J (1906) Peristaltic movements of the rabbit's cecum and their inhibition, with demonstration. Proc Soc Exp Biol Med 4:37–40
85. Messenger JP, Furness JB (1992) Distribution of enteric nerve cells that project to the coeliac ganglion of the Guinea-pig. Cell Tissue Res 269:119–132
86. Messenger JP, Furness JB (1993) Distribution of enteric nerve cells projecting to the superior and inferior mesenteric ganglia of the Guinea-pig. Cell Tissue Res 271:333–339
87. Miller SM, Szurszewski JH (1997) Colonic mechanosensory afferent input to neurons in the mouse superior mesenteric ganglion. Am J Physiol-Gastroint Liver Physiol 272:G357–G366
88. Miller SM, Szurszewski JH (2002) Relationship between colonic motility and cholinergic mechanosensory afferent synaptic input to mouse superior mesenteric ganglion. Neurogastroenterol Motil 14:339–348
89. Miller SM, Szurszewski JH (2003) Circumferential, not longitudinal, colonic stretch increases synaptic input to mouse prevertebral ganglion neurons. Am J Physiol 285:G1129–G1138
90. Mizutani M, Neya T, Nakayama S (1990) Capsaicin-sensitive afferents activate a sympathetic intestinointestinal inhibitory reflex in dogs. J Physiol 425:133–144
91. Morin G, Vial J (1934) Sur les voies et les centres du réflexe inhibiteur intestino-intestinal. CR Soc Biol(Paris) 116:536–538
92. Muller PA, Schneeberger M, Matheis F, Wang P, Kerner Z, Ilanges A, Pellegrino K, Del Mármol J, Castro TBR, Furuichi M, Perkins M, Han W, Rao A, Picard AJ, Cross JR, Honda K, de Araujo I, Mucida D (2020) Microbiota modulate sympathetic neurons via a gut-brain circuit. Nature 583:441–446
93. Neuhuber WL, Appelt M, Polak JM, Baier-Kustermann W, Abelli L, Ferri GL (1993) Rectospinal neurons: cell bodies, pathways, immu-

nocytochemistry and ultrastructure. Neuroscience 56:367–378
94. Olsson C, Costa M, Brookes SJ (2004) Neurochemical characterization of extrinsic innervation of the Guinea pig rectum. J Comp Neurol 470:357–371
95. Palmer G, Hibberd TJ, Roose T, Brookes SJ, Taylor M (2016) Measurement of strains experienced by viscerofugal nerve cell bodies during mechanosensitive firing using digital image correlation. Am J Physiol Gastrointest Liver Physiol 311:G869–G879
96. Parkman HP, Ma RC, Stapelfeldt WH, Szurszewski JH (1993) Direct and indirect mechanosensory pathways from the colon to the inferior mesenteric ganglion. Am J Physiol 265:G499–G505
97. Pearcy JF, Liere EJV (1926) Studies on the visceral nervous system. Am J Physiol 78:64–73
98. Pflüger EFW (1857) Ueber das Hemmungs-Nervensystem für die peristaltischen Bewegungen der Gedärme. A. Hirschwald, Berlin
99. Pilipenko V (1956) Contributions to the functional morphology of the peripheral nervous system part II: function nature of Dogiel's type II cells. Bull Exp Biol Med 41:449–452
100. Quinson N, Furness JB (2002) Intestino-intestinal secretomotor reflexes in the rat. Proc Aust Neurosci Soc 13:102
101. Ross JG (1958) On the presence of centripetal fibres in the superior mesenteric nerves of the rabbit. J Anat 92:8
102. Schemann M, Mai G, Costa M, Enck P (2020) Translating the seminal findings of Carl Lüderitz: a description in English of his extraordinary studies of gastrointestinal motility accompanied by a historical view of peristalsis. Neurogastroenterol Motil 33:e13995
103. Schofield GC (1960) Experimental studies on the innervation of the mucous membrane of the gut. Brain 83:490–514
104. Semba T (1954) Intestino-intestinal inhibitory reflexes. Jpn J Physiol 4:241
105. Sharkey KA, Lomax AE, Bertrand PP, Furness JB (1998) Electrophysiology, shape, and chemistry of neurons that project from Guinea pig colon to inferior mesenteric ganglia. Gastroenterology 115:909–918
106. Smith-Edwards KM, Edwards BS, Wright CM, Schneider S, Meerschaert KA, Ejoh LL, Najjar SA, Howard MJ, Albers KM, Heuckeroth RO, Davis BM (2021) Sympathetic input to multiple cell types in mouse and human colon produces region-specific responses. Gastroenterology 160:1208–1223
107. Spencer N, McCarron SL, Smith TK (1999) Sympathetic inhibition of ascending and descending interneurones during the peristaltic reflex in the isolated Guinea-pig distal colon. J Physiol 519(Pt 2):539–550
108. Spencer NJ, Bywater RA, Klemm MF (1998) Effects of sympathetic nerve stimulation on membrane potential in the circular muscle layer of mouse distal colon. Neurogastroenterol Motil 10:543–552
109. Spencer NJ, Hennig GW, Smith TK (2002) A rhythmic motor pattern activated by circumferential stretch in Guinea-pig distal colon. J Physiol 545:629–648
110. Spencer NJ, Smith TK (2004) Mechanosensory S-neurons rather than AH-neurons appear to generate a rhythmic motor pattern in Guinea-pig distal colon. J Physiol 558:577–596
111. Spencer NJ, Hibberd TJ, Travis L, Wiklendt L, Costa M, Hu H, Brookes SJ, Wattchow DA, Dinning PG, Keating DJ, Sorensen J (2018) Identification of a rhythmic firing pattern in the enteric nervous system that generates rhythmic electrical activity in smooth muscle. J Neurosci 38:5507–5522
112. Spencer NJ, Costa M, Hibberd TJ, Wood JD (2021) Advances in colonic motor complexes in mice. Am J Physiol Gastrointest Liver Physiol 320:G12–g29
113. Stebbing MJ, Bornstein JC (1993) Electrophysiological analysis of the convergence of peripheral inputs onto neurons of the coeliac ganglion in the Guinea pig. J Auton Nerv Syst 46:93–105
114. Szurszewski J, Miller S (1994) Physiology of prevertebral ganglia. In: Johnson LR (ed) Physiology of the gastrointestinal tract, vol 1. Raven Press, New York, pp 795–877
115. Szurszewski JH, Weems WA (1976) A study of peripheral input to and its control by post-ganglionic neurones of the inferior mesenteric ganglion. J Physiol 256:541–556
116. Szurszewski JH, Miller SM (2006) Physiology of prevertebral sympathetic ganglia. In: Johnson LR (ed) Physiology of the gastrointestinal tract, vol 1. Academic Press, Burlington, p 603-VIII
117. Tassicker BC, Hennig GW, Costa M, Brookes SJ (1999) Rapid anterograde and retrograde tracing from mesenteric nerve trunks to the Guinea-pig small intestine in vitro. Cell Tissue Res 295:437–452
118. Ungvary GY, Leranth CS (1970) Termination in the prevertebral abdominal sympathetic ganglia of axons arising from the local (terminal) vegetative plexus of visceral organs. Z Zellforsch 110:185–191
119. von Haller A (1755) A dissertation on the sensible and irritable parts of animals. Bull Inst History Med 4:651–699
120. Wattchow D, Heitmann P, Smolilo D, Spencer NJ, Parker D, Hibberd T, Brookes SSJ, Dinning PG, Costa M (2020) Postoperative ileus-an ongoing conundrum. Neurogastroenterol Motil 33:e14046
121. Weems WA, Szurszewski JH (1977) Modulation of colonic motility by peripheral neural inputs to neurons of the inferior mesenteric ganglion. Gastroenterology 73:273–278

122. Wood JD (1973) Electrical activity of the intestine of mice with hereditary megacolon and absence of enteric ganglion cells. Dig Dis Sci 18:477–488
123. Youmans W, Karstens A, Aumann K (1942) Nervous pathways for the reflex regulation of intestinal pressure. Am J Physiol--Legacy Content 135:619–627
124. Youmans WB, Meek WJ (1937) Reflex and humoral gastro-intestinal inhibition in unanesthetized dogs during rectal stimulation. Am J Physiol-Legacy Content 120:750–756
125. Zagorodnyuk VP, Brookes SJ (2000) Transduction sites of vagal mechanoreceptors in the Guinea pig esophagus. J Neurosci 20:6249–6255
126. Zagorodnyuk VP, Chen BN, Brookes SJ (2001) Intraganglionic laminar endings are mechano-transduction sites of vagal tension receptors in the guinea-pig stomach. J Physiol 534:255–268

Embryonic Development of Motility: Lessons from the Chicken

10

Nicolas R. Chevalier

Abstract

I outline here the development of intestinal motility in the chicken embryo. The first contractile events are circular smooth muscle driven calcium waves (E6), that gain a clock-like regularity when interstitial cells of Cajal become electrically active (E14). Soon after longitudinal smooth muscle contractions become prominent (E14), the enteric nervous system starts controlling motility (E16) by coupling the longitudinal and circular contractions via inhibitory neurotransmission. It gives rise to circular-longitudinal antagonism, to the migrating motor complex, and to the polarized ascending contraction-descending relaxation pressure response known as the "law of the intestine". The kinetics of gut development in the chicken appears to follow faithfully that of humans by simply converting embryonic days of chicken development into embryonic weeks of human development.

Keywords

Peristalsis · Embryo · Intestine · Enteric nervous system

N. R. Chevalier (✉)
Laboratoire Matière et Systèmes Complexes, Université Paris Cité, CNRS UMR 7057, Paris, France
e-mail: nicolas.chevalier@u-paris.fr

Newborn babies are often called "stomach on legs" because their foremost activity during this early period of life is to absorb vast quantities of milk. It is therefore vital that their gastrointestinal machinery be functional at birth so that the transition from placental to intestinal feeding occurs as smoothly as possible. Other less "urgent" faculties – vision, motor skills, hearing, smell, etc. – will develop fully only months or years after birth. The pressing need for functional motor reflexes of the gut entails precocious development in the womb. It is most interesting to study the embryonic history of these reflex movements for several reasons. First, these movements and their underlying causes go in the embryo from the easy to the more complex. Adult motility is remarkably complex, involving at least three vastly different cell types (circular and longitudinal smooth muscle, ICC, enteric neurons), chemical and mechanical inputs, a panoply of different neurons and neurotransmitters [16], and a rich phenomenology of different movements [22]. I believe that looking into the ontogenesis of this complexity holds a key to deciphering its essential building blocks, to read the "manual" of how the gut works in reverse. A second point is that many babies are born with intestinal dysmotility problems [12] and that ontogenic studies hold the promise of identifying the origin of these deficits to better treat them. Finally, because ontogeny does to some extent recapitulate phylogeny, studying the emergence of digestive

N. J. Spencer et al. (eds.), *The Enteric Nervous System II*, Advances in Experimental Medicine and Biology 1383, https://doi.org/10.1007/978-3-031-05843-1_10

movements gives us a way to understand how a vital reflex spontaneously emerged from the broth of evolution, and to fathom the digestive activity of more primitive species [11, 20].

Three animal models have contributed to this field of developmental physiology: the transparent zebrafish whose regular intestinal contractions can be seen without opening up the animal [13], the mouse [19], and the chicken. The first embryonic motility study in chicken can be traced back to Newgreen et al. [18]. It is noteworthy that, as far as the GI tract is concerned, the chronology of developmental events in chicken and humans is the same, in days for the former and weeks for the latter [6]. The chicken is of course phylogenetically more distant from humans than mice, but the ease with which embryonic samples of different ages can be obtained in the lab makes it of invaluable help for detailed experimental scrutiny. I will outline here some of the more salient results we obtained on this animal model over the past years.

10.1 The First Digestive Movements Are Just Calcium Waves in a Tube of Circular Smooth Muscle

The first digestive movements can be recorded as soon as circular smooth muscle differentiates [5], which occurs at slightly different timepoints in the different parts of the gut: E5 (E13.5 in mice) in the midgut, E6 (E14.5 in mice) in the hindgut, E9 (E16.5 in mice) in the cecal appendix. We have provided evidence [4] that motility at this early stage boils down to calcium waves that propagate across the circular smooth muscle layer. Enoxolone and heptanol halt propagation of these waves, showing the implication of gap junctions in the propagation mechanism; removing calcium from the medium totally halts motility; general Ca^{2+} channel blockers like $CoCl_2$ strongly perturb motility, as do nicardipine [15] and nifedipine [19], demonstrating the need for L-type channels for these early contractile events. W/W^v mice lacking ICCs display essentially the same motility as wild-type mice in the period E13.5–E16.5, indicating that ICCs are not necessary to trigger calcium waves at this stage [19]. Early mesenchymal cells of the gut thus appear to present both contractile and self-initiating depolarization properties; these two roles will eventually be partitioned to smooth muscle and ICC cells respectively, as they differentiate. We have provided the first videos of calcium wave propagation across the early gut smooth musculature by working on living transverse gut sections (Fig. 10.1a, b), so that the calcium indicator Fluo4-AM could more readily penetrate inside the smooth muscle cells. The longitudinal propagation of the calcium waves in embryonic chicken gut has recently been beautifully imaged by electroporating the smooth muscle with a GCamp-type calcium indicator [14]. IRM performed on human embryos fixed immediately after abortion shows undulations of the GI tract [23] at 6–9 weeks, which are most probably frozen-in, spontaneous contraction waves as seen in model animal species. Importantly, we found that the calcium waves can be stimulated mechanically, by applying local pressure with a pipette (pinching): this gives rise to two contractile waves traveling in opposite directions, from the point where pressure was applied (Fig. 10.1c). As we will see, this behavior is an immature (because unpolarized) form of the "law of the intestine" formulated by Bayliss and Starling [1], also called the barometric or peristaltic reflex.

10.2 Early Smooth Muscle Contractility Is Essential for Anisotropic Longitudinal Growth of the Gut

Calcium waves spontaneously circulate in the smooth muscle syncytium from E6 through E12 without significant qualitative change. Their speed gradually increases (from ~10 to 50 μm/s), as does their frequency (0.2 cpm/day) and the amplitude of the contractions they give rise to (from 2 to ~25% diameter decrease). We wondered whether these contractile waves had any functional role, because they do, after all, represent a significant energy expenditure for the

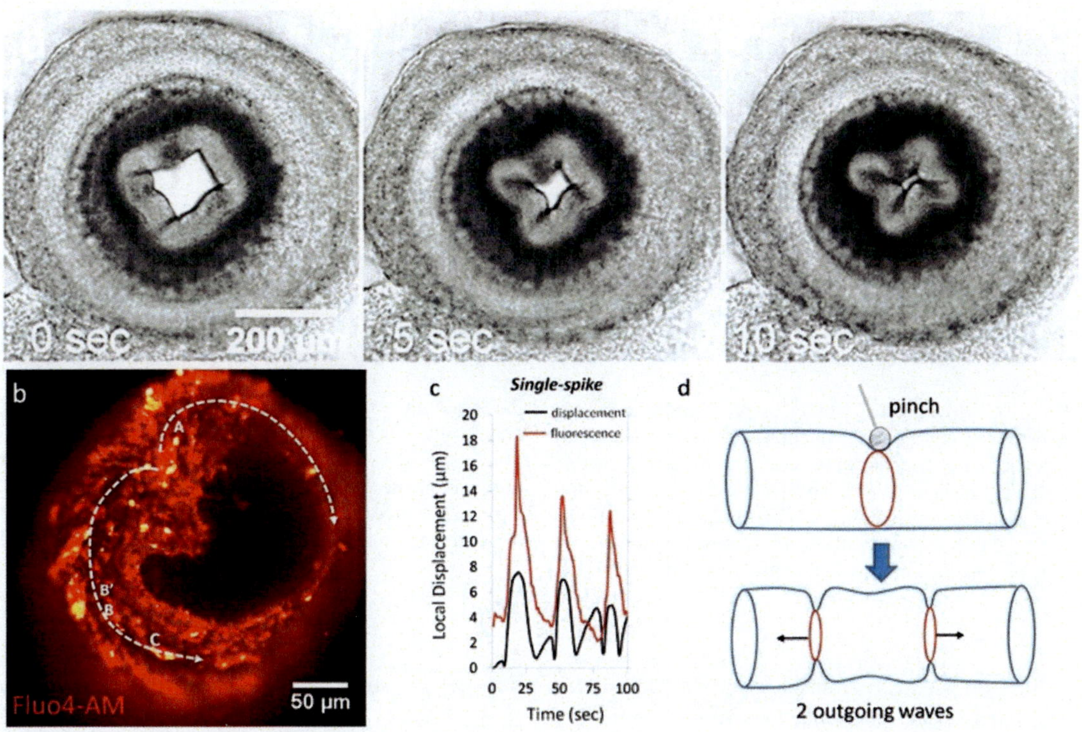

Fig. 10.1 Smooth muscle calcium waves and myogenic reflex in the early embryonic gut. (**a**) Transverse 1-mm-thick slice preparation of E9 chick hindgut still presents cyclic, lumen-obliterating contractions. (**b**) This transverse slice preparation can be used to ease Fluo4-AM calcium indicator penetration inside the smooth musculature, revealing propagating circumferential calcium waves (dashed arrows) that precede the contraction (**c**). (**d**) Mechanosensitivity in the early gut is characterized by the generation of two contractile calcium waves that propagate away from the point where pressure is applied (pinching with a pipette)

embryo. My colleague Sandrine Faure (INSERM, Montpellier) suggested that the mostly rostro-caudal propagation in the proximal gut and caudo-rostral propagation in the hind and distal guts [5] could retain bile and other intestinal secretions in the lumen of the central portion of the gut, preventing them from leaking out into the amniotic bag. Recently, we have found that the tonic and phasic contractility of the circular smooth muscle belt plays a major role in shaping the developing GI tract. Adolphe Bloch recognized in his 1904 manuscript on gut morphometrics across species that the dimensions of the intestine vary significantly depending on its contractile state [3]. We have developed software to accurately determine gut length, diameter, and volume and found that an active E10 gut is 30% longer and has ~25% smaller diameter than the same gut in which contractions have been relaxed after blocking Ca^{2+} channels [15]. This happens because the contracted circular muscle belt squeezes the tissue of the whole intestine along the only direction it is free to deform along—its length (Fig. 10.2c). Could the repeated, accumulated deformation resulting from smooth muscle activity during early embryonic development lead to irreversible, plastic, longitudinal deformation of the proliferating organ, that is, longitudinal growth? We first found that providing extra oxygen by continuous carbogen bubbling could partially resume intestinal growth in culture (60% lengthening and mass increase in 2 days, Fig. 10.2b). We also showed that continuous longitudinal stretch that mimicked in vivo tension applied on the early gut loop by the vitelline duct could also induce growth (Fig. 10.2a). To our knowledge, these are the first rigorous demonstrations that guts can be grown in vitro;

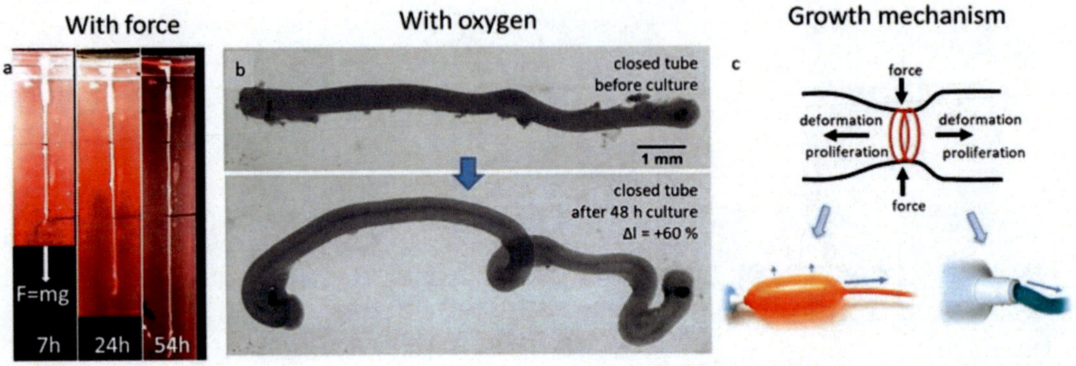

Fig. 10.2 Mechanical forces direct embryonic gut growth. (**a**) A constant stretching force (small weight) mimicking the mechanical tension of the vitelline duct on an E8 gut induces growth. (**b**) Culturing guts in a carbogen atmosphere (95% O_2, 5% CO_2) induces elongational growth that is due to circular smooth muscle contractions, as inhibiting the latter with a longitudinal cut or nicardipine results in isotropic growth (i.e., equal thickening and lengthening). (**c**) Schematic of the mechanical deformations coupled to proliferation that result in longitudinal growth. The mechanism is akin to longitudinal inflation of a rubber balloon reinforced along the circumferential direction longitudinal, or to extrusion of paste from a tube

culture on porous membrane had been advocated previously [10], but in our experience, flattening of the gut due to meniscus pressure in this setup can confound the actual volume increase of the organ; moreover, controlling the contractile state is essential when assessing morphometric parameters as smooth muscle contractions can induce very high (~100%), purely elastic deformations, that do not necessarily mean growth per se. We found that halting motility (tone and phasic contractions) with nicardipine changed the growth pattern from elongation to radial growth, that is, the gut got thicker; preventing the "self-squeezing" effect by cutting one of the gut wall had the same effect. Longitudinal smooth muscle differentiation at E13 and the resulting contractions along this direction have the opposite effect: they temporarily slow down elongation and favor an increase in diameter. The high-aspect ratio of the intestine (7 m in the human adult, only a few cm in diameter) is therefore a direct consequence of the early contractile forces that are exerted by the spontaneously contracting smooth muscle layers [15]. The intestine is "sculpted" by smooth muscle forces, a fact that had already been appreciated for villi formation of the epithelium [8, 21] or for branching alveola in the developing lung [17]. Our study suggests a potential causal link between dysmotility and abnormal growth in Short Bowel Syndrome [24]. Recently, the same smooth-muscle contraction-induced growth mechanism has been demonstrated [25] by the group of K. Mostov in the mouse gut and shown to be mediated by the mechanosensitive YAP pathway.

10.3 The Interstitial Cell of Cajal Transition

A sharp increase of contractile wave speed (from 25 to 125 μm/s) takes place between E12 and E14. Comparing spatiotemporal maps at these two stages shows obvious differences, with irregular and spaced-out contractions at E12, that gain a clock-like regularity at E14 (Fig. 10.3a). We recently showed that this transition is the result of the functional implication of the interstitial cells of Cajal [7]. ICCs are present early on in development: TMEM-16A is expressed in the whole intestine at E7, c-KIT at E9 (E14.5 in the mouse [19]). However, similarly to enteric nerves, ICCs are not functionally active until later in development. We showed that the first distinct calcium transients in Cajal cells can only be recorded as from E14 (Fig. 10.3b), and that imatinib, a known

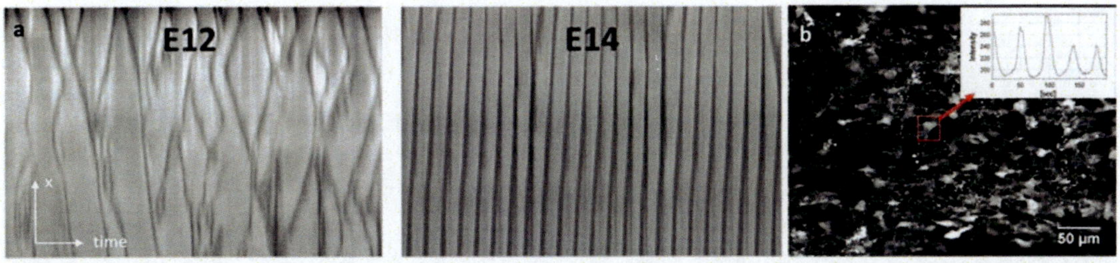

Fig. 10.3 A transition to ICC driven motility occurs between E12 and E14. (**a**) Contractions at E14 become rhythmic, with prolonged periods of propagation in the same direction. (**b**) Fluo4-AM confocal image of E14 duodenum, the bright cells are MP-ICCs that exhibit rhythmic, nicardipine-resistant electrical activity starting at E14

blocker of the c-kit receptor present at the surface of Cajal cells, degrades the ICC network and inhibits the physiological transition to ICC-driven motility between E12 and E14. The gut therefore undergoes a transition from myogenic to ICC-driven waves during development; applying the day-week equivalence between chicken and humans predicts that this transition occurs in the human embryo to around 12–14 weeks of development. A similar "ICC transition" should occur in mice around E18.5, time around which the first slow waves have been recorded [2, 19].

10.4 Early Enteric Nervous System Activity

The first influence of the enteric nervous system on motility occurs at E16, shortly after longitudinal muscle smooth muscle differentiation (E13) and after the first longitudinal contractions can be recorded (at E14). This order of events is common to all model species examined so far. At E16, the ENS forms two distinct interconnected plexuses and neurites extend from ganglia to innervate the smooth muscle layer as well as the epithelial villi (Fig. 10.4a). Pharmacological inhibition shows that the first active neurons of the ENS in the chicken are nitrinergic, they relax the circular smooth muscle; upon application of tetrodotoxin or NOLA, the tone of the circular muscle layer strongly increases, revealing this inhibitory, NO-mediated effect [6]. This neural inhibition is mechanosensitive and descending (Fig. 10.4c): when an E16 duodenum is pinched (local pressure application), the area 1–2 mm distal to the point where pressure was applied becomes "numb," that is, it does not present circular contractility anymore. Calcium waves travel through this numbed region but do not trigger a contraction. Applying tetrodotoxin reverts the reaction to pinching to what it was at earlier myogenic stages: two-waves traveling symmetrically away from the point of pinching. This reveals the embryonic make-up of the "law of the intestine": the myogenic reflex is responsible for a symmetric (ascending and descending) contraction (wave propagation), half of which is inhibited by neural circuitry (descending inhibition). The neural asymmetry is likely caused by rostrocaudally projecting neurons, a topological feature, which we speculate could be traced back to the rostro-caudal migration of neural crest cells along the gut tract.

At E14, when neurons are not yet active, circular and longitudinal contractions occur independently, each with its own distinct frequency (3 cpm for circular contractions, 0.5 cpm for longitudinal contractions). In contrast, at E16, when neurons and glia become active, a distinct correlation of longitudinal and circular contractions can be measured: circular contraction amplitude is notably less important during longitudinal contractions, so that the two muscle layers start working antagonistically. This antagonism is neurally driven as it vanishes in the presence of TTX. When this antagonism occurs locally, that is, when the longitudinal contraction only spans a

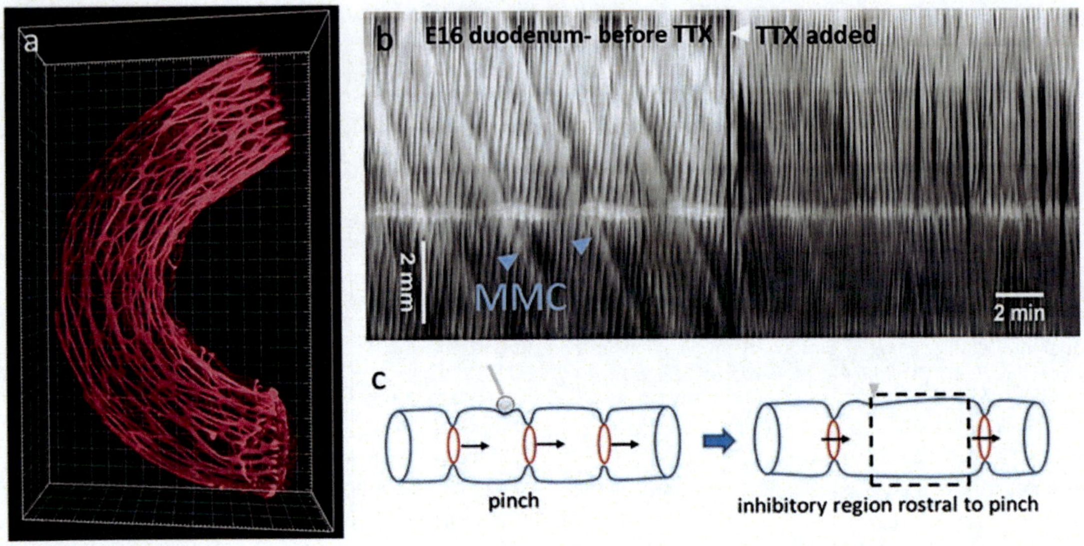

Fig. 10.4 Neurogenic motility. (**a**) Tuj1 immunohistochemistry of a transparized E16 duodenum, the first stage at which neurogenic motility events can be detected. (**b**) Tetrodotoxin-sensitive migrating motor complexes (white diagonal streaks pointed at with blue arrowheads) can be detected on a background of higher-frequency myogenic waves. (**c**) The descending inhibition—ascending contraction reflex appears at this neurogenic stage: two myogenic waves are generated after mechanical stimulation (see Fig. 10.1c), but the descending one is inhibited by neural nitrinergic input. This reveals the embryonic makeup of the "law of the intestine" described by Bayliss and Starling

finite length of the gut segment, it gives rise to a new wave propagation pattern, the migrating motor complex (MMC). MMCs are clearly neurogenic as they are abolished by tetrodotoxin (Fig. 10.4b). We found that the MMCs are initiated by longitudinal contractions, as they have exactly the same frequency as the cyclic longitudinal contractions. The local longitudinal contraction is able to trigger the same descending inhibition of the circular muscle as upon pinching. However, unlike with local pressure application, the MMC propagates, at a low velocity (40 μm/s), in the rostro-caudal direction. It is in fact the first rostro-caudally directed motor pattern in the developing embryo. This propagation may either result from electrotonic propagation of the longitudinal contraction (e.g., through gap junctions), or from a phenomenon where the pressure exerted by the liquid bolus accumulated in the lumen at the site of the MMC becomes the mechanical trigger for further descending neural inhibition: the properties of the MMC would then be dependent on the displacement and flow properties of the bolus itself. This possibility is called the neuromechanical loop and has received confirmation in adult gut motility studies in recent years [9]. Based on our observations, we therefore suggest the following definition of an MMC: a local, descending inhibitory signal propagating along the intestine, which can be stimulated by cyclic contractions of the longitudinal muscle layer [6]. It remains to be elucidated whether this definition holds in other species as well.

10.5 Outlook

We have outlined how intestinal motility evolved from simple, purely circular smooth muscle–driven calcium waves, to ICC-driven waves. Soon after the longitudinal layer differentiates, the enteric nervous system becomes active by coupling the longitudinal and circular contractions: this coupling is mechanosensitive, inhibitory, descending, and gives rise to circular-longitudinal antagonism and to the migrating motor complex. What happens beyond this neurally mediated migrating motor complex, at E16 and later? When can motility be considered "functional"? What about cholinergic neurons? What mechanical triggers activate the ENS? Many questions remain to be elucidated, which we shall tackle in the forthcoming years.

References

1. Bayliss WM, Starling EH (1899) The movements and innervation of the small intestine. J Physiol 24:99–143
2. Beckett EAH, Ro S, Bayguinov Y, Sanders KM, Ward SM (2007) Kit signaling is essential for development and maintenance of interstitial cells of Cajal and electrical rhythmicity in the embryonic gastrointestinal tract. Dev Dyn 236:60–72
3. Bloch A (1904) Des variations de longueur de l'intestin. Bull Mem Soc Anthropol Paris 5:160–197
4. Chevalier NR (2018) The first digestive movements in the embryo are mediated by mechanosensitive smooth muscle calcium waves. Philos Trans R Soc B Biol Sci. https://doi.org/10.1098/rstb.2017.0322
5. Chevalier NR, Fleury V, Dufour S, Proux-Gillardeaux V, Asnacios A (2017) Emergence and development of gut motility in the chicken embryo. PLoS One 12:e0172511
6. Chevalier NR, Dacher N, Jacques C, Langlois L, Guedj C, Faklaris O (2019) Embryogenesis of the peristaltic reflex. J Physiol 597:2785
7. Chevalier N, Ammouche Y, Gomis A, Teyssaire C, de Santa Barbara P, Faure S (2020) Shifting into high gear: how interstitial cells of Cajal change the motility pattern of the developing intestine. Am J Physiol Gastrointest Liver Physiol 319:519
8. Coulombres A, Coulombres J (1958) Intestinal development: morphogenesis of the villi and musculature. J Embryol Exp Morph 3:403–411
9. Dinning PG, Wiklendt L, Omari T, Arkwright JW, Spencer NJ, Brookes SJH et al (2014) Neural mechanisms of peristalsis in the isolated rabbit distal colon: a neuromechanical loop hypothesis. Front Neurosci 8:1–15. https://doi.org/10.3389/fnins.2014.00075
10. Duh G, Mouri N, Warburton D, Thomas DW (2000) EGF regulates early embryonic mouse gut development in chemically defined organ culture. Pediatr Res 48:794–802. https://doi.org/10.1203/00006450-200012000-00016
11. Furness JB, Stebbing MJ (2018) The first brain: species comparisons and evolutionary implications for the enteric and central nervous systems. Neurogastroenterol Motil 30:1–6. https://doi.org/10.1111/nmo.13234
12. Gfroerer S, Rolle U (2015) Pediatric intestinal motility disorders. World J Gastroenterol. https://doi.org/10.3748/wjg.v21.i33.9683
13. Holmberg A, Schwerte T, Fritsche R, Pelster B, Holmgren S (2003) Ontogeny of intestinal motility in correlation to neuronal development in zebrafish embryos and larvae. J Fish Biol 63:318–331. https://doi.org/10.1046/j.1095-8649.2003.00149.x
14. Huycke TR, Miller BM, Gill HK, Nerurkar NL, Sprinzak D, Mahadevan L et al (2019) Genetic and mechanical regulation of intestinal smooth muscle development. Cell 179:90–105.e21
15. Khalipina D, Kaga Y, Dacher N, Chevalier N (2019) Smooth muscle contractility causes the gut to grow anisotropically. J R Soc Interface 16:20190484
16. McConalogue K, Furness J (1994) Gastrointestinal neurotransmitters. Bailliere Clin Endocrinol Metab 8:51–76
17. Nelson CM, Gleghorn JP, Pang M-F, Jaslove JM, Goodwin K, Varner VD et al (2017) Microfluidic chest cavities reveal that transmural pressure controls the rate of lung development. Development 144:4328–4335. https://doi.org/10.1242/dev.154823
18. Newgreen DF, Jahnke I, Allan IJ, Gibbins IL (1980) Differentiation of sympathetic and enteric neurons of the fowl embryo in grafts to the chorio-allantoic membrane. Cell Tissue Res 208:1–19. https://doi.org/10.1007/BF00234168
19. Roberts RR, Ellis M, Gwynne RM, Bergner AJ, Lewis MD, Beckett EA et al (2010) The first intestinal motility patterns in fetal mice are not mediated by neurons or interstitial cells of Cajal. J Physiol 588:1153–1169
20. Shimizu H, Koizumi O, Fujisawa T (2004) Three digestive movements in hydra regulated by the diffuse nerve net in the body column. J Comp Physiol A Neuroethol Sens Neural Behav Physiol 190:623–630. https://doi.org/10.1007/s00359-004-0518-3
21. Shyer AE, Tallinen T, Nerurkar NL, Wei Z, Gil ES, Kaplan DL et al (2013) Villification: how the gut gets its villi. Science 342:212–218. https://doi.org/10.1126/science.1238842
22. Spencer NJ, Dinning PG, Brookes SJ, Costa M (2016) Insights into the mechanisms underlying colonic motor patterns. J Physiol 594:4099–4116. https://doi.org/10.1113/JP271919
23. Ueda Y, Yamada S, Uwabe C, Kose K, Takakuwa T (2016) Intestinal rotation and physiological umbilical herniation during the embryonic period. Anat Rec 299:197–206. https://doi.org/10.1002/ar.23296
24. van der Werf CS, Halim D, Verheij JBGM, Alves MM, Hofstra RMW (2015) Congenital short bowel syndrome: from clinical and genetic diagnosis to the molecular mechanisms involved in intestinal elongation. Biochim Biophys Acta Mol basis Dis 1852:2352–2361. https://doi.org/10.1016/j.bbadis.2015.08.007
25. Yang Y, Paivinen P, Xie C, Krup AL, Makela TP, Mostov KE et al (2020) Ciliary signaling-patterned smooth muscle drives tubular elongation. bioRxiv. https://doi.org/10.1101/2020.08.31.276295

11 Activation of ENS Circuits in Mouse Colon: Coordination in the Mouse Colonic Motor Complex as a Robust, Distributed Control System

Bradley B. Barth, Nick J. Spencer, and Warren M. Grill

Abstract

The characteristic motor patterns of the colon are coordinated by the enteric nervous system (ENS) and involve enterochromaffin (EC) cells, enteric glia, smooth muscle fibers, and interstitial cells. While the fundamental control mechanisms of colonic motor patterns are understood, greater complexity in the circuitry underlying motor patterns has been revealed by recent advances in the field. We review these recent advances and new findings from our laboratories that provide insights into how the ENS coordinates motor patterns in the isolated mouse colon. We contextualize these observations by describing the neuromuscular system underling the colonic motor complex (CMC) as a robust, distributed control system. Framing the colonic motor complex as a control system reveals a new perspective on the coordinated motor patterns in the colon. We test the control system by applying electrical stimulation in the isolated mouse colon to disrupt the coordination and propagation of the colonic motor complex.

B. B. Barth · W. M. Grill (✉)
Department of Biomedical Engineering, Duke University, Durham, NC, USA
e-mail: warren.grill@duke.edu

N. J. Spencer
College of Medicine and Public Health, Flinders University, Adelaide, SA, Australia

Keywords

Enteric nervous system · Colonic migrating motor complex · Control system · Enteric circuitry

11.1 Introduction

A variety of propulsive and nonpropulsive neurogenic motor patterns have been identified in the large intestine of vertebrates [48]. The characteristics of these different patterns of activity can be modulated by external stimuli, such as mechanical compression or distension, endogenous circulating hormones, and neurotransmitters released from intrinsic or extrinsic nerves. Two of the most commonly described neurogenic motor patterns in the colon are peristalsis and colonic motor complexes. Our understanding of the fundamental intrinsic control mechanisms that underlie the generation of neurogenic peristalsis and motor complexes is conceptual; distension initiates a polarized neural reflex of ascending excitation and descending inhibition in the smooth muscle. Recently, electrophysiology, spatiotemporal diameter mapping, and calcium

N. J. Spencer et al. (eds.), *The Enteric Nervous System II*, Advances in Experimental Medicine and Biology 1383, https://doi.org/10.1007/978-3-031-05843-1_11

imaging in the isolated whole mouse colon revealed great complexity in motor complex initiation and propagation (Fig. 11.1). This chapter reviews some of the recent advances in this field and some novel findings from our laboratories that provide new insights into how the ENS coordinates the colonic motor complex in the isolated mouse colon are presented. We discuss historical approaches for evoking and studying the colonic motor complex, the current understanding underlying network synchronization and coordination of the motor complex, and the effect of disrupting the coordinated onset of the motor complex with nonphysiological stimulation. Our findings suggest that the neuromuscular system underlying the colonic motor complex can be described as a robust, distributed control system.

For clarity and consistency, the terminologies for motor patterns used in this work are as follows: colonic motor complex (CMC) refers to a coordinated period of smooth muscle contraction that is dependent upon the ENS and can propa-

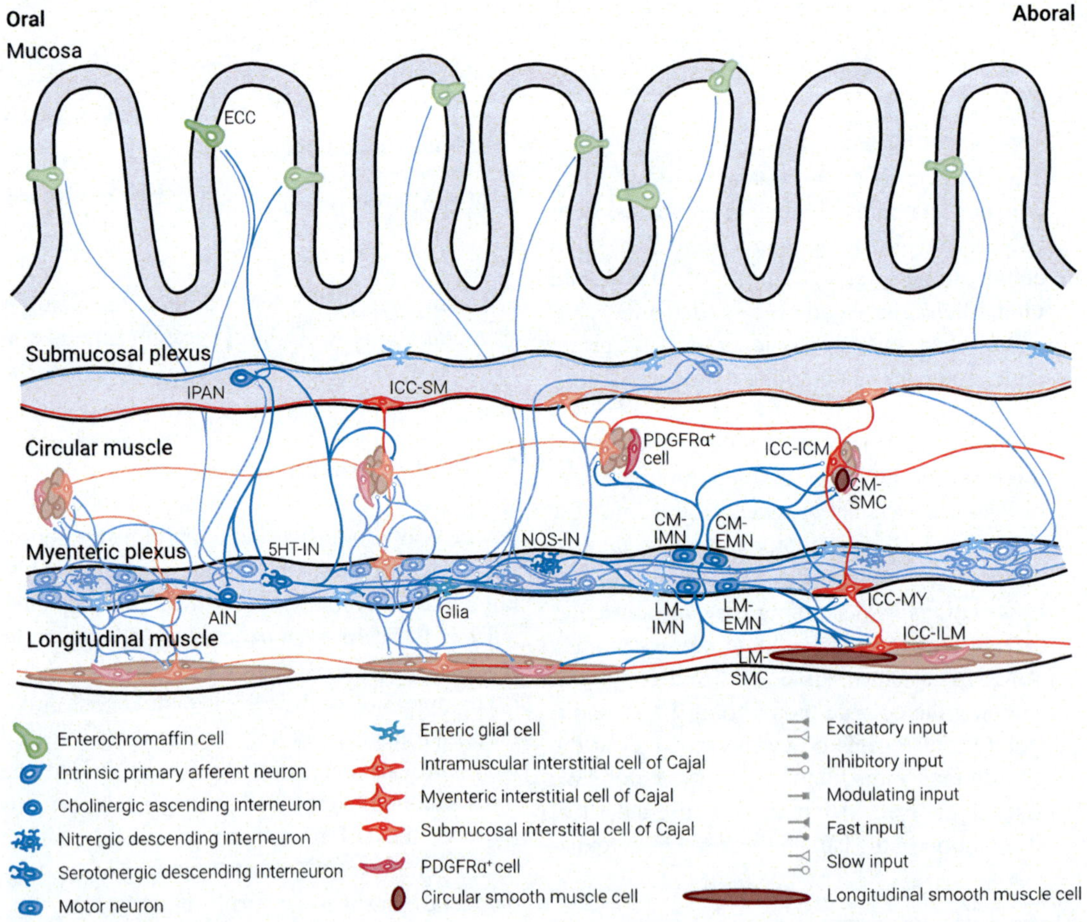

Fig. 11.1 Anatomical representation of the neuromuscular system as a robust, distributed control system. Abbreviations: enterochromaffin cell (ECC), intrinsic primary afferent neuron (IPAN), cholinergic ascending interneuron (AIN), nitrergic descending interneuron (NOS-IN), serotonergic descending interneuron (5HT-IN), circular muscle inhibitory motor neuron (CM-IMN), circular muscle excitatory motor neuron (CM-EMN), longitudinal muscle inhibitory motor neuron (LM-IMN), longitudinal muscle excitatory motor neuron (LM-EMN), circular intramuscular interstitial cell of Cajal (ICC-ICM), longitudinal intramuscular interstitial cell of Cajal (ICC-ILM), myenteric interstitial cell of Cajal (ICC-MY), submucosal interstitial cell of Cajal (ICC-SM), platelet-derived growth factor receptor alpha-positive (PDGFRα$^+$) cell, circular smooth muscle cell (CM-SMC), and longitudinal smooth muscle cell (LM-SMC)

gate in both anterograde and retrograde directions [11]. CMCs are typically measured by force transducers or intraluminal pressure sensors. The myoelectric complex refers to the electrical corollary of the CMC and is often associated with muscle action potentials detected by intracellular or extracellular electrodes. Traditionally, the colonic migrating motor complex had been defined as an event that propagates over at least half the length of the colon [40], either in the oral or aboral direction. However, we now know that the mechanisms that generate CMCs that propagate over less than half the length of colon are not different from those that propagate over the full length of colon [23]. Contraction wavefront refers to the leading edge of a contraction, propagating orally or aborally, and is typically identified by a banded decrease in colon diameter or an increase in intraluminal pressure.

11.2 History of Evoked Colonic Migrating Motor Complexes

A variety of stimuli can elicit a premature colonic motor complex. For example, mechanical compression of the colon is a reliable stimulus responsible for initiating a colonic motor complex. However, the means by which mechanical deformation initiates neurogenic peristalsis is contested, and can include involvement from enterochromaffin cells [19, 28], mechanosensitive enteric neurons [44, 54], and the smooth muscle cell – interstitial cell of Cajal – PDGFRα$^+$ cell (SIP) syncytium [15, 24, 31]. As such, a number of methods have been used to initiate motor complexes, including physical stimuli such as dynamic or constant distension and brushing or stimulating the mucosa. Originally, it was postulated that 5-HT release from the mucosa was critical for CMCs to occur, based on the finding that removing the mucosa abolished CMCs [19]. However, subsequent studies that carefully removed the mucosa found that CMCs persisted without the mucosa. Only then was it realized that too vigorous dissection to remove the mucosa can destroy the intricate neural pathways underlying CMCs [27]. As the mechanisms and downstream effects of mucosal stimulation are difficult to isolate and distinguish [20], we will focus on distension-evoked motor complexes. Fluid distension is perhaps the most reliable method to evoke a migrating motor complex or neurogenic peristalsis in the colon. It has been demonstrated across multiple species, including mice, rats, guinea pigs, and rabbits [12, 16]. Rapid fluid distension most frequently evokes a characteristic "all-or-none" type of CMC, and these were referred to as long-distance contractions [9]. Rapid distension of the entire length of colon engages a coordinated response in which the motor complex propagates the entire length, typically in the aboral direction.

Alternatively, maintained distension in the colon evokes repetitive complexes referred to as cyclical motor complexes. Previously, these patterns were described as "spontaneous" motor complexes [16]. However, it is now known that CMCs are highly sensitive to mechanical distension. For example, in unstimulated colon preparations, CMCs rarely occur, and if they do, they occur at a low frequency. The tissue hooks that were commonly used to measure the force of contraction from the colon themselves potently increase CMC frequency. Mounting the colon from tissue hooks in organ baths typically applies 1–20 mN of tension, and may sufficiently stretch the colon to the point of mechanical activation [16]. In the isolated mouse colon with endogenous fecal pellets, CMCs occur approximately every 30 seconds, when tissue hooks were applied to the same preparation, CMCs readily occurred approximately every 2 minutes [3], similar to the spontaneous events reported by others [16], whereas in the unstimulated colon, they occur only approximately once every 5.5 minutes. Nevertheless, maintained distension is now well accepted as a potent stimulus to generate reliably cyclical motor complexes.

In addition to physical stimuli, electrical and optogenetic stimulation techniques have been used to initiate migrating motor complexes. For example, transmural electrical stimulation delivered to the proximal, mid, or distal colon evoked premature complexes in a stretched isolated colon with cyclical motor complexes endogenously occurring at 0.4 events per minute [46]. Interestingly, complexes evoked in the proximal colon failed to migrate.

More recently, optogenetic stimulation was used to evoke motor complexes in the whole colon isolated from transgenic mice. Hibberd and colleagues used mice expressing channelrhodopsin via a *calb2* promoter and Cre-recombinase to activate selectively calretinin-positive neurons in the mouse colon. In isolated whole transgenic mouse colon, focal optogenetic activation of calretinin-expressing neurons evoked myoelectric complexes [22]. The evoked myoelectric activity increased endogenous fecal pellet output in the isolated preparation. Importantly, Hibberd and colleagues went on to replicate the in vitro observations and observed increased fecal pellet output with in vivo optogenetic stimulation in freely moving mice transgenic mice.

11.3 Coordination in the Colonic Migrating Motor Complex

11.3.1 Subthreshold Rapid Oscillations in the Smooth Muscle

Subthreshold rapid oscillations in colonic smooth muscle are observed in a number of species, including mice, cats, and humans [10, 38, 53]. The frequency of the oscillations is reportedly much faster than slow waves, on the order of 1.5–2 Hz. In mice, these rapid oscillations were found to be neural in origin as they were blocked by atropine or hexamethonium [8]. Clearly, the mechanism underlying their generation was distinct from that of canonical myogenic electrical slow waves. More recent studies have demonstrated that the cholinergic rapid oscillations represent a series of repetitive excitatory junction potentials, generated by coordinated temporal firing of a large ensemble of cholinergic motor neurons [50].

The rapid oscillations are often most prominent during the onset of a myoelectric complex underlying the CMC, and the action potentials of the motor complex occur during the peak of the oscillations. Intracellular recordings revealed that rapid oscillations have greater amplitude during the onset of the slow depolarization associated with a myoelectric complex than during the repolarization phase [8].

11.3.2 Myenteric Potential Oscillations and the Role of Interstitial Cells

In addition to the neurogenic cholinergic rapid oscillations, rapid oscillations initially described in the canine colon occur at 20–40 cycles per minute [43]. Termed myenteric potential oscillations, these oscillations appear similar to the neurogenic rapid oscillations described by Bywater et al. [8]. While the source of the myenteric potential oscillation is thought to be generated by myenteric interstitial cells of Cajal, the underlying mechanisms are unclear, and the myenteric potential oscillation may originate intrinsically from the interstitial cells of Cajal, be mediated by a smooth muscle fiber mechanism, or be driven by excitatory motor neurons [24].

To identify the role of myenteric interstitial cells of Cajal in rapid oscillations, Bayguinov et al. [6] recorded intracellular calcium transients via fluo-4 calcium indicator in c-Kit-positive myenteric interstitial cells of Cajal. In the presence of tetrodotoxin, calcium waves occurred at 38.1 ± 2.9 cycles per minute, consistent with the frequency of myenteric potential oscillations [6]. However, in the absence of tetrodotoxin, during a colonic migrating motor complex, intracellular calcium waves in the myenteric interstitial cells of Cajal increased in frequency to 62.9 ± 1.4 cycles per minute. Calcium transients in the myenteric interstitial cells of Cajal endogenously occur between 20 and 40 cycles per minute, but they increase in frequency during motor complexes by tetrodotoxin-sensitive mechanisms. In this same preparation, myenteric neurons that had taken up fluo-4 increased their firing during the motor complex, with calcium transients occurring at 2.06 ± 0.09 Hz in nerve varicosities, consistent with the frequency of rapid oscillations in the smooth muscle reported by Bywater et al. [8].

Cholinergic rapid oscillations are potently abolished by hexamethonium [8]. Hence, it is clear that fast nicotinic synaptic inputs in large populations of enteric cholinergic neurons are essential for the rapid oscillations to occur. This was later confirmed by neuronal imaging of the ENS during cholinergic rapid oscillations underlying each CMC [50]. Together, these data suggest myenteric neurons directly drive cholinergic rapid oscillations in the smooth muscle and possibly indirectly drive myenteric potential oscillations by increasing the frequency of myogenic calcium transients in the myenteric interstitial cells of Cajal (ICC-MY). However, it is important to acknowledge that in mutant mice deficient in ICC-MY and lacking slow waves, migrating motor complexes still occur and propagate along the intestine [52]. At present, the role of ICC-MY and slow waves in the mouse colon remains unclear.

11.3.3 Enteric Neuron Synchronization During Motor Complex Initiation

Although myoelectric complexes do not usually occur simultaneously along the length of the colon, subthreshold rapid oscillations in smooth muscle can synchronize in time over large spatial distances [49]. During simultaneous intracellular recordings from multiple smooth muscle fibers, inhibitory junction potentials and neurogenic cholinergic rapid oscillations synchronized during the colonic migrating motor complex [49]. The junction potentials were not synchronized during the interval between complexes, suggesting that their temporal synchronization during the CMC was not due to electrotonic coupling in the syncytium alone.

Although blocking cholinergic transmission with hyoscine occludes the cholinergic rapid oscillations, activating cholinergic receptors alone is not sufficient to drive rapid oscillations. If acetylcholine were to serve as a gate or switch to enable a pattern mediated by interstitial cells of Cajal, then reapplying cholinergic agonists in the presence of hyoscine would rescue the cholinergic oscillations. However, applying carbachol, a muscarinic agonist, in the presence of hyoscine, does not rescue cholinergic rapid oscillations [32], suggesting that prolonged exposure of the muscle to a muscarinic agonist was not the underlying cause of the rapid oscillations. Clearly, the repetitive, pulsatile release of acetylcholine from intrinsic motor neurons is fundamental to the cholinergic rapid oscillations.

During simultaneous intracellular recordings from smooth muscle cells, synchronized cholinergic oscillations occurred in smooth muscle cells 15 mm apart from one another [49]. Enteric motor neuron projections reach up to 3–5 mm in the mouse, suggesting that the synchronization likely relies on interneurons [49]. Calcium imaging in multiple myenteric neurons simultaneously revealed increased firing rates in individual neurons and their varicosities, supporting a role for interneurons in coordinating cholinergic oscillations via motor neurons [5, 6]. Evaluating the calcium transients in populations of cholinergic and nitrergic myenteric neurons revealed synchronization periods during the onset of the colonic migrating motor complex [50]. These observations suggest a coordinating role for interneurons that drive excitatory and inhibitory motor neurons in synchrony over large spatial distances in the colon, and that this manifests as cholinergic rapid oscillations in the smooth muscle and increased calcium transient frequency in myenteric interstitial cells of Cajal.

11.3.4 Coordination in a Robust, Distributed Control System

The colonic migrating motor complex invokes a hardwired coordination from a heterogenous network. The heterogenous network acts as a robust, distributed control system to respond to stochastic inputs, such as distension, stretch, or luminal content, and generate an output motor function, such as propulsion or segmentation (Fig. 11.2).

In the robust, distributed control system, the transducer is the enterochromaffin cells, which respond to mucosal stimulation, including physical and nutritional stimuli [7]. Enterochromaffin

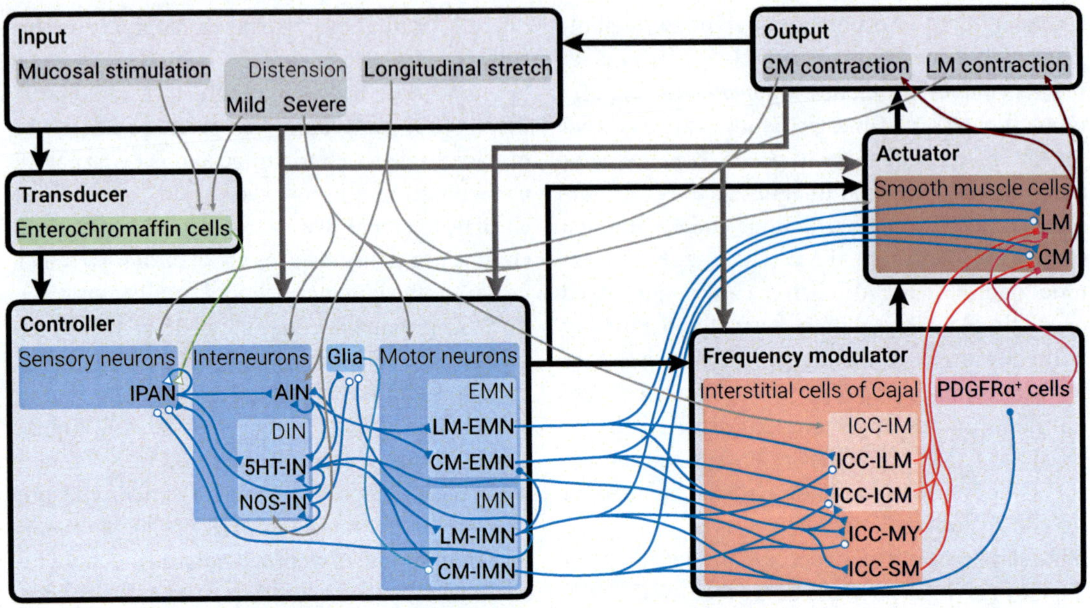

Fig. 11.2 Block model of the neuromuscular system as a robust, distributed control system. Black arrows indicate primary pathways, and gray arrows indicate secondary pathways. Abbreviations: intrinsic primary afferent neuron (IPAN), cholinergic ascending interneurons (AIN), serotonergic (5HT-IN) or nitrergic (NOS-IN) descending interneurons (DIN), excitatory (EMN) and inhibitory motor neurons (IMN), longitudinal muscle excitatory motor neuron (LM-EMN), longitudinal muscle inhibitory motor neuron (LM-IMN), circular muscle excitatory motor neuron (CM-EMN), circular muscle inhibitory motor neuron (CM-IMN), intramuscular interstitial cells of Cajal (ICC-IM), longitudinal intramuscular interstitial cell of Cajal (ICC-ILM), circular intramuscular interstitial cell of Cajal (ICC-ICM), interstitial cells of Cajal in the myenteric plexus (ICC-MY) and submucosal plexus (ICC-SM), platelet-derived growth factor receptor alpha-positive (PDGFRα$^+$) cell, longitudinal muscle cells (LM), and circular muscle cells (CM)

cells are also sensitive to stretch (more sensitive than mechanosensitive enteric neurons), and they release serotonin to activate the controller through intrinsic primary afferent neurons [2]. The controller is comprised of, in addition to intrinsic primary afferent neurons, cholinergic ascending interneurons, serotonergic descending interneurons, nitrergic descending interneurons, excitatory and inhibitory motor neurons, and enteric glial cells. The controller outputs from motor neurons and serotonergic descending "command" interneurons, acting on interstitial cells to modulate their intrinsic frequency and on smooth muscle cells [37].

In this model, PDGFRα$^+$ cells and interstitial cells of Cajal receive input from the controller to modulate myoelectric rhythms. Submucosal and intramuscular interstitial cells of Cajal are distinct pacemakers that generate independent electrical rhythmicity [25, 43]. The myenteric interstitial cells of Cajal are directly modulated by motor neurons and serotonergic descending interneurons [6, 18]. Collectively, the SIP syncytium maintains intrinsic, myogenic rhythms with frequency modulation by enteric neurons and mechanical deformation. The frequency modulation enables dynamic frequency patterns in the smooth muscle as observed in Bayguinov et al. [6]. Please note that the previous notion of a serotonergic "command" neuron in the mouse ENS as a driver of CMC activity is tenuous, based on more recent studies, since depletion of 5-HT from enteric neurons in mouse colon has little to no effect on CMC activity [51].

Distribution within the control system is provided primarily at the level of the controller through interneurons, which have projections that reach up to 15 mm in the mouse colon [30,

49]. Subpopulations of interneurons, including cholinergic ascending interneurons, and descending interneurons, form individual chains that distribute polarized responses and coordinate activity over large spatial distances. In addition, the control system is distributed over shorter distances. Intrinsic primary afferent neurons have overlapping receptive fields and form local networks with other intrinsic primary afferent neurons spanning roughly 3 mm [29]. Indeed, in the guinea pig small intestine, approximately five intrinsic primary afferent neurons innervate a common intestinal villus and ten intestinal villi are innervated by a common intrinsic primary afferent neuron [45]. Further, smooth muscle cells, interstitial cells of Cajal, and PDGFRα^+ cells are electrically coupled in a spatially extended syncytium [39]. The multiple scales of distribution in the control system enable independent global and local coordination.

The robust, distributed controller incorporates redundant mechanisms for mechanosensation via intrinsic primary afferent neurons, interneurons, motor neurons, and interstitial cells of Cajal [1, 33]. This redundancy enables error calculation and a robust response to stochastic inputs by incorporating a neuromuscular feedback signal from smooth muscle cell contractions distinct from tensile or distension signals [41]. This model provides a mechanism consistent with the neuromechanical loop hypothesis as well as myogenic bursting activity [13, 14, 21].

11.4 Perturbing the Control System and the Colonic Migrating Motor Complex

11.4.1 Initiating a Colonic Migrating Motor Complex by Nonphysiological Stimulation

One of the primary physiological stimuli to induce or initiate colonic motor patterns is luminal distension, which can be detected by enterochromaffin cells or enteric neurons directly. Recent studies suggest enterochromaffin (EC) cells act on extrinsic vagal afferent endings in the mucosa [26]. It is likely that many classes of enteric neurons are mechanosensitive and can initiate CMCs in response to circumferential stretch [1, 33, 54]. Indeed, the non–EC cell mediated mechanisms for initiating a motor complex are consistent with the model of a robust or redundant, distributed control system model.

Although electrical stimulation has been widely used to evoke premature motor complexes in the intact colon, it is unclear which cells need to be activated by electrical stimulation to initiate a CMC. Heredia et al. [19] demonstrated that electrical stimulation continued to evoke colonic migrating motor complexes even after removing the mucosa to abolish spontaneous motor complexes [19], suggesting enterochromaffin cells are not the necessary target of electrical stimulation. Smith and Koh [42] suggest a key role for serotonergic descending interneurons; however, their postulate would be better supported with cell-specific pharmacological blockade or optogenetic stimulation or inhibition during electrical stimulation. A remaining concern with the proposed role for serotonergic interneurons in CMC generation and/or propagation is the lack of evidence of serotonin as a neurotransmitter in the ENS of the mouse colon. The two detailed intracellular electrophysiological studies from myenteric neurons in the mouse colon only found evidence of acetylcholine as the major excitatory fast neurotransmitter [17, 36].

11.4.2 Disrupting Coordination in the Colonic Migrating Motor Complex

In the model of a robust, distributed control system, interneurons coordinate synchronization at the level of the controller and the frequency modulator. The interneurons are thought to be responsible for synchronizing increased firing rates in other interneurons and motor neurons [42, 50]. Here, we address the question of how the characteristics of CMC generation and propagation are altered when the ENS circuits downstream, in the mid colon, are perturbed. We seek to understand whether long descending and/or ascending pathways contribute to the circuits in the CMC wavefront that underlies the CMC contraction.

In our recent experiments, we investigated the effects of delivering electrical nerve stimulation (400 μs, 50 V, 20 Hz bipolar stimulation) along the mouse colon to determine effects on ENS activity that underlies cholinergic rapid oscillations. We delivered a repetitive train of electrical pulses for 5 seconds in the mid colon in isolated whole mouse colon preparations [4]. The stimulus was delivered when cholinergic rapid oscillations were first detected in the proximal colon, as harbingers of a CMC evoked by fluid distension. Cholinergic rapid oscillations and myoelectric complexes were detected by extracellular suction electrodes, and propagation was confirmed via spatiotemporal diameter mapping. Delivering the stimulus to disrupt the rapid oscillations did not block, delay, or observably alter the myoelectric complex or contraction wavefront propagation (Fig. 11.3a, b).

However, when brief trains of electrical pulses were delivered with a slight delay, the contraction wavefront propagation was temporarily halted or delayed for the duration of the stimulus (Fig. 11.3c, d). Across repeated trials, it became clear that the timing was essential to delay successfully contraction propagation. Interfering electrical stimulation delivered to the site of contraction as the contraction wavefront arrived robustly halted the contraction wavefront. When the stimulus was terminated, the contraction wavefront proceeded as if it had not been prevented or perturbed, sometimes with an increased contraction velocity after stimulation compared to before stimulation. We hypothesize the interfering electrical stimulation activates chains or assemblies of excitatory cholinergic interneurons that feed into inhibitory motor neurons or nitrergic interneurons that prevent contraction at the site of the stimulation electrode, because the interfering electrical stimulation did not appear to drive novel contraction sites. Incorporating pharmacological compounds to the assay will be necessary to confirm or refute this potential mechanism.

Temporarily halting a contraction wavefront by delivering interfering brief trains of electrical pulses occurs by disrupting the robust, distributed control system. Activating the controller at precisely the right time effectively causes destructive interference in the frequency modulator and actuator, hence the temporary delay in propagation. This model explains the importance of the time and location of the electrical stimulation to interfere with a migrating motor complex. Interpreting the neuro-neuronal and neuromuscular apparatus as a robust, distributed control system offers new insights into the mechanisms underlying CMC initiation, coordination, and propagation.

11.5 Concluding Remarks

We discussed the coordinating events underlying the motor complex in the isolated mouse colon, including historical approaches for evoking and studying the motor complex, as well as underlying network synchronization. We proposed that the neuromuscular system underlying the coordination and onset of the CMC can be described as a robust, distributed control system. The model offers a means to describe frequency modulation and tuning in myenteric potential oscillations and cholinergic rapid oscillations. It includes mechanisms for primary signal transducers through the mucosa and enterochromaffin cells, and it includes secondary sensor systems for feedback control and error detection via mechanosensation in enteric neurons, smooth muscle cells, and interstitial cells. The model incorporates serotonergic descending interneurons with possible "command" roles as described by Smith and colleagues [37, 42], and a potential role for disinhibition of smooth muscle cells by presynaptic inhibition of inhibitory motor neurons [47]. This model also allows for CMC generation in the absence of serotonergic descending interneurons, to reconcile findings that depletion of neuronal 5-HT in the mouse colon (i.e., all interneurons and motor neurons) does not prevent CMC generation in mice [51]. Further, the model includes cooperation of distinct populations of interstitial cells of Cajal as pacemakers, and frequency modulation in the SIP syncytium [6, 24, 25].

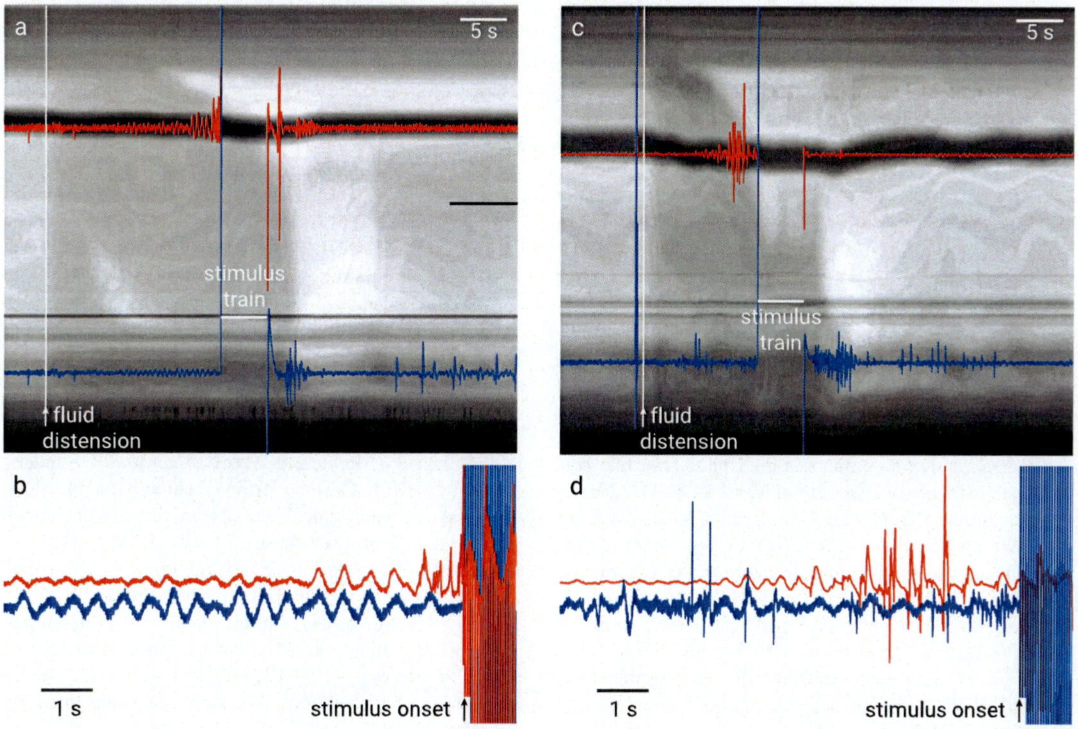

Fig. 11.3 Spatiotemporal diameter maps and myoelectric recordings from the isolated whole mouse colon. Fluid distension evoked colonic migrating motor complex that could be modulated by exogenous electrical stimulation. (**a**, **b**) Interfering transmural electrical stimulation was delivered after the onset of cholinergic rapid oscillations prior to smooth muscle action potentials, and it did not observably interfere with the colonic migrating motor complex. However, when the stimulus was delivered, the contraction wavefront reached the site of stimulation (**c**, **d**), the interfering stimulus temporarily halted the contraction wavefront until the stimulus was terminated

There are limitations and potential inconsistencies in the model. The details of the robust, distributed model appear to vary between species. For example, in the presence of tetrodotoxin, the muscarinic agonist carbachol evoked myogenic contractions in rabbits and rats but not in mice or guinea pigs [12]. The distribution of mechanosensitive neurons and their neurochemical code, for example, also vary across species [34, 35]. These discrepancies between species do not invalidate the robust, distributed control system model, but rather suggest that the details and tuning of the model vary across species. As well, the majority of the observations used to develop the model were from the ex vivo isolated colon. It remains to be determined how well these observations and models translate to the in vivo murine colon, and the in vivo colon of other species, including humans. Across species, the colonic motor complex is coordinated by a robust, distributed control system that routinely and reliably generates specific output motor function in response to stochastic and varied inputs.

Acknowledgments We acknowledge the financial support of the Department of Biomedical Engineering at Duke University, Durham, North Carolina, USA, and the National Institute of Diabetes and Digestive and Kidney Diseases (R01DK119795). Experiments in this study were performed in the Visceral Neurophysiology Laboratory at Flinders University, Adelaide, South Australia, AU, with support from the National Health and Medical Research Council (1156416) to NJS.

References

1. Alcaino C, Farrugia G, Beyder A (2017) Mechanosensitive piezo channels in the gastrointestinal tract. Curr Top Membr 79:219–244

2. Alcaino C, Knutson KR, Treichel AJ, Yildiz G, Strege PR, Linden DR, Li JH, Leiter AB, Szurszewski JH, Farrugia G, Beyder A (2018) A population of gut epithelial enterochromaffin cells is mechanosensitive and requires Piezo2 to convert force into serotonin release. Proc Natl Acad Sci 115:E7632
3. Barnes KJ, Beckett EA, Brookes SJ, Sia TC, Spencer NJ (2014) Control of intrinsic pacemaker frequency and velocity of colonic migrating motor complexes in mouse. Front Neurosci 8:96–96
4. Barth B, Travis L, Spencer NJ, Grill WM (2021) Control of colonic motility using electrical stimulation to modulate enteric neural activity. Am J Physiol-Gastrointest Liver Physiol 320(4):G675–G687
5. Bayguinov PO, Broadhead MJ, Okamoto T, Hennig GW, Smith TK (2012) Activity in varicosities within the myenteric plexus between and during the colonic migrating motor complex in the isolated murine large intestine. Neurogastroenterol Motil 24:e185–e201
6. Bayguinov PO, Hennig GW, Smith TK (2010) Ca2+ imaging of activity in ICC-MY during local mucosal reflexes and the colonic migrating motor complex in the murine large intestine. J Physiol 588:4453–4474
7. Bellono NW, Bayrer JR, Leitch DB, Castro J, Zhang C, O'Donnell TA, Brierley SM, Ingraham HA, Julius D (2017) Enterochromaffin cells are gut chemosensors that couple to sensory neural pathways. Cell 170:185–198.e16
8. Bywater RA, Small RC, Taylor GS (1989) Neurogenic slow depolarizations and rapid oscillations in the membrane potential of circular muscle of mouse colon. J Physiol 413:505–519
9. Chen JH, Zhang Q, Yu Y, Li K, Liao H, Jiang L, Hong L, Du X, Hu X, Chen S, Yin S, Gao Q, Yin X, Luo H, Huizinga JD (2013) Neurogenic and myogenic properties of pan-colonic motor patterns and their spatiotemporal organization in rats. PLoS One 8:e60474
10. Christensen J, Caprilli R, Lund GF (1969) Electric slow waves in circular muscle of cat colon. Am J Phys 217:771–776
11. Corsetti M, Costa M, Bassotti G, Bharucha AE, Borrelli O, Dinning P, Di Lorenzo C, Huizinga JD, Jimenez M, Rao S, Spiller R, Spencer NJ, Lentle R, Pannemans J, Thys A, Benninga M, Tack J (2019) First translational consensus on terminology and definitions of colonic motility in animals and humans studied by manometric and other techniques. Nat Rev Gastroenterol Hepatol 16:559–579
12. Costa M, Dodds KN, Wiklendt L, Spencer NJ, Brookes SJH, Dinning PG (2013) Neurogenic and myogenic motor activity in the colon of the guinea pig, mouse, rabbit, and rat. Am J Physiol Gastrointest Liver Physiol 305:G749–G759
13. Costa M, Keightley LJ, Wiklendt L, Hibberd TJ, Arkwright JW, Omari T, Wattchow DA, Zagorodnyuk V, Brookes SJH, Dinning PG, Spencer NJ (2019) Roles of three distinct neurogenic motor patterns during pellet propulsion in guinea-pig distal colon. J Physiol 597:5125–5140
14. Dinning PG, Wiklendt L, Omari T, Arkwright JW, Spencer NJ, Brookes SJH, Costa M (2014) Neural mechanisms of peristalsis in the isolated rabbit distal colon: a neuromechanical loop hypothesis. Front Neurosci 8:75–75
15. Drumm BT, Rembetski BE, Baker SA, Sanders KM (2019) Tonic inhibition of murine proximal colon is due to nitrergic suppression of Ca(2+) signaling in interstitial cells of Cajal. Sci Rep 9:4402
16. Fida R, Lyster DJ, Bywater RA, Taylor GS (1997) Colonic migrating motor complexes (CMMCs) in the isolated mouse colon. Neurogastroenterol Motil 9:99–107
17. Furukawa K, Taylor GS, Bywater RA (1986) An intracellular study of myenteric neurons in the mouse colon. J Neurophysiol 55:1395–1406
18. Hennig GW, Spencer NJ, Jokela-Willis S, Bayguinov PO, Lee HT, Ritchie LA, Ward SM, Smith TK, Sanders KM (2010) ICC-MY coordinate smooth muscle electrical and mechanical activity in the murine small intestine. Neurogastroenterol Motil 22:e138–e151
19. Heredia DJ, Dickson EJ, Bayguinov PO, Hennig GW, Smith TK (2009) Localized release of serotonin (5-hydroxytryptamine) by a fecal pellet regulates migrating motor complexes in murine colon. Gastroenterology 136:1328–1338
20. Heredia DJ, Gershon MD, Koh SD, Corrigan RD, Okamoto T, Smith TK (2013) Important role of mucosal serotonin in colonic propulsion and peristaltic reflexes: in vitro analyses in mice lacking tryptophan hydroxylase 1. J Physiol 591:5939–5957
21. Hibberd TJ, Costa M, Travis L, Brookes SJH, Wattchow DA, Feng J, Hu H, Spencer NJ (2017) Neurogenic and myogenic patterns of electrical activity in isolated intact mouse colon. Neurogastroenterol Motil 29:1–12
22. Hibberd TJ, Feng J, Luo J, Yang P, Samineni VK, Gereau RWT, Kelley N, Hu H, Spencer NJ (2018) Optogenetic induction of colonic motility in mice. Gastroenterology 155(2):514–528e6
23. Hu H, Spencer NJ (2018) Enteric nervous system structure and neurochemistry related to function and neuropathology. In: SAID HM (ed) Physiology of the gastrointestinal tract, 6th edn. Elsevier, London
24. Huizinga JD, Martz S, Gil V, Wang X-Y, Jimenez M, Parsons S (2011) Two independent networks of interstitial cells of Cajal work cooperatively with the enteric nervous system to create colonic motor patterns. Front Neurosci 5:93
25. Huizinga JD, Parsons SP, Chen JH, Pawelka A, Pistilli M, Li C, Yu Y, Ye P, Liu Q, Tong M, Zhu YF, Wei D (2015) Motor patterns of the small intestine explained by phase-amplitude coupling of two pacemaker activities: the critical importance of propagation velocity. Am J Physiol Cell Physiol 309:C403–C414
26. Kaelberer MM, Buchanan KL, Klein ME, Barth BB, Montoya MM, Shen X, Bohórquez DV (2018) A gut-brain neural circuit for nutrient sensory transduction. Science 361(408):eaat5236

27. Keating DJ, Spencer NJ (2010) Release of 5-hydroxytryptamine from the mucosa is not required for the generation or propagation of colonic migrating motor complexes. Gastroenterology 138:652–659
28. Kunze WA, Bertrand PP, Furness JB, Bornstein JC (1997) Influence of the mucosa on the excitability of myenteric neurons. Neuroscience 76:619–634
29. Kunze WA, Furness JB, Bornstein JC (1993) Simultaneous intracellular recordings from enteric neurons reveal that myenteric AH neurons transmit via slow excitatory postsynaptic potentials. Neuroscience 55:685–694
30. Li Z, Hao MM, Van Den Haute C, Baekelandt V, Boesmans W, Vanden Berghe P (2019) Regional complexity in enteric neuron wiring reflects diversity of motility patterns in the mouse large intestine. elife 8:e42914
31. Lies B, Beck K, Keppler J, Saur D, Groneberg D, Friebe A (2015) Nitrergic signalling via interstitial cells of Cajal regulates motor activity in murine colon. J Physiol 593:4589–4601
32. Lyster DJ, Bywater RA, Taylor GS (1995) Neurogenic control of myoelectric complexes in the mouse isolated colon. Gastroenterology 108:1371–1378
33. Mazzuoli-Weber G, Schemann M (2015) Mechanosensitivity in the enteric nervous system. Front Cell Neurosci 9:408–408
34. Mazzuoli G, Schemann M (2009) Multifunctional rapidly adapting mechanosensitive enteric neurons (RAMEN) in the myenteric plexus of the guinea pig ileum. J Physiol 587:4681–4694
35. Mazzuoli G, Schemann M (2012) Mechanosensitive enteric neurons in the myenteric plexus of the mouse intestine. PLoS One 7:e39887
36. Nurgali K, Stebbing MJ, Furness JB (2004) Correlation of electrophysiological and morphological characteristics of enteric neurons in the mouse colon. J Comp Neurol 468:112–124
37. Okamoto T, Barton MJ, Hennig GW, Birch GC, Grainger N, Corrigan RD, Koh SD, Sanders KM, Smith TK (2014) Extensive projections of myenteric serotonergic neurons suggest they comprise the central processing unit in the colon. Neurogastroenterol Motil 26:556–570
38. Provenzale L, Pisano M (1971) Methods for recording electrical activity of the human colon in vivo – clinical applications. Am J Dig Dis 16:712–722
39. Sanders KM, Ward SM, Koh SD (2014) Interstitial cells: regulators of smooth muscle function. Physiol Rev 94:859–907
40. Sarna SK (1985) Cyclic motor activity; migrating motor complex: 1985. Gastroenterology 89:894–913
41. Schemann M, Mazzuoli G (2010) Multifunctional mechanosensitive neurons in the enteric nervous system. Auton Neurosci 153:21–25
42. Smith TK, Koh SD (2017) A model of the enteric neural circuitry underlying the generation of rhythmic motor patterns in the colon: the role of serotonin. Am J Physiol Gastrointest Liver Physiol 312:G1–G14
43. Smith TK, Reed JB, Sanders KM (1987) Interaction of two electrical pacemakers in muscularis of canine proximal colon. Am J Phys 252:C290–C299
44. Smith TK, Spencer NJ, Hennig GW, Dickson EJ (2007) Recent advances in enteric neurobiology: mechanosensitive interneurons. Neurogastroenterol Motil 19:869–878
45. Song Z-M, Brookes SJH, Costa M (1994) All calbindin-immunoreactive myenteric neurons project to the mucosa of the guinea-pig small intestine. Neurosci Lett 180:219–222
46. Spencer NJ, Bywater RA (2002) Enteric nerve stimulation evokes a premature colonic migrating motor complex in mouse. Neurogastroenterol Motil 14:657–665
47. Spencer NJ, Bywater RA, Taylor GS (1998) Disinhibition during myoelectric complexes in the mouse colon. J Auton Nerv Syst 71:37–47
48. Spencer NJ, Dinning PG, Brookes SJ, Costa M (2016) Insights into the mechanisms underlying colonic motor patterns. J Physiol 594:4099–4116
49. Spencer NJ, Hennig GW, Dickson E, Smith TK (2005) Synchronization of enteric neuronal firing during the murine colonic MMC. J Physiol 564:829–847
50. Spencer NJ, Hibberd TJ, Travis L, Wiklendt L, Costa M, Hu H, Brookes SJ, Wattchow DA, Dinning PG, Keating DJ, Sorensen J (2018) Identification of a rhythmic firing pattern in the enteric nervous system that generates rhythmic electrical activity in smooth muscle. J Neurosci 38:5507–5522
51. Spencer NJ, Nicholas SJ, Sia TC, Staikopoulos V, Kyloh M, Beckett EA (2013) By what mechanism does ondansetron inhibit colonic migrating motor complexes: does it require endogenous serotonin in the gut wall? Neurogastroenterol Motil 25:677–685
52. Spencer NJ, Sanders KM, Smith TK (2003) Migrating motor complexes do not require electrical slow waves in the mouse small intestine. J Physiol 553:881–893
53. Wood JD (1973) Electrical activity of the intestine of mice with hereditary megacolon and absence of enteric ganglion cells. Am J Dig Dis 18:477–488
54. Zagorodnyuk VP, Spencer NJ (2011) Localization of the sensory neurons and mechanoreceptors required for stretch-evoked colonic migrating motor complexes in mouse colon. Front Physiol 2:98–98

12 Colonic Response to Physiological, Chemical, Electrical and Mechanical Stimuli; What Can Be Used to Define Normal Motility?

Phil G. Dinning

Abstract

The colon plays an important functional role in the bacterial fermentation of carbohydrates, transmural exchange of fluid and short-chain fatty acids, and the formation, storage and evacuation of faeces and gaseous contents. Coordinated colonic motor patterns are essential for these functions to occur. Our understanding of human colonic motor patterns has largely come through the use of various forms of colonic manometry catheters, combined with a range of stimuli, both physiological and artificial. These stimuli are used in patients with colonic disorders such as constipation, irritable bowel syndrome and faecal incontinence to understand the pathophysiology mechanisms that may cause the disorder and/or the associated symptoms. However, our understanding of a "normal" colonic response remains poor. This review will assess our understanding of the normal colonic response to commonly used stimuli in short duration studies (<8 hrs) and the mechanisms that control the response.

P. G. Dinning (✉)
Gastroenterology Unit, Flinders Medical Centre & College of Medicine and Public Health, Flinders University, Adelaide, SA, Australia
e-mail: phil.dinning@flinders.edu.au

Keywords

Colonic manometry · Meal · Colon motility · Laxatives · Distension

12.1 Protocols and Catheter Types

Colonic manometry is a technique in which a small diameter, highly flexible catheter, incorporating an array of pressure sensors is placed into the colon to record contractile activity. A detailed description of colonic manometry catheter types, placement techniques and protocols have been published elsewhere [1]. Here, a brief summary is provided.

Manometry catheters in use today include water perfused, solid state and fibre-optic. These catheters are classified as being high resolution if they have an array of sensors (typically 36–72) spaced at 10–30 mm intervals. In adults, all studies in the past ten years have placed catheters into the colon through the anus with the aid of a colonoscope. This placement technique involves some form of bowel preparation (the complete or partial removal of faecal material) and sedation. As such, most current studies record colonic contractile activity in a colon devoid of faecal content. This differs from many of the previous colonic manometry studies, using low-resolution (sensors spaced at >7 cm) manometry. Those studies were mostly recorded over periods of

N. J. Spencer et al. (eds.), *The Enteric Nervous System II*, Advances in Experimental Medicine and Biology 1383, https://doi.org/10.1007/978-3-031-05843-1_12

12–24 hrs, which allowed the colon to fill again. The colonic manometry protocols vary between research groups and the design is very much dependent upon the aims of the research group. However, most studies will have a recording period of unstimulated colonic motility, followed by a meal and/or chemical/mechanical stimulation.

12.2 Colonic Motor Patterns

Over the years, there have been a wide range of colonic motor patterns described, and in many instances, different terminology was applied to the same motor pattern. In 2019, a consensus statement was published to unify nomenclature [2]. A brief description of the main motor patterns used in this chapter are as follows:

(i) *High-amplitude propagating contractions (HAPC)*. As the name suggests, this motor pattern consists of high amplitude pressure waves, usually >75 mmHg that travel over at least 30 cm of the colon. They mostly originate in the proximal colon and are associated with "mass movement" of colonic content and defecation.

(ii) *The cyclic motor pattern*. Repetitive propagating pressure events with a cyclic frequency of 2–8/min in either a retrograde or antegrade direction, or aligned synchronously across ≥3 sensors. Low-resolution manometry labelled this activity as non-propagating, segmental or, if recorded in the rectum, periodic rectal motor activity.

(iii) *Low-amplitude propagating contractions*. Propagating contraction with pressure waves below that used to define HAPCs.

(iv) *Colonic pressurizations*. Synchronous pressure increase across all manometry channels temporally associated with relaxation of the anal sphincter.

12.3 Colonic Response to a Meal

Of all the physiological influences on colonic motility, the meal is arguably the most studied. In 1910, Hertz and Frowde wrote, after viewing X-ray images of the human colon, that they "*discovered that the taking of food is the most powerful of all stimuli to the motor activity of the colon in man*" [3]. Meals were reported to be a key stimulus for "mass movements", [4] and the term "gastro-colonic reflex" was being used by 1913 [5]. Recordings with low-resolution water perfused manometry in healthy adults showed that a 1000 kCal meal rapidly increased contractile activity throughout the colon and induced an increase in low-amplitude contractions and HAPC [6–13].

With the introduction of high-resolution manometry, the shorter duration studies in a prepared colon have changed our interpretation of the meal response. For example, while meals in healthy adults have typically been associated with an increase in the number of HAPCs, in high-resolution colonic manometry studies, in the prepared colon, post-prandial HAPCs are absent in many recordings [14–18]. A potential stimulus of HAPC is distension caused by colonic content (see Sect. 12.5) and presumably its removal during a bowel preparation has a negative impact upon their initiation. Of the other motor patterns, one study in healthy adults recorded a significant increase the cyclic motor pattern after a meal (Fig. 12.1), [19] while another did not [20]. Colonic pressurizations have been shown to increase during a meal period in several studies, [15–18] leading one study to suggest that these motor patterns may be a suitable biomarker for defining dysmotility in patient populations [16]. However, subsequent studies in healthy adults have failed to show a postprandial increase in this motor pattern [21]. It is important to note that the meals, catheter types and analysis techniques differed amongst all studies reporting a meal response and this may account for the varied results.

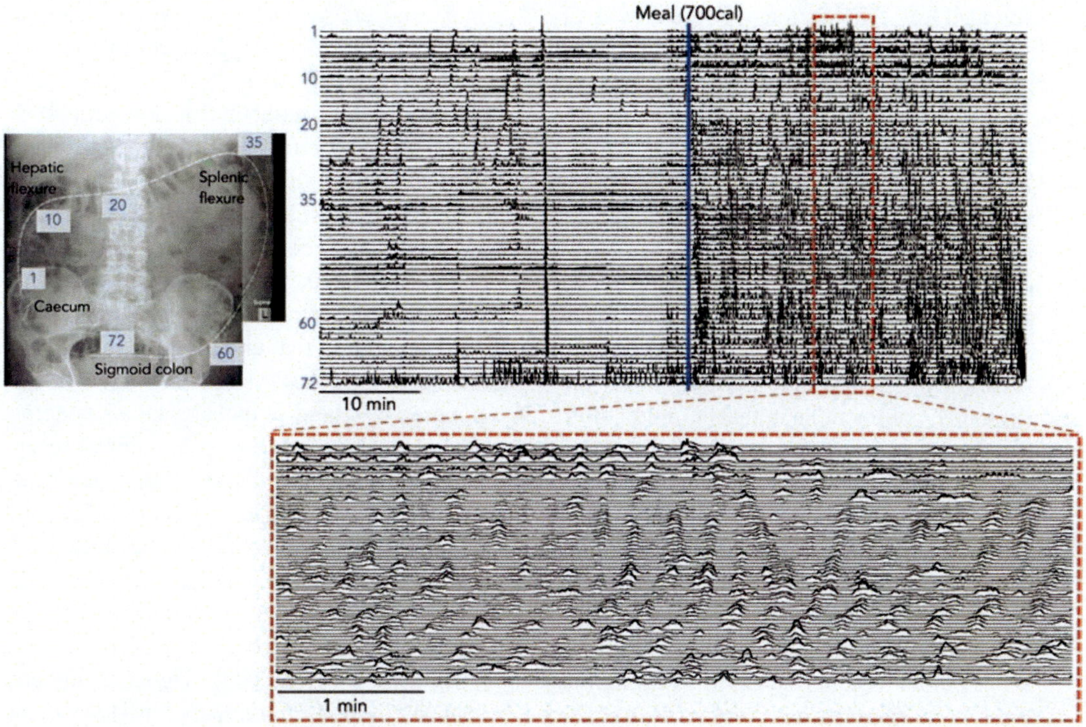

Fig. 12.1 Fibre-optic high-resolution manometry recording in a healthy adult volunteer. The numbers on the X-ray image represent the sensor number. These numbers can also be seen on the manometry trace to the right of the X-ray image, providing the location of where each sensor is located within the colon. An increase in colonic motor patterns can be seen when the meal is given (vertical blue line). The area within the red hatch rectangle have been expanded and shown at the bottom of the figure. The expanded image shows the cyclic motor pattern initiated by the meal

Mechanisms Controlling the Meal Response The colonic response to a meal is rapid and the level of the response may be dependent upon the calories content [22] and the type of food consumed [23]. However, as the increase in colonic activity is evident within seconds of starting to eat, [19] clearly a large calories intake is not required for its initiation. The speed of the response and its inhibition by pre-treatment with the anticholinergic drug clidinium bromide [22] suggest the initial meal response is neurally mediated. Therefore, a failed colonic response to meal in a patient population may indicate an abnormality of the neural inputs to the colon. However, given some healthy adults can also fail to show an increase in any of the defined motor pattern after a meal, a pathophysiological diagnosis based upon a meal response alone should not be made.

Conclusion The colonic meal response in healthy adults remains poorly defined and as yet there are no post-prandial motor patterns that can clearly define normal from abnormal motility.

12.4 Colonic Response to Chemical Stimulation

Over many years, chemicals of various kinds have been used to stimulate or inhibit colonic activity in the humans. Here, these studies have been divided into two sections: (i) the effects of chemicals on general colonic motility and low-amplitude propagating contraction; and (ii) the effects of chemicals on HAPCs.

(i) *The effects of chemicals on general colonic motility and low-amplitude propagating contraction.* In the early 1950s, the "antispasmodic" post-ganglionic parasympathetic antagonist Banthine (100 mg) was given orally to subjects without bowel complaints. It resulted in profound inhibition of the wave-like contractions in the sigmoid colon. [24] In 1960, recording motility from a single location in the caecum and one in the descending colon, Fink and Friedman tested the effects of several drugs [25]. All doses (5–12.5 mg) of the cholinergic receptor agonist Mecholyl increased colonic contractions in the proximal colon of all subjects, while in the sigmoid colon, increased contractions were only recorded with higher doses. The cholinergic agonists neostigmine and urecholine increased phasic activity in the proximal and distal colon in all subjects. The 5-hydroxytryptamine (5-HT) neurotransmitter, serotonin, increased the phasic activity of the proximal colon and inhibited distal colonic motility.

   In 14 healthy adults, the anticholinergic drug, clidinium bromide blocked the normal rapid increase in sigmoid phasic activity in response to a high calories meal [22]. A rectal infusion (150 ml) of the bile acid, chenodeoxycholic acid (1 mmol), increased proximal colonic propagating motor patterns in ten healthy adults [26]. An intravenous infusion of the cholinergic stimulant, edrophonium chloride (10 mg), significantly increased both proximal and distal colonic contractile activity in eight healthy adults [27]. An IV infusion of the nitric oxide inhibitor, N(G)-monomethyl-L-arginine (L-NMMA; 3 mg.kg-1), increased the number of proximal colonic propagating contractions in six health adults [28]. In three healthy adults, acetylcholinesterase inhibitor, prostigmine (0.5 mg; IV), significantly increased the number of colonic pressurizations [15], whilst, oral administration of polyethylene glycol (2 × 13.8 g satchels) increased the number of low-amplitude propagating contraction in ten healthy subjects [29].

(ii) *The effects of chemicals on HAPCs.* In 11 healthy volunteers, an oral dose (0.0125–0.5 mg/kg) of the adrenergic $\alpha 2$ antagonist, yohimbine, resulted in an increase in HAPCs, [30] suggesting that under normal conditions the central nervous system may play a role in suppressing these propulsive motor patterns. Prucalopride, a 5-hydroxytryptamine4 receptor agonist, increases HAPC frequency when recordings were made 12 hours after oral intake [31]. However, in short duration studies, no HAPC increase was noted [18, 29, 32]. The most used chemical to induce HAPC is biasocodyl. In healthy adults, bisacodyl (3 mg in 10 ml saline) infused through a manometry catheter into the proximal transverse colon resulted in a rapid and significant increase in the number of HAPCs [33]. Similar increases in HAPCs have been recorded with rectal infusion (10 mg) (Fig. 12.2) [18] and oral dosage (10 mg), albeit with a delayed response for the latter [29].

Mechanisms Controlling the Chemical Response The mechanisms underpinning the colonic response to chemicals will depend upon the chemicals used. For the majority of manometry studies, bisacodyl is the drug of choice. Bisacodyl is converted to an active form, desacetyl-bisacodyl ("des-bisacodyl") in the lumen of the gut by mucosal or bacterial enzymes [34]. With oral administration, bisacodyl and des-bisacodyl are poorly absorbed through the gut wall. There is a poor correlation between the laxative effect (increase in HAPC) and the blood concentrations of bisacodyl [35, 36]. This indicates that it works on the luminal surface of the mucosa. Pre-treatment of lignocaine to the mucosa of the human colon blocks the effects of bisacodyl [37]. Stimulation of sensory nerves within the mucosa, by bisacodyl is then likely to

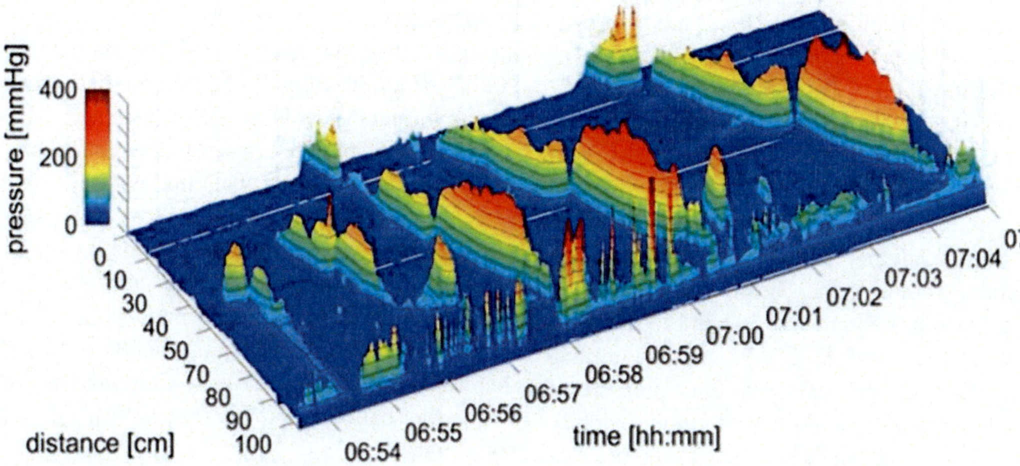

Fig. 12.2 A spatio-temporal colour plot of HAPC initiated by a rectal infusion of bisacodyl in a healthy adult. The Y-axis shows the distance in cm from the anal canal, the X-axis shows time (h:m) from the time of rectal infusion. The Z-axis shows the amplitude of the pressure waves in the HAPC. Note that while the bisacodyl was infused into the rectum, the site of origin of the HAPC is the proximal colon. (Reproduced with permission from [18])

trigger extrinsic vagal nerves to initiate the HAPCs in proximal colon. This response appears to be an all or nothing one – either HAPCs are initiated or they are not. In children there appears to be little increase in any other type of motor pattern [38]. Once initiated, the progression of the HAPC along the colon is also a good indication of normal colonic musculature and enteric neural circuits. [18, 39]

Conclusion From a clinical perspective, in short duration studies, Bisacodyl is the most commonly used stimulant, particularly in paediatric studies [38]. As a rectal infusion of bisacodyl can induce HAPC starting in the proximal colon [18], this would appear to be the simplest test to induce these propulsive motor patterns. However, bisacodly can also be considered a blunt tool, and most patients with severe constipation will still display a "normal" colonic response to this chemical stimulus [38]. Therefore, while it may help isolate those few patients who fail to respond, it is unlikely to help subtype patients beyond that. In addition, it must be noted that while patients may respond to a chemical like bisacodyl, it does not mean their colon will respond to the normal physiological chemical and mechanical stimulation provided by the patients' normal colonic content [39].

12.5 Colonic Response to Distension

Direct observation using X-ray images noted that as caecal filling and subsequent distension and elongation of the region occurred, propulsive motor patterns were initiated to move content towards the rectum [40, 41]. This suggests that distending the colon may be a suitable stimulus for initiating colonic motor patterns. While not commonly performed, there are some colonic manometry studies that have conducted colonic balloon distension during the recordings, the results of which have been highly variable. The studies below detail "colonic" distension, as opposed to rectal distension, which has been shown to inhibit proximal colonic motor propagating pressure waves in healthy adults [26].

In healthy adults, distension with a 5 cm long balloon (maximum of 150 ml [~6 cm diameter] or until the patient reported pain) in the transverse, descending and sigmoid colon induced low-amplitude propagating pressure waves in 8/15 healthy adults [42]. In another study, in the descending colon of 32 healthy volunteers, a 10 cm barostat balloon was inflated to an intra-balloon pressure of 8, 16, 24 and 32 mmHg greater than the minimal respiratory pressure of the colon (distension diameters not provided). The inflation remained at each pressure for 1 minute, before a 1-minute rest, and then the next inflation [30, 43]. Within 5 minutes of balloon distension, just four HAPCs were recorded [30]. That study did not provide details on which of four balloon distensions were associated with the HAPCs, and in how many volunteers those four HAPCs were recorded; however, it is safe to assume that no post-inflation HAPCs were recorded in at least 28/32 adults. In a more recent study, distension of a 10 cm balloon in the proximal colon, to each subject's maximal tolerable volume (250–400 ml) for 2–3 minutes was shown to stimulate HAPCs in 17 of 19 healthy adults [18].

Mechanisms Controlling the Distension Response In humans, the pathways involved in the colonic response to distension are poorly defined. It is likely to involve both intrinsic and extrinsic pathways. As normal distension of the caecum and ascending colon with colon content appears to initiate propulsive motor patterns [40, 41], it is likely that extrinsic vagal neurons play a role. The progression of the HAPC along the colon also indicates an intact enteric nervous system.

Conclusion The colonic response to balloon stimulation is likely to be dependent upon the site of stimulation, diameter of the balloon and duration of the stimulation. Current evidence suggests that proximal ascending colonic distension is good intervention for inducing HAPC, with a low probability of non-responders [18]. However, given that the proximal colon distension induced HAPCs required 2–3 min of balloon inflation at a maximal tolerable volume, the usefulness of the protocol may be limited in certain patient populations. For example, in patients with irritable bowel syndrome, maintaining uncomfortable/painful stimuli for prolonged periods would be unfeasible. The volume required to induce a maximal tolerable volume in such patients may also be much lower than in healthy adults [44] and therefore a failed response may simply be due to the smaller balloon diameter, rather than a motor abnormality.

12.6 Colonic Response to Gas Insufflation

Intestinal gas in the bowel can cause significant discomfort or pain. However, the effects of gas upon colonic motor patterns have rarely been assessed. A recent study demonstrated that small amounts of gas insufflated directly into the sigmoid colon (30 ml over 2 minutes) could reliably trigger the cyclic motor patterns in nine healthy adults [21]. Further studies are required to confirm the reproducibility of this response to establish normal ranges and to determine whether or not the stimulus can be used in patients with suspected colonic hypersensitivity.

12.7 Summary, Current Limitations and Future Directions

While the colon in healthy adults responds to many different types of chemical, mechanical and physiological stimuli, with the exception of bisacodyl, the "normal" response remains poorly defined. Despite most groups around the world now using high-resolution colonic manometry, there remains little to no standardization of the procedure. Catheter types, the number of sensors, the spacing between sensors, protocols used, definitions of motor patterns and analysis of manometry traces all differ amongst institutions [1, 45, 46]. There are also very few studies performed in healthy adults and most publications will report data on <20 subjects [18, 29, 47]. The small sample size, combined with all the other variables associated with the procedure, makes it very difficult to establish normal ranges.

Based upon the current evidence, a rectal infusion of bisacodyl or proximal colonic balloon inflation would appear to give the most reproducible results. However, the usefulness of either technique in helping to subtype patients, understand associated motor abnormalities or guide/predict their treatment outcomes remains undetermined. With colonic manometry, it is also highly unlikely that one test will be sufficient for all colonic motility disorders. Prolonged balloon distension in a patient with constipation-predominant IBS may not be feasible, while in a patient with faecal incontinence or diarrhoea, a rectal infusion of Bisacodyl may not be appropriate.

References

1. Dinning PG, O'Grady G (2020) Colonic Manometry. In: Kuipers EJe (ed) Encyclopedia of gastroenterology, 2nd edn. Academic Press, Oxford, pp 618–626
2. Corsetti M, Costa M, Bassotti G et al (2019) First translational consensus on terminology and definitions of colonic motility in animals and humans studied by manometric and other techniques. Nat Rev Gastroenterol Hepatol 16(9):559–579
3. Hertz A, Frowde H (1910) Constipation and allied intestinal disorders. In: The British Institute of Radiology
4. Barclay AE (1912) Note on the movements of the large intestine. Arch Roentgen Ray 16:422–424
5. Hertz AF, Newton A (1913) The normal movements of the colon in man. 319–330. J Phisiol 47(1–2):57–65
6. Narducci F, Bassotti G, Gaburri M, Morelli A (1987) Twenty four hour manometric recording of colonic motor activity in healthy man. Gut 28(1):17–25
7. Bassotti G, Gaburri M (1988) Manometric investigation of high-amplitude propagated contractile activity of the human colon. Am J Phys 255(5 Pt 1):G660–G664
8. Bassotti G, Imbimbo BP, Gaburri M, Daniotti S, Morelli A (1987) Transverse and sigmoid colon motility in healthy humans: effects of eating and of cimetropium bromide. Digestion 37(1):59–64
9. Bassotti G, Betti C, Imbimbo BP, Pelli MA, Morelli A (1989) Colonic motor response to eating: a manometric investigation in proximal and distal portions of the viscus in man. Am J Gastroenterol 84(2):118–122
10. Bassotti G, Clementi M, Antonelli E, Pelli MA, Tonini M (2001) Low-amplitude propagated contractile waves: a relevant propulsive mechanism of human colon. Dig Liver Dis 33(1):36–40
11. Bampton PA, Dinning PG, Kennedy ML, Lubowski DZ, Cook IJ (2001) Prolonged multi-point recording of colonic manometry in the unprepared human colon: providing insight into potentially relevant pressure wave parameters. Am J Gastroenterol 96(6):1838–1848
12. Rao SS, Sadeghi P, Beaty J, Kavlock R, Ackerson K (2001) Ambulatory 24-h colonic manometry in healthy humans. Am J Physiol Gastrointest Liver Physiol 280(4):G629–G639
13. Dinning PG, Zarate N, Hunt LM et al (2010) Pancolonic spatiotemporal mapping reveals regional deficiencies in, and disorganization of colonic propagating pressure waves in severe constipation. Neurogastroenterol Motil 22(12):e340–e349
14. Dinning PG, Wiklendt L, Omari T et al (2014) Neural mechanisms of peristalsis in the isolated rabbit distal colon: a neuromechanical loop hypothesis. Front Neurosci 8:75
15. Corsetti M, Pagliaro G, Demedts I et al (2017) Pan-Colonic pressurizations associated with relaxation of the anal Sphincter in health and disease: a new Colonic motor pattern identified using high-resolution manometry. Am J Gastroenterol 112(3):479–489
16. Chen JH, Parsons SP, Shokrollahi M et al (2018) Characterization of simultaneous pressure waves as biomarkers for Colonic motility assessed by high-resolution Colonic manometry. Front Physiol 9:1248
17. Chen JH, Yu Y, Yang Z et al (2017) Intraluminal pressure patterns in the human colon assessed by high-resolution manometry. Sci Rep 7:41436
18. Milkova N, Parsons SP, Ratcliffe E, Huizinga JD, Chen JH (2020) On the nature of high-amplitude propagating pressure waves in the human colon. Am J Physiol Gastrointest Liver Physiol 318(4):G646–G660
19. Dinning PG, Wiklendt L, Maslen L et al (2014) Quantification of in vivo colonic motor patterns in healthy humans before and after a meal revealed by high-resolution fiber-optic manometry. Neurogastroenterol Motil 26(10):1443–1457
20. Pervez M, Ratcliffe E, Parsons SP, Chen JH, Huizinga JD (2020) The cyclic motor patterns in the human colon. Neurogastroenterol Motil 32:e13807
21. Heitmann PT, Mohd Rosli R, Maslen L et al (2021) High-resolution impedance manometry characterizes the functional role of distal colonic motility in gas transit. Neurogastroenterol Motil In Press:e14178
22. Snape WJ, Wright SH, Battle WM, Cohen S (1979) The gastrocolonic response: evidence for a neural mechanism. Gastroenterology 77:1235–1240

23. Rao SS, Kavelock R, Beaty J, Ackerson K, Stumbo P (2000) Effects of fat and carbohydrate meals on colonic motor response. Gut 46(2):205–211
24. Kern F Jr, Almy TP, Stolk NJ (1951) Effects of certain antispasmodic drugs on the intact human colon, with special reference to banthine (beta-diethylaminoethyl xanthene-9-carboxylate methobromide). Am J Med 11(1):67–74
25. Fink S, Friedman G (1960) The differential effect of drugs on the proximal and distal colon. Am J Med 28:534–540
26. Bampton PA, Dinning PG, Kennedy ML, Lubowski DZ, Cook IJ (2002) The proximal colonic motor response to rectal mechanical and chemical stimulation. Am J Physiol Gastrointest Liver Physiol 282(3):G443–G449
27. Bassotti G, Imbimbo BP, Betti C, Erbella GS, Pelli MA, Morelli A (1991) Edrophonium chloride for testing colonic contractile activity in man. Acta Physiol Scand 141(3):289–293
28. Dinning PG, Szczesniak MM, Cook IJ (2006) Removal of tonic nitrergic inhibition is a potent stimulus for human proximal colonic propagating sequences. Neurogastroenterol Mot 18(1):37–44
29. Corsetti M, Thys A, Harris A et al (2021) High-resolution manometry reveals different effect of polyethylene glycol, bisacodyl, and prucalopride on colonic motility in healthy subjects: an acute, open label, randomized, crossover, reader-blinded study with potential clinical implications. Neurogastroenterol Motil 33(5):e14040
30. Malcolm A, Camilleri M (2000) Coloanal motor coordination in association with high-amplitude colonic contractions after pharmacological stimulation. Am J Gastroenterol 95(3):715–719
31. Miner PB Jr, Koltun WD, Wiener GJ et al (2017) A randomized phase III clinical trial of Plecanatide, a Uroguanylin analog, in patients with chronic idiopathic constipation. Am J Gastroenterol 112(4):613–621
32. De Schryver AMP, Andriesse GI, Samsom M, Smout AJPM, Gooszen HG, Akkermans LMA (2002) The effects of the specific 5HT4 receptor agonist, prucalopride on coloinc motility in healthy volunteers. Aliment Pharmacol Ther 2002:603–612
33. De Schryver AM, Samsom M, Smout AI (2003) Effects of a meal and bisacodyl on colonic motility in healthy volunteers and patients with slow-transit constipation. Dig Dis Sci 48(7):1206–1212
34. Jauch R (1975) Laxoberal-sodium picosulphate. Drug Ther Bull 25:1796
35. Roth W, Beschke K (1988) Pharmacokinetics and laxative effect of bisacodyl following administration of various dosage forms. Arzneimittelforschung 38(4):570–574
36. Flig E, Hermann T, Zabel M (2000) Is bisacodyl absorbed at all from suppositories in man? Int J Pharm 196(1):11–20
37. Hardcastle JD, Mann CV (1968) Study of large bowel peristalsis. Gut 9:512–520
38. Heitmann PT, Wiklendt L, Thapar N et al (2020) Characterization of the colonic response to bisacodyl in children with treatment-refractory constipation. Neurogastroenterol Motil 32:e13851
39. Chen JH, Huizinga JD (2018) High-pressure tactic: Colonic manometry in chronic constipation. Dig Dis Sci 63(11):2820–2822
40. Hertz AF (1907) The passage of food along the human alimentary canal. Guys Hosp Rep 61:389–427
41. Todd TW (1930) Behavior patterns of the alimentary tract. Williams & Wilkins
42. Bassotti G, Gaburri M, Imbimbo BP, Morelli A, Whitehead WE (1994) Distension-stimulated propagated contractions in human colon. Dig Dis Sci 39(9):1955–1960
43. Bharucha AE, Camilleri M, Zinsmeister AR, Hanson RB (1997) Adrenergic modulation of human colonic motor and sensory function. Am J Phys 273(5 Pt 1):G997–1006
44. Whitehead WE, Holtkotter B, Enck P et al (1990) Tolerance for rectosigmoid distention in irritable bowel syndrome. Gastroenterology 98(5 Pt 1):1187–1192
45. Scott SM (2003) Manometric techniques for the evaluation of colonic motor activity: current status. Neurogastroenterol Motil 15(5):483–513
46. Dinning PG, Smith TK, Scott SM (2009) Pathophysiology of colonic causes of chronic constipation. Neurogastroenterol Motil 21 Suppl 2(s2):20–30
47. Dinning PG (2018) A new understanding of the physiology and pathophysiology of colonic motility? Neurogastroenterol Motil 30(11):e13395

13 New Insights on Extrinsic Innervation of the Enteric Nervous System and Non-neuronal Cell Types That Influence Colon Function

Kimberly A. Meerschaert, Brian M. Davis, and Kristen M. Smith-Edwards

Abstract

The enteric nervous system not only innervates the colon to execute various functions in a semi-autonomous manner but also receives neural input from three extrinsic sources, (1) vagal, (2) thoracolumbar (splanchnic), and (3) lumbosacral (pelvic) pathways, that permit bidirectional communication between the colon and central nervous system. Extrinsic pathways signal sensory input via afferent fibers, as well as motor autonomic output via parasympathetic or sympathetic efferent fibers, but the shared and unique roles for each pathway in executing sensory-motor control of colon function have not been well understood. Here, we describe the recently developed approaches that have provided new insights into the diverse mechanisms utilized by extrinsic pathways to influence colon functions related to visceral sensation, motility, and inflammation. Based on the cumulative results from anatomical, molecular, and functional studies, we propose pathway-specific functions for vagal, thoracolumbar, and lumbosacral innervation of the colon.

Keywords

Visceral pain · Motility · Autonomic nervous system · Spinal cord · Vagus nerve · Splanchnic nerve · Pelvic nerve

13.1 Introduction

The colon is responsible for the formation, storage, and expulsion of fecal contents, while also maintaining a homeostatic environment with respect to the microbiome, epithelial barrier, and immune status. To execute these integrated functions, the colon is innervated by intrinsic neurons, enteric nervous system (ENS) that initiate reflexes and generate activity patterns that produce propulsive smooth muscle contractions and

K. A. Meerschaert
Department of Immunology, Harvard Medical School, Boston, MA, USA

B. M. Davis
Department of Neurobiology, University of Pittsburgh, Pittsburgh, PA, USA

K. M. Smith-Edwards (✉)
Department of Physiology and Biomedical Engineering, Mayo Clinic, Rochester, MN, USA
e-mail: edwards.kristen@mayo.edu

N. J. Spencer et al. (eds.), *The Enteric Nervous System II*, Advances in Experimental Medicine and Biology 1383, https://doi.org/10.1007/978-3-031-05843-1_13

relaxation. Mechanical and/or chemical stimuli at the luminal surface activate sensory terminals of intrinsic primary afferent neurons (IPANs), either directly or indirectly via enteroendocrine cells in the colon epithelium. IPANs then initiate activity in ascending ENS pathways that cause oral contraction, as well as descending ENS pathways that cause anal relaxation. This pattern of neuro-mechanical activity facilitates forward movement of fecal matter toward the anus [1]. Although the ENS can produce motor patterns autonomously, the colon also receives neural input from three distinct extrinsic sources: (1) vagal, (2) thoracolumbar (TL) spinal, and (3) lumbosacral (LS) spinal pathways. Each of these pathways contains sensory input via afferent fibers, as well as autonomic motor output via parasympathetic (vagal and LS, pelvic) or sympathetic (TL, splanchnic) efferent fibers, permitting bidirectional communication between the colon and brain. Adding to the complexity and diversity of mechanisms that execute colon functions, intrinsic and extrinsic neurons interact with non-neuronal cell populations, including interstitial cells of Cajal, enteric glia, immune cells, and epithelial cells, that have been shown to shape motility patterns, regulate neuroinflammation and visceral pain, and maintain a protective epithelial barrier [2–7]. Many labs have focused their attention on understanding extrinsic *afferent* signaling and sensory input to the central nervous system (CNS; see [8, 9] for reviews). It is also critical to determine how input from vagal, TL, and LS pathways is integrated within the central nervous system (CNS) to, in turn, send output commands that influence colon function. Much of what is known about extrinsic *efferent* nerve pathways to the colon and their cellular targets comes from studies that have either used in vivo or in vitro motility assays combined with pharmacology and/or nerve stimulation (see [10] for review). However, in the last decade, there has been an explosion of new tools (e.g., optogenetics, genetically encoded calcium indicators, single cell RNA sequencing) that allow for high-throughput comprehensive, yet also detailed, investigations of anatomical, molecular [11–15], and functional characteristics of the ENS and extrinsic pathways. For example, single cell RNA sequencing has been used to identify molecularly unique subpopulations of myenteric neurons [11–14], as well as extrinsic sympathetic [11] and sensory neurons [11, 15] that innervate the colon. Studies using opto- and chemogenetic stimulation or inhibition have begun to attribute certain motility and visceral pain behaviors to specific neuronal (and non-neuronal) subtypes [16–18]. Of particular importance for mapping functional connectivity, genetically encoded calcium indicators, such as GCaMP, have been used for direct recordings of network activity with single-cell and millisecond resolution in more intact ex vivo colon preparations [17, 19] and in vivo models [19, 20]. This mini-review describes how recently developed approaches have painted a more holistic picture of the diverse mechanisms utilized by extrinsic pathways to influence colon function, and pathway-specific functions are proposed for LS, TL, and vagal innervation.

13.2 Distinct ENS Organization and Function in Proximal and Distal Colon

To fully understand how extrinsic neural innervation influences colon function, it is important to first highlight recent advances in our knowledge of the differences between proximal and distal colon regions regarding structure and function. This has been a major topic of ENS neurobiology in the last few years, and a comprehensive discussion is unfortunately not within the scope of the present review (but for references, see [12, 21, 22, 23]). Briefly, it is generally agreed upon that the increased complexity of neural circuits in proximal regions corresponds to the more diverse motor patterns that this area of colon must generate in order to mix and process luminal contents, absorb water, and shape into fecal pellets. By contrast, the neural circuits in distal colon regions are relatively simpler, reflecting the peristaltic contractions, driven by neuro-mechanical feed-forward interactions [1], that this area produces in order to transport fecal contents to the anus for excretion. Therefore, to achieve more precise control, the mechanisms utilized by extrinsic neural pathways that modulate colon function may also exhibit region-specific characteristics (e.g., spatial distribution of innervation, cellular targets, neurotransmitters/receptors), and this concept will be discussed in further detail below as it relates to the three major extrinsic pathways that innervate the colon.

13.3 Lumbosacral (LS) Pathway

In the true spirit of the colon, we will start at the "bottom" with pathways that travel from LS spinal cord segments to the colon via the pelvic nerve and are traditionally associated with both afferent input and parasympathetic output (but see [24]). Teased fiber recordings in ex vivo colon preparations indicate an even distribution of LS afferents sensitive to different types of mechanical stimuli applied to the colon (e.g., stretch, probe, and mucosal brushing) [9, 25, 26]. Molecular studies have revealed that a major distinction between LS and other extrinsic colon afferents is the expression of Piezo-2 [27], a channel responsible for converting mechanical input to electrical activity, indicating an important role for LS afferents in monitoring the mechanical state of the colon. Afferents in the LS pathway respond to both noxious and innocuous distension, and activate second-order neurons in the dorsal horn of the spinal cord [19, 28, 29]. However, during conditions of inflammation in rodents, afferent input to the LS spinal cord remains unchanged [28] (also Meerschaert et al., *manuscript in preparation*), indicating that LS afferents are involved in mechanical sensation and perhaps acute visceral pain, but inflammatory pain is not likely mediated through this pathway.

Recent work performed in our lab using dual-color back-labeling from the colon showed that LS afferents almost exclusively project to distal regions (98% of colon-projecting LS afferents), whereas afferents of other pathways innervate both the proximal and distal colon to a similar extent (below) [27]. Therefore, despite also receiving sensory innervation from TL and vagal pathways, the distal colon is innervated by a spatially restricted group of afferents completely dedicated to monitoring the mechanical state of this particular colon region, regardless of the inflammatory state. Spinal segregation of LS-specific input coming from distal colon likely helps the CNS manage or compartmentalize the large amounts of sensory information it continuously processes, but more importantly, probably corresponds to distinct functions specifically carried out by the LS pathway.

In the late 1970s, in vivo electrophysiology was used in cats to show that the LS parasympathetic reflexes to the large intestine are mediated via a spinal pathway and have an essential role in the initiation of propulsive activity during defecation [30]. More recent studies have confirmed and extended our understanding of the neural circuits involved. In an ex vivo preparation that keeps the colon intact with the LS pathway, GCaMP calcium imaging was used to monitor the activity in myenteric neurons and colon motility produced by stimulating afferent and efferent fibers of the LS pathway (i.e., L6 DRG/dorsal root and ventral root) [19]. Activating preganglionic parasympathetic neurons via ventral root stimulation increased calcium signals in myenteric neurons and evoked contractions, regardless of whether spinal cord circuitry was intact. By contrast, capsaicin and electrical stimulation applied to LS afferents caused a similar effect on myenteric neurons and motility only when the spinal cord was present in continuity. In vivo, optogenetic stimulation of TRPV1+ afferents by application of blue light to the colon lumen was associated with increased activation of autonomic nuclei in LS spinal cord segments [19]. In follow-up experiments, electrical stimulation of the pelvic nerve activated myenteric neurons in the distal colon and also produced a delayed, time-locked activation of myenteric neurons in the proximal colon that consistently initiated a colonic motor complex (Meerschaert and Smith-Edwards, unpublished). The organization of sensory input, response patterns to mechanical stimulation in LS afferents, combined with the engagement of ENS circuits that produce contractions, are all consistent with the primary function of the LS pathway as mediating defecation or expulsion of formed fecal contents by monitoring the mechanical status of the distal colon.

13.4 Thoracolumbar (TL) Pathway

The neural circuits of the TL pathway to the colon, associated with afferent input and sympathetic output, are more complex in anatomy, cellular targets, and functional effects on colon motility patterns. As a population, TL colon afferents do not respond well to stretch or mucosal brushing; rather, the majority are activated by blunt probing of their receptive fields, 50% of which are located within or adjacent to the mesenteries that hold the colon in place within the abdominal cavity [9, 25, 26]. This has led to the hypothesis that TL afferents detect twisting or torsion of the colon and are involved in producing sharp, transient pain associated with spasm or distension. During conditions of inflammation induced by mustard oil [28] as well as a 5-day treatment of dextran sulfate sodium (DSS; Meerschaert et al., *manuscript in preparation*), input from colon afferents to TL spinal cord neurons was significantly increased. Neurogenic inflammation is mediated via sensory neurons that release neuropeptides from their peripheral terminals, and a high proportion of TL afferents have a peptidergic molecular profile, with expression of calcitonin gene-related polypeptide (Calca), preprotachykinin-1 (Tac1), and growth factor receptor 3 (Gfra3) [27]. Therefore, one would predict that TL afferents have a large role in mediating and transducing signals of inflammation and inflammatory pain.

From the colon, TL afferents course through sympathetic prevertebral ganglia on their way to TL spinal cord segments, and likely provide direct sensory input to sympathetic postganglionic neurons (SPNs) via axon collaterals along the way. Sensory neurons within the colon (intestinofugal afferent neurons, IFANs) also have axons that project and form functional synapses onto SPNs [31–34]. Therefore, direct sensory-to-sympathetic communication occurs outside of the spinal cord (i.e., CNS) on a regular basis, and based on evidence described in the above paragraph, it appears that TL spinal cord neurons become more involved during inflammatory or noxious conditions. There are three different levels of sympathetic prevertebral ganglia that contain colon-innervating SPNs [17]. The celiac ganglion is the largest and most rostral of the SPG, and houses SPNs that project to proximal regions of the colon [17] as well as many other visceral organs. By contrast, the inferior mesenteric ganglion (IMG) and pelvic hypogastric plexus (PHP) each have much fewer SPNs in total, but nearly 60% innervate the colon, half going to proximal regions and the other half to distal regions [17]. Therefore, whereas the celiac ganglion mediates sympathetic output to numerous visceral organs, the IMG and PHP are more restricted to colon, and likely the bladder and reproductive organs.

Unlike sympathetic innervation of the vasculature, the effects of sympathetic input on colon motility occur via the enteric nervous system (ENS) rather than directly on smooth muscle cells. Early studies using distal colon muscle recordings, sympathetic nerve stimulation, and neuronal blockers reported that sympathetic input has an inhibitory effect on ascending ENS reflex responses to mucosal stimulation [35]; sympathetic stimulation also affected the resting inhibitory tone of colon smooth muscles [35] and altered muscle activity in a context-dependent manner by acting on the ENS [36]. More recent work has provided evidence that SPNs also target non-neuronal cells that help to shape motility patterns (e.g., ICC, glia), regulate neuroinflammation (e.g., glia, immune cells), mediate responses to infection (e.g., immune cells), and make up the ever-important epithelial barrier (e.g., epithelial cells) [17, 37–39]. Interestingly, sympathetic input has been shown to have region-specific effects on activity in myenteric neurons, ICC and epithelial cells, causing the inhibition of propulsive contractions in distal colon, and facilitating motor patterns for mixing and processing fecal contents in proximal colon [17]. Overall, these findings are consistent with a primary function of the TL pathway as moderating neuro- and myogenic motility, as well as coordinating the timing and propulsion of movement from proximal to more distal colon regions, which must adapt to inflammatory conditions in order to reestablish homeostasis. Therefore, whereas the LS pathway mediates motility based on mechanical input specifically from the distal colon, the TL pathway mediates motility based on mechanical and inflammatory input from the entire length of colon, and likely other regions of the GI tract.

13.5 Vagal Pathway

Several labs have used optogenetic, electrophysiological, and tracing approaches to map functional connections between vagal neurons and visceral organs, but few reports have included vagal innervation of the colon. The extent to which the vagus nerve innervates the colon has been debated throughout history [40–44], with more recent studies finding that contrary to previous thought, extrinsic neurons of the vagus nerve innervate and form functional synapses throughout the entire length of colon, even distal regions [45–49]. In mice, back-labeling from proximal and distal colon suggest a relatively similar extent of vagal innervation in the two regions [27]. Interestingly, vagal colon afferents were molecularly distinct from spinal colon afferents, mainly due to higher mRNA expression of the purinergic receptors, P2X2 and P2X3, and low expression of Calca and Tac1, suggesting that vagal afferents innervating the colon are non-peptidergic and sensitive to metabolites, such as ATP [27]. Although data are currently limited, these findings are consistent with the vagal pathway as being a major source of sensory input from both proximal and distal colon.

Vagovagal reflexes refer to the vagal afferent, brainstem, vagal efferent neuron circuits that coordinate the functions of visceral organs. Appetite, satiety, gastric volume, and secretion of pancreatic enzymes are all functions thought to be mediated by vagovagal reflexes [50, 51]. Information from the vagus nerve converges within the nucleus tractus solitarius (NTS) of the brainstem, which projects to other brain structures that are involved in setting autonomic control, metabolism, satiety, and affect (e.g., limbic system, hypothalamus). Therefore, vagal afferents appear to be responsible for providing the brain with the information required for major "top-down" decision-making regarding intestinal digestion and motility, either via vagal efferents or descending pathways to the spinal cord. It is estimated that only 20% of neural fibers in the vagus nerve are parasympathetic efferents, and these are distributed across several other visceral organs besides the colon. Therefore, future research should focus efforts to determine whether the vagal pathway mediates colon function via vagovagal reflexes, or if vagal input from the colon is distributed across multiple brain regions, and effector functions occur via descending spinal cord pathways.

13.6 Summary

Based on recent studies using new approaches to investigate extrinsic sensory and autonomic innervation of the colon, the following pathway-specific functions are proposed:

- The lumbosacral pathway to the colon mediates motility reflexes for defecation or expulsion of fecal pellets by specifically monitoring the mechanical state of distal colon regions.
- The thoracolumbar pathway mediates neuro- and myogenic motility to coordinate the timing of fecal movement from proximal to distal regions; it does this by integrating input regarding the mechanical and inflammatory state of the entire length of colon.
- The vagal pathway is a major source of sensory input from both proximal and distal colon, providing the brain information required for "top-down" control of intestinal

digestion and motility. Colon-specific effector functions need further investigation.

References

1. Spencer NJ, Dinning PG, Brookes SJ et al (2016) Insights into the mechanisms underlying colonic motor patterns. J Physiol 594:4099–116. PMC4967752
2. Drumm BT, Rembetski BE, Baker SA et al (2019) Tonic inhibition of murine proximal colon is due to nitrergic suppression of Ca(2+) signaling in interstitial cells of Cajal. Sci Rep 9:4402. PMC6416298
3. McClain J, Grubisic V, Fried D et al (2014) Ca2+ responses in enteric glia are mediated by connexin-43 hemichannels and modulate colonic transit in mice. Gastroenterology 146:497–507 e1. PMC3935238
4. McClain JL, Fried DE, Gulbransen BD (2015) Agonist-evoked Ca(2+) signaling in enteric glia drives neural programs that regulate intestinal motility in mice. Cell Mol Gastroenterol Hepatol 1:631–645. PMC4673674
5. Schneider S, Wright CM, Heuckeroth RO (2019) Unexpected roles for the second brain: enteric nervous system as master regulator of bowel function. Annu Rev Physiol 81:235–259
6. Ward SM, Beckett EA, Wang X et al (2000) Interstitial cells of Cajal mediate cholinergic neurotransmission from enteric motor neurons. J Neurosci 20:1393–403. PMC6772355
7. Yoo BB, Mazmanian SK (2017) The enteric network: interactions between the immune and nervous systems of the gut. Immunity 46:910–926. PMC5551410
8. Brookes SJ, Spencer NJ, Costa M et al (2013) Extrinsic primary afferent signalling in the gut. Nat Rev Gastroenterol Hepatol 10:286–296
9. Brierley SM, Hibberd TJ, Spencer NJ (2018) Spinal afferent innervation of the colon and rectum. Front Cell Neurosci 12:467. PMC6288476
10. Browning KN, Travagli RA (2014) Central nervous system control of gastrointestinal motility and secretion and modulation of gastrointestinal functions. Compr Physiol 4:1339–68. PMC4858318
11. Zeisel A, Hochgerner H, Lonnerberg P et al (2018) Molecular architecture of the mouse nervous system. Cell 174:999–1014 e22. PMC6086934
12. Drokhlyansky E, Smillie CS, Van Wittenberghe N et al (2020) The human and mouse enteric nervous system at single-cell resolution. Cell 182:1606–1622 e23
13. Wright CM, Schneider S, Smith-Edwards KM et al (2021) scRNA-Seq reveals new enteric nervous system roles for GDNF, NRTN, and TBX3. Cell Mol Gastroenterol Hepatol 11:1548
14. May-Zhang AA, Tycksen E, Southard-Smith AN et al (2021) Combinatorial transcriptional profiling of mouse and human enteric neurons identifies shared and disparate subtypes in situ. Gastroenterology 160:755–770 e26. PMC7878294
15. Hockley JRF, Taylor TS, Callejo G et al (2019) Single-cell RNAseq reveals seven classes of colonic sensory neuron. Gut 68:633–644. PMC6580772
16. Najjar SA, Davis BM, Albers KM (2020) Epithelial-neuronal communication in the colon: implications for visceral pain. Trends Neurosci 43:170–181. PMC7047572
17. Smith-Edwards KM, Edwards BS, Wright CM et al (2021) Sympathetic input to multiple cell types in mouse and human colon produces region-specific responses. Gastroenterology 160:1208–1223 e4. PMC7956113
18. Hibberd TJ, Feng J, Luo J et al (2018) Optogenetic induction of colonic motility in mice. Gastroenterology 155:514–528 e6. PMC6715392
19. Smith-Edwards KM, Najjar SA, Edwards BS et al (2019) Extrinsic primary afferent neurons link visceral pain to colon motility through a spinal reflex in mice. Gastroenterology 157:522–536 e2. PMC6995031
20. Rakhilin N, Barth B, Choi J et al (2016) Simultaneous optical and electrical in vivo analysis of the enteric nervous system. Nat Commun 7:11800. PMC4899629
21. Li Z, Hao MM, Van den Haute C et al (2019) Regional complexity in enteric neuron wiring reflects diversity of motility patterns in the mouse large intestine. elife 8:e42914. PMC6391068
22. Costa M, Keightley LJ, Hibberd TJ et al (2020) Characterization of alternating neurogenic motor patterns in mouse colon. Neurogastroenterol Motil 33:e14047
23. Nestor-Kalinoski A, Smith-Edwards KM, Meerschaert K et al (2022) Unique neural circuit connectivity of mouse proximal middle and distal colon defines regional colonic motor patterns. Cell Mol Gastroenterol Hepatol 13(1):309–337. e3. S2352345X21001831. https://doi.org/10.1016/j.jcmgh.2021.08.016
24. Espinosa-Medina I, Saha O, Boismoreau F et al (2016) The sacral autonomic outflow is sympathetic. Science 354:893–897. PMC6326350
25. Lynn PA, Blackshaw LA (1999) In vitro recordings of afferent fibres with receptive fields in the serosa, muscle and mucosa of rat colon. J Physiol 518:271–82. PMC2269405
26. Brierley SM, Jones RC 3rd, Gebhart GF et al (2004) Splanchnic and pelvic mechanosensory afferents signal different qualities of colonic stimuli in mice. Gastroenterology 127:166–178
27. Meerschaert KA, Adelman PC, Friedman RL et al (2020) Unique molecular characteristics of visceral afferents arising from different levels of the Neuraxis: location of afferent Somata predicts function and stimulus detection modalities. J Neurosci 40:7216–7228. PMC7534907
28. Traub RJ, Murphy A (2002) Colonic inflammation induces fos expression in the thoracolumbar spinal cord increasing activity in the spinoparabrachial pathway. Pain 95:93–102

29. Harrington AM, Caraballo SG, Maddern JE et al (2019) Colonic afferent input and dorsal horn neuron activation differs between the thoracolumbar and lumbosacral spinal cord. Am J Physiol Gastrointest Liver Physiol 317:G285–G303
30. De Groat WC, Krier J (1978) The sacral parasympathetic reflex pathway regulating colonic motility and defaecation in the cat. J Physiol 276:481–500. PMC1282439
31. Miller SM, Szurszewski JH (1997) Colonic mechanosensory afferent input to neurons in the mouse superior mesenteric ganglion. Am J Phys 272:G357–G366
32. Miller SM, Szurszewski JH (2002) Relationship between colonic motility and cholinergic mechanosensory afferent synaptic input to mouse superior mesenteric ganglion. Neurogastroenterol Motil 14:339–348
33. Miller SM, Szurszewski JH (2003) Circumferential, not longitudinal, colonic stretch increases synaptic input to mouse prevertebral ganglion neurons. Am J Physiol Gastrointest Liver Physiol 285:G1129–G1138
34. Sharkey KA, Lomax AE, Bertrand PP et al (1998) Electrophysiology, shape, and chemistry of neurons that project from guinea pig colon to inferior mesenteric ganglia. Gastroenterology 115:909–918
35. Spencer N, McCarron SL, Smith TK (1999) Sympathetic inhibition of ascending and descending interneurones during the peristaltic reflex in the isolated guinea-pig distal colon. J Physiol 519(Pt 2):539–50. PMC2269523
36. Spencer NJ, Bywater RA, Klemm MF (1998) Effects of sympathetic nerve stimulation on membrane potential in the circular muscle layer of mouse distal colon. Neurogastroenterol Motil 10:543–552
37. Gulbransen BD, Bains JS, Sharkey KA (2010) Enteric glia are targets of the sympathetic innervation of the myenteric plexus in the guinea pig distal colon. J Neurosci 30:6801–9. PMC6632550
38. Bellono NW, Bayrer JR, Leitch DB et al (2017) Enterochromaffin cells are gut Chemosensors that couple to sensory neural pathways. Cell 170:185–198 e16. PMC5839326
39. Matheis F, Muller PA, Graves CL et al (2020) Adrenergic signaling in muscularis macrophages limits infection-induced neuronal loss. Cell 180:64–78 e16. PMC7271821
40. Bayliss WM, Starling EH (1900) The movements and the innervation of the large intestine. J Physiol 26:107–18. PMC1540523
41. Dennis C, Eddy FD, Frykman HM et al (1948) The response to Vagotomy in idiopathic ulcerative colitis and regional enteritis. Ann Surg 128:479–93. PMC1514076
42. Gray GW, Hendershot LC, Whitrock RM et al (1955) Influence of the parasympathetic nerves and their relation to the action of atropine in the ileum and colon of the dog. Am J Phys 181:679–687
43. Hulten L, Jodal M, Lundgren O (1969) The effects of graded electrical stimulation or reflex activation of the sympathetic and the parasympathetic nerve supply on the regional blood flow in cat colon. Bibl Anat 10:312–315
44. Lannon J, Weller E (1947) The parasympathetic supply of the distal colon. Br J Surg 34:373–378
45. Altschuler SM, Escardo J, Lynn RB et al (1993) The central organization of the vagus nerve innervating the colon of the rat. Gastroenterology 104:502–509
46. Berthoud HR, Carlson NR, Powley TL (1991) Topography of efferent vagal innervation of the rat gastrointestinal tract. Am J Phys 260:R200–R207
47. Dapoigny M, Cowles VE, Zhu YR et al (1992) Vagal influence on colonic motor activity in conscious nonhuman primates. Am J Phys 262:G231–G236
48. Larauche M, Wang Y, Wang PM et al (2020) The effect of colonic tissue electrical stimulation and celiac branch of the abdominal vagus nerve neuromodulation on colonic motility in anesthetized pigs. Neurogastroenterol Motil 32:e13925. PMC7606494
49. Tong WD, Ridolfi TJ, Kosinski L et al (2010) Effects of autonomic nerve stimulation on colorectal motility in rats. Neurogastroenterol Motil 22:688–93. PMC2952396
50. Furness JB, Callaghan BP, Rivera LR et al (2014) The enteric nervous system and gastrointestinal innervation: integrated local and central control. Adv Exp Med Biol 817:39–71
51. Travagli RA, Hermann GE, Browning KN et al (2003) Musings on the wanderer: what's new in our understanding of vago-vagal reflexes? III. Activity-dependent plasticity in vago-vagal reflexes controlling the stomach. Am J Physiol Gastrointest Liver Physiol 284:G180–7. PMC3055655

The Emerging Role of the Gut–Brain–Microbiota Axis in Neurodevelopmental Disorders

14

S. Hosie, T. Abo-Shaban, C. Y. Q. Lee, S. M. Matta, A. Shindler, R. Gore, S. S. Sharna, M. Herath, P. J. Crack, A. E. Franks, and E. L. Hill-Yardin

Abstract

Autism spectrum disorder (ASD; autism) is a prevalent neurodevelopmental disorder associated with changes in gut-brain axis communication. Gastrointestinal (GI) symptoms are experienced by a large proportion of individuals diagnosed with autism. Several mutations associated with autism modify cellular communication via neuronal synapses. It has been suggested that modifications to the enteric nervous system, an intrinsic nervous system of the GI tract, could contribute to GI dysfunction. Changes in gut motility, permeability, and the mucosal barrier as well as shifts in the large population of microbes inhabiting the GI tract could contribute to GI symptoms. Preclinical research has demonstrated that mice expressing the well-studied R451C missense mutation in *Nlgn3* gene, which encodes cell adhesion protein neuroligin-3 at neuronal synapses, exhibit GI dysfunction. Specifically, $NL3^{R451C}$ mice show altered colonic motility and faster small intestinal transit. As well as dysmotility, macrophages located within the gut-associated lymphoid tissue of the $NL3^{R451C}$ mouse caecum show altered morphology, suggesting that neuro-inflammation pathways are modified in this model. Interestingly, $NL3^{R451C}$ mice maintained in a shared environment demonstrate fecal microbial dysbiosis indicating a role for the nervous system in regulating gut microbial populations. To better understand host-microbe interactions, further clarification and comparison of clinical and animal model profiles of dysbiosis should be obtained, which in turn will provide better insights into the efforts taken to design personalized microbial therapies. In addition to changes in neurophysiological measures, the mucosal component of the GI barrier may contribute to GI dysfunction more broadly in individuals diagnosed with a wide range of neurological disorders. As the study of GI dysfunction

S. Hosie · T. Abo-Shaban · C. Y. Q. Lee · S. M. Matta
R. Gore · S. S. Sharna
School of Health & Biomedical Sciences, STEM College, RMIT University, Bundoora, VIC, Australia

A. Shindler · A. E. Franks
Department of Physiology, Anatomy and Microbiology, La Trobe University, Bundoora, VIC, Australia

M. Herath
Department of Anatomy & Physiology, The University of Melbourne, Parkville, VIC, Australia

P. J. Crack
Department of Biochemistry and Pharmacology, The University of Melbourne, Parkville, VIC, Australia

E. L. Hill-Yardin (✉)
School of Health & Biomedical Sciences, STEM College, RMIT University, Bundoora, VIC, Australia

Department of Anatomy & Physiology, The University of Melbourne, Parkville, VIC, Australia
e-mail: Elisa.hill@rmit.edu.au

N. J. Spencer et al. (eds.), *The Enteric Nervous System II*, Advances in Experimental Medicine and Biology 1383, https://doi.org/10.1007/978-3-031-05843-1_14

advances to encompass multiple components of the gut-brain-microbiota axis, findings will help understand future directions such as microbiome engineering and optimisation of the mucosal barrier for health.

Keywords

Autism · Enteric nervous system · Synapse · Mouse models · Gut–brain axis · Microbes

14.1 Introduction

Autism spectrum disorder (ASD; autism) is a highly prevalent neurodevelopmental disorder diagnosed based on the presence of impaired social communication skills alongside repetitive and/or restricted behavioural patterns. Recent reports reveal that approximately 1 in 54 school-aged children in the United States are diagnosed with ASD [1]. A significant proportion of people on the autism spectrum experience major challenges interacting in the community as well as reduced quality of life due to profound behavioural impairments. Surprisingly, gastrointestinal symptoms are commonly reported by individuals diagnosed with neurodevelopmental disorders such as autism. Since these gastrointestinal symptoms potentially impact quality of life, effective treatment options for these issues may provide a broader benefit for people with autism. It is now well established that genetics are a major contributor to behavioural profiles included in the ASD diagnostic criteria. In addition to a large number of gene mutations affecting neuronal communication in the brain, emerging research shows that several autism-associated mutations also impact the enteric nervous system (ENS). In preclinical models, these changes are evident in both the myenteric and submucosal neuronal plexuses of the ENS, which are arranged in parallel adjacent to the muscular layers of the gut wall, suggesting that both gastrointestinal motility and gut permeability as well as fluid and enzyme secretion may be impacted in individuals with autism. A role for the mucus environment, which lines the gastrointestinal tract and forms part of the mucosal barrier to pathogen invasion, is also being investigated in light of gastrointestinal disorders in autism. In the last decade, a renewed interest in understanding how the gastrointestinal tract communicates with the brain has revealed novel avenues for therapeutic design and optimising health. Multiple components of the gut–brain axis pathway appear to be dysregulated in autism, including inflammation pathways and neuronal signalling in the brain, intestinal inflammation and an increased prevalence of gastrointestinal disorders as well as alterations in the composition of the microbiome [2]. Although microbes undoubtedly interact with the host physiology [3], whether microbial populations in the gastrointestinal tract can be modulated to improve gut function in addition to mood and behaviour in individuals with autism remains unclear. In this chapter, we review current understandings around how genetic mutations resulting in impaired synaptic function alter the gut–brain-microbiota axis to influence gastrointestinal function in autism.

14.2 Gastrointestinal Symptoms in Autism

Gastrointestinal dysfunction is commonly experienced by individuals diagnosed with neurodevelopmental disorders such as autism [4–8]. For example, individuals with autism are approximately four times more likely to be hospitalised for gastrointestinal problems [9]. Symptoms are wide-ranging and include alternating diarrhoea and constipation, bloating, abdominal pain, reflux and postprandial vomiting. In addition to an increased occurrence of gastrointestinal symptoms, individuals with ASD exhibit a higher prevalence for inflammatory bowel disease compared to neurotypical controls [10], suggesting that subtle changes in the functioning of the nervous system also impact immune pathways.

Many components within the multitude of pathways connecting microbes, the gastrointestinal tract and the brain may be dysregulated in individuals with autism. Several preclinical studies report disruptions of the ENS, including altered proportions of neuronal populations and dysmotility [11, 12], as well as gastrointestinal symptoms in ASD clinical populations [11, 13, 14]. In addition, these changes likely influence the distribution and structure of microbial populations located in the gastrointestinal tract. Given that gut microbes play an important role in influencing mental state and behaviour [15–19], a better understanding of host-microbe relationships is needed for tackling gastrointestinal symptoms and behavioural issues in neurodevelopmental disorders. The gut–brain–microbiota axis comprises both the enteric and central nervous systems (including vagal and spinal nerves), the mucosal barrier including the mucus biofilm, and microbial populations located in the lumen of the gastrointestinal tract. Currently, an active area of investigation is assessing for changes in the mucus biofilm in preclinical models of neurodevelopmental and other disorders since this plays an important role in supporting the colonisation of commensal microbes and components of the immune system.

14.3 Genetic Contributions to Autism

The aetiology of ASD is heterogeneous in nature, with a strong genetic influence. Currently, over 1000 ASD-implicated mutations that are inherited or of de novo origin are registered in the Simons Foundation Autism Research Initiative (SFARI) gene database (https://gene.sfari.org/). These mutations predominantly occur in genes involved in development, chromatin remodelling and neuronal communication. Many gene mutations implicated in ASD impact proteins located at the synapse and therefore affect neuronal communication, cognition and brain function [20–29]. It has been proposed that alterations in the central nervous system similarly impact the gastrointestinal tract [11, 30, 31] and studies supporting this theory continue to emerge. It is therefore important to understand the complex interactions between the ENS and bacteria in health and disease states.

14.4 Microbial Dysbiosis

Several studies report an altered profile (dysbiosis) of gut microbial populations in individuals diagnosed with ASD [32–34]. These changes may contribute to and exacerbate gastrointestinal symptoms in the context of autism. For example, microbes can modify the absorption and digestion of dietary nutrients [35], levels of inflammation in the periphery and potentially in brain tissue, and influence anxiety-like behaviour in rodents [36]. Before we can shift microbial populations with the aim to improve gut health, however, it is critical to characterise the biological pathways involved and identify targets that will enable us to modulate physiological outcomes. Modifying microbial populations may enhance resilience or resistance to infection (for example, in response to Sars-CoV-2 infection), optimise gut health and overall quality of life, including mental health indicators.

14.5 Preclinical Studies

Studies in animal models enable factors such as diet, housing and even host genetics to be maintained, so that the impact of autism-associated gene mutations can be carefully examined. In addition to these advantages, complex behavioural disorders such as autism can be modelled (including as endophenotypes, or subsets of behaviours) in rodents [37]. Behavioural testing batteries for these preclinical models of autism are well established [38] and multiple transgenic mouse models expressing gene mutations identified in individuals with ASD demonstrate autism-relevant phenotypes. These tools hold significant value for understanding the role of the ENS in neurodevelopmental disorders and the use of preclinical models hold the potential to clarify biological mechanisms relevant to clinical cohorts.

14.6 Gastrointestinal Dysfunction in Patients and Mice Expressing the Autism-Associated R451C Mutation in Neuroligin-3

The first gene mutations identified in individuals with autism were shown to impact the neuroligin family of postsynaptic adhesion proteins [24]. Specifically, a truncation mutation in the *Nlgn4* gene encoding neuroligin-4 was identified in two children with ASD. In the same study, a missense mutation (R451C) in *Nlgn3*, which causes an arginine to be replaced by a cysteine at position 451 of the neuroligin 3 protein, was identified in a second family, also with two children as the probands. Cellular studies indicated that as a result of the missense mutation, the majority of the neuroligin-3 protein is retained intracellularly within the Golgi complex with only 10% of typical levels reaching the cell membrane (reviewed in [39]). Clinical reports of the two brothers expressing the R451C mutation described gastrointestinal dysfunction, including diarrhoea and faecal incontinence, post-meal regurgitation, oesophageal inflammation, chronic intestinal pain, and significantly delayed bladder and bowel control [11]. In addition to these symptoms, one brother has epilepsy and one has problematic aggressive behaviour [24].

Mice expressing the R451C mutation are arguably the best studied preclinical model of autism. This work has clearly demonstrated that $Nlgn3^{R451C}$ mice show autism-relevant behavioural traits and have altered cellular structure and function in the brain [22, 26, 28, 40–43]. We have demonstrated that $Nlgn3^{R451C}$ mice bred on a mixed genetic background have faster small intestinal transit and increased sensitivity to $GABA_A$ receptor modulators in the colon (Fig. 14.1; [11]). Further investigation also revealed altered neuronal numbers in the myenteric plexus of the small intestine, but not the colon in these mice. Significantly, the jejunum exhibits an increase in the proportion of nitric oxide synthase-labelled neurons per ganglion (Fig. 14.2; [11]). These neurons release nitric oxide, the major inhibitory neurotransmitter of the gastrointestinal tract. Furthermore, Sharna et al. [12] observed that the myenteric and submucosal plexuses of the caecum in mutant mice contain more total neurons and a higher proportion of nitric oxide-producing neurons per ganglia compared to wild type littermates [12]. Although our work using an ex vivo assay approach showed that $Nlgn3^{R451C}$ mice have gastrointestinal dysmotility, it is also important to examine if these changes are present in vivo (i.e. using techniques such as X-ray imaging of motility [44]). Additional research suggests that ablation of the *Nlgn*3 gene in mice also impacts gastrointestinal function. For example, neuroligin-3 knockout mice have an increased colonic diameter as measured in an ex vivo assay as well as a subtle increase in the velocity of colonic migrating complexes [45]. Overall, recent work has indicated that the *Nlgn3* gene plays a role in gastrointestinal structure and function in mice and may therefore also be relevant to gut dysfunction in individuals expressing this mutation. More broadly, changes in the function of proteins at neuronal synapses may contribute to gastrointestinal dysfunction in autism populations.

In addition to the R451C mutation impacting the neuroligin-3 protein, other well-studied mutations associated with autism in clinical studies include *CHD8* (chromodomain helicase DNA binding protein 8) mutations and SHANK3 (SH3 and multiple ankyrin repeat domains 3, also known as proline-rich synapse-associated protein 2; ProSAP2). Individuals diagnosed with autism and expressing mutations in the CHD8 gene, which is involved in chromatin remodelling, commonly report symptoms of gastrointestinal dysfunction. In a study of 15 individuals, each carrying a single nucleotide variant in the CHD8 gene, 80% experienced recurrent and consistent problems with constipation [46]. Interestingly, further investigation into the role of the CHD8 gene in development showed that suppression of the CHD8 gene in zebrafish caused a decrease in

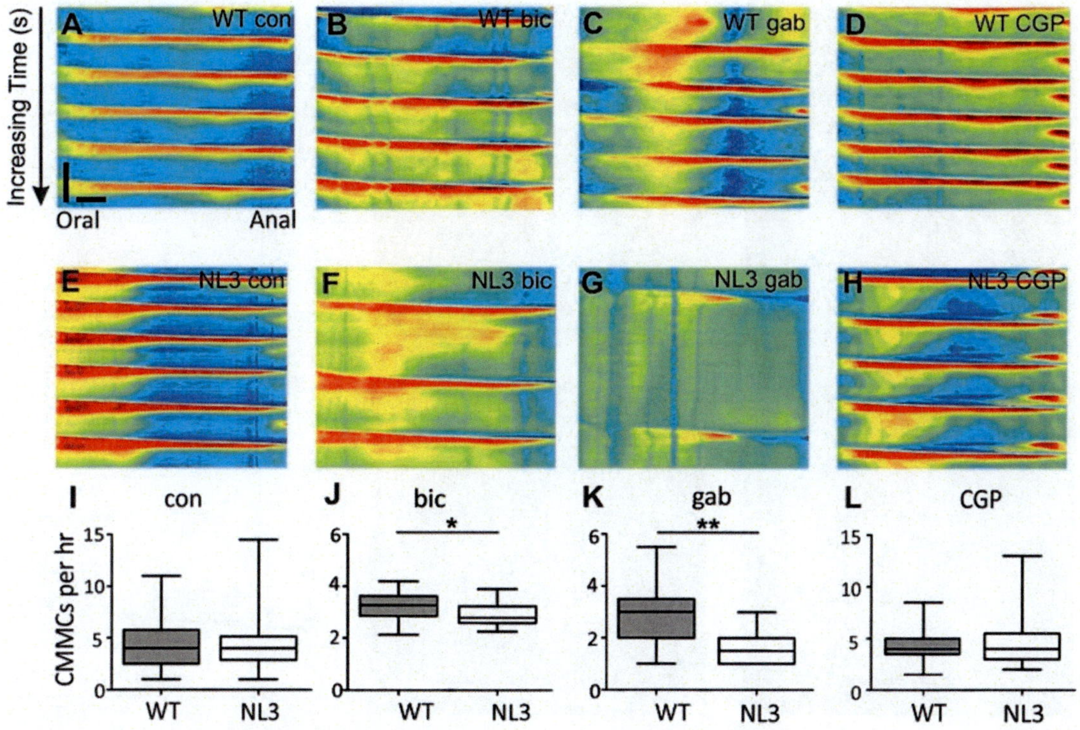

Fig. 14.1 Mice expressing the R451C mutation in neuroligin-3 show increased sensitivity in the colon to $GABA_A$ but not $GABA_B$ receptor antagonists. Representative spatiotemporal maps derived from video files of mouse colon motility: (**a–d**) WT mice; (**e–h**) $NL3^{R451C}$ mutant mice. Group data indicating (**i**) no change in the number of CMMCs under baseline conditions; reduced numbers of contractions in $NL3^{R451C}$ mice compared to WT in response to (**j**) 10 uM bicuculline, (**k**) 10 uM gabazine, but not (**l**) the $GABA_B$ antagonist CGP-35348. (CMMCs: colonic migrating motor complexes, also known as Colonic Motor Complexes; CMCs;). (Reproduced with permission from [11]))

gastrointestinal motility [46]. In a separate study of symptoms of individuals with Phelan-McDermid Syndrome which causes intellectual delay and autism symptoms, gastrointestinal problems including constipation, diarrhoea, gastro-oesophageal reflux and dysphagia were identified [47]. Preclinical studies also indicate a role for SHANK3 mutations in gastrointestinal dysfunction, whereby zebrafish showed intestinal dysmotility and a reduction in the number of serotonin-positive enteroendocrine cells [48]. Phelan-McDermid Syndrome is caused by copy number variations occurring on chromosome 22q13.3 and commonly involves the SHANK3 gene [49] which, similar to genes in the neuroligin family, is important for synaptic transmission.

14.7 Neuroinflammation in Autism

Neuroinflammation (i.e. abnormal and chronic brain inflammation) as well as altered synapse function has been observed in brain tissue derived from individuals with autism (reviewed in Matta, Hill-Yardin and Crack [50]) and may result from prolonged microglial and astrocyte activation. Families of individuals with autism show a higher incidence of inflammatory disorders [51] and alterations in inflammation pathways are commonly reported alongside an autism diagnosis [50, 52]. Serum, plasma, peripheral blood mononuclear cells, and even amniotic fluid samples from individuals with autism reveal elevated levels of pro-inflammatory cytokines and chemo-

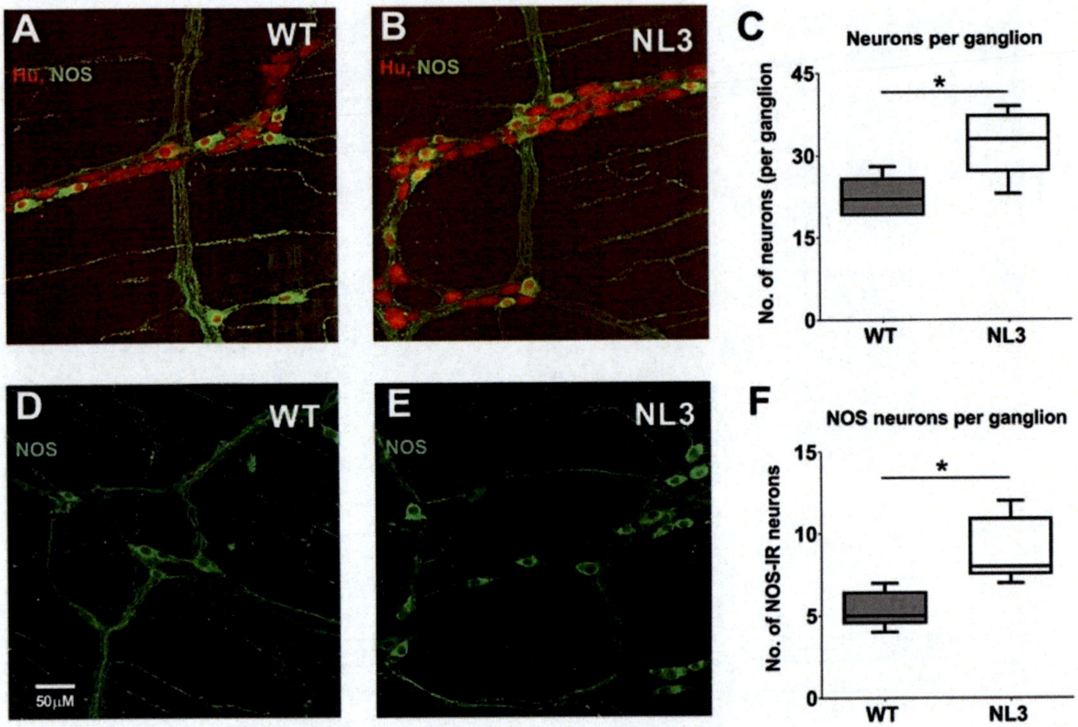

Fig. 14.2 The autism-associated R451C mutation in the *Nlgn-3* gene increases myenteric neuronal numbers in the mouse jejunum. Immunofluorescence labelling in (**a**) wild-type and (**b**) Neuroligin-3^{R451C} mutant mice show increased total myenteric neurons (**c**). Jejunal neurons labelled for NOS in wild-type (**d**) and neuroligin-3^{R451C} mutant mice (**e**) show increased numbers of NOS-immunoreactive neurons per myenteric ganglion in neuroligin-3^{R451C} mutant mice (**f**). Hu: pan-neuronal marker, NOS: Nitric Oxide Synthase, IR immunoreactive. (Reproduced with permission from Hosie et al. [11])

kines. Similarly, brain and cerebrospinal fluid from individuals diagnosed with ASD show dysregulation of inflammatory mediators [50]. These inflammatory mediators are released by immune cells, including microglia (brain-resident macrophages) and astrocytes, as well as B lymphocytes, T lymphocytes and mast cells. Immune cells undertake constant immunosurveillance of the tissue environment for foreign invaders and promote the recruitment or proliferation of other immune cells during an immune response. This activity can be beneficial in fighting infection, but harmful in the absence of a threat or injury. Since microglia and astrocytes play important roles in synapse maintenance and pruning under physiological conditions, neuroinflammation can alter synaptic properties in neural networks. Microglia and astrocytes proliferate and undergo phenotypic changes when in a pro-inflammatory state. In response to inflammatory stimuli, the ramified branches of microglial cells retract and thicken, whereas cell somata elongate to enable greater motility. Astrocytes, in contrast, undergo cell hypertrophy and extend their processes under these conditions. These phenotypic changes can interfere with their homeostatic roles in synapse maintenance, resulting in changes in the neuronal circuitry. Central and peripheral increases in inflammation in people with autism may therefore contribute to altered brain function and cognitive and behavioural changes.

Altered immune function has also been reported in a number of animal models expressing autism-relevant gene mutations (reviewed in [50]) and environmental risk factors [53, 54]. Neuroligin-3^{R451C} mice, for example, display subtle changes in astrocyte and microglial cell populations in the hippocampus [43]. Astrocytes are reduced in size and have fewer processes, whereas an increased density of microglia in the dentate gyrus is observed in $NL3^{R451C}$ mice [43]. Similarly, $NL3^{R451C}$ mice exhibit changes in the density and morphology of gastrointestinal macrophages in the caecum of the gastrointestinal tract [12]. Intestinal macrophages tolerate higher levels of inflammation compared to conventional macrophages as the gut is constantly in an inflammatory state. They play a complex role in immune defence due to the need to distinguish between foreign bacteria, viruses and microbes, and the commensal microbiota of the host. The function and localisation of intestinal macrophages is highly heterogeneous and different subpopulations (including monocyte-derived mature macrophages, monocyte-derived inflammatory macrophages and self-maintaining macrophages) play varying roles in gastrointestinal disorders [55]. Overall, dysregulation of immune homeostasis in the gut can disrupt gastrointestinal motility and secretion, as well as CNS function via the gut-brain-microbiota axis. Studies in germ-free mice show disrupted blood-brain permeability, altered synaptic plasticity and neurotransmitter expression, as well as immature microglia and impaired immune function [13, 56] which reinforce the critical role that microbial populations play in health. These studies demonstrate that the microbiome significantly impacts the nervous system and immune function in particular. These findings provide an impetus to assess the potential for targeting dysbiosis and gastrointestinal inflammation as a viable avenue to treat neuroimmune dysfunction occurring in the brain in autism.

14.7.1 Altered Caecal Neuroimmune Interactions in Mouse Models of Autism

Surprisingly, the neuroligin-3 R451C mutation causes a robust gross anatomical phenotype in the caecum of mice, further suggesting neuronal interactions with the immune system. In short, mice expressing this autism-associated mutation have reduced caecal weight. This phenotype was observed in mice bred on two different genetic strains and persisted when these mice were maintained in three different animal facilities [12] indicating a likely impact of the gene mutation itself on host physiology (and not genetic background or environmental effects such as diet and housing conditions) as the main cause of this phenotype. The gastrointestinal tract is covered in a layer of mucus which serves as lubrication, physical protection from luminal contents, and a nutrient-rich biofilm enabling microorganisms to thrive. Although as yet unconfirmed, the difference observed in caecal weight may be due to a reduction in the mucus content within the caecum. This phenomenon has been observed in a preclinical model of infection and is associated with accelerated secretion of mucin-filled vacuoles from the caecal epithelium in response to inflammatory signals [57, 58]. Within the apex of the caecum, mice typically have one or two lymphoid aggregates of gut-associated lymphoid tissue (GALT), known as the caecal patch. Interestingly, $NL3^{R451C}$ mice showed alterations in the morphology and density of ionised calcium binding adaptor molecule-1 (Iba-1) immunoreactive enteric macrophages (Fig. 14.3). Specifically, mice expressing the autism-associated mutation in the Nlgn3 gene had a higher density of Iba-1 positive cells, which were smaller in volume and more spherical in shape compared to those seen in the caecal patch of controls [12]. Further investigation of caecal function is therefore warranted. Since the ENS is a significant regulator of gastrointestinal function, it is important to investigate how the changes in enteric neuronal populations affect caecal motility, secretion and mucosal barrier function. It is anticipated that functional changes in the caecum will affect the churning of microbial content and digesta, the composition of the microbial community and metabolome produced at various gastrointestinal regions as well as digestion and nutrient uptake.

14.8 The Gastrointestinal Mucus Environment and Implications for Neurodevelopmental Disorders

Changes in the nervous system associated with neurodevelopmental disorders or neurological disease may also impact mucus production. The

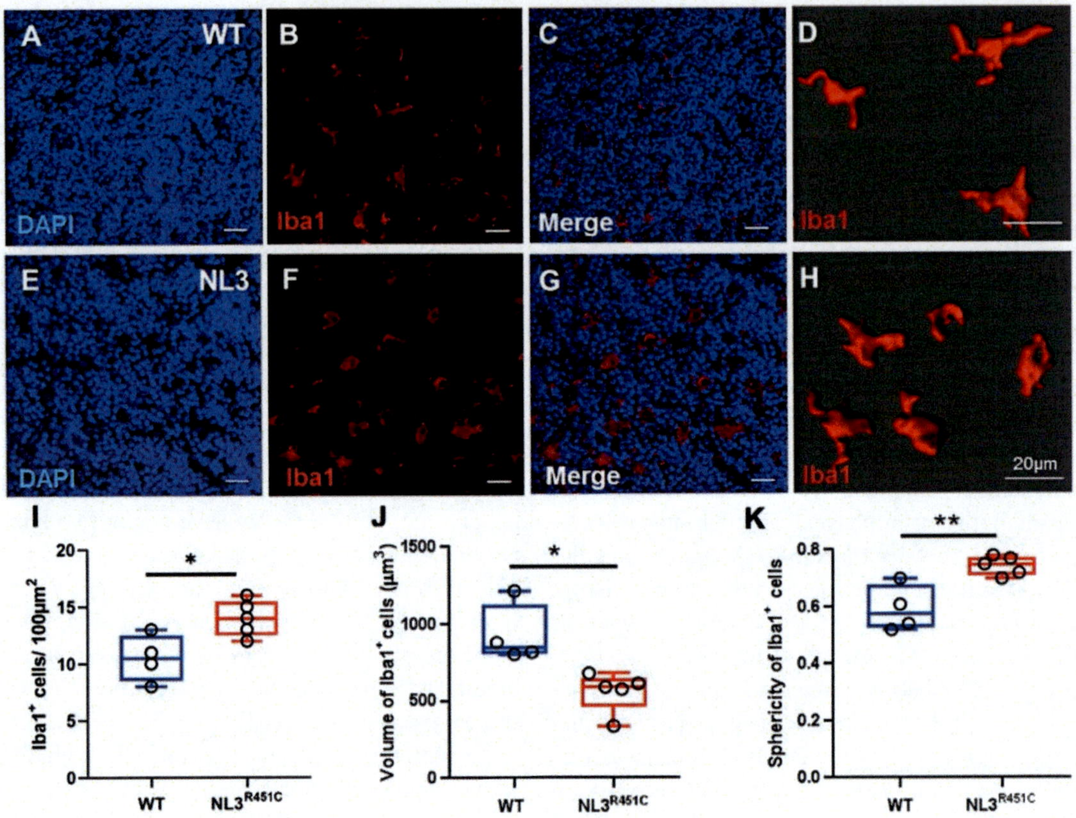

Fig. 14.3 Altered gut-associated lymphoid tissue in the caecal patch in the neuroligin-3^{R451C} mouse model of autism. All cells (labelled with DAPI; blue) and gastrointestinal macrophages (labelled with Iba-1; red) (**a–d**) in a representative WT sample; (**e–h**) representative images of NL3 mutant tissue. (**i–k**) NL3 mutant mice show increased density, decreased volume and a more spherical morphology of Iba-1 positive macrophages. (DAPI: 4′,6-diamidino-2-phenylindole nuclear counterstain; Iba-1: ionised calcium binding adaptor molecule 1; $*p < 0.05$, $**p < 0.01$. (Image reproduced with permission from Sharna et al. [12])

mucus biofilm lining the wall of the gastrointestinal tract provides an energy source and a structural niche for commensal bacteria. Modifications to this biofilm could therefore contribute to dysbiosis and aberrant inflammatory responses to exacerbate gut dysfunction in ASD. More broadly, patients with neurological disorders, including Parkinson's disease, Alzheimer's disease and multiple sclerosis, frequently report gastrointestinal symptoms in advance of diagnostic symptoms, but whether modification of the mucus may also impact disease progression for these individuals is unknown [59–69].

We propose that changes in mucosal barrier function are highly likely to occur in individuals with ASD due to altered activity of the ENS (Fig. 14.4; [70]). Firstly, it is expected that altered ENS activity can modify stem cell proliferation (changing the physical packing density of mucosal epithelial cells) and the density of mucus-producing goblet cells. Secondly, since neuronal activity also broadly drives mucus production from goblet cells, subtle changes in enteric neuronal activity could also modify the mucus biofilm that supports gut microbial communities. Thirdly, modified gastrointestinal motility has been demonstrated in the presence of autism-associated gene mutations that alter the ENS (e.g [11, 45, 71].). Because gut motility plays a role in flushing mucus along the gastrointestinal tract,

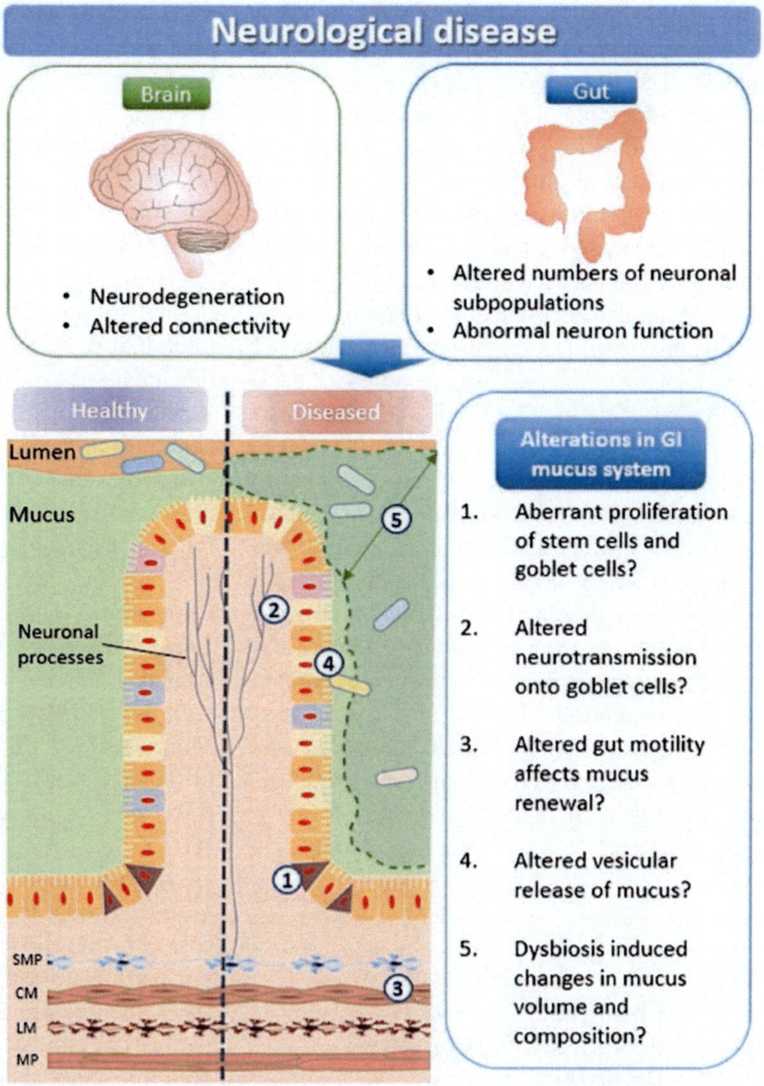

Fig. 14.4 Potential mechanisms underlying changes in the gastrointestinal mucus system in neurological disease. Neurological diseases that are established in the brain may be associated with gastrointestinal issues due to a shift in enteric neuron subpopulations and/or function. These changes could also impact the mucus system that lines the gastrointestinal tract and result in dysbiosis and aberrant gut function. SMP submucosal plexus; CM circular muscle; MP myenteric plexus; LM, longitudinal muscle. (Reproduced from Herath et al. [70] with permission)

such changes could impact mucus renewal (i.e., via reduced motility that may result in an excess of mucus debris within the lumen, and excess motility potentially resulting in thinning of the mucus layer). Fourth, mucus production by goblet cells may be impaired due to changes in the ENS at the molecular level, since much of the biological machinery involved in the vesicular release of mucus overlaps with that utilised for classical neurotransmission. Fifth, alterations in the ENS such as those emerging in the context of preclinical models of autism are also anticipated to modify mucus volume and composition to cause dysbiosis and exacerbate disease-mediated changes.

14.9 Region-Specific Motility Patterns

The ENS regulates spontaneous motility patterns in the gastrointestinal tract and current techniques enable changes in the enteric neural circuitry to be assessed in a quantitative manner. Ex vivo video imaging methods enable the measurement of multiple parameters such as gut diameter, contractile length as well as the velocity, frequency and magnitude of contractions in preclinical models. This approach has been predominantly used in guinea pigs [72–75] and mice [76–79], and has more recently been applied to transgenic mouse models of autism [11, 45, 71]. Since a large component of these studies have provided detailed characterisations of colonic motility, current work is focused on enhancing experimental protocols and edge detection software reported by Swaminathan et al. [80] to provide the research community with an open-source software package for investigating motility in other regions of the gastrointestinal tract. These approaches are useful for identifying subtle changes in motility profiles due to altered enteric neural circuitry associated with neurodevelopmental disorders. This is important because distinct motility patterns are generated in different anatomical regions of the gastrointestinal tract and in combination, these regional functions enable optimal nutrient absorption, digestion and expulsion of luminal content. Alterations in these processes commonly correspond with microbial dysbiosis and dysregulation of inflammation, relevant to symptoms commonly reported by people with ASD.

14.10 Examining Microbial Changes in Neurodevelopmental Disorders

Gastrointestinal microbial dysbiosis is often correlated with the presence of neurodevelopmental disorders, including ASD [81]. For example, an increased abundance of pathogenic bacteria such as *Clostridium*, *Lactobacillus*, *Corynebacterium* and *Sutterella* have been demonstrated in microbial samples obtained from children with autism [34, 81–84]. Alterations to the Firmicutes: Bacteroidetes ratio (two phyla that combined contribute to 90% of human gastrointestinal microbial bacterial populations; [85]) have also been observed in patients with ASD, with bacterial species from the Firmicutes phylum being more prevalent [81, 86–88]. Interestingly, faecal transplant of microbial samples from people with ASD into germ-free mice induced autism-relevant behaviours in the offspring of the mice [19], further suggesting that host microbial interactions are important in behavioural profiles.

Several studies in preclinical rodent models of autism have examined for gastrointestinal microbial dysbiosis. For instance, mice lacking the autism-associated SHANK3 gene as well as Black and Tan BRachyury (BTBR) mice (which show ASD-relevant behaviour) exhibit microbial gastrointestinal variations compared to wild-type controls [90–92]. Multiple studies report microbial dysbiosis in SHANK3 mouse models [92–94] and reversal of dysfunctional social behavioural traits following administration of *Lactobacillus reuteri* has even been demonstrated in SHANK3 knockout and BTBR mice as well as mice treated with the antiepileptic drug, Valproate [93]. In 2017, Golubeva and others observed an increased abundance in *Akkermansia*, *Bilophila*, and *Desulfovibrio* in BTBR mice compared to controls [91]. Clarification of the similarity and differences in gastrointestinal dysbiosis observed in clinical cases of ASD and preclinical models is emerging but further information in this area is needed [95]. Such a comparison of microbial dysbiosis patterns in clinical cases of autism and relevant preclinical models should reveal important information about underlying host-microbial communication and could identify shared patterns of dysbiosis across species.

The expression and role of ASD-implicated genes in the gastrointestinal tract are not well understood, an issue that stems primarily from a lack of suitable experimental tools. Advances in techniques to label nucleic acids (mainly mRNA) or proteins of interest are therefore needed to shed light on expression patterns and potential effects on the underlying ENS circuitry. These approaches will assist in determining the role of these genes in regulating neurally mediated gastrointestinal physiology in neurodevelopmental disorders. Given that altered gastrointestinal permeability and secretion can impact mucus composition and subsequently contribute to intestinal dysbiosis, a next step is to begin to examine microbial spatial distribution patterns to highlight how host-microbial interactions are altered in neurodevelopmental disorders. Although sequencing data show changes in microbial abundance and diversity, labelling of microbes using fluorescence in situ hybridisation can provide precise maps of the 'microbial biogeography' within the gastrointestinal tract [96, 97]. Implementation of approaches to quantify microbial distribution profiles in preclinical models is essential to determine host-microbial interactions in neurodevelopmental disorders.

14.11 Conclusion

Studying animal models of neurodevelopmental disorders provides important insights into how bidirectional communication via the gut-brain-microbiota axis influences gastrointestinal tract function, immune function and behaviour. Evidence suggests that changes in the structure and function of the ENS as well as altered interactions with microbial populations in the gut lumen occur in mice expressing gene mutations associated with ASD. Accordingly, it is crucial to further develop preclinical knowledge around these interactions with the aim of translation to clinical practice and providing evidence-based therapeutic options for gastrointestinal issues in autism. To achieve this goal, a multidisciplinary

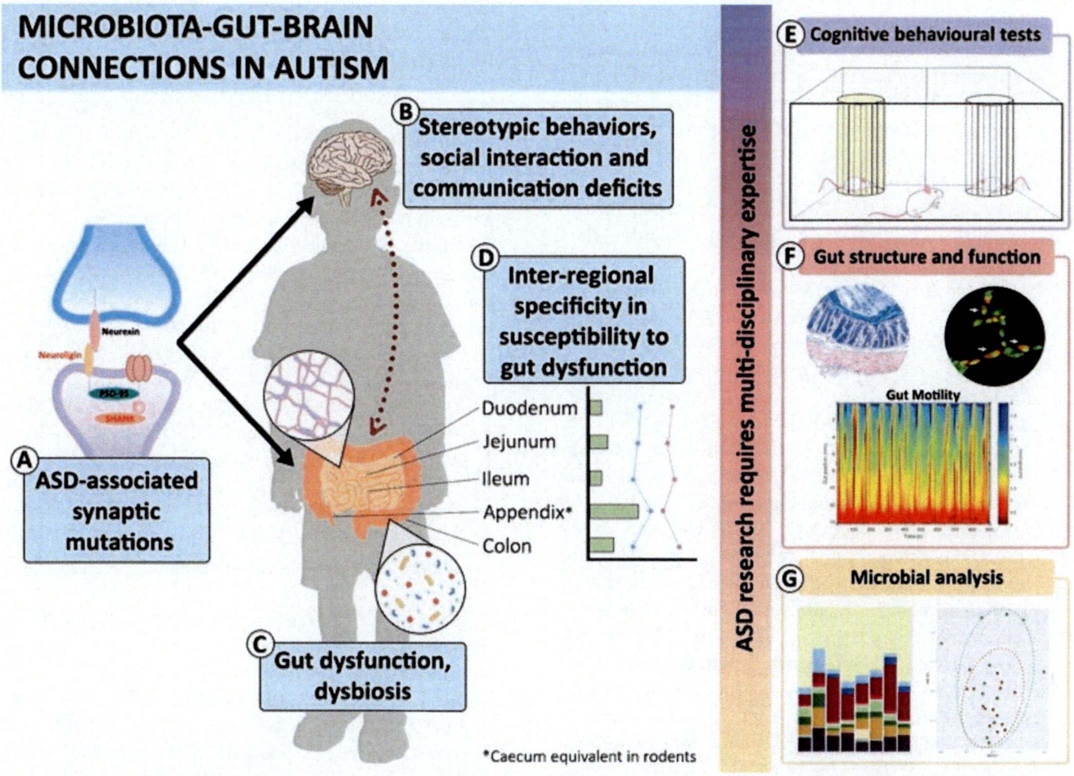

Fig. 14.5 Autism-associated mutations located at neuronal synapses contribute to alterations in the microbiota–gut–brain axis. (**a–c**) Gene mutations identified in autism can impact brain and gut neuronal circuitry and contribute to gut dysfunction and potentially, dysbiosis. (**d**) Different anatomical regions of the gastrointestinal tract contribute distinct functions and may show varying susceptibility to dysfunction in the context of neurodevelopmental disorders. (**e–g**) Given the need for expertise in genetics, microbiology, physiology, neuroscience and behaviour in addition to core clinical expertise, gaining an in-depth understanding of how microbiota–gut–brain interactions are altered in ASD will require multidisciplinary research approaches. (Figure included with permission, [89])

research approach across the fields of genetics, microbiology, physiology, neuroscience and behaviour with core clinical understanding is needed (Fig. 14.5). Based on mounting evidence that the microbiome plays an important role in host physiology, the use of novel approaches to shift microbial populations and optimise the gastrointestinal environment for individuals diagnosed with ASD needs further investigation.

References

1. Maenner MJ, Shaw KA, Baio J, Washington A, Patrick M, DiRienzo M, Christensen DL, Wiggins LD, Pettygrove S, Andrews JG, Lopez M, Hudson A, Baroud T, Schwenk Y, White T, Rosenberg CR, Lee L-C, Harrington RA, Huston M, Hewitt A, Esler A, Hall-Lande J, Poynter JN, Hallas-Muchow L, Constantino JN, Fitzgerald RT, Zahorodny W, Shenouda J, Daniels JL, Warren Z, Vehorn A, Salinas A, Durkin MS, Dietz PM (2020) Prevalence of Autism Spectrum Disorder among children aged 8 years — autism and developmental disabilities monitoring network, 11 sites, United States, 2016. MMWR Surveill Summ 69(4):1–12
2. Vuong HE, Hsiao EY (2017) Emerging roles for the gut microbiome in Autism Spectrum Disorder. Biol Psychiatry 81(5):411–423
3. Buffington SA, Dooling SW, Sgritta M, Noecker C, Murillo OD, Felice DF, Turnbaugh PJ, Costa-Mattioli M (2021) Dissecting the contribution of host genetics and the microbiome in complex behaviors. Cell 184:1740
4. Gorrindo P, Williams KC, Lee EB, Walker LS, McGrew SG, Levitt P (2012) Gastrointestinal dysfunction in autism: parental report, clinical evaluation, and associated factors. Autism Res 5(2):101–108
5. Parracho HM, Bingham MO, Gibson GR, McCartney AL (2005) Differences between the gut microflora of children with autistic spectrum disorders and that of healthy children. J Med Microbiol 54(10):987–991
6. Restrepo B, Angkustsiri K, Taylor SL, Rogers SJ, Cabral J, Heath B, Hechtman A, Solomon M, Ashwood P, Amaral DG, Nordahl CW (2020) Developmental–behavioral profiles in children with Autism Spectrum Disorder and co-occurring gastrointestinal symptoms. Autism Res 13(10):1778–1789
7. Thulasi V, Steer RA, Monteiro IM, Ming X (2019) Overall severities of gastrointestinal symptoms in pediatric outpatients with and without Autism Spectrum Disorder. Autism 23(2):524–530
8. Valicenti-McDermott M, McVicar K, Rapin I, Wershil BK, Cohen H, Shinnar S (2006) Frequency of gastrointestinal symptoms in children with autistic spectrum disorders and association with family history of autoimmune disease. J Dev Behav Pediatr 27(2):S128–S136
9. McElhanon BO, McCracken C, Karpen S, Sharp WG (2014) Gastrointestinal symptoms in Autism Spectrum Disorder: a meta-analysis. Pediatrics 133(5):872–883
10. Lee M, Krishnamurthy J, Susi A, Sullivan C, Gorman GH, Hisle-Gorman E, Erdie-Lalena CR, Nylund CM (2018) Association of Autism Spectrum Disorders and inflammatory bowel disease. J Autism Dev Disord 48(5):1523–1529
11. Hosie S, Ellis M, Swaminathan M, Ramalhosa F, Seger GO, Balasuriya GK, Gillberg C, Råstam M, Churilov L, McKeown SJ, Yalcinkaya N, Urvil P, Savidge T, Bell CA, Bodin O, Wood J, Franks AE, Bornstein JC, Hill-Yardin EL (2019) Gastrointestinal dysfunction in patients and mice expressing the autism-associated R451C mutation in neuroligin-3. Autism Res 12(7):1043–1056
12. Sharna SS, Balasuriya GK, Hosie S, Nithianantharajah J, Franks AE, Hill-Yardin EL (2020) Altered caecal neuroimmune interactions in the Neuroligin-3R451C mouse model of Autism. Front Cell Neurosci 14:85
13. Cryan JF, O'Riordan KJ, Cowan CSM, Sandhu KV, Bastiaanssen TFS, Boehme M, Codagnone MG, Cussotto S, Fulling C, Golubeva AV, Guzzetta KE, Jaggar M, Long-Smith CM, Lyte JM, Martin JA, Molinero-Perez A, Moloney G, Morelli E, Morillas E, O'Connor R, Cruz-Pereira JS, Peterson VL, Rea K, Ritz NL, Sherwin E, Spichak S, Teichman EM, Wouw M, Ventura-Silva AP, Wallace-Fitzsimons SE, Hyland N, Clarke G, Dinan TG (2019) The microbiota-gut-brain axis. Physiol Rev 99(4):1877–2013
14. Rao M, Gershon MD (2016) The bowel and beyond: the enteric nervous system in neurological disorders. Nat Rev Gastroenterol Hepatol 13(9):517–528
15. Cryan JF, Dinan TG (2012) Mind-altering microorganisms: the impact of the gut microbiota on brain and behaviour. Nat Rev Neurosci 13(10):701–712
16. Hsiao EY, McBride SW, Hsien S, Sharon G, Hyde ER, McCue T, Codelli JA, Chow J, Reisman SE, Petrosino JF, Patterson PH, Mazmanian SK (2013) Microbiota modulate behavioral and physiological abnormalities associated with neurodevelopmental disorders. Cell 155(7):1451–1463
17. Mayer EA (2011) Gut feelings: the emerging biology of gut–brain communication. Nat Rev Neurosci 12(8):453–466
18. Mayer EA, Tillisch K, Gupta A (2015) Gut/brain axis and the microbiota. J Clin Invest 125(3):926–938
19. Sharon G, Cruz NJ, Kang D-W, Gandal MJ, Wang B, Kim Y-M, Zink EM, Casey CP, Taylor BC, Lane CJ, Bramer LM, Isern NG, Hoyt DW, Noecker C, Sweredoski MJ, Moradian A, Borenstein E, Jansson JK, Knight R, Metz TO, Lois C, Geschwind DH, Krajmalnik-Brown R, Mazmanian SK (2019) Human gut microbiota from Autism Spectrum Disorder promote behavioral symptoms in mice. Cell 177(6):1600–1618.e1617

20. Alarcón M, Abrahams BS, Stone JL, Duvall JA, Perederiy JV, Bomar JM, Sebat J, Wigler M, Martin CL, Ledbetter DH, Nelson SF, Cantor RM, Geschwind DH (2008) Linkage, association, and gene-expression analyses identify CNTNAP2 as an autism-susceptibility gene. Am J Hum Genet 82(1):150–159
21. Bourgeron T (2009) A synaptic trek to autism. Curr Opin Neurobiol 19(2):231–234
22. Burrows EL, Laskaris L, Koyama L, Churilov L, Bornstein JC, Hill-Yardin EL, Hannan AJ (2015) A neuroligin-3 mutation implicated in autism causes abnormal aggression and increases repetitive behavior in mice. Mol Autism 6(1):62
23. Delorme R, Ey E, Toro R, Leboyer M, Gillberg C, Bourgeron T (2013) Progress toward treatments for synaptic defects in autism. Nat Med 19(6):685–694
24. Jamain S, Quach H, Betancur C, Råstam M, Colineaux C, Gillberg IC, Soderstrom H, Giros B, Leboyer M, Gillberg C, Bourgeron T, Gillberg C, Råstam M, Gillberg C, Nydén A, Söderström H, Leboyer M, Betancur C, Philippe A, Giros B, Colineaux C, Cohen D, Chabane N, Mouren-Siméoni M-C, Brice A, Sponheim E, Spurkland I, Skjeldal OH, Coleman M, Pearl PL, Cohen IL, Tsiouris J, Zappella M, Menchetti G, Pompella A, Aschauer H, Van Maldergem L, Paris Autism Research International Sibpair, S (2003) Mutations of the X-linked genes encoding neuroligins NLGN3 and NLGN4 are associated with autism. Nat Genet 34(1):27–29
25. Peça J, Feliciano C, Ting JT, Wang W, Wells MF, Venkatraman TN, Lascola CD, Fu Z, Feng G (2011) Shank3 mutant mice display autistic-like behaviours and striatal dysfunction. Nature 472(7344):437–442
26. Rothwell PE, Fuccillo MV, Maxeiner S, Hayton SJ, Gokce O, Lim BK, Fowler SC, Malenka RC, Südhof TC (2014) Autism-associated Neuroligin-3 mutations commonly impair striatal circuits to boost repetitive behaviors. Cell 158(1):198–212
27. Schmeisser MJ, Ey E, Wegener S, Bockmann J, Stempel AV, Kuebler A, Janssen A-L, Udvardi PT, Shiban E, Spilker C, Balschun D, Skryabin BV, Dieck S, Smalla K-H, Montag D, Leblond CS, Faure P, Torquet N, Le Sourd A-M, Toro R, Grabrucker AM, Shoichet SA, Schmitz D, Kreutz MR, Bourgeron T, Gundelfinger ED, Boeckers TM (2012) Autistic-like behaviours and hyperactivity in mice lacking ProSAP1/Shank2. Nature 486(7402):256–260
28. Tabuchi K, Blundell J, Etherton MR, Hammer RE, Liu X, Powell CM, Südhof TC (2007) A Neuroligin-3 mutation implicated in autism increases inhibitory synaptic transmission in mice. Science 318(5847):71–76
29. Vicidomini C, Ponzoni L, Lim D, Schmeisser MJ, Reim D, Morello N, Orellana D, Tozzi A, Durante V, Scalmani P, Mantegazza M, Genazzani AA, Giustetto M, Sala M, Calabresi P, Boeckers TM, Sala C, Verpelli C (2017) Pharmacological enhancement of mGlu5 receptors rescues behavioral deficits in SHANK3 knock-out mice. Mol Psychiatry 22(5):689–702
30. Gershon MD, Ratcliffe EM (2004) Developmental biology of the enteric nervous system: pathogenesis of Hirschsprung's disease and other congenital dysmotilities. Semin Pediatr Surg 13(4):224–235
31. Niesler B, Rappold GA (2021) Emerging evidence for gene mutations driving both brain and gut dysfunction in Autism Spectrum Disorder. Mol Psychiatry 26(5):1442–1444
32. Gondalia SV, Palombo EA, Knowles SR, Cox SB, Meyer D, Austin DW (2012) Molecular characterisation of gastrointestinal microbiota of children with autism (with and without gastrointestinal dysfunction) and their neurotypical siblings. Autism Res 5(6):419–427
33. Kang D-W, Park JG, Ilhan ZE, Wallstrom G, LaBaer J, Adams JB, Krajmalnik-Brown R (2013) Reduced incidence of Prevotella and other fermenters in intestinal microflora of autistic children. PLoS One 8(7):e68322
34. Luna RA, Oezguen N, Balderas M, Venkatachalam A, Runge JK, Versalovic J, Veenstra-VanderWeele J, Anderson GM, Savidge T, Williams KC (2017) Distinct microbiome-neuroimmune signatures correlate with functional abdominal pain in children with Autism Spectrum Disorder. Cell Mol Gastroenterol Hepatol 3(2):218–230
35. Chen RY, Mostafa I, Hibberd MC, Das S, Mahfuz M, Naila NN, Islam MM, Huq S, Alam MA, Zaman MU, Raman AS, Webber D, Zhou C, Sundaresan V, Ahsan K, Meier MF, Barratt MJ, Ahmed T, Gordon JI (2021) A microbiota-directed food intervention for undernourished children. N Engl J Med 384(16):1517–1528
36. Luczynski P, McVey Neufeld K-A, Oriach CS, Clarke G, Dinan TG, Cryan JF (2016) Growing up in a bubble: using germ-free animals to assess the influence of the gut microbiota on brain and behavior. Int J Neuropsychopharmacol 19(8):pyw020
37. Argyropoulos A, Gilby K, Hill-Yardin E (2013) Studying autism in rodent models: reconciling endophenotypes with comorbidities. Front Hum Neurosci 7:417
38. Silverman JL, Yang M, Lord C, Crawley JN (2010) Behavioural phenotyping assays for mouse models of autism. Nat Rev Neurosci 11(7):490–502
39. Südhof TC (2008) Neuroligins and neurexins link synaptic function to cognitive disease. Nature 455(7215):903–911
40. Etherton M, Földy C, Sharma M, Tabuchi K, Liu X, Shamloo M, Malenka RC, Südhof TC (2011) Autism-linked neuroligin-3 R451C mutation differentially alters hippocampal and cortical synaptic function. Proc Natl Acad Sci 108(33):13764–13769
41. Földy C, Malenka RC, Südhof TC (2013) Autism-associated neuroligin-3 mutations commonly disrupt tonic endocannabinoid signaling. Neuron 78(3):498–509
42. Hosie S, Malone DT, Liu S, Glass M, Adlard PA, Hannan AJ, Hill-Yardin EL (2018) Altered amygdala excitation and CB1 receptor modulation of aggressive behavior in the Neuroligin-3R451C mouse model of Autism. Front Cell Neurosci 12:234
43. Matta SM, Moore Z, Walker FR, Hill-Yardin EL, Crack PJ (2020) An altered glial phenotype in

the NL3R451C mouse model of autism. Sci Rep 10(1):14492
44. McQuade RM, Stojanovska V, Donald E, Abalo R, Bornstein JC, Nurgali K (2016) Gastrointestinal dysfunction and enteric neurotoxicity following treatment with anticancer chemotherapeutic agent 5-fluorouracil. Neurogastroenterol Motil 28(12):1861–1875
45. Leembruggen AJL, Balasuriya GK, Zhang J, Schokman S, Swiderski K, Bornstein JC, Nithianantharajah J, Hill-Yardin EL (2019) Colonic dilation and altered ex vivo gastrointestinal motility in the neuroligin-3 knockout mouse. Autism Res 13:691
46. Bernier R, Golzio C, Xiong B, Stessman HA, Coe BP, Penn O, Witherspoon K, Gerdts J, Baker C, Silfhout V-v, Anneke T, Schuurs-Hoeijmakers JH, Fichera M, Bosco P, Buono S, Alberti A, Failla P, Peeters H, Steyaert J, Vissers LELM, Francescatto L, Mefford HC, Rosenfeld JA, Bakken T, O'Roak BJ, Pawlus M, Moon R, Shendure J, Amaral DG, Lein E, Rankin J, Romano C, de Vries BBA, Katsanis N, Eichler EE (2014) Disruptive CHD8 mutations define a subtype of autism early in development. Cell 158(2):263–276
47. De Rubeis S, Siper PM, Durkin A, Weissman J, Muratet F, Halpern D, Trelles MP, Frank Y, Lozano R, Wang AT, Holder JL, Betancur C, Buxbaum JD, Kolevzon A (2018) Delineation of the genetic and clinical spectrum of Phelan-McDermid syndrome caused by SHANK3 point mutations. Mol Autism 9(1):31
48. James DM, Kozol RA, Kajiwara Y, Wahl AL, Storrs EC, Buxbaum JD, Klein M, Moshiree B, Dallman JE (2019) Intestinal dysmotility in a zebrafish (Danio rerio) shank3a;shank3b mutant model of autism. Mol Autism 10(1):3
49. Tabet A-C, Rolland T, Ducloy M, Lévy J, Buratti J, Mathieu A, Haye D, Perrin L, Dupont C, Passemard S, Capri Y, Verloes A, Drunat S, Keren B, Mignot C, Marey I, Jacquette A, Whalen S, Pipiras E, Benzacken B, Chantot-Bastaraud S, Afenjar A, Héron D, Le Caignec C, Beneteau C, Pichon O, Isidor B, David A, El Khattabi L, Kemeny S, Gouas L, Vago P, Mosca-Boidron A-L, Faivre L, Missirian C, Philip N, Sanlaville D, Edery P, Satre V, Coutton C, Devillard F, Dieterich K, Vuillaume M-L, Rooryck C, Lacombe D, Pinson L, Gatinois V, Puechberty J, Chiesa J, Lespinasse J, Dubourg C, Quelin C, Fradin M, Journel H, Toutain A, Martin D, Benmansour A, Leblond CS, Toro R, Amsellem F, Delorme R, Bourgeron T (2017) A framework to identify contributing genes in patients with Phelan-McDermid syndrome. NPJ Genom Med 2(1):32
50. Matta SM, Hill-Yardin EL, Crack PJ (2019) The influence of neuroinflammation in Autism Spectrum Disorder. Brain Behav Immun 79:75–90
51. Atladóttir HÓ, Pedersen MG, Thorsen P, Mortensen PB, Deleuran B, Eaton WW, Parner ET (2009) Association of family history of autoimmune diseases and Autism Spectrum Disorders. Pediatrics 124(2):687–694
52. Siniscalco D, Schultz S, Brigida AL, Antonucci N (2018) Inflammation and neuro-immune dysregulations in Autism Spectrum Disorders. Pharmaceuticals (Basel, Switzerland) 11(2):56
53. Hsiao EY, McBride SW, Chow J, Mazmanian SK, Patterson PH (2012) Modeling an autism risk factor in mice leads to permanent immune dysregulation. Proc Natl Acad Sci 109(31):12776–12781
54. Lucchina L, Depino AM (2014) Altered peripheral and central inflammatory responses in a mouse model of autism. Autism Res 7(2):273–289
55. Yip JLK, Balasuriya GK, Spencer SJ, Hill-Yardin EL (2021) The role of intestinal macrophages in gastrointestinal homeostasis: heterogeneity and implications in disease. Cell Mol Gastroenterol Hepatol 12(5):1701-1718
56. Erny D, Hrabě de Angelis AL, Jaitin D, Wieghofer P, Staszewski O, David E, Keren-Shaul H, Mahlakoiv T, Jakobshagen K, Buch T, Schwierzeck V, Utermöhlen O, Chun E, Garrett WS, McCoy KD, Diefenbach A, Staeheli P, Stecher B, Amit I, Prinz M (2015) Host microbiota constantly control maturation and function of microglia in the CNS. Nat Neurosci 18(7):965–977
57. Bergstrom KSB, Kissoon-Singh V, Gibson DL, Ma C, Montero M, Sham HP, Ryz N, Huang T, Velcich A, Finlay BB, Chadee K, Vallance BA (2010) Muc2 protects against lethal infectious colitis by disassociating pathogenic and commensal bacteria from the Colonic Mucosa. PLoS Pathog 6(5):e1000902
58. Forstner G (1995) Signal transduction packaging and secretion of mucins. Annu Rev Physiol 57(1):585–605
59. Berer K, Gerdes LA, Cekanaviciute E, Jia X, Xiao L, Xia Z, Liu C, Klotz L, Stauffer U, Baranzini SE, Kümpfel T, Hohlfeld R, Krishnamoorthy G, Wekerle H (2017) Gut microbiota from multiple sclerosis patients enables spontaneous autoimmune encephalomyelitis in mice. Proc Natl Acad Sci 114(40):10719–10724
60. Cantarel BL, Waubant E, Chehoud C, Kuczynski J, DeSantis TZ, Warrington J, Venkatesan A, Fraser CM, Mowry EM (2015) Gut microbiota in Multiple Sclerosis. Possible Influence Immunomodulators 63(5):729–734
61. Cattaneo A, Cattane N, Galluzzi S, Provasi S, Lopizzo N, Festari C, Ferrari C, Guerra UP, Paghera B, Muscio C, Bianchetti A, Volta GD, Turla M, Cotelli MS, Gennuso M, Prelle A, Zanetti O, Lussignoli G, Mirabile D, Bellandi D, Gentile S, Belotti G, Villani D, Harach T, Bolmont T, Padovani A, Boccardi M, Frisoni GB (2017) Association of brain amyloidosis with pro-inflammatory gut bacterial taxa and peripheral inflammation markers in cognitively impaired elderly. Neurobiol Aging 49:60–68
62. Cekanaviciute E, Yoo BB, Runia TF, Debelius JW, Singh S, Nelson CA, Kanner R, Bencosme Y, Lee YK, Hauser SL, Crabtree-Hartman E, Sand IK, Gacias M, Zhu Y, Casaccia P, Cree BAC, Knight R, Mazmanian SK, Baranzini SE (2017) Gut bacteria from multiple sclerosis patients modulate human T cells and exacer-

bate symptoms in mouse models. Proc Natl Acad Sci 114(40):10713–10718

63. Heintz-Buschart A, Pandey U, Wicke T, Sixel-Döring F, Janzen A, Sittig-Wiegand E, Trenkwalder C, Oertel WH, Mollenhauer B, Wilmes P (2018) The nasal and gut microbiome in Parkinson's disease and idiopathic rapid eye movement sleep behavior disorder. Mov Disord 33(1):88–98
64. Jangi S, Gandhi R, Cox LM, Li N, von Glehn F, Yan R, Patel B, Mazzola MA, Liu S, Glanz BL, Cook S, Tankou S, Stuart F, Melo K, Nejad P, Smith K, Topçuolu BD, Holden J, Kivisäkk P, Chitnis T, De Jager PL, Quintana FJ, Gerber GK, Bry L, Weiner HL (2016) Alterations of the human gut microbiome in multiple sclerosis. Nat Commun 7(1):12015
65. Keshavarzian A, Green SJ, Engen PA, Voigt RM, Naqib A, Forsyth CB, Mutlu E, Shannon KM (2015) Colonic bacterial composition in Parkinson's disease. Mov Disord 30(10):1351–1360
66. Petrov VA, Saltykova IV, Zhukova IA, Alifirova VM, Zhukova NG, Dorofeeva YB, Tyakht AV, Kovarsky BA, Alekseev DG, Kostryukova ES, Mironova YS, Izhboldina OP, Nikitina MA, Perevozchikova TV, Fait EA, Babenko VV, Vakhitova MT, Govorun VM, Sazonov AE (2017) Analysis of gut microbiota in patients with Parkinson's disease. Bull Exp Biol Med 162(6):734–737
67. Scheperjans F, Aho V, Pereira PAB, Koskinen K, Paulin L, Pekkonen E, Haapaniemi E, Kaakkola S, Eerola-Rautio J, Pohja M, Kinnunen E, Murros K, Auvinen P (2015) Gut microbiota are related to Parkinson's disease and clinical phenotype. Mov Disord 30(3):350–358
68. Vogt NM, Kerby RL, Dill-McFarland KA, Harding SJ, Merluzzi AP, Johnson SC, Carlsson CM, Asthana S, Zetterberg H, Blennow K, Bendlin BB, Rey FE (2017) Gut microbiome alterations in Alzheimer's disease. Sci Rep 7(1):13537
69. Zhuang Z-Q, Shen L-L, Li W-W, Fu X, Zeng F, Gui L, Lü Y, Cai M, Zhu C, Tan Y-L, Zheng P, Li H-Y, Zhu J, Zhou H-D, Bu X-L, Wang Y-J (2018) Gut microbiota is altered in patients with Alzheimer's disease. J Alzheimers Dis 63:1337–1346
70. Herath M, Hosie S, Bornstein JC, Franks AE, Hill-Yardin EL (2020) The role of the gastrointestinal mucus system in intestinal homeostasis: implications for neurological disorders. Front Cell Infect Microbiol 10:248
71. Margolis KG, Li Z, Stevanovic K, Saurman V, Israelyan N, Anderson GM, Snyder I, Veenstra-VanderWeele J, Blakely RD, Gershon MD (2016) Serotonin transporter variant drives preventable gastrointestinal abnormalities in development and function. J Clin Invest 126(6):2221–2235
72. Costa M, Keightley LJ, Wiklendt L, Hibberd TJ, Arkwright JW, Omari T, Wattchow DA, Brookes SJH, Dinning PG, Spencer NJ (2019) Identification of multiple distinct neurogenic motor patterns that can occur simultaneously in the guinea pig distal colon. Am J Physiol Gastrointest Liver Physiol 316(1):G32–G44
73. Gwynne RM, Bornstein JC (2007) Local inhibitory reflexes excited by mucosal application of nutrient amino acids in guinea pig jejunum. Am J Physiol Gastrointest Liver Physiol 292(6):G1660–G1670
74. Gwynne RM, Thomas EA, Goh SM, Sjövall H, Bornstein JC (2004) Segmentation induced by intraluminal fatty acid in isolated guinea-pig duodenum and jejunum. J Physiol 556(2):557–569
75. Hennig GW, Costa M, Chen BN, Brookes SJH (1999) Quantitative analysis of peristalsis in the guinea-pig small intestine using spatio-temporal maps. J Physiol 517(2):575–590
76. Hung LY, Parathan P, Boonma P, Wu Q, Wang Y, Haag A, Luna RA, Bornstein JC, Savidge TC, Foong JPP (2020) Antibiotic exposure postweaning disrupts the neurochemistry and function of enteric neurons mediating colonic motor activity. Am J Physiol Gastrointest Liver Physiol 318(6):G1042–G1053
77. Neal KB, Parry LJ, Bornstein JC (2009) Strain-specific genetics, anatomy and function of enteric neural serotonergic pathways in inbred mice. J Physiol 587(3):567–586
78. Roberts RR, Ellis M, Gwynne RM, Bergner AJ, Lewis MD, Beckett EA, Bornstein JC, Young HM (2010) The first intestinal motility patterns in fetal mice are not mediated by neurons or interstitial cells of Cajal. J Physiol 588(7):1153–1169
79. Spencer NJ, Hibberd TJ, Travis L, Wiklendt L, Costa M, Hu H, Brookes SJ, Wattchow DA, Dinning PG, Keating DJ, Sorensen J (2018) Identification of a rhythmic firing pattern in the enteric nervous system that generates rhythmic electrical activity in smooth muscle. J Neurosci 38(24):5507–5522
80. Swaminathan M, Hill-Yardin E, Ellis M, Zygorodimos M, Johnston LA, Gwynne RM, Bornstein JC (2016) Video imaging and spatiotemporal maps to analyze gastrointestinal motility in mice. J Vis Exp 108:53828–53828
81. Strati F, Cavalieri D, Albanese D, De Felice C, Donati C, Hayek J, Jousson O, Leoncini S, Renzi D, Calabrò A, De Filippo C (2017) New evidences on the altered gut microbiota in Autism Spectrum Disorders. Microbiome 5(1):24
82. Bennet SMP, Böhn L, Störsrud S, Liljebo T, Collin L, Lindfors P, Törnblom H, Öhman L, Simrén M (2018) Multivariate modelling of faecal bacterial profiles of patients with IBS predicts responsiveness to a diet low in FODMAPs. Gut 67(5):872–881
83. Finegold SM, Summanen PH, Downes J, Corbett K, Komoriya T (2017) Detection of Clostridium perfringens toxin genes in the gut microbiota of autistic children. Anaerobe 45:133–137
84. Williams BL, Hornig M, Parekh T, Lipkin WI (2012) Application of novel PCR-based methods for detection, quantitation, and phylogenetic characterization of Sutterella species in intestinal biopsy samples from children with autism and gastrointestinal disturbances. MBio 3(1):e00261
85. The Human Microbiome Project Consortium, Huttenhower C, Gevers D, Knight R, Abubucker

S, Badger JH, Chinwalla AT, Creasy HH, Earl AM, FitzGerald MG, Fulton RS, Giglio MG, Hallsworth-Pepin K, Lobos EA, Madupu R, Magrini V, Martin JC, Mitreva M, Muzny DM, Sodergren EJ, Versalovic J, Wollam AM, Worley KC, Wortman JR, Young SK, Zeng Q, Aagaard KM, Abolude OO, Allen-Vercoe E, Alm EJ, Alvarado L, Andersen GL, Anderson S, Appelbaum E, Arachchi HM, Armitage G, Arze CA, Ayvaz T, Baker CC, Begg L, Belachew T, Bhonagiri V, Bihan M, Blaser MJ, Bloom T, Bonazzi V, Paul Brooks J, Buck GA, Buhay CJ, Busam DA, Campbell JL, Canon SR, Cantarel BL, Chain PSG, Chen IMA, Chen L, Chhibba S, Chu K, Ciulla DM, Clemente JC, Clifton SW, Conlan S, Crabtree J, Cutting MA, Davidovics NJ, Davis CC, DeSantis TZ, Deal C, Delehaunty KD, Dewhirst FE, Deych E, Ding Y, Dooling DJ, Dugan SP, Michael Dunne W, Scott Durkin A, Edgar RC, Erlich RL, Farmer CN, Farrell RM, Faust K, Feldgarden M, Felix VM, Fisher S, Fodor AA, Forney LJ, Foster L, Di Francesco V, Friedman J, Friedrich DC, Fronick CC, Fulton LL, Gao H, Garcia N, Giannoukos G, Giblin C, Giovanni MY, Goldberg JM, Goll J, Gonzalez A, Griggs A, Gujja S, Kinder Haake S, Haas BJ, Hamilton HA, Harris EL, Hepburn TA, Herter B, Hoffmann DE, Holder ME, Howarth C, Huang KH, Huse SM, Izard J, Jansson JK, Jiang H, Jordan C, Joshi V, Katancik JA, Keitel WA, Kelley ST, Kells C, King NB, Knights D, Kong HH, Koren O, Koren S, Kota KC, Kovar CL, Kyrpides NC, La Rosa PS, Lee SL, Lemon KP, Lennon N, Lewis CM, Lewis L, Ley RE, Li K, Liolios K, Liu B, Liu Y, Lo C-C, Lozupone CA, Dwayne Lunsford R, Madden T, Mahurkar AA, Mannon PJ, Mardis ER, Markowitz VM, Mavromatis K, McCorrison JM, McDonald D, McEwen J, McGuire AL, McInnes P, Mehta T, Mihindukulasuriya KA, Miller JR, Minx PJ, Newsham I, Nusbaum C, O'Laughlin M, Orvis J, Pagani I, Palaniappan K, Patel SM, Pearson M, Peterson J, Podar M, Pohl C, Pollard KS, Pop M, Priest ME, Proctor LM, Qin X, Raes J, Ravel J, Reid JG, Rho M, Rhodes R, Riehle KP, Rivera MC, Rodriguez-Mueller B, Rogers Y-H, Ross MC, Russ C, Sanka RK, Sankar P, Fah Sathirapongsasuti J, Schloss JA, Schloss PD, Schmidt TM, Scholz M, Schriml L, Schubert AM, Segata N, Segre JA, Shannon WD, Sharp RR, Sharpton TJ, Shenoy N, Sheth NU, Simone GA, Singh I, Smillie CS, Sobel JD, Sommer DD, Spicer P, Sutton GG, Sykes SM, Tabbaa DG, Thiagarajan M, Tomlinson CM, Torralba M, Treangen TJ, Truty RM, Vishnivetskaya TA, Walker J, Wang L, Wang Z, Ward DV, Warren W, Watson MA, Wellington C, Wetterstrand KA, White JR, Wilczek-Boney K, Wu Y, Wylie KM, Wylie T, Yandava C, Ye L, Ye Y, Yooseph S, Youmans BP, Zhang L, Zhou Y, Zhu Y, Zoloth L, Zucker JD, Birren BW, Gibbs RA, Highlander SK, Methé BA, Nelson KE, Petrosino JF, Weinstock GM, Wilson RK, White O, The Human Microbiome Project, C (2012) Structure, function and diversity of the healthy human microbiome. Nature 486(7402):207–214

86. Macfabe DF (2012) Short-chain fatty acid fermentation products of the gut microbiome: implications in Autism Spectrum Disorders. Microb Ecol Health Dis 23:10.3402/mehd.v23i0.19260
87. Song Y, Liu C, Finegold SM (2004) Real-time PCR quantitation of Clostridia in feces of autistic children. Appl Environ Microbiol 70(11):6459–6465
88. Williams BL, Hornig M, Buie T, Bauman ML, Cho Paik M, Wick I, Bennett A, Jabado O, Hirschberg DL, Lipkin WI (2011) Impaired carbohydrate digestion and transport and mucosal dysbiosis in the intestines of children with autism and gastrointestinal disturbances. PLoS One 6(9):e24585
89. Lee CYQ, Franks AE, Hill-Yardin EL (2020) Autism-associated synaptic mutations impact the gut-brain axis in mice. Brain Behav Immun 88:275–282
90. Cristiano C, Pirozzi C, Coretti L, Cavaliere G, Lama A, Russo R, Lembo F, Mollica MP, Meli R, Calignano A, Mattace Raso G (2018) Palmitoylethanolamide counteracts autistic-like behaviours in BTBR T+tf/J mice: contribution of central and peripheral mechanisms. Brain Behav Immun 74:166–175
91. Golubeva AV, Joyce SA, Moloney G, Burokas A, Sherwin E, Arboleya S, Flynn I, Khochanskiy D, Moya-Pérez A, Peterson V, Rea K, Murphy K, Makarova O, Buravkov S, Hyland NP, Stanton C, Clarke G, Gahan CGM, Dinan TG, Cryan JF (2017) Microbiota-related changes in bile acid & tryptophan metabolism are associated with gastrointestinal dysfunction in a mouse model of autism. EBioMedicine 24:166–178
92. Sauer AK, Bockmann J, Steinestel K, Boeckers TM, Grabrucker AM (2019) Altered intestinal morphology and microbiota composition in the Autism Spectrum Disorders associated SHANK3 mouse model. Int J Mol Sci 20(9):2134
93. Sgritta M, Dooling SW, Buffington SA, Momin EN, Francis MB, Britton RA, Costa-Mattioli M (2019) Mechanisms underlying microbial-mediated changes in social behavior in mouse models of Autism Spectrum Disorder. Neuron 101(2):246–259.e246
94. Tabouy L, Getselter D, Ziv O, Karpuj M, Tabouy T, Lukic I, Maayouf R, Werbner N, Ben-Amram H, Nuriel-Ohayon M, Koren O, Elliott E (2018) Dysbiosis of microbiome and probiotic treatment in a genetic model of Autism Spectrum Disorders. Brain Behav Immun 73:310–319
95. Alamoudi MU, Hosie S, Shindler AE, Wood JL, Franks AE, Hill-Yardin EL (2022) Comparing the Gut Microbiome in Autism and Preclinical Models: A Systematic Review. Frontiers in Cellular and Infection Microbiology, 889
96. Welch JLM, Hasegawa Y, McNulty NP, Gordon JI, Borisy GG (2017) Spatial organization of a model 15-member human gut microbiota established in gnotobiotic mice. Proc Natl Acad Sci 114(43):E9105–E9114
97. Woo M (2018) Inner workings: mapping the microbiome location helps elucidate its role. Proc Natl Acad Sci 115(48):12078–12080

15 Interaction of the Microbiota and the Enteric Nervous System During Development

Jaime Pei Pei Foong

Abstract

The gastrointestinal tract contains the enteric nervous system within its walls and a large community of microbial symbionts (microbiota) in its lumen. In recent years, studies have shown that these two systems that lie adjacent to each other interact. This review will summarize new data using mouse models demonstrating the concurrent development of the enteric nervous system and microbiota during key pre- and postnatal stages. It will also discuss the possible roles that microbiota play on influencing enteric nervous system development and implications of antibiotic exposure during developmental windows.

Keywords

Enteric nervous system · Microbiota · Antibiotics · Development

The gastrointestinal tract is home to several hundred species of microbes collectively referred to as the microbiota. In humans, the colon harbours the largest microbial population (10^{14} bacterial cells) compared to the upper gut and the rest of the body. Although the symbiotic relationship between the host and intestinal microbiota developed over millions of years, we have only recently begun to understand its importance to the overall well-being of the host, instead of being regarded merely as pathogens. The luminal microbiota lie near the enteric nervous system (ENS) embedded in the gut wall; many studies in recent years have shown that there is crosstalk between microbiota and the ENS [4, 5, 8, 25, 26, 28, 32, 39, 45, 57]; however, when this interaction begins in life and the health implications of its disruption during critical developmental windows is a new research frontier.

15.1 The ENS and Microbiota Develop Concurrently

Mice have been pivotal for studying the physiological significance of ENS-microbiota interactions. Almost 99% of mouse genes are shared with humans and microbiota of mice and humans have many core similarities [11, 30]. Furthermore, the mouse is the best-studied model for the anatomical and functional development of the ENS. This enables us to create a detailed timeline for key developmental milestones which show the concurrent maturation of the ENS and microbiota within the gut (Fig. 15.1). Thus, in this chapter, I will be focusing on gut research conducted using

J. P. P. Foong (✉)
Department of Anatomy & Physiology, University of Melbourne, Parkville, VIC, Australia
e-mail: j.foong@unimelb.edu.au

N. J. Spencer et al. (eds.), *The Enteric Nervous System II*, Advances in Experimental Medicine and Biology 1383, https://doi.org/10.1007/978-3-031-05843-1_15

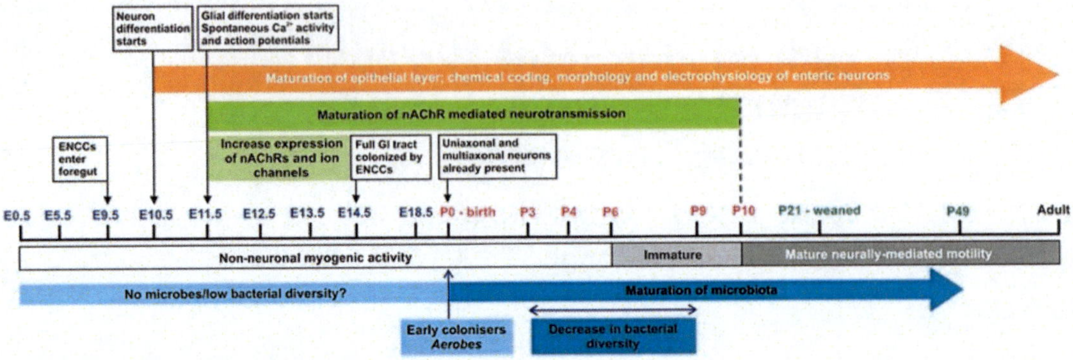

Fig. 15.1 Developmental timeline for the mouse colon. This timeline summarizes and simplifies the developmental milestones for ENS and microbiota development during embryogenesis (E0.5–E18.5), early postnatal (P0–P10) period, post-weaning period (P21–P49) and adulthood

the murine model which has revealed three key developmental windows broadly referred to as embryogenesis, early postnatal and post weaning periods.

Embryogenesis Substantial development of the ENS begins and occurs during embryogenesis [21, 22, 50, 55]. The ENS mainly derives from the neural crest and during embryogenesis, a complex array of molecular and cellular mechanisms orchestrates the migration of neural crest cells to and within the developing gastrointestinal tract, proliferation of these precursors and differentiation into neuron and glial subtypes [24, 43]. The mature ENS comprises a large variety of enteric neuronal and glial subtypes which are differentiated by various properties, including subtype specific function, electrophysiology, neurochemical coding and morphology [16–18]. But, each of these identifying properties seems to have a staggered appearance during development, and when each neuronal/glial subtype has completed the collection of their unique properties remains unclear [21, 53]. Furthermore, enteric cells can express certain properties such as morphology and neurochemistry only transiently during development. It was only discovered in the last decade that young enteric neurons are electrically active early during embryogenesis long before coordinated neurally mediated gut motility has commenced. These embryonic neurons have begun to form functional connections with each other through neurotransmitters such as acetylcholine and 5-hydroxytryptamine (5-HT) and their early communication is postulated to affect the survival of later born enteric neurons [14, 19, 20, 23, 37].

Opposed to the decades of dedicated research which gave rise to a generally unified understanding of ENS development. There has been intense debate over 100 years about whether the foetal environment is sterile and whether there are microbiota transferred from mother to foetus in utero [48]. Some studies report that in normal pregnancies, the placenta, amniotic fluid and first stool of the infant (meconium) contains microorganisms that are not harmful to the foetus, thus raising the possibility that the foetus may have already encountered bacteria [1, 27, 31]. Nonetheless, the vast majority of research indicates that a foetus normally develops in a sterile environment, notably a recent heroic study examined placenta samples from 537 women to demonstrate that there are indeed no bacteria in healthy placenta [7, 48]. Yet, this was immediately challenged by another study identifying viable bacteria from murine and foetal tissues [59]. Hence, this issue remains contentious and an area of active research.

Early Postnatal Period In an earlier book chapter [12], I discussed the emerging research on the postnatal development of the murine ENS. The first ten days of life (postnatal day, P0–10) in the mouse, which represents the early postnatal period, is a critical time for ENS development. The motility patterns of the colon and its main underlying neuronal circuitry within the myenteric plexus are still maturing [52]. Significant numbers of myenteric neurons are still being born (exiting the cell cycle), acquiring their neurochemistry and undergoing substantial maturation in their morphology and electrophysiological properties [3, 13, 14, 34, 49, 54]. Little is known about the maturation of the other division of the ENS, the submucous plexus, but, what we do know is that submucosal neurons are still developing, and that they tend to differentiate later than myenteric neurons [35, 38, 49, 54].

Early postnatal life is the critical period for colonization and establishment of microbiota. At birth, the intestines of infants are rapidly colonized by bacteria from their immediate environment which is the mother's vagina or skin, depending on the mode of delivery [42]. Like humans born of natural birth, the microbiome of the newborn mouse colon has a similar composition to that of their mother's vagina. An initial bloom of Streptococcus occurs after birth and is replaced by Lactobacillus after postnatal day (P) 3 [46]. Then, many other environmental factors contribute to the development of microbiota, including whether babies are fed breast milk or formula [42].

Post-weaning Period The post-weaning period is defined as the period during and immediately after weaning. Mice have a shorter and more accelerated early life compared to humans, and adolescence or puberty can begin in mice as early as postnatal day (P) 18–28, when juvenile mice are weaned from the female dam [11, 36]. The vital components of the ENS would have been acquired during embryogenesis and early postnatal life. Post-weaning development is likely to involve continued formation of synapses in the enteric network that commenced in earlier stages and fine-tuning of the circuit connections that underpin maturation of gastrointestinal functions. Indeed, the electrophysiological properties and synaptic profile of enteric neurons, particularly those that are characteristic of the intrinsic sensory neurons of the circuitry, are still immature at P10 [13]. We have recently shown that between pre-weaning and post-weaning periods, there is significant maturation in the somata size of enteric neurons and numbers of synaptophysin+ varicosities closely apposed to their cell bodies. The architecture of the enteric plexi is still maturing as the ganglia containing neurons and glia are stretched further apart during development. While there is no appreciable change in neurochemistry of myenteric neurons, substantial maturation in submucosal neurochemistry still occurs during the post-weaning period [47]. Further, Schwann cell-derived enteric neurons in the distal parts of the gastrointestinal tract, and S100β + glia that are found in the intestinal mucosa are all still developing during the post-weaning period [28, 54].

In humans, major shifts in microbiota occur during and after weaning, due to the transition from mother's milk to solid food [30]. Unsurprisingly, in mice, we observe significant increases in abundance and communities of microbiota between P10 and P42–49 (6-weeks of age) [26].

15.2 Role of Microbiota on the Developing ENS

Although the ENS and microbiota develop concurrently within the gastrointestinal tract especially during the early postnatal period, the crosstalk between the two systems and how the microbiota contributes to the development of the ENS remains unclear. A study using germfree mice showed that the ENS in the small intestine of these mice is abnormal at P3 [5]. Whether disruption to the developing ENS is due to the lack of microbiota is unclear as other systems are significantly perturbed in germ-free mice [58]. Moreover, as the germfree mice were only examined at P3, thus it may be possible that the ENS has been disrupted earlier during embryogenesis. Recent work from my group shows that oral administration of the antibiotic, vancomycin, to neonatal mice daily from birth to P10 disrupts their colonic microbiota (dysbiosis), motility, myenteric neurochemistry and activity [25]. While this supports the view that microbiota is important in supporting ENS development, the possibility that the antibiotic itself could have toxic effects directly on the ENS cannot be excluded.

There are only a couple of studies from other groups that have examined the role of microbiota in ENS development, and they studied later stages of postnatal development. Microbiota dysbiosis has been shown to impair gut neuromuscular function in 3-week-old male mice [4]. Another study showed that gut microbiota controls the influx of enteric glial cells into the lamina propria of the small intestine by comparing adult germ-free mice with those raised in a specific pathogen-free environment [28]. More recently, we exposed mice to vancomycin during the post-weaning period and showed that the significant shifts in microbiota communities were accompanied by disruptions to neurochemistry and function of myenteric and submucosal neurons leading to dysmotility in mice [26]. Interestingly, compared to our work with the same antibiotic exposed in a different developmental window [25], we found that the impact of vancomycin appeared to be greater when administered during the early postnatal compared to the post-weaning period.

There is very little understanding of the mechanisms in which microbiota influences ENS development. Work on the adult ENS has shown that microbes signal to the ENS by modifying 5-HT metabolism and signalling from the mucosa [8, 51, 57]. Although our work on neonatal mice ENS seems to be in agreement to the adult studies by showing antibiotic-induced perturbation in the mucosal levels of the 5-HT metabolite, 5-HIAA [25], more studies investigating how the microbes signal to the developing ENS via 5-HT or other processes is warranted. Adult studies have also revealed other mechanisms by which microbiota signal to the ENS, including toll-like receptors, microvesicles, transcription and neurotrophic factors, various microbial metabolites and mediators released from enteroendocrine cells and immune system that could serve roles in the developing microbiota-ENS crosstalk [15]. Further, while collective studies so far have supported the view that microbiota plays important roles mediating ENS development, the possibility that antibiotics can have direct toxic effects on the ENS cannot be excluded [10].

15.3 Implications of Antibiotic Exposure During Critical Developmental Windows

Antibiotics were developed and used as medicines around the 1940s, and they soon earned the title of "miracle cure" by effectively treating infections, thereby leading to a dramatic decrease in death rates and serious illnesses. Although antibiotics are necessary in many circumstances, they are not harmless and may have negative health consequences especially when used during critical developmental windows.

Healthy pregnancies are accompanied by increased numbers of bacteria and dramatic changes in gut microbiota composition from the first to the third trimester, which can persist till at least one month after delivery [29, 44]. Several antibiotics are considered safe for use during pregnancy, and they account for 80% of all prescribed medication to pregnant women [33]. Recent startling findings show that maternal exposure to antibiotics, even before pregnancy, is associated with an increased risk of childhood hospitalized infections, including gastroenteritis [40]. Moreover, maternal antibiotics have been linked to childhood obesity [41, 44]. More recently, maternal microbiota and their metabolites have been shown to significantly impact the development of the foetal brain [56]. However, the impact of maternal antibiotics during pregnancy on the growing foetus, especially its ENS, is massively understudied.

Infants and young children have the highest antibiotic exposures globally [2]. Exposure to antibiotics early in life has been linked to increased susceptibility to several diseases such as functional gastrointestinal disorders, obesity, metabolic dysfunction and allergies later in life

[6, 9, 42]. Yet, the lasting impact of early life antibiotic exposure on host physiology, and how antibiotics given during critical developmental windows may predispose the host to gastrointestinal disorders is currently unknown.

15.4 Conclusions and Future Directions

The gastrointestinal tract is an organ where multiple systems coexist, and it is not a coincidence that the microbiota and ENS develop concurrently. Future studies should aim to provide mechanistic insights into the crosstalk between the microbiota and ENS during various developmental windows. Antibiotics are important drugs especially for their life-saving qualities. Identifying how antibiotic usage during critical developmental windows affects the host would advance antibiotic therapy by revealing preventative measures against its unwanted side effects and improve the short- and long-term health and well-being of our next generation.

References

1. Aagaard K, Ma J, Antony KM, Ganu R, Petrosino J, Versalovic J (2014) The placenta harbors a unique microbiome. Sci Transl Med 6:237ra65
2. Anderson H, Vuillermin P, Jachno K, Allen KJ, Tang ML, Collier F, Kemp A, Ponsonby AL, Burgner D (2017) Prevalence and determinants of antibiotic exposure in infants: a population-derived Australian birth cohort study. J Paediatr Child Health 53:942–949
3. Bergner AJ, Stamp LA, Gonsalvez DG, Allison MB, Olson DP, Myers MG Jr, Anderson CR, Young HM (2014) Birthdating of myenteric neuron subtypes in the small intestine of the mouse. J Comp Neurol 522:514–527
4. Caputi V, Marsilio I, Filpa V, Cerantola S, Orso G, Bistoletti M, Paccagnella N, de Martin S, Montopoli M, Dall'acqua S, Crema F, Di Gangi IM, Galuppini F, Lante I, Bogialli S, Rugge M, Debetto P, Giaroni C, Giron MC (2017) Antibiotic-induced dysbiosis of the microbiota impairs gut neuromuscular function in juvenile mice. Br J Pharmacol 174:3623–3639
5. Collins J, Borojevic R, Verdu EF, Huizinga JD, Ratcliffe EM (2014) Intestinal microbiota influence the early postnatal development of the enteric nervous system. Neurogastroenterol Motil 26:98–107
6. Cox LM, Blaser MJ (2015) Antibiotics in early life and obesity. Nat Rev Endocrinol 11:182–190
7. De Goffau MC, Lager S, Sovio U, Gaccioli F, Cook E, Peacock SJ, Parkhill J, Charnock-Jones DS, Smith GCS (2019) Human placenta has no microbiome but can contain potential pathogens. Nature 572:329–334
8. De Vadder F, Grasset E, Manneras Holm L, Karsenty G, Macpherson AJ, Olofsson LE, Backhed F (2018) Gut microbiota regulates maturation of the adult enteric nervous system via enteric serotonin networks. Proc Natl Acad Sci U S A 115(25):6458–6463
9. De Vroey B, De Cassan C, Gower-Rousseau C, Colombel JF (2010) Editorial: antibiotics earlier, IBD later? Am J Gastroenterol 105:2693–2696
10. Delungahawatta T, Amin JY, Stanisz AM, Bienenstock J, Forsythe P, Kunze WA (2017) Antibiotic driven changes in gut motility suggest direct modulation of enteric nervous system. Front Neurosci 11:588
11. Dutta S, Sengupta P (2016) Men and mice: relating their ages. Life Sci 152:244–248
12. Foong JP (2016) Postnatal development of the mouse enteric nervous system. Adv Exp Med Biol 891:135–143
13. Foong JP, Nguyen TV, Furness JB, Bornstein JC, Young HM (2012) Myenteric neurons of the mouse small intestine undergo significant electrophysiological and morphological changes during postnatal development. J Physiol 590:2375–2390
14. Foong JP, Hirst CS, Hao MM, Mckeown SJ, Boesmans W, Young HM, Bornstein JC, Vanden Berghe P (2015) Changes in nicotinic neurotransmission during enteric nervous system development. J Neurosci 35:7106–7115
15. Foong JPP, Hung LY, Poon S, Savidge TC, Bornstein JC (2020) Early life interaction between the microbiota and the Enteric Nervous System. Am J Physiol Gastrointest Liver Physiol 319(5):G541–G548
16. Fung C, Vanden Berghe P (2020) Functional circuits and signal processing in the enteric nervous system. Cell Mol Life Sci 77:4505–4522
17. Grundmann D, Loris E, Maas-Omlor S, Huang W, Scheller A, Kirchhoff F, Schäfer KH (2019) Enteric glia: S100, GFAP, and beyond. Anat Rec (Hoboken) 302:1333–1344
18. Gwynne RM, Bornstein JC (2007) Synaptic transmission at functionally identified synapses in the enteric nervous system: roles for both ionotropic and metabotropic receptors. Curr Neuropharmacol 5:1–17
19. Hao MM, Boesmans W, van den Abbeel V, Jennings EA, Bornstein JC, Young HM, Vanden Berghe P (2011) Early emergence of neural activity in the developing mouse enteric nervous system. J Neurosci 31:15352–15361
20. Hao MM, Lomax AE, Mckeown SJ, Reid CA, Young HM, Bornstein JC (2012) Early development of electrical excitability in the mouse enteric nervous system. J Neurosci 32:10949–10960
21. Hao MM, Bornstein JC, Vanden Berghe P, Lomax AE, Young HM, Foong JP (2013) The emergence of neural

activity and its role in the development of the enteric nervous system. Dev Biol 382:365–374
22. Hao MM, Foong JP, Bornstein JC, Li ZL, Vanden Berghe P, Boesmans W (2016) Enteric nervous system assembly: functional integration within the developing gut. Dev Biol 417:168–181
23. Hao MM, Bergner AJ, Hirst CS, Stamp LA, Casagranda F, Bornstein JC, Boesmans W, vanden Berghe P, Young HM (2017) Spontaneous calcium waves in the developing enteric nervous system. Dev Biol 428:74–87
24. Hirst CS, Foong JP, Stamp LA, Fegan E, Dent S, Cooper EC, Lomax AE, Anderson CR, Bornstein JC, Young HM, Mckeown SJ (2015) Ion channel expression in the developing enteric nervous system. PLoS One 10:e0123436
25. Hung LY, Boonma P, Unterweger P, Parathan P, Haag A, Luna RA, Bornstein JC, Savidge TC, Foong JPP (2019) Neonatal antibiotics disrupt motility and enteric neural circuits in mouse colon. Cell Mol Gastroenterol Hepatol 8:298–300.e6
26. Hung LY, Parathan P, Boonma P, Wu Q, Wang Y, Haag A, Luna RA, Bornstein JC, Savidge TC, Foong JPP (2020) Antibiotic exposure postweaning disrupts the neurochemistry and function of enteric neurons mediating colonic motor activity. Am J Physiol Gastrointest Liver Physiol 318:G1042–g1053
27. Jimenez E, Marin ML, Martin R, Odriozola JM, Olivares M, Xaus J, Fernandez L, Rodriguez JM (2008) Is meconium from healthy newborns actually sterile? Res Microbiol 159:187–193
28. Kabouridis PS, Lasrado R, Mccallum S, Chng SH, Snippert HJ, Clevers H, Pettersson S, Pachnis V (2015) Microbiota controls the homeostasis of glial cells in the gut lamina propria. Neuron 85:289–295
29. Koren O, Goodrich JK, Cullender TC, Spor A, Laitinen K, Backhed HK, Gonzalez A, Werner JJ, Angenent LT, Knight R, Backhed F, Isolauri E, Salminen S, Ley RE (2012) Host remodeling of the gut microbiome and metabolic changes during pregnancy. Cell 150:470–480
30. Kostic AD, Howitt MR, Garrett WS (2013) Exploring host-microbiota interactions in animal models and humans. Genes Dev 27:701–718
31. Kovalovszki L, Villanyi Z, Pataki I, Veszelowvsky I, Nagy ZB (1982) Isolation of aerobic bacteria from the placenta. Acta Paediatr Acad Sci Hung 23:357–360
32. Kunze WA, Mao YK, Wang B, Huizinga JD, Ma X, Forsythe P, Bienenstock J (2009) Lactobacillus reuteri enhances excitability of colonic AH neurons by inhibiting calcium-dependent potassium channel opening. J Cell Mol Med 13:2261–2270
33. Kuperman AA, Koren O (2016) Antibiotic use during pregnancy: how bad is it? BMC Med 14:91
34. Laranjeira C, Sandgren K, Kessaris N, Richardson W, Potocnik A, Vanden Berghe P, Pachnis V (2011) Glial cells in the mouse enteric nervous system can undergo neurogenesis in response to injury. J Clin Invest 121:3412–3424
35. Lasrado R, Boesmans W, Kleinjung J, Pin C, Bell D, Bhaw L, Mccallum S, Zong H, Luo L, Clevers H, Vanden Berghe P, Pachnis V (2017) Lineage-dependent spatial and functional organization of the mammalian enteric nervous system. Science 356:722–726
36. Laviola G, Macri S, Morley-Fletcher S, Adriani W (2003) Risk-taking behavior in adolescent mice: psychobiological determinants and early epigenetic influence. Neurosci Biobehav Rev 27:19–31
37. Li Z, Chalazonitis A, Huang YY, Mann JJ, Margolis KG, Yang QM, Kim DO, Cote F, Mallet J, Gershon MD (2011) Essential roles of enteric neuronal serotonin in gastrointestinal motility and the development/survival of enteric dopaminergic neurons. J Neurosci 31:8998–9009
38. Mckeown SJ, Chow CW, Young HM (2001) Development of the submucous plexus in the large intestine of the mouse. Cell Tissue Res 303:301–305
39. Mcvey Neufeld KA, Perez-Burgos A, Mao YK, Bienenstock J, Kunze WA (2015) The gut microbiome restores intrinsic and extrinsic nerve function in germ-free mice accompanied by changes in calbindin. Neurogastroenterol Motil 27:627–636
40. Miller JE, Wu C, Pedersen LH, De Klerk N, Olsen J, Burgner DP (2018) Maternal antibiotic exposure during pregnancy and hospitalization with infection in offspring: a population-based cohort study. Int J Epidemiol 47:561–571
41. Mueller NT, Whyatt R, Hoepner L, Oberfield S, Dominguez-Bello MG, Widen EM, Hassoun A, Perera F, Rundle A (2015) Prenatal exposure to antibiotics, cesarean section and risk of childhood obesity. Int J Obes 39:665–670
42. Munyaka PM, Khafipour E, Ghia JE (2014) External influence of early childhood establishment of gut microbiota and subsequent health implications. Front Pediatr 2:109
43. Nagy N, Goldstein AM (2017) Enteric nervous system development: a crest cell's journey from neural tube to colon. Semin Cell Dev Biol 66:94–106
44. Nuriel-Ohayon M, Neuman H, Koren O (2016) Microbial changes during pregnancy, birth, and infancy. Front Microbiol 7:1031
45. Obata Y, Castano A, Boeing S, Bon-Frauches AC, Fung C, Fallesen T, De Aguero MG, Yilmaz B, Lopes R, Huseynova A, Horswell S, Maradana MR, Boesmans W, Vanden Berghe P, Murray AJ, Stockinger B, Macpherson AJ, Pachnis V (2020) Neuronal programming by microbiota regulates intestinal physiology. Nature 578(7794):284–289
46. Pantoja-Feliciano IG, Clemente JC, Costello EK, Perez ME, Blaser MJ, Knight R, Dominguez-Bello MG (2013) Biphasic assembly of the murine intestinal microbiota during early development. ISME J 7:1112–1115
47. Parathan P, Wang Y, Leembruggen AJ, Bornstein JC, Foong JP (2020) The enteric nervous system undergoes significant chemical and synaptic maturation during adolescence in mice. Dev Biol 458(1):75–87

48. Perez-Munoz ME, Arrieta MC, Ramer-Tait AE, Walter J (2017) A critical assessment of the "sterile womb" and "in utero colonization" hypotheses: implications for research on the pioneer infant microbiome. Microbiome 5:48
49. Pham TD, Gershon MD, Rothman TP (1991) Time of origin of neurons in the murine enteric nervous system: sequence in relation to phenotype. J Comp Neurol 314:789–798
50. Rao M, Gershon MD (2018) Enteric nervous system development: what could possibly go wrong? Nat Rev Neurosci 19:552–565
51. Reigstad CS, Salmonson CE, Rainey JF 3rd, Szurszewski JH, Linden DR, Sonnenburg JL, Farrugia G, Kashyap PC (2015) Gut microbes promote colonic serotonin production through an effect of short-chain fatty acids on enterochromaffin cells. FASEB J 29:1395–1403
52. Roberts RR, Murphy JF, Young HM, Bornstein JC (2007) Development of colonic motility in the neonatal mouse-studies using spatiotemporal maps. Am J Physiol Gastrointest Liver Physiol 292:G930–G938
53. Sasselli V, Pachnis V, Burns AJ (2012) The enteric nervous system. Dev Biol 366:64–73
54. Uesaka T, Nagashimada M, Enomoto H (2015) Neuronal differentiation in Schwann cell lineage underlies postnatal neurogenesis in the enteric nervous system. J Neurosci 35:9879–9888
55. Uesaka T, Young HM, Pachnis V, Enomoto H (2016) Development of the intrinsic and extrinsic innervation of the gut. Dev Biol 417:158–167
56. Vuong HE, Pronovost GN, Williams DW, Coley EJL, Siegler EL, Qiu A, Kazantsev M, Wilson CJ, Rendon T, Hsiao EY (2020) The maternal microbiome modulates fetal neurodevelopment in mice. Nature 586(7828):281–286
57. Yano JM, Yu K, Donaldson GP, Shastri GG, Ann P, Ma L, Nagler CR, Ismagilov RF, Mazmanian SK, Hsiao EY (2015) Indigenous bacteria from the gut microbiota regulate host serotonin biosynthesis. Cell 161:264–276
58. Yi P, Li L (2012) The germfree murine animal: an important animal model for research on the relationship between gut microbiota and the host. Vet Microbiol 157:1–7
59. Younge N, Mccann JR, Ballard J, Plunkett C, Akhtar S, Araújo-Pérez F, Murtha A, Brandon D, Seed PC (2019) Fetal exposure to the maternal microbiota in humans and mice. JCI Insight 4:e127806

16 Comparative and Evolutionary Aspects of the Digestive System and Its Enteric Nervous System Control

John B. Furness

Abstract

All life forms must gain nutrients from the environment and from single cell organisms to mammals a digestive system is present. Components of the digestive system that are recognized in mammals can be seen in the sea squirt that has had its current form for around 500my. Nevertheless, in mammals, the organ system that is most varied is the digestive system, its architecture being related to the dietary niche of each species. Forms include those of foregut or hindgut fermenters, single or multicompartment stomachs and short or capacious large intestines. Dietary niches include nectarivores, folivores, carnivores, etc. The human is exceptional in that, through food preparation (>80% of human consumption is prepared food in modern societies), humans can utilize a wider range of foods than other species. They are cucinivores, food preparers. In direct descendants of simple organisms, such as sponges, there is no ENS, but as the digestive tract becomes more complex, it requires integrated control of the movement and assimilation of its content. This is achieved by the nervous system, notably the enteric nervous system (ENS) and an array of gut hormones. An ENS is first observed in the phylum cnidaria, exemplified by hydra. But hydra has no collections of neurons that could in any way be regarded as a central nervous system. All animals more complex than hydra have an ENS, but not all have a CNS. In mammals, the ENS is extensive and is necessary for control of movement, enteric secretions and local blood flow, and regulation of the gut immune system. In animals with a CNS, the ENS and CNS have reciprocal connections. From hydra to human, an ENS is essential to life.

Keywords

Evolution · Digestion · *Paramecium* · Diet · Cucinivores · Cnidaria · Enteric nervous system

16.1 Nutritional Strategies of Simple Life Forms

Cells, the basic units of all complex life forms, emerged some 3.5 billion years ago (bya). The first cells (the prokaryotes, that include bacteria and archaea) had a surrounding membrane, but nothing obvious to separate cellular tasks.

J. B. Furness (✉)
Digestive Physiology and Nutrition Laboratories, Florey Institute of Neuroscience and Mental Health, Parkville, VIC, Australia

Department of Anatomy & Physiology, University of Melbourne, Parkville, VIC, Australia
e-mail: j.furness@unimelb.edu.au

N. J. Spencer et al. (eds.), *The Enteric Nervous System II*, Advances in Experimental Medicine and Biology 1383, https://doi.org/10.1007/978-3-031-05843-1_16

Prokaryotes gain nutrient molecules from the environment through diffusion, membrane transporters and specialized pores. About 1.6–2.1 bya, eukaryote cells arose, in which functions are separated into compartments that are commonly surrounded by membranes within the cell, for example, genetic machinery was concentrated in the nucleus and energy production in the mitochondria. The single-celled eukaryote, paramecium, exhibits the first structurally identifiable digestive apparatus, including a mouth region and an anal region [4] (Fig. 16.1). These organisms engulf small particles from the aqueous environment in which they live and transport them into food vacuoles where the particles are exposed to digestive enzymes.

Multicellular organisms developed from single cells some 1.2 bya. The earliest multicellular members of the animal kingdom are the sponges (*porifera*), the oldest of which appeared some 540–650 mya. These animals have a very simple body plan; their bodies are made of two surface cell layers with a jelly-like substance in between, which contains some individual cells derived from the surface. The two cell layers define these as diploblasts. The pores from which the sponge gets its name (*porifera* = pore bearing) are of two types, pores through which water bearing nutrients flow in and pores through which water flows out [58]. The entrance pores are connected by canals to chambers where absorption of nutrients occurs, and the chambers open into broader exit pores (Fig. 16.1). The direction of flow is determined by active pulsatile movements of the canals and the movements of brush-like processes of cells in the walls of the pores and chambers. These animals have no nervous system.

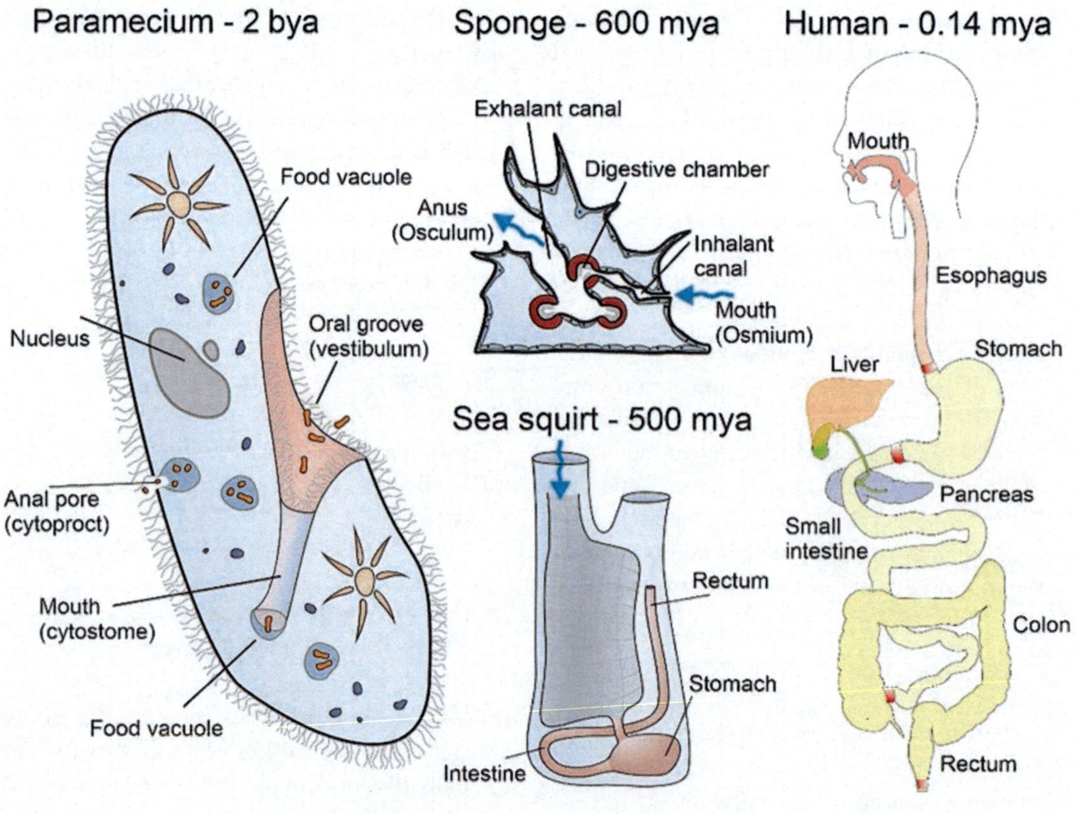

Fig. 16.1 Digestive systems: from paramecium to human. The single celled paramecium has an identifiable mouth part and anal excretory mechanism. The sponge, shown in cross section, has mouth, anus and a digestive chamber. The sea squirt that has maintained its form for 500 my, has equivalents of mouth, stomach, intestine and rectum. The pattern common to most mammals is seen in modern humans

A next level of digestive system complexity was reached in the phylum *Cnidaria*, about 500 mya. This phylum includes Hydra and the corals, *Anthazoa* [57], the first multicellular species with readily identifiable neurally regulated digestive systems. Hydra has a digestive tube which contracts under the influence of nerves and can in an organized way admit food and expel waste [59, 60].

Multicellularity provides a challenge for access of cells to nutrients as animals became bigger and more complex. This was solved by the evolution of the three-layered embryo, in which internal surfaces could be developed to accommodate a circulatory system. The epithelium of a diploblast is a continuous layer around the outside, as in the sponge or Hydra (Fig. 16.1), whereas the triploblast has an external epithelium and several internal epithelia. It is in triploblasts, the sea squirts (e.g. *Ciona intestinalis*), that we encounter the most ancient animals to have a digestive tract with identifiable homologues of the mammalian digestive tract: mouth, oesophagus, stomach, intestine, rectum and anus (Fig. 16.1). *Ciona* have an enteric nervous system.

Thus, a differentiated gastrointestinal tract whose structure and function has resemblance to the mammalian gastrointestinal tract can be traced back more than 400 million years. Also the primitive placazoans (Tricoplax) are flat multicellular organisms that absorb nutrients through their surfaces and truly lack a tubular digestive system.

16.2 Design Features of the Vertebrate Digestive System

16.2.1 Nutrient Exchange

By having a cardiovascular system that can deliver nutrients to all the cells and organs of the body, vertebrates can attain the enormous sizes of extant whales and elephants or the prehistoric brontosaurus. We must now concern ourselves with how the animal gets the nutrients to the insides of the blood vessels from the external environment, and eventually from the blood to the tissues. Vertebrates obtain nutrition through the lungs and digestive system, each of which has a highly permeable external epithelium (derived from ectoderm) and only a very thin layer of mesoderm separating this surface from the endoderm that lines the smallest blood vessels, the capillaries. For the lungs, the external surface is bathed in air, from which oxygen is taken up into the blood capillaries. For the digestive system, the lining of the intestine is bathed in a watery solution containing digested nutrients derived from the food. It is relevant that the lungs develop as outgrowths of the digestive tract.

The main nutrient absorptive surface within our digestive system is found in the small intestine. The lining epithelium has only a fragile single layer, beneath which are the absorptive vessels of the vascular system, lymphatics and blood capillaries. Between the external epithelium (digestive tract lining) and the internal epithelium (capillary or lymphatic vessel lining) there is very little, an extracellular protein-rich gel and a small number of connective tissue cells. To be able to absorb all the necessary nutrients, the human small intestine presents a surface membrane area of about 60 m^2 to its contents, making it the largest vulnerable surface of the body; the total surface of the gastrointestinal tract that faces the external environment is about 100 m^2, with some estimates up to 400 m^2, compared to 2 m^2 of skin [24]. The surface that faces the external environment is substantially increased by the presence of villi and crypts in the small intestine and by microvilli at the surfaces of the cells making up the lining.

16.3 Comparisons of Digestive Strategies in Mammals

Different mammalian species have adapted to consume different types of food [20]. Conversely, adaptations to different foods have constrained the digestive systems of species, which is reflected in digestive tract anatomy (Fig. 16.2). That digestive physiology and anatomy imposes dietary restriction is exemplified by ruminant dependence on low protein fibrous plant material, the cat being an obligate carnivore, specialist feeders being reliant on nectar, and the koala subsisting on a diet almost exclusively of eucalyptus leaves. In the wild, a modern human could not exist on the diet of a sheep, koala or panda. And a koala or cat could not exist on the diet of a deer or a cow. The differences in the diets that animals prefer or are obliged to eat have led to the classification of animals by their dietary specialization as carnivore, omnivore, herbivore, folivore, frugivore and so on.

Mammals lack cellulases and thus bacteria are essential to the breakdown of plant cell walls in the diet. Most herbivores solve the problem of digesting plant cells by holding food in capacious gut reservoirs (Fig. 16.2) and enlisting the aid of cellulolytic, pectinolytic and xylanolytic bacteria that are adapted to breakdown of plant cell walls. Foregut fermenters include ruminants, such as cattle and sheep, that have multichambered stomachs, and kangaroos, colobus monkey and

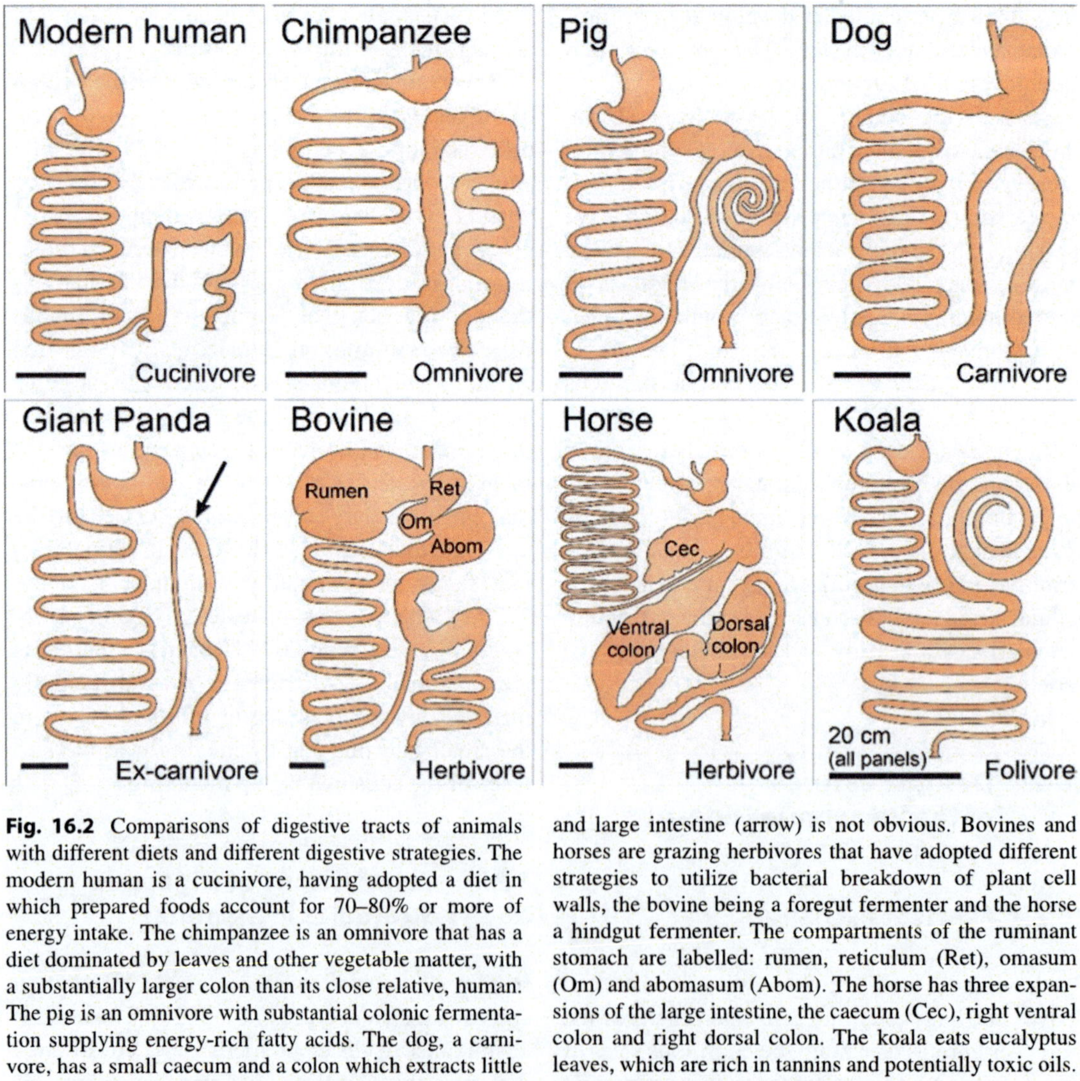

Fig. 16.2 Comparisons of digestive tracts of animals with different diets and different digestive strategies. The modern human is a cucinivore, having adopted a diet in which prepared foods account for 70–80% or more of energy intake. The chimpanzee is an omnivore that has a diet dominated by leaves and other vegetable matter, with a substantially larger colon than its close relative, human. The pig is an omnivore with substantial colonic fermentation supplying energy-rich fatty acids. The dog, a carnivore, has a small caecum and a colon which extracts little energy from food. The giant panda has retained a carnivore gut form, despite having adopted a diet of leaves and shoots. It has no caecum, and the junction between small and large intestine (arrow) is not obvious. Bovines and horses are grazing herbivores that have adopted different strategies to utilize bacterial breakdown of plant cell walls, the bovine being a foregut fermenter and the horse a hindgut fermenter. The compartments of the ruminant stomach are labelled: rumen, reticulum (Ret), omasum (Om) and abomasum (Abom). The horse has three expansions of the large intestine, the caecum (Cec), right ventral colon and right dorsal colon. The koala eats eucalyptus leaves, which are rich in tannins and potentially toxic oils. It has an extensive large bowel, and very slow transit times. Its large bowel harbours tannin-metabolizing bacteria

other species that have an enlarged single chambered forestomach. Hindgut fermenters include horses, most rodents, rabbits and rhinoceros. In addition to ruminants, non-ruminant foregut fermenters and hindgut fermenters, a fourth herbivore specialization is to possess a very long small intestine in which microbial digestion can proceed. Such an intestine occurs in the black bear. Sheep have both a ruminant stomach and an exceedingly long small intestine, and bovines have both a rumen and an enlarged colon (Fig. 16.2).

16.3.1 Ruminant Foregut Fermenters

Ruminants are able to utilize one of the most abundant sources of carbohydrate on the planet, grasses, because the rumen holds large populations of bacteria that are adapted to breakdown of plant cells (cellulolytic and xylanolytic and pectinolytic bacteria). A cost of microbial reliance for herbivorous species, especially foregut fermenters, is that the anaerobic bacteria within the rumen or other fermentation reservoirs utilize almost all the plant carbohydrate, including simple sugars. From this carbohydrate, ruminal bacteria produce short chain fatty acids (SCFA), notably acetate, propionate and butyrate, plus ethanol, hydrogen, methane and CO_2. Bacterially produced SCFA must be absorbed across the gut wall and converted enzymatically to glucose and other products, a net energy cost to the host [61]. Thus, ruminants are more prone than other groups to enter into negative glucose balance, for example, during lactation, soon after giving birth [29]. Foregut fermenters are also advantaged by being able to readily utilize vitamins produced by bacteria, particularly B vitamins [62]. For this reason, ruminants do not require B vitamins in the diet. It is thought that eating faeces (coprophagy) by hind-gut fermenters, such as rabbits, provides access to such vitamins. Foregut fermentation also contributes to detoxification, for which hindgut fermenters and primarily enzymatic digesters rely mainly on the liver. For humans detoxification is additionally achieved by cooking food.

It is notable that all mammals lack cellulases, and that in animals with only hind-gut fermentation, and short hind-guts, such as cats, cell walls are essentially indigestible. By contrast, cellulases are present in some invertebrates [66].

The capability of ruminal bacteria to digest protein is limited. High amounts of protein putrefy in the rumen and produce potentially toxic by-products, including ammonia, phenols, indoles, amines and amino acids, poisoning the host [6, 8]. For humans, it has been suggested that high protein diets, particularly high protein/low carbohydrate diets, can result in excess protein reaching the colon where it ferments to produce injurious sulphides, ammonia, phenols and indoles [70].

Non-ruminant foregut fermenters, such as kangaroos, are less efficient than ruminants for two reasons, the lack of rumination and the absence of efficient sorting that allows digested material to be moved on while undigested material is more or less retained [10].

16.3.2 Autoenzyme Digesters: Carnivores, Omnivores and Cucinivores

By autoenzyme digesters is meant those species that have considerable reliance on fore and mid gut digestion, utilizing enzymes that they themselves produce, rather than relying on bacterial enzymes. Examples are humans, pigs, practically all carnivores and rodents. These species commence the digestive process by mastication in the presence of salivary enzymes, primarily amylases. They have simple stomachs that subject the masticated food to acid hydrolysis and exposure to enzymes, primarily proteases. Gastric acidity and proteases also kill the majority of bacteria in the food. As in ruminants, the stomach has a reservoir function and, as in ruminants, the stomach sorts digested from non-digested foods, in this case by trituration. In humans, antral contractions massage and push the contents towards the pyloric sphincter, and triturated aspirates containing only small particles, less than about 1 mm in diameter, are propelled into duodenum [41].

Further autoenzymatic digestion occurs in the small intestine, utilizing enzymes released from the exocrine pancreas and those in the gut wall, especially in the glycocalyx of enterocytes. The pancreatointestinal enzymes include a range of proteases (trypsins, chymotrypsins, carboxypeptidases), carbohydrases (amylases, disaccharidases) and lipases. The effectiveness of lipases is enhanced by emulsification of fats by bile salts. None of these enzymes is able to digest plant walls or plant wall components, such as pectin and cellulose; pectinases and cellulases being absent. On the other hand, indigestible plant carbohydrate can influence the digestion of other dietary components. Pectin binds cholesterol, tends to reduce plasma cholesterol and may slow glucose absorption by trapping carbohydrates [9].

The large intestines of autoenzyme digesters differ considerably between omnivores, in which the large intestine can be capacious, as in the pig, and include a significant caecum, as in the rat and guinea pig. It is not until the substantially digested food reaches the caecum and colon that significant populations of bacteria are encountered in omnivores. The populations of hind gut bacterial colonies express digestive enzymes similar to those found in the rumen, and in the caecum and colon of hind gut fermenters. Carnivores have relatively short hindguts (Fig. 16.2), which are often continuous with the small intestine, and in external features and diameter are similar to the midgut [62]. Most carnivores have little need or ability to handle carbohydrate.

16.3.3 Cucinivores

Modern humans have escaped the restrictions they would encounter in the wild to become cucinivores, that is consumers of cooked or otherwise prepared foods [20, 22]. It is estimated that 80–90% of food that is cooked in modern societies is pre-processed [64]. Before they were able to prepare and cook foods, many of the major nutrient sources used by modern humans were less readily available or unavailable [14]. The most obvious example is grains, which only became a common food source for humans about 10,000–30,000 years ago, with the development of grinding to create flour, and the use of cooking [23]. Grains, that were formerly poorly accessible food sources, account for up to 70% of dietary intake in modern societies [12]. Mechanical break-up and cooking also provided access to the starches of underground storage organs (USOs), such as potatoes and yams.

Cooking, food preparation and storage have contributed to extending the environmental range of human habitation and, it can be argued, have allowed humans the time to develop technologies and eventually modern civilization. Humans have consumed cooked foods for around 300,000–400,000 years, perhaps 12,000 generations, and no groups of humans who live without cooking have been recorded [67, 69]. Humans also initiate food breakdown through storage, such as by hanging meat, drying in the sun, fermentation, or prolonged marination (e.g. civet de lièvre), and by pounding and grinding. Cooking changes the palatability, digestibility and texture of food, and removes toxins. This history may have influenced humans to have alimentary tracts that are quite different to that expected of an anthropoid primate. In humans, the colon represents only 20% of the total volume of the digestive tract, whereas in hominid apes, it is about 50% [48, 49]. The sizeable colons of most large-bodied primates permit fermentation of low-quality plant fibres, allowing for extraction of energy in the form of SCFAs [45]. Thus, humans are relatively poor amongst autoenzyme-dependent omnivores in digesting uncooked plant fibre. The human large intestine lies somewhere between that of the pig, an omnivore with similar digestive apparatus to human, and the dog, a carnivore capable of consuming an omnivore diet that has a very small caecum and short colon, like human (Fig. 16.2). Evidence for this trend is that hind gut absorption of SCFA accounts for 2% of maintenance energy for dogs, 6–9% for human and 10–31% for pigs [62]. In horses, it provides 46% of maintenance energy requirements. A further adaptation to softer foods is the smaller molars and reduced bite strength in humans compared to other primates [17]. The smaller gastrointestinal tract of humans requires smaller abdominal and pelvic cavities [2], which may be an advantage to a mammal that stands erect.

The extreme changes in diet in the last 50–60 years, especially foods high in animal fats, high in simple readily assimilated sugars, high in salt and high in foods cooked at high temperature have arguably been detrimental to human health [23].

16.3.4 Extremes of Diversity

The koala, which eats a diet exclusively of eucalyptus leaves, has a number of anatomical, physiological and microbial hosting adaptations [7]. The koala caecum is the largest of any mammal in relation to body size (Fig. 16.2), and mean gastrointestinal retention time is the longest known amongst mammals. Eucalyptus leaves contain high levels of tannins, and the pure eucalyptus leaf diet would be toxic for other mammals. However, the koala's large intestine is colonized by tannin-digesting bacteria, including *Lonepinella koalarum*, discovered in the koala and perhaps unique to this species [7]. Mother koalas are observed to feed their young a fecal paste that is presumably a mechanism to ensure colonization with bacteria appropriate to their diet.

The giant and red pandas diverged separately from earlier carnivores, to become species that consume an almost exclusively (99%) plant diet [35, 38, 52]. In parallel with the sweet taste receptor being a pseudogene in the cat, an obligate carnivore, in the herbivorous giant and red pandas the umami receptor for the savoury taste of meat is a pseudogene [35, 71]. The giant and red pandas belong to different carnivore families (*Ursidae* and *Ailuridae*, respectively) and it seems certain that the pseudogenes arose independently, as the mutations in the two species are distinct. We can conclude that pseudogenization of taste receptors is related to disuse in cats and pandas. Despite its diet, the giant panda has a simple stomach and short gastrointestinal tract typical of carnivores (Fig. 16.2) [15, 56], and it needs to eat large amounts of bamboo to survive because of the inefficiency of its carnivore digestive tract to digest bamboo. The giant panda alimentary tract clearly has not adapted during an estimated period of 2–2.4 my of exclusive bamboo diet [38], although there has been evolution of dentition to a herbivore type [15].

16.4 The Enteric Nervous System

The gastrointestinal tract has dual, extensive systems to control its functions and to balance its activities in relation to the whole body, the ENS, reviewed below and the gut enteroendocrine system, a collection of hormone secreting cells, also in the gut wall [30]. The gut enteroendocrine system comprises tens of thousands of cells dispersed through the lining of the gastrointestinal tract. These release hormone messengers that subserve three major roles: control of the ingestion of food; control of the digestion of the food; and control of metabolic functions. To do this, gut hormones act locally, on other regions of the digestive tract and on other organs, notably the brain, pancreas and liver.

16.5 Essential Nature of the ENS

The ENS has been called the brain in the gut, or the second brain [28], although in an evolutionary sense, it can be regarded as the first brain [25]. It contains entire neural circuits for physiological control, the gastrointestinal tract being the only organ with its own extensive intrinsic nervous system (there are also intrinsic ganglia in the airways and heart).

The ENS is essential to normal life, as indicated by the high morbidity or mortality associated with congenital or acquired dysfunction of this system. In Hirschsprung disease, the ganglia of the ENS fail to develop in the distal bowel, but all other tissue components are intact and functional. No propulsive activity occurs in the aganglionic bowel, and the newborn child will die if untreated. Rescue of these children requires removal of the section of bowel without enteric neurons and a re-joining of the ends of the intact bowel. The ENS of hydra, like that of human, is essential to life and neurons must be intact for adequate peristaltic movements of the gut tube to occur [59]. Without an ENS, the hydra will die, but can be kept alive by placing fresh food into the body cavity using a fine pipette. Degeneration of enteric neurons of the colon in Chagas disease, which is precipitated by infection with the protozoan *Trypanosoma cruzi*, causes colorectal propulsion to fail and megacolon to develop in adults, which is similar to this problem associated with Hirschsprung disease in children. The bowel with a deficient enteric nervous system needs to be removed.

Failure of the ENS to control fluid movement between the intestinal lumen and body fluid compartments is also life-threatening. Fluid movement across the epithelium lining the gut is controlled by enteric neurons that are pathologically activated by certain infective agents or their products. These pathogens, including *Vibrio cholerae* (which secretes cholera toxin) and rotavirus (which secretes an enterotoxin), act directly on enteric neurons and the epithelium itself, triggering hypersecretion and subsequent diarrhea. Infectious diarrhea causes over 1.5 million deaths a year, primarily in underdeveloped tropical countries [55].

16.5.1 Evolution of the Enteric Nervous System

The animals with a nervous system whose form can be traced back furthest are cnidarians, exemplified by the genus *Hydra* (500 mya). The cnidarian ENS is distributed as several hundreds of neurons that form a meshwork within the wall of the gut tube, but there is no evidence of the formation of specialized aggregations of neurons (head ganglia) that might be a first stage of brain formation [31, 32]. The neuronal plexus extends into the tentacles that sweep water and nutrients into the gut tube. The individual neurons of the cnidarian nervous system contain neurotransmitters, including catecholamines, 5-HT and neuropeptides, with synaptic specializations between the neurons [27, 39, 68].

Propulsive and mixing movements are observed in normal hydra in response to nutrients, but not to non-nutrients placed within the hydra digestive cavity. Thus it is clear that hydra have nutrient detectors and that neurons of the hydra enteric nervous system react to the introduction of food [59]. Therefore, like mammals, the enteric nervous system of hydra has sensory and motor neuron functions, which, given that the hydra enteric nervous system contains synaptically connected neurons, suggests that there may be separate sensory and motor neurons, and perhaps interneurons. In normal hydra after food has been digested (6–9 hours after feeding), circumferential and longitudinal movement of the body wall expels the residuum (cnidarian feces) into the environment. This suggests that, directly or indirectly, the hydra ENS detects products of digestion and the decline in nutrient concentration. Thus the cnidarian ENS, like the mammalian ENS, is programmed to direct a range of movements: peristalsis, mixing movements and expulsion. Lack of an ENS is incompatible with life in this most primitive of animals that has a nervous system, as it is in humans, who cannot survive the congenital lack of an enteric nervous system.

Other invertebrates, including echinoderms [26], acorn worms [40], cephalopods [3], gastropod molluscs [37, 53] and annelid worms [36] also have an ENS, some of which have architectures reminiscent of that of vertebrates, including human (Fig. 16.3). However, the ENS of insects has diverged from that of other invertebrates, in that enteric neurons are in ganglia and along nerve strands on the surface of the gut, rather than being embedded in its wall [13, 50]. In snails [1, 46], two layers of ganglia, identified as myenteric and submucosal ganglia, as is observed in other invertebrates and in mammals (Fig. 16.3), have been described. Nerve tracing studies indicate that ganglionic clusters of molluscs, including the stomatogastric ganglia, project axons to the digestive tube, perhaps making connections with enteric neurons [47].

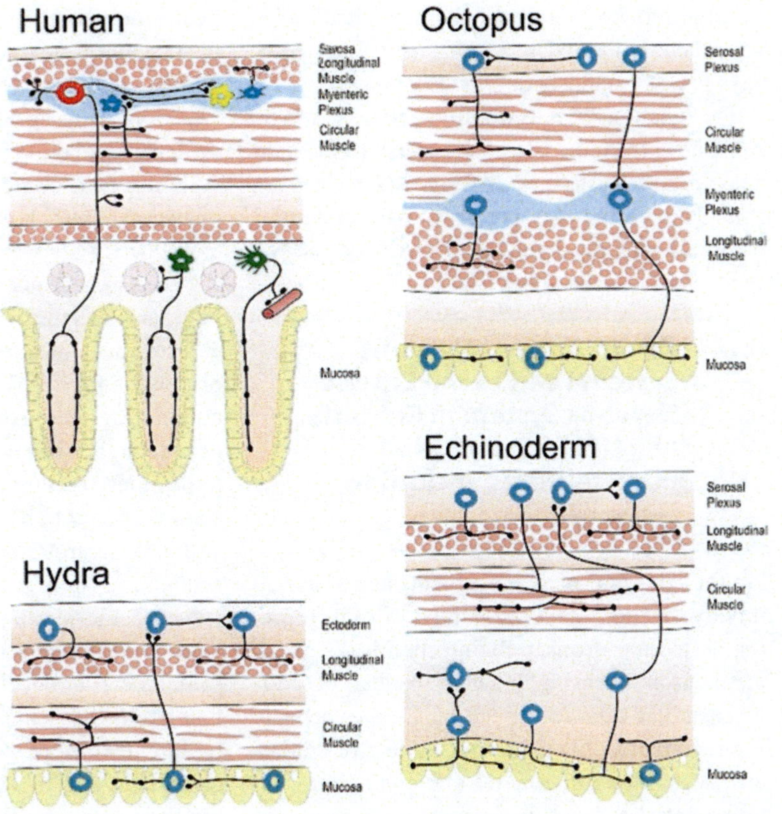

Fig. 16.3 Comparison of the ENS of a hydra and of an echinoderm, animals that do not have a CNS, with human and octopus, both of which have a CNS. Each diagram represents the gut wall in transverse section, with nerve cell bodies shown with their processes in black. In each of the four species there is an ENS that provides innervation of the muscle of the gut wall. There is detailed knowledge of the projections and connections of enteric neurons in human and other mammals, but detailed circuitry has not been determined in the other animals illustrated here

16.6 Reciprocal and Convergent Connections of the ENS and CNS

The ENS has outwardly projecting neurons, that, like those of the CNS, influence the functions of their targets [25]. In this sense, the CNS and ENS mirror each other, both giving rise to efferent pathways and both receiving afferent input. Both also contain intrinsic integrative circuitry. The innervation of peripheral organs other than the gut is connected with the CNS through efferent control pathways that emanate from the CNS and through primary afferent neurons with cell bodies in dorsal root and cranial ganglia close to the CNS, such as the trigeminal, nodose and jugular ganglia.

Neurons that project from the ENS to the CNS are entero-bulbar neurons with cell bodies in the stomach and proximal duodenum that project to the lower brain stem [34] and neurons that project from the rectum to the lumbo-sacral spinal cord [16, 51]. The recto-spinal neurons have typical enteric neuron, type I morphology and receive synapses, presumably from other enteric neurons. The cell bodies of gastro-bulbar and duodeno-bulbar neurons are contacted by vagal efferent neurons with cell bodies in the lower brain stem [34].

Other pathways that are peculiar to the ENS are the intestinofugal pathways, in which neurons with cell bodies in the intestine innervate the trachea [18], biliary system and gallbladder [42, 54], pancreas [43] and the cell bodies of sympathetic post-ganglionic neurons within prevertebral ganglia [21]. The axons of intestinofugal neurons that project to prevertebral ganglia converge on neurons that also receive synaptic input from the CNS. These intestinofugal neurons are parts of the afferent limbs of entero-enteric reflexes, through which one region of intestine causes inhibition of other, generally more proximal, regions [19, 63]. These peripheral reflexes that pass through abdominal sympathetic ganglia usually remain unmentioned in textbooks. Nevertheless, the existence of such reflex pathways is well established.

The functions of intestinofugal neurons whose axons arise from the ENS and project to the pancreas, biliary system or the airways have been little explored but they are presumably involved in coordinating the activities of these organs.

16.7 Did an Ancient Nervous System Lead to the Enteric Nervous System in Cnidaria and the ENS and CNS in Vertebrates, Including Human?

Our thoughts, our body movements and a myriad of body functions depend on the flow of information between neurons, utilizing chemical signals, one neuron releasing a small packet of neurotransmitter chemical to act on another neuron. It is remarkable that neurotransmitters of cnidarians, which have an enteric nervous system but not a central nervous system, are in many cases identical, and in other cases very similar, to the neurotransmitters that occur in the enteric nervous system and the central nervous system of mammals. These transmitters include nitric oxide [11] and oxytocin/vasopressin-like peptides [31, 59] and ACh [27]. The mammalian ENS and CNS neurotransmitter, serotonin (5-HT), occurs in the invertebrate ENS of a wide range of species, including hydra, acorn worms and insects [40, 44, 68]. It seems unlikely that the same chemicals, and the enzyme pathways and transporters needed to form and store them, evolved independently in *cnidaria* and mammals. More likely, the ancient enteric nervous system of hydras and corals was an evolutionary precursor of the central nervous system of mammals. It is possible that there was a common ancestor with a type of nervous system that gave rise to both. It is difficult to imagine what such an ancient nervous system would be doing if it was not controlling digestion. Thus, however the lineage is considered, the ENS appears to be the most evolutionary ancient nervous system, and should probably be regarded as the first nervous system (Fig. 16.4).

Previous considerations of the evolutionary origins of the CNS suggest that it arose in the common ancestors of the bilateria, urbilateria [5], even though it is generally agreed that some deuterostomes (echinoderms and hemichordates) cannot be said to have a CNS [33]. If the conservation of neurotransmitter systems is taken into account, as well as similarities in neuronal and synaptic morphologies between cnidarian and other invertebrate and vertebrate nervous systems [68], an earlier development of a nervous system than urbilateria can be proposed to have occurred in 'ureumetazoa', postulated ancestors of both cnidarians and bilateria [65]. This would place the ureumetazoan ENS as a likely precursor of both the modern ENS and the CNS. The proposal that the ENS is the 'first brain' is consistent with the ENS and CNS of mammals both having afferent and efferent connections with other organs, and having reciprocal connections with each other [25].

16.8 Conclusions

Diffusion of nutrients through surface membranes or endocytosis by single cells, and flow of fluid through penetrating channels in sponges, was sufficient for nutrition. As animals became

	Bilateria							Cnidaria
	Protostomes		Deuterostomes					
	Insects	Annelids	Echinoderms	Hemichordates	Amphioxis	Ascidians	Vertebrates	Hydra
CNS	Ventral	Ventral	None	None	Dorsal	Dorsal	Dorsal	None
ENS	Yes	Yes	Yes	Yes	Yes	Yes	Yes	Yes

Fig. 16.4 Animals with an ENS and no CNS (ventral or dorsal), or with both

larger and more complex, a nervous system, the ENS, was needed to direct the assimilation of nutrients. This improved, neutrally controlled digestive apparatus provided the energy flow that could sustain a central nervous system. Animals became dependent on an ENS, without which they could not survive.

References

1. Ábrahám A (1940) Die Innervation des Darmkanals der Gastropoden. Cell Tissue Res 30:273–296
2. Aiello LC, Wheeler P (1995) The expensive-tissue hypothesis: the brain and the digestive system in human and primate evolution. Curr Anthropol 36:199–221
3. Alexandrowicz J-S (1928) Notes sur l'innervation du tube digestif des céphalopodes. Arch Zool Exp 67:69–90
4. Allen RD (1974) Food vacuole membrane growth with microtubule-associated membrane transport in *Paramecium*. J Cell Biol 63:904–922
5. Arendt D, Denes AS, Jékely G, Tessmar-Raible K (2008) The evolution of nervous system centralization. Philos Trans R Soc Lond B Biol Sci 363:1523–1528
6. Bach A, Calsamiglia S, Stern MD (2005) Nitrogen metabolism in the rumen. J Dairy Sci 88:E9–E21
7. Barker CJ, Gillett A, Polkinghorne A, Timms P (2013) Investigation of the koala *(Phascolarctos cinereus)* hindgut microbiome via 16S pyrosequencing. Vet Microbiol 167:554–564
8. Bernalier-Donadille A (2010) Fermentative metabolism by the human gut microbiota. Gastroenterol Clin Biol 34:S16–S22
9. Brouns F, Theuwissen E, Adam A, Bell M, Berger A, Mensink RP (2012) Cholesterol-lowering properties of different pectin types in mildly hypercholesterolemic men and women. Eur J Clin Nutr 66:591–599
10. Clauss M, Hume ID, Hummel J (2010) Evolutionary adaptations of ruminants and their potential relevance for modern production systems. Animal 4:979–992
11. Colasanti M, Venturini G, Merante A, Musci G, Lauro GM (1997) Nitric oxide involvement in *Hydra vulgaris* very primitive olfactory-like system. J Neurosci 17:493–499
12. Copeland L, Blazek J, Salman H, Tang MC (2009) Form and functionality of starch. Food Hydrocoll 23:1527–1534
13. Copenhaver PF (2007) How to innervate a simple gut: familiar themes and unique aspects in the formation of the insect enteric nervous system. Dev Dyn 236:1841–1864
14. Cordain L, Eaton SB, Sebastian A, Mann N, Lindeberg S, Watkins BA, O'Keefe JH, Brand-Miller J (2005) Origins and evolution of the Western diet: health implications for the 21st century. Am J Clin Nutr 81:341–354
15. Dierenfeld ES, Hintz HF, Robertson JB (1982) Utilization of bamboo by the giant panda. J Nutr 112:636–641
16. Doerffler-Melly J, Neuhuber WL (1988) Rectospinal neurons: evidence for a direct projection from the enteric to the central nervous system in the rat. Neurosci Lett 92:121–125
17. Eng CM, Lieberman DE, Zink KD, Peters MA (2013) Bite force and occlusal stress production in hominin evolution. Am J Phys Anthropol 151:544–557
18. Fischer A, Canning BJ, Undem BJ, Kummer W (1998) Evidence for an esophageal origin of VIP-IR and NO synthase-IR nerves innervating the guinea pig trachealis: a retrograde neuronal tracing and immunohistochemical analysis. J Comp Neurol 394:326–334
19. Furness JB (2006) The enteric nervous system. Blackwell, Oxford
20. Furness JB, Bravo DM (2015) Humans as cucinivores: comparisons with other species. J Comp Physiol B 185:825–834
21. Furness JB, Callaghan B, Rivera LR, Cho HJ (2014) The enteric nervous system and gastrointestinal innervation: integrated local and central control. Adv Exp Med Biol 817:39–71
22. Furness JB, Cottrell JJ, Bravo DM (2015) Comparative physiology of digestion. J Anim Sci 93:485–491
23. Furness JB, Fakhry J, Gajewski J, Boyle EK, Fothergill L (2019) The digestive system in evolutionary medicine. In: Brüne M, Schiefenhövel W (eds) Oxford handbook of evolutionary medicine. Oxford University Press, pp 531–562
24. Furness JB, Rivera LR, Cho H-J, Bravo DM, Callaghan B (2013) The gut as a sensory organ. Nat Rev Gastroenterol Hepatol 10:729–740
25. Furness JB, Stebbing MJ (2018) The first brain: species comparisons and evolutionary implications for the enteric and central nervous systems. Neurogastroenterol Motil 30:e13234
26. García-Arrarás JE, Rojas-Soto M, Jiménez LB, Díaz-Miranda L (2001) The enteric nervous system of echinoderms: unexpected complexity revealed by neurochemical analysis. J Exp Biol 204:865–873
27. Giez C, Klimovich A, Bosch TC (2021) Neurons interact with the microbiome: an evolutionary-informed perspective. e-Neuroforum. https://doi.org/10.1515/nf-2021-0003
28. Gershon MD (1999) The enteric nervous system: a second brain. Hosp Pract 34:31–32
29. Gordon JL, LeBlanc SJ, Duffield TF (2013) Ketosis treatment in lactating dairy cattle. Vet Clin North Am Food Anim Pract 29:433–445
30. Gribble FM, Reimann F, Roberts GP (2018) Gastrointestinal hormones. In: Said HM (ed) Physiology of the gastrointestinal tract 6edn. Academic Press, New York, pp 31–70
31. Grimmelikhuijzen CJP, Dierickx K, Boer GJ (1982) Oxytocin/vasopressin-like immunoreactivity is

present in the nervous system of hydra. Neuroscience 7:3191–3199
32. Grimmelikhuijzen CJP, Spencer AN, Carré D (1986) Organization of the nervous system of physonectid siphonophores. Cell Tissue Res 246:463–479
33. Holland LZ, Carvalho JE, Escriva H, Laudet V, Schubert M, Shimeld SM, Yu J-K (2013) Evolution of bilaterian central nervous systems: a single origin? EvoDevo 4:27
34. Holst MC, Kelly JB, Powley TL (1997) Vagal preganglionic projections to the enteric nervous system characterized with phaseolus vulgaris-leucoagglutinin. J Comp Neurol 381:81–100
35. Hu Y, Wu Q, Ma S, Ma T, Shan L, Wang X, Nie Y, Ning Z, Yan L, Xiu Y, Wei F (2017) Comparative genomics reveals convergent evolution between the bamboo-eating giant and red pandas. Proc Natl Acad Sci 114:1081–1086
36. Ierusalimsky VN, Balaban PM (2006) Immunoreactivity to molluskan neuropeptides in the central and stomatogastric nervous systems of the earthworm, *lumbricus terrestris* L. Cell Tissue Res 325:555–565
37. Ito S, Kurokawa M (2007) Coordinated peripheral neuronal activities among the different regions of the digestive tract in Aplysia. Zool Sci 24:714–722
38. Jin C, Ciochon RL, Dong W, Hunt RM, Liu J, Jaeger M, Zhu Q (2007) The first skull of the earliest giant panda. Proc Natl Acad Sci U S A 104:10932–10937
39. Kass-Simon G, Pierobon P (2007) Cnidarian chemical neurotransmission, an updated overview. Comp Biochem Physiol 146:9–25
40. Kaul-Strehlow S, Urata M, Minokawa T, Stach T, Wanninger A (2015) Neurogenesis in directly and indirectly developing enteropneusts: of nets and cords. Org Divers Evol 15:405–422
41. Kelly KA (1980) Gastric emptying of liquids and solids: roles of proximal and distal stomach. Am J Phys 239:G71–G76
42. Kennedy AL, Mawe GM (1998) Duodenal sensory neurons project to sphincter of Oddi ganglia in guinea pig. J Neurosci 18:8065–8073
43. Kirchgessner AL, Gershon MD (1990) Innervation of the pancreas by neurons in the gut. J Neurosci 10:1626–1642
44. Klemm N, Hustert R, Cantera R, Nässel DR (1986) Neurons reactive to antibodies against serotonin in the stomatogastric nervous system and in the alimentary canal of locust and crickets (Orthoptera, Insecta). Neuroscience 17:247–261
45. Leonard WR, Snodgrass JJ, Robertson ML (2007) Effects of brain evolution on human nutrition and metabolism. Annu Rev Nutr 27:311–327
46. Martinez-Pereira MA, Franceschi RDC, Antunes GDF, Coelho BP, Achaval M, Zancan DM (2013) General morphology and innervation of the midgut and hindgut of *Megalobulimus abbreviatus* (Gastropoda, Pulmonata). Zool Sci 30:319–330
47. Martinez-Pereira MA, Franceschi RDC, Coelho BP, Zancan DM (2017) The stomatogastric and enteric nervous system of the pulmonate snail *Megalobulimus abbreviatus*: a neurochemical analysis. Zool Sci 34:300–311
48. Milton K (2003) The critical role played by animal source foods in human (*Homo*) evolution. J Nutr 133:3886S–3892S
49. Milton K, Demment MW (1988) Digestion and passage kinetics of chimpanzees fed high and low fiber diets and comparison with human data. J Nutr 118:1082–1088
50. Nässel DR, Eckert M, Muren JE, Penzlin H (1998) Species-specific action and distribution of tachykinin-related peptides in the foregut of the cockroaches *Leucophaea maderae* and *Periplaneta americana*. J Exp Biol 201:1615–1626
51. Neuhuber WL, Appelt M, Polak JM, Baier Kustermann W, Abelli L, Ferri GL (1993) Rectospinal neurons: cell bodies, pathways, immunocytochemistry and ultrastructure. Neuroscience 56:367–378
52. Nie Y, Zhang Z, Raubenheimer D, Elser JJ, Wei W, Wei F (2015) Obligate herbivory in an ancestrally carnivorous lineage: the giant panda and bamboo from the perspective of nutritional geometry. Funct Ecol 29:26–34
53. Okamoto T, Kurokawa M (2010) The role of the peripheral enteric nervous system in the control of gut motility in the snail Lymnaea stagnalis. Zool Sci 27:602–610
54. Padbury RTA, Furness JB, Baker RA, Toouli J, Messenger JP (1993) Projections of nerve cells from the duodenum to the sphincter of Oddi and gallbladder of the Australian possum. Gastroenterology 104:130–136
55. Podewils LJ, Mintz ED, Nataro JP, Parashar UD (2004) Acute, infectious diarrhea among children in eveloping countries. Semin Pediatr Infect Dis 15:155–168
56. Raven HC (1936) Notes on the anatomy of the viscera of the giant panda (*Ailuropoda melanoleuca*). Am Mus Novit 877:1–23
57. Raz-Bahat M, Douek J, Moiseeva E, Peters EC, Rinkevich B (2017) The digestive system of the stony coral *Stylophora pistillata*. Cell Tissue Res 368:311–323
58. Renard E, Gazave E, Fierro-Constain L, Schenkelaars Q, Ereskovsky A, Vacelet J, Borchiellini C (2013) Porifera (sponges): recent knowledge and new perspectives. eLS. https://doi.org/10.1002/9780470015902.a0001582.pub2
59. Shimizu H, Koizumi O, Fujisawa T (2004) Three digestive movements in *Hydra* regulated by the diffuse nerve net in the body column. J Comp Physiol 190:623–630
60. Shimizu H, Takaku Y, Zhang X, Fujisawa T (2007) The aboral pore of hydra: evidence that the digestive

tract of hydra is a tube not a sac. Dev Genes Evol 217:563–568
61. Stevens CE, Hume ID (1995) Comparative physiology of the vertebrate digestive system. Cambridge University Press, Cambridge, New York
62. Stevens CE, Hume ID (1998) Contributions of microbes in vertebrate gastrointestinal tract to production and conservation of nutrients. Physiol Rev 78:303–427
63. Szurszewski JH, Ermilov LG, Miller SM (2002) Prevertebral ganglia and intestinofugal afferent neurones. Gut 51:i6–i10
64. van Boekel M, Fogliano V, Pellegrini N, Stanton C, Scholz G, Lalljie S, Somoza V, Knorr D, Rao Jasti P, Eisenbrand G (2010) A review on the beneficial aspects of food processing. Mol Nutr Food Res 54:1215–1247
65. Watanabe H, Fujisawa T, Holstein TW (2009) Cnidarians and the evolutionary origin of the nervous system. Develop Growth Differ 51: 167–183
66. Watanabe H, Tokuda G (2001) Animal cellulases. Cell Mol Life Sci 58:1167–1178
67. Weaver TD (2012) Did a discrete event 200,000—100,000 years ago produce modern humans? J Hum Evol 63:121–126
68. Westfall JA, Elliott SR, MohanKumar PS, Carlin RW (2000) Immunocytochemical evidence for biogenic amines and immunogold labeling of serotonergic synapses in tentacles of *Aiptasia pallida* (Cnidaria, Anthozoa). Invertebr Biol 119:370–378
69. Wrangham R, Conklin-Brittain N (2003) Cooking as a biological trait. Comp Biochem Physiol A Mol Integr Physiol 136:35–46
70. Yao CK, Muir JG, Gibson PR (2016) Review article: insights into colonic protein fermentation, its modulation and potential health implications. Aliment Pharmacol Ther 43:181–196
71. Zhao H, Yang J-R, Xu H, Zhang J (2010) Pseudogenization of the umami taste receptor gene Tas1r1 in the giant panda coincided with its dietary switch to bamboo. Mol Biol Evol 27:2669–2673

Enteric Glia and Enteric Neurons, Associated

17

Giorgio Gabella

Abstract

Peripheral neurons are never found alone and are invariably accompanied by glial cells, with which they are intimately associated in compact, highly deformable structures.

Myenteric ganglia of the guinea-pig, examined *in situ* by electron microscopy, show that in their neuropil (axons and dendrites, and glial cells and processes) the glia constitutes almost half of the volume and almost half of membrane extent.

In the glia, the expanse of the cell membrane predominates over that of their cytoplasm, the opposite being the case with the neural elements.

The profile of the glial elements is passive and is dictated by the surrounding elements, mainly the axons, and hence it is predominantly concave.

The enteric glia is widely developed; however, it is not sufficient to form a full wrapping around all neurons and around all axons (unlike what is found in other autonomic ganglia).

Glial processes are radially expanding laminae, irregularly tapering, branching, and penetrating between axons.

Some processes have a specialized termination attached to the basal lamina of the ganglion.

Mitochondria are markedly more abundant in neural element that in the glia (up to a factor of 2).

Many expanded axons, laden with vesicles clustered beneath membrane sites, abut on glial cells and processes, while these show no matching structural specializations.

Keywords

Enteric glia · Myenteric ganglia · Mitochondria · Axon-glia contacts

The symbiosis of neurons and glia is a remarkable aspect of the nervous tissue. In living bodies, neurons never live alone, or move alone, or develop alone, and conversely glial cells are not found that are unattached to a neuron. In the periphery too axons are not without a glial process around them (except at the very end of some autonomic axons or in discrete areas of enteric ganglia). In living tissues no glial cell has been found that is unattached to a nerve cell or to a neurite.

The intramural nerves of the gut (the enteric nervous system) are no exception, and for its glia the term 'enteric glial cells' was introduced

G. Gabella (✉)
University College London, London, UK
e-mail: g.gabella@ucl.ac.uk

N. J. Spencer et al. (eds.), *The Enteric Nervous System II*, Advances in Experimental Medicine and Biology 1383, https://doi.org/10.1007/978-3-031-05843-1_17

50 years ago [5]. The fine structural features [3, 5–7, 14] and some immunochemical markers [1, 4, 13] of enteric glial cells were quickly characterized. Chemical markers, such as glial fibrillary acidic protein (GFAP) and S100b have become key tools to investigate these cells in vivo [12] and far more frequently in vitro; the transcription factor Sox8/9/10, localized in the nucleus, has also been useful for cell counts [12]. Comprehensive review articles on the subject have appeared regularly from as early as 30 years ago and they have maintained their usefulness (e.g. [8, 18]) and include a full historical review [9].

Our understanding of the structure of enteric glia has progressed slowly; here, a summary account of enteric glial morphology is presented, one mainly based on data from electron microscopy and founded on the supposition that structural evidence is critical for correct attribution of functional properties.

In addition, although outside the scope of this article, a large population of enteric glial cells, mainly located in the mucosa, has come to light. These cells are attracting interest because of suggestive evidence of a role in immune processes in the mucosal epithelium and an interaction with the gut microbe. A direct role has been suggested for these cells in the peristaltic reflex and in inflammation, and these properties are examined in several studies (e.g. [2, 19]; Gubisic and Gulbransen 2017). However, the present review is focused on documented ultrastructural evidence of ganglia and intramural nerves studied in situ, and the studies of the particular enteric glial cells in the mucosa are beyond its scope.

17.1 Cytology of Glia

Any micrograph of enteric ganglia shows a distinct population of neurons – however variable in size, shape and other structural features – that are intermingled with an unambiguously different population of glial cells. Unidentified cells may be found, even by electron microscopy, but they are quite rare.

There is no uncertainty in distinguishing nerve cell bodies and neurites (that is dendrites and axons) from glial cells and glial processes (Figs. 17.1a, b and 17.2). Whether there are different types of enteric glial cells is a difficult question. Even with supportive data (e.g. [10]) there is still no conclusive evidence. Beyond the physiologic variability in the engagement of these cells, there may be different sources and lineages [18]. Enteric glial cells are much smaller than the neurons and are rather uniform in size; their nucleus is slightly elongated, electron-dense, non-globular, and often presents deep indentations.

Some immunochemical preparations [1] and microinjected preparations [10] document the exceptional branching out of the glial processes over a relatively short distance from the cell body. Branches constitute to larger part of these cells. Glial laminar processes are common that taper away regularly from the cell body; they form large expanses whose cytoplasm is progressively reduced in thickness until it is less than one twentieth of a micrometre and becomes devoid of any organelles. Some large processes extend to the ganglion's surface (Fig. 17.1b), and there are glial cells that reach both surfaces of a myenteric ganglion (that is facing both muscle layers). At least half of the surface of a myenteric ganglion is made by the glia. A single basal lamina covers the entire ganglion and the nerve strands; it is markedly more evident in the myenteric than in the submucosal plexus.

Adherens junctions are found between glial processes or between glia and neuritis, and gap junctions are noted between glial processes, although neither type of junction is frequent. An extensive laminar covering by a glial laminar process is found around some neurons. However, in all ganglia studied, some large areas of neuronal cell bodies lie directly beneath the ganglionic basal lamina.

A typical component in the majority of glial cells are 10-nm filaments (gliofilaments), gathered in thin bundles which appear to lie in all directions (Fig.17.1c). They are more abundant in the myenteric than in the submucosal glial

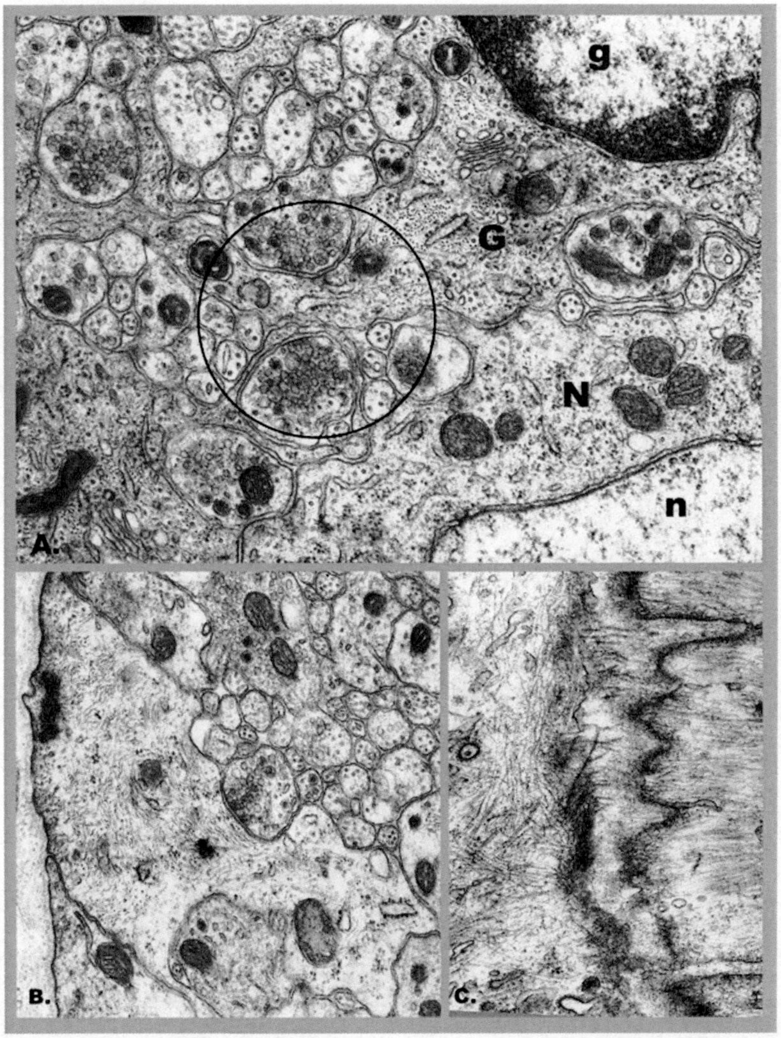

Fig. 17.1 (**a**) Electron micrograph of a myenteric ganglion of a guinea pig ileum. To the right, at top, a glia cell (G) with its nucleus (g), and, at bottom, a neuron (N) with its nucleus (n). The circle highlights two axonal profiles (varicosities), packed with vesicles and with a membrane density associated with vesicles; both axons are apposed to a process issued from the glial cell. At these points of contact between axons and glia, the glial membrane shows no specialization, that is, no structural features, and does not suggest the presence of a junction. At the left, the microscopic field is occupied by many axons, in transverse section, roundish or oval in profile, containing microtubules, some clusters of vesicles and an occasional mitochondrion. (**b**) In this micrograph of another similar ganglion, a large glial process expands and reaches the surface of the ganglion at the left edge. Small groups of gliofilaments lie in all directions and some are inserted in the part of the glial membrane that is in contact with the basal lamina (gliofilaments are not inserted on any other area of the cell membrane). A membrane-bound dense body projects into the glial cytoplasm and is reached by many gliofilaments. A vesicle-loaded axon makes contact with this glial process, at the centre of the image. (**c**) In this image from the surface of a myenteric ganglion of a guinea pig ileum, the different patterns of three types of filamentous structures are in evidence. To the right, myofilaments are oriented parallel and many of them are attached to dense bodies beneath the surface of a smooth muscle cell. To the left, a glial process in a myenteric ganglion shows gliofilaments lying in all directions and inserting on densities beneath the glial cell membrane, obliquely sectioned

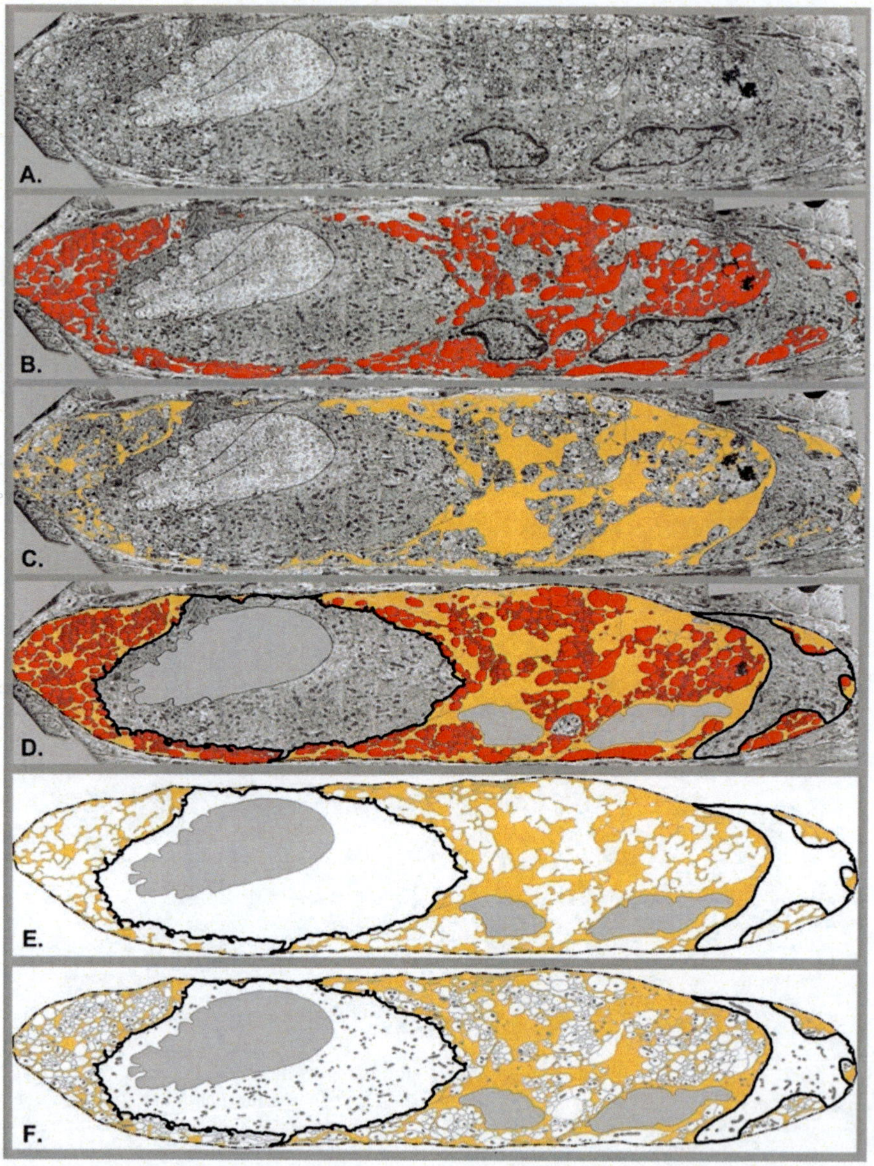

Fig. 17.2 (**a**) Myenteric ganglion from a guinea pig. Several electron micrographs of a thin section are assembled in a montage, showing the entire profile of the ganglion at low resolution. A large neuron occupies the left part of the ganglion, and it shows its pale nucleus; the corner of another neuron appears at the far right. The nuclei of two glial cells are visible at bottom right. Morphometric data on this ganglion are in Table 17.2. (**b**) In the same ganglion, all the axons (axonal profiles) are traced digitally (with the software FreeHand from Micromedia) with a thin dark line, and filled in red. There are 1208 axons. (**c**) The membrane of all the glial elements is traced digitally with a thin dark line, and filled in yellow. (**d**) The tracings of axons (as in B.) and of glia (as in C.) are added, highlighting the interdigitation of the two components. The neuronal nucleus is outlined and shaded in grey. The outline of the neurons is traced with a thick black line. (**e**) Only the outline of the neural cell and all the glia are shown here. The glial nuclei are shaded in grey. (**f**) On the image in E, the tracing of all the mitochondria is added

cells, and in poor evidence in the glial cells of nerve strands. Some gliofilament bundles are anchored to the cell membrane that is apposed to the basal lamina: dense bodies, some projecting deeply into the cytoplasm, are characteristic mostly in the glial processes facing the longitudinal muscle (Fig. 17.1c).

An intriguing observation, not uncommon and already noted in earlier studies [7], is well documented in guinea pig ganglia, both myenteric and submucosal: apposed to a glial cell or a glial process are axonal varicosities with vesicles (electron lucent, of 40–45 nm diameter) clustered against a dense patch of the axolemma (the cell membrane) in an arrangement similar to the axonal side of a synapse; no specializations appear on the glial side, except for an extended apposition with these axons. (A matter of concern is that these structures have not been mentioned by other authors.) Here, these intriguing structures are referred to as axon-glia contacts (or appositions), which might facilitate an influence of the axon on the glia, but are not regarded as junctions because we lack evidence of some membrane specialization on the glial side (Fig. 17.1a, b). Conservatively, the micrographs simply show that axons can develop varicosities, with clustering of vesicles and with a membrane density usually associated with a synapse, even in the absence of a potential post-junctional element. If these axonal expansions (varicosities) with their vesicle clustering and membrane densities are sites of some chemical release, then they suggest the possibility of a non-synaptic release of axonal neurotransmitters inside a ganglion.

17.2 Glial Populations

The enteric glial population is vast. Anne Rühl and colleagues [12], by histochemistry, found about six times as many glial cells as neurons in human myenteric ganglia. Earlier studies on guinea pigs, based on histologic sections, showed a preponderance of glial cells over neurons in many enteric ganglia [7] (Table 17.1). In the ileum, for example, myenteric ganglia have a substantial numerical predominance of glial cells over neurons; the values are reproduced between individuals and therefore must be regulated to some extent – with an average of 250:100, or two and a half glial cells for every neuron. There are no clear differences in other regions of the small intestine, and the values are similar – and probably the same – in the stomach and large intestine (Table 17.1).

A relevant cytological parameter is the extent of the neuropil, expressed as its percentage volume within a ganglion (volumes are calculated morphometrically from surface areas in thin sections). Neuropil is the collection of axons, dendrites, glial cell bodies, and glial processes, that is, all the components of a ganglion except the nerve cell bodies. The neuropil is thus the part of the plexus where impulses are transmitted and delivered, as opposed to the neuronal cell bodies

Table 17.1 Morphometric data on myenteric and submucosal ganglia of some mammalian species

		Myenteric ganglia		Submucosal ganglia	
		Neuron to glia	Neuropil	Neuron to glia	Neuropil
Mouse	Ileum	100:109	51%	100:64	38%
Guinea pig	Stomach	100:235	63%	–	–
	Duodenum	100:224	66%	100:96	50%
	Ileum	100:266	65%	100:71	51%
	Colon	100:217	62%	100:77	53%
	Rectum	100:228	67%	100:78	54%
Rabbit	Ileum	100:259	70%	100:109	44%
Sheep	Ileum	100:466	74%	100:150	44%

The figures are the number of glial cells per 100 ganglion neurons, counted on tangential sections of ganglia, and corrected for the different size of nerve and glial cells

The second numerical column from the left and the last column on the right give the percentage of the space occupied by the neuropil

Table 17.2 An example of morphometric data obtained by measuring the area of all the profiles forming the ganglion (nerve cells, axons, glial cells, glial processes, as illustrated in Fig. 17.1 and in the same ganglion as in Fig. 17.2 from a guinea pig ileum) in a thin section under the electron microscope and the length of the cell membranes of all those profiles

GP ileum AFJ myenteric	Cellular areas	Percentage areas	Cellular perimeters	Membrane percentage	Mitochondria number area % area (excl. nuclei)			
Ganglion	753 μm^2	93%	134 μm	[2435 μm]	603	30.19 μm^2		
2 nerve cells	312 μm^2	41.4%	111 μm	4.6%	249	14.69 μm^2	6.34%	6.43%
1196 neurites	197 μm^2	26.2%	1507 μm	61.9%	275	12.93 μm^2	6.55%	
3 glial cells	91.7 μm^2	12.2%	188 μm	7.7%	28	1.29 μm^2	2.36%	1.67%
Glial processes	98.7 μm^2	13.1%	629 μm	25.8%	51	1.28 μm^2	1.30%	

peri/extracellular ~7%

The three glial cells and all the glial processes occupy a quarter of the sectional area of this ganglion (this includes the three glial nuclei)

The linear extent of all the cell membranes within the section is more than 18 times the perimeter of the ganglion, and one third of all the membrane is contributed by the glia

The columns to the right present the total number of mitochondria in the section of this ganglion and their distribution in the nerve cell bodies, the neurites (mainly axons), the glial cell bodies and the glial processes. More than 6% of the neuronal cytoplasm, in cell and processes, is occupied by mitochondria, while less than 2% is the corresponding figures for the glia

where impulses are generated (and, also, impulses are received and integrated).

In myenteric ganglia of guinea pigs, the neuropil amounts consistently to about two-thirds of the ganglion volume, the remaining one-third being made up of nerve cell bodies (Table 17.1). In contrast, the submucosal plexus in guinea pigs has a neuropil amounting to only about half the ganglion volume; glial cells are less abundant than in myenteric ganglia and not more numerous than neurons (except in the duodenum) (Table 17.1).

While these figures have limited practical use, they might be handy in comparative studies, and to monitor development, maintenance state and ageing of these structures. In addition, figures of this type have, in theory, a use in exploring the morphogenesis of the ganglia, that is, those processes becoming visible as the endogenous emergence of forms that are not pre-existing.

Taking a single myenteric ganglion of a guinea pig ileum as an example (Fig. 17.2), of its sectional area of 750 square microns about a quarter is glia, while the cell membranes of all the glia amount to at least a third of the total of 2400 microns of cell membrane (Table 17.1).

Applying the same histologic probe to a few other species (not many, out of over 5000 mammalian species in existence), substantial differences are found as well as some patterns (Table 17.1); for example, in mice, fewer glial cells and a lesser extent of the neuropil, whereas in sheep, a much greater proportion of glial cells to neurons and a more extensive neuropil.

The difference between myenteric and submucosal ganglia remains noticeable in all species. On average, neurons of the submucosal ganglia are smaller and more tightly packed than in the corresponding myenteric ganglia, and glial cells less numerous. Variation in size (at least the nuclear size) in glial cells is minimal between ganglia and between species; in contrast, the range of neuronal sizes and the average neuronal size vary between species with trends somehow related to the size of the body or of the organ.

These data may suggest, therefore, some vague correlation between body size, neuronal populations and amount of glia. However, quantitative differences of this type are not indications of variations of function of the glia in different species (as has been suggested). They may be related, in large species, to the longer distances covered by the axons, to the larger innervation territories, to differences in size-related external mechanical effects, and to other factors.

Therefore one should be prudent when considering factors that may influence these relationships. Glia and neurons are so intimately associated that one can expect as much influence of neurons on glia as from glia on neurons. The two cell types (and their subtypes) cannot be set apart or investigated independently, and only partial views are obtained when they are torn apart. Neurons and glia coexist, having reached some kind of working equilibrium and compatibility, including relatively stable structural parameters.

17.3 Relative Extent of Glia

From the comparative data a critical point emerges (Table 17.1). In spite of their large numbers and the vast extent of the glial domain, the enteric glial cells are relatively scarce, scarce by relation to the associated neurons, or at least the scarcer that is found in the nervous tissue in general. The enteric glial cells seem to be in some limited supply, one might say they are inadequately developed, however surprising the notion may sound. The enteric glial population is indeed vast; however, the ratio glia-to-neuron is markedly lower in the gut than in most parts of the nervous system. Everywhere in the peripheral autonomic nervous system, glial cells are more abundant, relative to neurons, than in the enteric nervous system.

The point on the limited extent of enteric glia is documented by ultrastructural observations.

The predominant rule in autonomic nerves is that each axon is fully wrapped by a glial process and it has an individual mesaxon (Fig. 17.3a). In contrast, many axons in enteric ganglia and strands are in direct contact with each other without any glia intervening or wrapped around them (Fig. 17.3b). This feature can be observed in all nerves and ganglia, and is seen also in ganglia where the neuropil, with its glia, is much extended, as in myenteric ganglia of sheep (Fig. 17.3c–e).

17.4 Glial Chondrioma

An aspect of the enteric glia that is shared widely across the nervous tissue is the amount of mitochondria (the chondrioma) (Fig. 17.2f).

On large montages of ganglia, mitochondria can be traced digitally in neurons, axons, glial cells and glial processes, generating accurate values on number, size, shape and localization of every mitochondrion, each easily identifiable by electron microscopy (Fig. 17.1a). Numbers and distributions are rather constant and therefore must be regulated. In addition – and probably of physiological significance – the number of mitochondria and their spatial density (that is, their overall percentage volume in a given structure) is far from similar in neuronal and glial elements (Fig. 17.4a–c). In all the preparations of enteric ganglia, the number of mitochondria and their percentage volume are markedly lower in the glia than in axons and nerve cells, as shown in the myenteric ganglion used as an example of the morphometric approach (Table 17.1) and visible in micrographs and tracings (Figs. 17.2f and 17.4a–c).

The discrepancy is conspicuous and is noted also in submucosal ganglia; it is found in the ganglia of all the other species examined, for example the sheep (Fig. 17.4). The volume of the chondrioma indicates the maximum metabolic rate available in a cell. In all cases the values strongly suggest that the metabolic rate in the glia is substantially lower than in neurons.

17.5 Research Limitations

The anatomical work on the enteric glia presented here has noticeable limitations, because it operates under the current view that biological structures are fully determined by genetic factors, a Mendelian perspective, and it may not account sufficiently for an influence of the environment on development of neural circuits (there are some strong data in a study by Dulac and Le Douarin, 1991), and for mechanisms acquired by trial and error or some form of learning. Surely, some

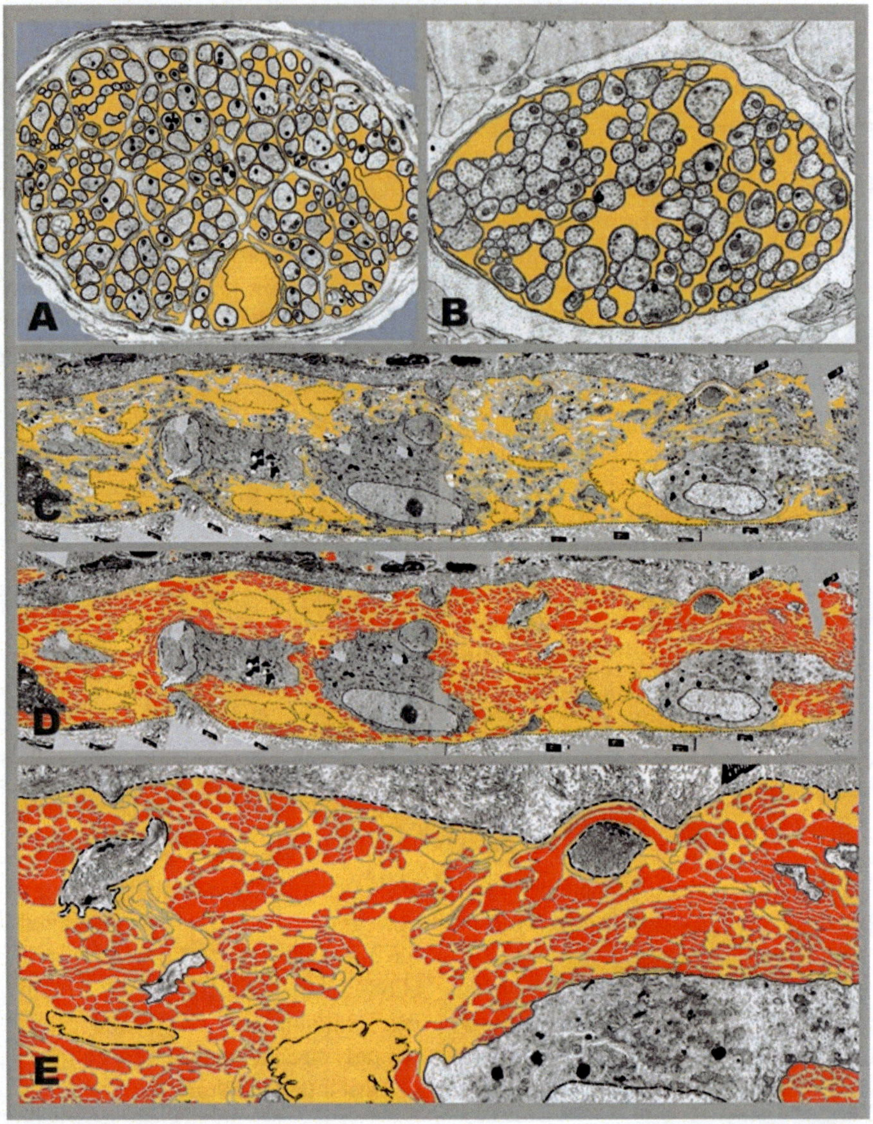

Fig. 17.3 (**a**) Transverse section of an intramural branch of the vagus nerve of a guinea pig, close to its point of entry in the stomach wall. The glial elements are traced with a dark line and filled in yellow. All the axons (numbering 232) are fully wrapped by glial processes, and none of the axons are in contact with each other. Two glial nuclei are shaded in dark grey. A perineurium surrounds this nerve trunk. (**b**) Transverse section of a nerve strand of the myenteric plexus of a guinea pig ileum. All the glia is traced and filled in yellow. Many of the 149 axons present are in direct contact with each other, without intervening glia. A thin basal lamina surrounds the strand and some muscle cells of the circular layer are at the top. (**c**) Myenteric ganglion of the ileum of a sheep. Several electron micrographs of a thin section are assembled in a montage, showing the profile of the ganglion at low resolution. Two large neurons, with nucleus, are visible below the centre and near the right bottom corner. All the glial elements, and the glial nuclei, are traced with a dark line and filled in yellow. The basal lamina of the ganglion is traced with a black dotted line. (**d**) Same as (**c**), with the added tracings of all the axons with a dark line and a red fill. (**e**) Detail of D. enlarged. Part of the nucleated neuron is at bottom right. Near the top surface of the ganglion (facing the circular muscle), there are deep invaginations of the periganglionic connective tissue penetrating into the ganglion

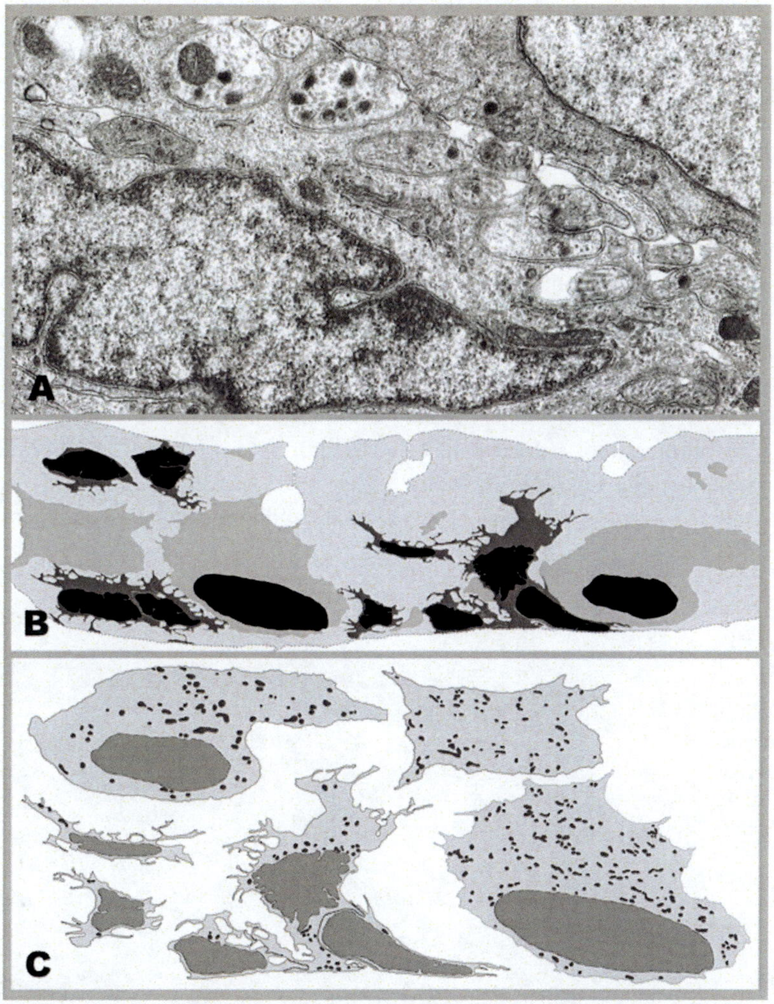

Fig. 17.4 (**a**) Myenteric ganglion of the ileum of a sheep, showing a detail from the bottom right area of Fig. 17.3a. A neuronal nucleus and a glial nucleus are visible. (**b**) Tracing of the same sheep ganglion (light grey), with tracings of the three neurons (grey) and of nine glial cells (dark grey). The nuclei are traced and filled in black and the non-ganglionic space is in white. (**c**) From the ganglion in Figs. 17.3a and 17.4b, three neurons (two of them nucleated) and five nucleated glial cell bodies are traced individually and filled in light grey; the nuclei are traced and filled in grey. All the mitochondria are also traced, to document their size, shape and distribution, and, above all, to show the difference in their spatial density in glial and neuronal elements. In this set of cells, mitochondria in nerve cells ($n = 536$) occupy 7.7% of the cytoplasm, in the glial cells ($n = 70$) they occupy 4.53% of the cytoplasm

functions of the enteric nerves are learned rather than simply inherited. In addition, there are many technical limitations in microscopy, hence legitimate doubts whether one is sufficiently critical in evaluating the reliability of the methods used.

Lastly, morphologists such as the author tend to fall prey to aesthetic considerations when they look down a microscope. Nevertheless, it cannot be denied that there is attraction, and maybe meaning, in the compactness of the enteric ganglia, their linear precision, the overdose of detail, and the absence of any available space. The minute interdigitations of neurons and glia produce a perfect construct of positive and negative form, and of reciprocal moulding. To say nothing of possible hidden patterns.

17.6 Ganglionic Dense Packing

Looking at a corner of a ganglion, and lifting axons and glia by separate tracings, the small amount of glia is apparent, with all the glial membrane deeply curved inward (Fig. 17.5). Except at the edge of extremely sharp folds, the glial membrane is always concave, with much reduction of the cytoplasm. In contrast, the surface of axons, whose transverse profiles are close to circular, approaches a minimum for the volume and is virtually always convex. A kind of turgidity of the axonal cytoplasm makes axons more rigid and incompressible and capable of always indenting into the glia and never the other way around.

The summing-up impression – granting the use of the word 'impression', or visual impression, which occasionally may be justified – is that axons minimize the axonal membrane and maximize the axonal cytoplasm; conversely, the glia seems to maximize its surface membrane, deploying as little cytoplasm as possible.

Across ganglia, the separation between nerve membrane and glial cell membrane is minimal and quite constant. The space, about 12 nm wide, is electron-lucent, and must be occupied by structured materials that maintain the exact distance and presumably offer a specific bondage of some strength between axons and between axons and glia. Enteric ganglia have no capsule (unlike other autonomic ganglia and nerves) and their

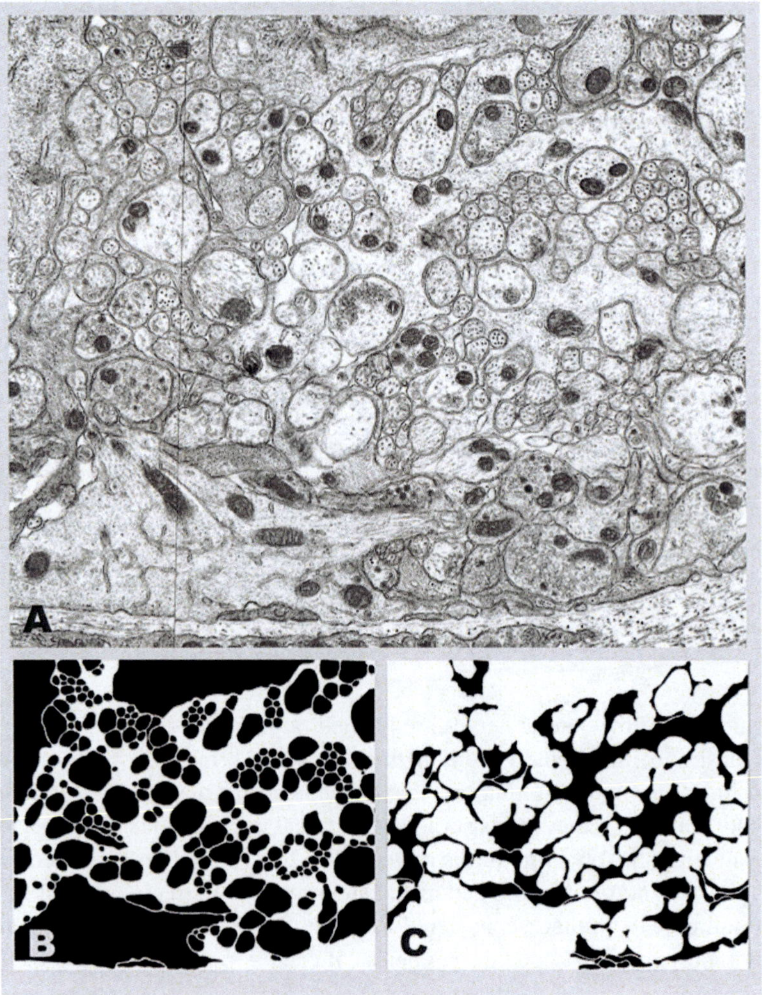

Fig. 17.5 (**a**) Myenteric ganglion of a guinea pig ileum. Several micrographs are assembled in a montage used for digital tracing and morphometry. The small detail displayed here shows the general appearance of the neuropil of the ganglion with innumerable processes both glial and neuronal, all approximately in transverse section. (**b**) In the same ganglionic area shown in the micrograph above, all the axons are traced and filled in in black. (**c**) In the same area as in A, all the glial processes are traced and filled in black. The two tracings are spatially complementary and when assembled together fill completely the space available, as seen in the original micrograph

components stay compactly together because of adhesion between the various elements, neuronal and glial. (Referring back to morphogenesis, the neuronal precursors colonizing the early embryonic gut are initially scattered around the wall; would they aggregate into a highly patterned plexus in the absence of glia?)

17.7 Dynamic Form of Ganglia

Several features picked up in micrographs of ganglia raise issues of possible mechanical relevance. The deformation of enteric ganglia and nerves produced by the contraction of the muscles is extraordinary, and is found nowhere else in the body except in the bladder wall. It occurs endlessly and fully reversibly. It involves stretch and compression in all three axes, with different patterns in the submucosal and the myenteric ganglia. Mechanical stress puts the enteric glia in a position that could not be more different from that of the CNS.

When the ileal wall contracts, going as far as occluding the lumen, the myenteric ganglia of that region become up to three times as thick as they are in the wall distended. The process involves elaborated changes in shape and relative position of all the ganglionic components, which take place in a matter of seconds and then are quickly and fully reversed (silently to the aware body). Similarly, transient changes are seen in the distal colon, where the gut diameter in the segment between two pellets is reduced to about a half that in the adjacent segments around faecal pellets, and the length of that segment is also slightly reduced (the contraction in diameter is limited by the occlusion of the luminal space). The volumes of the tissues are unchanged, hence both muscle layers become correspondingly thicker, and the myenteric ganglia, compressed mainly along their long (circumferential) axis acquire a roundish profile from the flattened shape they have in the distended segments.

The large and passive deformations in the ganglia are permitted, and are facilitated, by the glia, which minimizes the mechanical stress on neurons, axons and dendrites. The glia, particularly because of its laminar geometry, is much affected by the stretching and compression of the ganglia, and it plays a role in facilitating an orderly and reversible deformation of the ganglion, with extensive repositioning of the axons and shape change of the neuronal cells.

When the circular muscle contracts and shortens, an axon bundle (a strand of the myenteric plexus) is compressed lengthwise and it takes up a wavy course, clearly observable by microscopy. The axons, however, both in the nerve strands and in the ganglia, maintain a near circular profile, even when they compress longitudinally into a wavy or coiled shape, while at the same time changing their relative position, as allowed and facilitated by the glia.

17.8 Life Times

The need to take into account the temporal dimension having been referred to above, it should be reminded that glial cells of the myenteric ganglia are still dividing (by mitosis) in guinea pig foetuses at term, after the neurons have stopped dividing.

The process of colonization of the gut by cells from the neural crest and their differentiation is being intensively studied, without as yet firm conclusions. An eye is kept on different populations of precursors [11], on possible waves of appearance [16] and on different timing (dates of birth). The possibility is raised of new neurons being incorporated in adult mice [15].

In fact, in some places, for example, in the ileal myenteric ganglia of guinea pigs, glia continues to expand (not necessarily in number) in the course of post-natal life, wrapping around an increased number of axons, even if never more than a minority of them. Thus, under the microscope, enteric nerves and ganglia from aged animals can be distinguished from those of young adults.

Phillips et al. [17] provide good evidence of age-related loss of glia (and neurons) in myenteric ganglia of the rats, confirming the occurrence of structural changes in the enteric nervous system throughout life. The matter is quite complex given the number of factors involved and the large variability between animal species, organs and lifestyles.

The quantitative and spatial relationships between neurons and glia appear, on the one hand, to be constant and regulated, while on the other hand, they change over life time and are never still.

References

1. Björklund H, Dahl D, Seiger A (1984) Neurofilament and glial fibrillay acid protein-related immunoreactivity in rodent enteric nervous system. Neuroscience 12:277–287
2. Cabarrocas J, Savidge TC, Liblau RS (2003) Role of enteric glial cells in inflammatory bowel disease. Glia 41:81–93
3. Cook RD, Burnstock G (1976) The ultrastructure of Auerbach's plexus in the guinea-pig - II. Non-neuronal elements. J Neurocytol 5:195–206. https://doi.org/10.1007/BF01181656
4. Ferri GL, Probert L, Cocchia D, Marangos PJ, Polak JM (1982) Evidence for the presence of S-100 protein in the glial component of the human enteric nervous system. Nature 297:409–410
5. Gabella G (1971) Glial cells in the myenteric plexus. Z Naturforsch 26b:244–245
6. Gabella G (1972) Fine structure of the myenteric plexus in the guinea-pig ileum. J Anat 111:69–97
7. Gabella G (1981) Ultrastructure of the nerve plexuses of the mammalian intestine: the enteric glial cells. Neuroscience 6: 425–436 10.1016/0306-4522(81)90135-4
8. Gershon MD, Rothman TP (1991) Enteric glia. Glia 4:195–204
9. Gulbransen BD (2014) Enteric glia. In: Colloquium series on neuroglia in biology and medicine: from physiology to disease. Morgan and Claypool Publishers, pp 1–55. https://doi.org/10.4199/C00113ED1V01Y201407NGL002
10. Hanani M, Reichenbach A (1994) Morphology of horseradish peroxidase (HRP)-injected glial cells in the myenteric plexus of the guinea-pig. Cell Tissue Res 278:153–160. https://doi.org/10.1007/BF00305787
11. Hao MM, Young HM (2009) Development of enteric neuron diversity. J Cell Mol Med 13:1198–1210
12. Hoff S, Zeller F, Von Weyhern CWH, Wegner M, Schemann M, Michel K, Rühl A (2008) Quantitative assessment of glial cells in the human and guinea-pig enteric nervous system with an anti-Sox8/9/10 antibody. J Comp Neurol 509:356–371
13. Jessen KJ, Mirsky R (1980) Glial cells in the enteric nervous system contain fibrillary acidic protein. Nature 286:737–737
14. Komuro T, Baluk P, Burnstock G (1982) An ultrastructural study of neurons and non-neuronal cells in the myenteric plexus of the rabbit colon. Neuroscience 7:1797–1806. https://doi.org/10.1016/0306-4522(82)90037-9
15. Liu MT, Kuan YH, Wang J, Hen R, Gershon MD (2009) 5-HT4 receptor-mediated neuroprotection and neurogenesis in the enteric nervous system of adult mice. J Neurosci 29:9683–9699
16. Pham TD, Gershon MD, Rothman TP (1991) Time of origin of neurons in the murine enteric nervous system: sequence in relation to phenotype. J Comp Neurol 314(4):789–798
17. Phillips RJ, Kieffer EJ, Powley TL (2004) Loss of glial and neurons in the myenteric plexus of the aged Fischer 344 rat. Anat Embryol 209:19–30
18. Rühl A (2005) Glial cells in the gut. Neurogastroenterol Motil 17:777–790
19. Sharkey KA (2015) Emerging roles for enteric glia in gastrointestinal disorders. J Clin Invest 125:918–925

18 Circadian Control of Gastrointestinal Motility

Anita J. L. Leembruggen, Lincon A. Stamp,
Joel C. Bornstein, and Marlene M. Hao

Abstract

With the earth's 24-h rotation cycle, physiological function fluctuates in both diurnal and nocturnal animals, thereby ensuring optimal functioning of the body. The main regulator of circadian rhythm is the suprachiasmatic nucleus (SCN), which is considered the main pacemaker or "central clock" of the body. Located in the anterior hypothalamus, the SCN influences the activity of other brain regions, as well as peripheral organs, through the release of melatonin and corticosteroids. The SCN can be entrained by several cues, with light being the major cue. Light information from the retina is received by the SCN via the retinohypothalamic tract. Non-photic cues such as temperature and exercise can also entrain the SCN, while feeding time can entrain the "molecular clock" contained within peripheral tissues. This enables organs such as the gastrointestinal (GI) tract to coordinate function with environmental factors, such as food availability.

The GI tract, which has the main functions of receiving and digesting food, and expelling waste, also shows oscillations in its activity during the circadian cycle. While these changes are evident under normal conditions, GI function is affected when normal circadian rhythm is disrupted. Recent reviews have assessed interactions between the central clock and gut clock; as such, this review aims to focus on the presence of endogenous circadian rhythms in the GI tract, with particular focus to changes to gastrointestinal motility.

Keywords

Circadian rhythm · Enteric nervous system · Intestine · Hypothalamus

18.1 Clock Genes

Circadian rhythm is generated by a transcription-translation feedback loop driving expression of clock genes [17]. In both diurnal and nocturnal mammals, the molecular clock comprises two loops: the core loop and auxiliary loop. Circadian locomotor output cycles kaput (CLOCK) and aryl hydrocarbon receptor nuclear translocator-like protein 1 (ARNTL, also known as brain and muscle ARNT-like 1 (BMAL1)) are known as "positive elements" in the core loop [17, 18], which dimerize and stim-

A. J. L. Leembruggen · L. A. Stamp · J. C. Bornstein
M. M. Hao (✉)
Department of Anatomy & Physiology, School of Biomedical Sciences, Faculty of Medicine, Dentistry & Health Sciences, University of Melbourne, Parkville, VIC, Australia
e-mail: hao.m@unimelb.edu.au

N. J. Spencer et al. (eds.), *The Enteric Nervous System II*, Advances in Experimental Medicine and Biology 1383, https://doi.org/10.1007/978-3-031-05843-1_18

ulate the transcription of "negative elements" known as period genes (PER) and cryptochrome genes (CRY) while animals are awake (Fig. 18.1) [17]. During sleep, the accumulation of PER and CRY in the cytosol acts as a transcriptional repressor of the CLOCK/BMAL1 heterodimer (Fig. 18.1) [17, 19]. The auxiliary loop involves the regulation of the BMAL1 promoter region by ROR/REV-ERBα, which results in rhythmic transcription of BMAL1 [17, 19]. The expression of genes encoding positive elements and negative elements occurs approximately 12 h apart [20].

18.2 Circadian Cycle and the Gut

Overall, many components of gastrointestinal function are under circadian modulation. Clock genes have been identified in peripheral organs, including the colon and liver, in rodents, chickens, and humans [20, 23–27]. Expression of *Per1*, *Per2*, and *Bmal1* is evident throughout the gastrointestinal tract (Fig. 18.1), with *Per* and *Bmal* expressed in epithelial cells lining colonic crypts in mice [27] and rats [20], as well as in neurons of the myenteric plexus in the stomach and colon of mice [27]. Approximately 3.7% of genes in the mouse distal colon show rhythmic expression during the light/dark cycle [28]. Within the gut itself, regional variation in clock gene expression has been observed, with *Per1*, *Per2*, and *Bmal1* and *Rev-Erbα* phase-advanced in the duodenum compared to the distal colon [29].

Peripheral clock gene expression is not wholly dependent on cues from the SCN. Instead, peripheral clocks are also entrained by activities in the gut, most prominently, the timing of food intake [12]. Further, cues from the gut also entrain other peripheral organs, in particular, the liver and adipose tissue, to coordinate metabolic function with feeding times. Therefore, it is important to note that while circadian rhythms in the mammalian SCN occurs at the same astrophysical time across both nocturnal and diurnal animals, as assessed by the cycling of clock genes [30], neuropeptide expression [31], and metabolic activity [32]; in contrast, the circadian rhythms in the gut and peripheral organs show opposite phases between diurnal and nocturnal animals [30, 33].

18.2.1 Nutrient Absorption and Metabolism

Early studies in the 1970s showed that nutrient absorption in the intestine peaked at night in rats, when they are usually active and feeding [34–36]. Since then, a number of key pathways involved in digestion and nutrient absorption have been shown to display circadian rhythmicity. Within the gastrointestinal tract, these range from the components of saliva to intestinal enzyme activity and nutrient transporters (reviewed in [16]). In the intestinal epithelium, the expression of several genes involved in the absorption of sugars, fats, and peptides has been shown to fluctuate through the circadian cycle. These include the sodium-glucose co-transporter *Sglt1*, glucose transporters 2 (*Glut2*), and *Glut5* [37], the intestinal H+-coupled peptide cotransporter (*Pept1*) [38], and the microsomal triglyceride transfer protein *Mttp*, which is important for apoB-lipoprotein assembly [39]. Further, clock gene elements have been shown to target these genes and influence their expression levels, either directly [40] or indirectly via other transcription factors [41, 42].

Hence, the cycling of clock gene expression is key for the control of nutrient absorption. Interestingly, this system is led by bidirectional communication, as the availability of food and timing of food intake are, in turn, crucial controls for setting clock gene expression cycles in the gut and the periphery.

The original studies in the 1970s already identified that if laboratory animals were fed at restricted times instead of having ad libitum access to food, the peaks in nutrient absorption would also shift to different times of day. For example, when rats could only feed during the day, the pattern of glucose and amino acid absorption also changed to peak during the day [35, 36]. Firstly, this change in absorption logically follows the availability of food. However, it was also identified that the pattern of increased day-

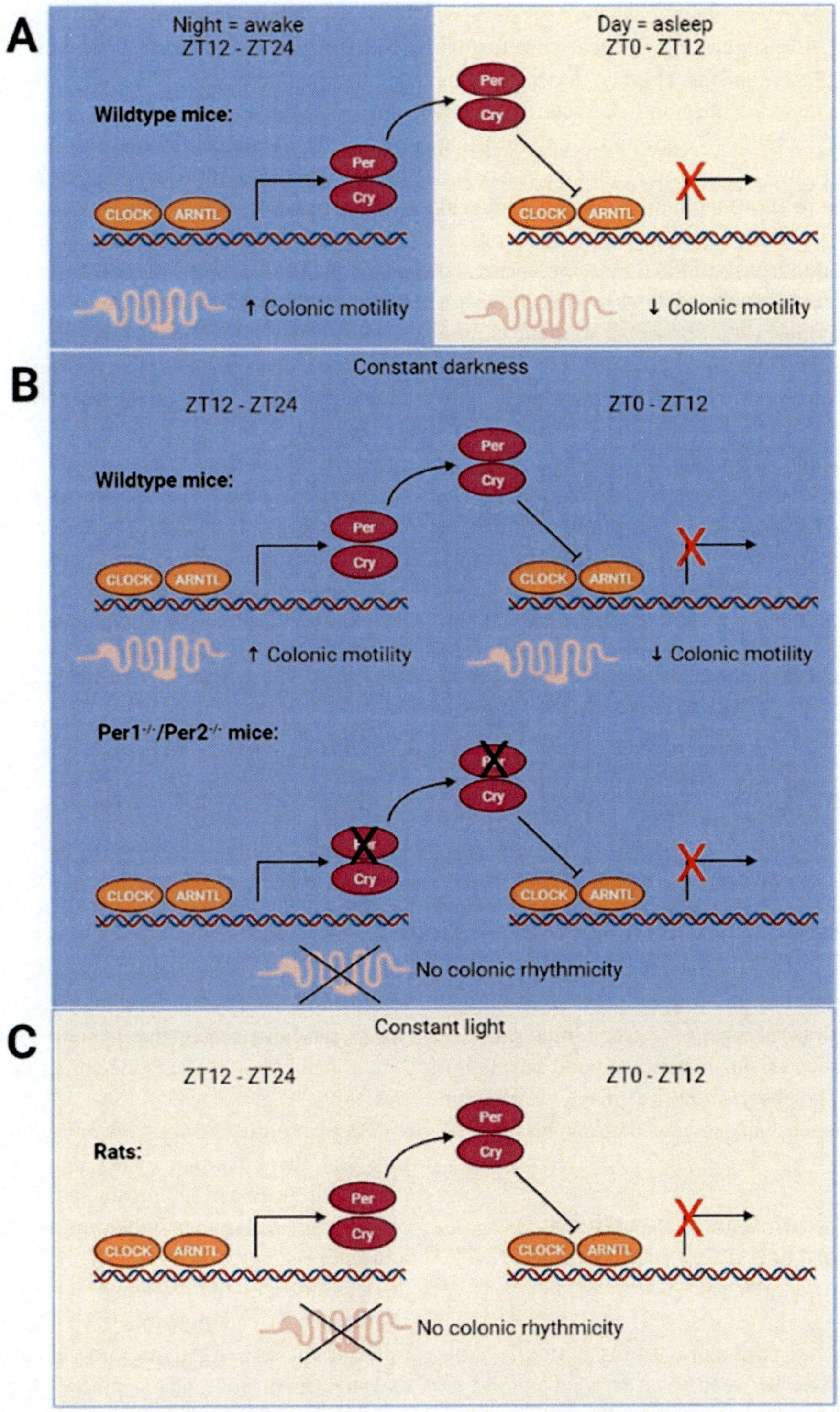

Fig. 18.1 Clock gene expression during the light/dark cycle in mice and correlation with gastrointestinal motility. (**a**) Changes to mouse clock gene expression in the gastrointestinal tract and corresponding changes to colonic motility during a standard 12 h light/12 h dark cycle [21]. (**b**) Changes to mouse clock gene expression in the gastrointestinal tract and corresponding changes to colonic motility during constant darkness for 48 h [21]. (**c**) Changes to clock gene expression in rats and the loss of gastrointestinal motility during constant light [22]. (Figure created with BioRender.com)

time absorption was maintained, even if rats were fasted for 24 h, suggesting that there are further underlying mechanisms at play. Many studies have gone on to identify and characterize "food anticipatory activities," a group of activities that have been shown to increase in anticipation of the availability of food, including changes in digestive enzyme activity, increased body temperature, increased duodenal contractions, and increased physical activity such as wheel-running [43, 44]. These activities were so named as their timing was dependent on the availability of food and changed when restricted feeding paradigms were set [45].

A key study in 2000 by Damiola et al. showed that if food availability was restricted to the daytime (the inactive phase) in mice, the cycling of clock genes in many peripheral tissues, including the gut, liver, and many major organs became shifted, cycling in opposition to their normal patterns [12]. Interestingly, clock gene cycling in the SCN remained unchanged, showing that timed feeding does not entrain the central pacemaker [12]. Since then, the importance of feeding times as an entrainer of peripheral circadian rhythms has become well established, and many studies have identified that food is an important zeitgeiber, an environmental cue that helps resynchronize clock genes [46, 47]. Changes in insulin levels and signaling via the insulin receptor have been shown to be a direct link between nutrient availability and the setting of clock gene expression in peripheral organs [48]. The importance of feeding times on our metabolism and in relation to obesity has been central to many studies and has been extensively reviewed elsewhere [49].

18.2.2 Regulation of Intestinal Epithelial Cell Proliferation and Cancer

The intestinal epithelium forms a tightly controlled barrier between the "external" world of the gut lumen and the "internal" body environment. Intestinal epithelial cells are generated at an immense rate of 10^{11} cells/day in humans [50], differentiating from proliferating stem cells and transiting amplifying cells in the intestinal crypts [51]. Diurnal fluctuations in the mitotic activity of intestinal crypt cells were shown over 50 years ago, with increased cell division observed at night [52]. Circadian fluctuations in the ability to induce cell apoptosis have also been demonstrated using ionizing radiation, with increased ability to induce apoptosis at the beginning of the day [53, 54]. However, there were also other earlier studies that failed to detect changes in proliferation rates [55, 56]. These could be due to differences in regions of the gut examined and on the methods used to measure proliferation.

Cycling clock gene expression has been observed in the intestinal epithelium, including the stem cell niche [27, 57]. Recently, mitosis in the crypts of mouse small intestine, as identified by phosophohistocine-3 immunohistochemistry, was found to be maintained at a constant level throughout the circadian cycle under homeostatic conditions [58]. However, stem cell proliferation was increased during the day following irradiation of the gut to damage the intestinal epithelium [58]. This circadian fluctuation in stem cell proliferation was absent in mice lacking *Bmal1*. In addition, in drosophila, intestinal epithelial regeneration following damage to the gut by the chemical detergent dextran sulfate sodium (DSS) is also diurnally regulated [59]. Some studies have suggested that without environmental inputs, intestinal stem cells by themselves exhibit lower circadian rhythmicity, but circadian fluctuations are enhanced by the presence of other differentiated epithelial cells via paracrine signaling [60].

There are clear links between cell proliferation and the circadian cycle, as expression of many cell cycle and apoptosis genes is regulated by clock gene elements, including several cyclins and tumor suppressor genes [61, 62]. Therefore, dysregulation of the circadian cycle can lead to aberrances in cell division [63]. This is highly relevant for investigations into cancer initiation and treatment [64, 65]. Epidemiology studies have shown that there are correlations between shift work, a known disrupter of the circadian cycle, and the prevalence of breast and prostate cancer. Different studies have provided conflicting evidence for whether there is a clear correlation between shift work and the incidence of colorectal cancer [66, 67]. Despite this, shift work has recently been classified as a risk factor for colorectal cancer [68]. The mechanisms underlying the relationship between circadian disruption and colorectal cancer remain to be investigated. As described above, there are clear implications for circadian regulation of intestinal stem cell division following injury; however, this is more opaque under normal homeostatic conditions. Studies in animal models have shown that disruptions to the peripheral circadian cycle by mistimed eating, together with alcohol consumption, promote carcinogenesis in the colon [69, 70]. These changes were shown to implicate the intestinal microbiota and immune cells.

18.2.3 Immune Function and Influence of the Microbiota

The human gastrointestinal tract is host to a vast ecosystem of approximately 10^{14} microorganisms, collectively known as the gut microbiota. The gut microbiota has been shown to play important roles in many aspects of body function, in particular, interacting with our metabolic and immune systems. Circadian changes in the gut microbiota and immune function have been extensively reviewed elsewhere [71–73]. Changes in the gut microbiota, both in terms of the composition of the different species and their gene expression patterns, have been shown to fluctuate with the circadian cycle in laboratory mice and humans [74]. These changes were dependent on the host's circadian cycle and feeding rhythms [75, 76]. Dysbiosis of the microbiome led to metabolic disruptions in the host mice, leading to increased obesity [74]. Changes in the microbiome composition and the microbiome metabolome were responsible for epigenetic changes in not only the host intestinal epithelium but also in the liver [77]. The link between the microbiome and the development of metabolic disease has been the focus of many different studies [78, 79].

In addition to influencing host metabolism, both the circadian cycle and the microbiome have important roles in influencing immune function. The presence of circadian oscillators in immune cells and immune function has been well documented [80, 81]. Clock genes are expressed by many immune cells, and the expression of many cytokines and immune cell receptors is regulated by clock elements. The most well-known example of circadian synchronization of infection is probably that of the malaria parasite, from the genus *Plasmodium*. *Plasmodium* parasites infect the hosts' blood cells and burst from them in a co-ordinated synchrony in multiples of 24 h, causing periodic fevers in patient [82]. Circadian fluctuations in hosts' melatonin levels have been identified as important for synchronizing the *Plasmodium* life cycle [83]; however, this is unlikely to be the only signal and factors such as blood glucose levels and circulating cytokines are also important [84]. How these interactions between the host and parasite evolved and what advantages they confer are less clear [85]. Nonetheless, in the gut, changes to epithelial barrier function, the prevalence of inflammatory bowel disease, and colon carcinogenesis are all linked to circadian disruption of immune function [71, 86]. These studies have also raised the issue of the importance of timing for providing medication to boost drug efficacy and decrease drug toxicity [87].

Neural-immune crosstalk has been shown to be critical immune function [88] and also plays an important role in the control of enteric nervous system (ENS) function and gut motility [89]. Whether these factors also impact on the circadian control of gut motility has not yet been investigated.

18.3 Gastrointestinal Motility and the Enteric Nervous System

The ENS plays a key role in the regulation of gastrointestinal motility. In response to a stimulus, such as distension due to the presence of luminal contents within the gut, the gut contracts orally due to ascending excitation and relaxes anally due to descending relaxation [90]. The gut is normally in a state of tonic neurogenic inhibition, driven by nitric oxide (NO), but this is interrupted periodically by bursts of activity in excitatory motor neurons leading to a variety of contractile patterns (for recent reviews, see [91, 92]).

The two main contractile patterns in the GI tract after a meal are segmentation and peristalsis, which occur in response to varying states of intestinal contents [93]. Segmentation involves localized constriction of circular muscle, thereby leading to a reduction in GI diameter and movement of content away from the constriction, ensuring that intestinal contents are mixed with gastric and other secretions, thereby enabling optimal digestion and absorption [94–97]. Peristalsis involves circular muscle constrictions that propagate varying distances from oral to anal along the gut, moving the contents along the length of the GI tract [98].

The arrival of food in the stomach stimulates relaxation of the stomach wall, allowing more food to enter the stomach, without increasing intragastric pressure. This is referred to as "accommodation" and is regulated by the vago-vagal reflex pathway, which involves interactions between the vagus and neurons of the myenteric plexus [99], specifically, activating nitrergic neurons and leading to the release of inhibitory neurotransmitters such as vasoactive intestinal peptide (VIP) and NO [100]. Peristaltic activity in the corpus and proximal antrum moves food toward the distal antrum and pyloric sphincter, where it is pulverized and mixed with gastric secretions to form chyme [101]. This mixing activity is achieved through specialized contractions known as slow waves, which are generated by interstitial cells of Cajal (ICCs, [93, 102, 103]).

The migrating myoelectric complex (MMC), a form of very slow peristalsis, typically occurs in a fasted state and clears mucosal debris and bacteria from the gut lumen [104]. MMCs occurring in the colon are referred to as colonic MCs (CMCs) and are useful for assessing changes to the neural circuits mediating GI motility [105, 106]. An example of regional differences in motility can be observed in the mouse colon [107]. In the proximal colon, both segmentation (which optimizes the absorption of water and electrolytes) and CMCs can be observed. In the mouse distal colon, however, CMCs are the predominant motor pattern observed [106, 108].

Current understandings of mechanisms in the ENS underlying motility in gut have recently been reviewed in detail [91, 92]; however, it is established that GI motility can be influenced by other cell types, such as enteric glia [109], interstitial cells of Cajal [110, 111], immune cells [112], as well as other factors such as the gut microbiome [113, 114], and circadian rhythm [14, 115].

18.3.1 Circadian Cycle and Gastrointestinal Motility

As the structure and function of the GI tract vary along its length, different changes in GI function associated with circadian rhythm can also be observed [14, 115]. The following sections will focus on changes to gastric and intestinal motility; however, it has been established that activity of the esophagus [116–118] and rectum [119–121] also oscillate during the circadian cycle.

In both diurnal and nocturnal animals, gastric motility increases several hours before a meal [97, 122–124]. In mice, this occurs with a corresponding decrease in the mechanosensitivity of gastric tension receptors during the active, night phase, which allowing more food to be consumed before satiety is reached [116]. This oscillation in gastric tension receptor mechanosensitivity is lost in mice on a high-fat diet, potentially due to these mice showing increased food intake during the light phase when they would normally be asleep [125]. Gastric emptying and slow wave contractions also vary during the circadian cycle [126–129].

Colonic motility also oscillates during the circadian cycle in rodents and humans [13, 122]. Humans show increased amplitude of colonic contractions and increased motility in the morning compared to night, as well as after consumption of a meal [121, 130–133]. In humans, different regions of the colon show varying activity during the day, with increased motility observed during the day in the descending colon compared to the rectosigmoid colon [121]. Similarly, mice show increased colonic motility and stool production during the night when they are awake and reduced colonic activity during sleep (Fig. 18.1) [21].

When the circadian cycle is disrupted (known as "chronodisruption"), gastrointestinal dysfunction such as constipation, diarrhea, abdominal pain, and bloating are common [134–136]. This is especially evident in nurses working rotating shifts, who experience increased functional bowel disorders such as irritable bowel syndrome (IBS) compared to the general population [137, 138]. Nurses working rotating shifts report more severe gastrointestinal symptoms compared to nurses working nightshifts only or dayshifts only, potentially due to an insufficient rest period between night shifts and day shifts as well as irregular mealtimes [139, 140]. However, other factors such as stress and difficulty of work could contribute to these symptoms as well [139–141]. This was highlighted in a recent study of ICU nurses in Wuhan, China, during the Covid-19 pandemic, with nurses experiencing GI symptoms such as nausea and diarrhea, which were exacerbated when risk factors (e.g., exposure to bodily fluids of infected patients) occurred [142].

Similarly, pilots and flight attendants have increased prevalence of upper GI symptoms compared to ground staff, likely due to reduced sleep and irregular meals [143–146]. Studies of the effects of jet-lag have also been performed in rodents; however, while no changes to food intake were observed in these mice [74], care should be taken when generalizing results of nocturnal animals to diurnal animals.

18.3.2 Molecular Mechanisms Underlying Circadian Control of Gut Motility

These changes to GI motility do not appear to be solely dependent on the presence of light. Mice kept in constant darkness have shown typical cycling of stool production with ad libitum access to food; however, this rhythmicity is abolished in *Per1* and *Per2* double knock-out mice (Fig. 18.1) [21]. This suggests that instead of light-driving peripheral clocks, the cycling of negative clock elements is important for this oscillation in colonic activity.

Conversely, rats exposed to constant light displayed a complete loss of circadian rhythmicity of food intake and activity (Fig. 18.1), despite the expression of *Per1*, *Per2*, and *Bmal1* persisting the duodenum [22]. $Bmal^{-/-}$ mice with ad libitum food access also show a loss of oscillation of food intake and ghrelin release during the light/dark cycle [147]. This has been supported by findings of decreased jejunum motility and contractility accompanied by an increased release of pancreatic enzyme output during the night in humans [131, 132]. It has also been shown that alterations to clock genes can also lead to disrupted GI function. In female adults, minor polymorphisms of *CLOCK* and *PER3* genes have been associated with reduced morning gastric motility [148].

Indeed, changes to neuronal nitric oxide synthase (*nNOS*) and *VIP* expression in the mouse distal colon fluctuates with the circadian cycle, with expression reduced at the start of the dark phase and increasing over the course of the dark phase in mice [28]. This pattern of expression persists in mice housed in constant darkness [28]. These studies also found that the circadian cycling of changes in colonic contractions was absent in $nNOS^{-/-}$ knockout mice [21], highlighting the importance of circadian neurotransmission for control of gastrointestinal motility.

Recent single-cell RNAseq studies have provided insights into the diversity in gene expression of the enteric neural population. Expression of calcitonin gene-related peptide (CGRP), which has roles in neuroimmune signaling in the GI tract [112, 149], is upregulated in the mouse colon during the day compared to night [23]. Cytoskeleton genes responsible for neuronal remodeling (*Tubb2a*, *Tubb3*, and *Prph*) are also upregulated during the day compared to the night in mice [23]. Additionally, genes involved in ENS signaling and neuroimmune signaling were expressed highly during the day, while transcription factors (*Nr1d2*, *Tef*, and *Dbp*) were upregulated in the evening [23]. These data suggest potential mechanisms for changes to enteric neural function that in turn could explain the changes to GI function during the light/dark cycle, which requires additional studies.

Further investigation of circadian changes in enteric neuronal plasticity can be informed by existing studies of the central nervous system (CNS). In addition to the SCN, there are significant changes to neuronal properties in many regions of the brain, for example in the cerebellum [150] and hippocampus [151–154]. Changes to neurotransmitter expression and activity in the CNS have been reviewed in detail, with glutamate, serotonin, acetylcholine, GABA, and VIP involved in signaling within the SCN (reviewed in [155]). Whether these neurotransmitters have similar roles in circadian rhythm in the ENS remain to be characterized.

18.4 Conclusions

It is clear that the circadian rhythm has profound effects on GI function within our daily 24-h cycles. This occurs through input from the SCN and, most importantly, regulation of the gut clock by molecular clock genes and stimuli such as time of food intake. Apart from changes to nNOS and VIP, neuronal plasticity of the ENS through the circadian cycle remains to be examined. This will allow better understanding of the molecular mechanisms underpinning the daily fluctuating changes in GI motility. The GI tract is a complex milieu of many interacting components, including the neural, immune, and microbial systems. Understanding the relationship between the circadian control of gut function will have important implications for individuals experiencing chronodisruption, leading to improved treatments for GI dysfunction.

Bibliography

1. Honma S (2018) The mammalian circadian system: a hierarchical multi-oscillator structure for generating circadian rhythm. J Physiol Sci 68:207–219
2. Reppert SM, Weaver DR (2001) Molecular analysis of mammalian circadian rhythms. Annu Rev Physiol 63:647–676
3. Reppert SM, Weaver DR (2002) Coordination of circadian timing in mammals. Nature 418:935–941
4. Saini C, Morf J, Stratmann M, Gos P, Schibler U (2012) Simulated body temperature rhythms reveal the phase-shifting behavior and plasticity of mammalian circadian oscillators. Genes Dev 26:567–580
5. Wright KP Jr, McHill AW, Birks BR, Griffin BR, Rusterholz T, Chinoy ED (2013) Entrainment of the human circadian clock to the natural light-dark cycle. Curr Biol 23:1554–1558
6. Hattar S, Kumar M, Park A, Tong P, Tung J, Yau KW, Berson DM (2006) Central projections of melanopsin-expressing retinal ganglion cells in the mouse. J Comp Neurol 497:326–349
7. Provencio I, Cooper HM, Foster RG (1998) Retinal projections in mice with inherited retinal degeneration: implications for circadian photoentrainment. J Comp Neurol 395:417–439
8. Rensing L, Ruoff P (2002) Temperature effect on entrainment, phase shifting, and amplitude of circadian clocks and its molecular bases. Chronobiol Int 19:807–864
9. Tahara Y, Aoyama S, Shibata S (2017) The mammalian circadian clock and its entrainment by stress and exercise. J Physiol Sci 67:1–10
10. Scheving LA, Russell WE (2007) It's about time: clock genes unveiled in the gut. Gastroenterology 133:1373–1376
11. Hamaguchi Y, Tahara Y, Kuroda H, Haraguchi A, Shibata S (2015) Entrainment of mouse peripheral circadian clocks to <24 h feeding/fasting cycles under 24 h light/dark conditions. Sci Rep 5:14207
12. Damiola F, Le Minh N, Preitner N, Kornmann BT, Fleury-Olela F, Schibler U (2000) Restricted feeding uncouples circadian oscillators in peripheral tissues from the central pacemaker in the suprachiasmatic nucleus. Genes Dev 14:2950–2961
13. Hoogerwerf WA (2010) Role of clock genes in gastrointestinal motility. Am J Physiol Gastrointest Liver Physiol 299:G549–G555
14. Duboc H, Coffin B, Siproudhis L (2020) Disruption of circadian rhythms and gut motility: an overview of underlying mechanisms and associated pathologies. J Clin Gastroenterol 54:405–414
15. Page AJ (2021) Gastrointestinal vagal afferents and food intake: relevance of circadian rhythms. Nutrients 13(3):844
16. Segers A, Depoortere I (2021) Circadian clocks in the digestive system. Nat Rev Gastroenterol Hepatol 18(4):239–251

References

17. Takahashi JS (2017) Transcriptional architecture of the mammalian circadian clock. Nat Rev Genet 18:164–179
18. Dunlap JC (1999) Molecular bases for circadian clocks. Cell 96:271–290
19. Bell-Pedersen D, Cassone VM, Earnest DJ, Golden SS, Hardin PE, Thomas TL, Zoran MJ (2005) Circadian rhythms from multiple oscillators: lessons from diverse organisms. Nat Rev Genet 6:544–556
20. Sladek M, Rybova M, Jindrakova Z, Zemanova Z, Polidarova L, Mrnka L, O'Neill J, Pacha J, Sumova A (2007) Insight into the circadian clock within rat colonic epithelial cells. Gastroenterology 133:1240–1249
21. Hoogerwerf WA, Shahinian VB, Cornelissen G, Halberg F, Bostwick J, Timm J, Bartell PA, Cassone VM (2010) Rhythmic changes in colonic motility are regulated by period genes. Am J Physiol Gastrointest Liver Physiol 298:G143–G150
22. Polidarová L, Sládek M, Soták M, Pácha J, Sumová A (2011) Hepatic, duodenal, and colonic circadian clocks differ in their persistence under conditions of constant light and in their entrainment by restricted feeding. Chronobiol Int 28:204–215
23. Drokhlyansky E, Smillie CS, Van Wittenberghe N, Ericsson M, Griffin GK, Eraslan G, Dionne D, Cuoco MS, Goder-Reiser MN, Sharova T, Kuksenko O, Aguirre AJ, Boland GM, Graham D, Rozenblatt-Rosen O, Xavier RJ, Regev A (2020) The human and mouse enteric nervous system at single-cell resolution. Cell 182(1606-22):e23
24. Hoogerwerf WA (2009) Role of biological rhythms in gastrointestinal health and disease. Rev Endocr Metab Disord 10:293–300
25. Ebert-Zavos E, Horvat-Gordon M, Taylor A, Bartell PA (2013) Biological clocks in the duodenum and the diurnal regulation of duodenal and plasma serotonin. PLoS One 8:e58477
26. Pardini L, Kaeffer B, Trubuil A, Bourreille A, Galmiche JP (2005) Human intestinal circadian clock: expression of clock genes in colonocytes lining the crypt. Chronobiol Int 22:951–961
27. Hoogerwerf WA, Hellmich HL, Cornelissen G, Halberg F, Shahinian VB, Bostwick J, Savidge TC, Cassone VM (2007) Clock gene expression in the murine gastrointestinal tract: endogenous rhythmicity and effects of a feeding regimen. Gastroenterology 133:1250–1260
28. Hoogerwerf WA, Sinha M, Conesa A, Luxon BA, Shahinian VB, Cornélissen G, Halberg F, Bostwick J, Timm J, Cassone VM (2008) Transcriptional profiling of mRNA expression in the mouse distal colon. Gastroenterology 135:2019–2029
29. Polidarova L, Sotak M, Sladek M, Pacha J, Sumova A (2009) Temporal gradient in the clock gene and cell-cycle checkpoint kinase Wee1 expression along the gut. Chronobiol Int 26:607–620
30. Mure LS, Le HD, Benegiamo G, Chang MW, Rios L, Jillani N, Ngotho M, Kariuki T, Dkhissi-Benyahya O, Cooper HM, Panda S (2018) Diurnal transcriptome atlas of a primate across major neural and peripheral tissues. Science 359:eaao0318
31. Dardente H, Menet JS, Challet E, Tournier BB, Pévet P, Masson-Pévet M (2004) Daily and circadian expression of neuropeptides in the suprachiasmatic nuclei of nocturnal and diurnal rodents. Brain Res Mol Brain Res 124:143–151
32. Schwartz WJ, Reppert SM, Eagan SM, Moore-Ede MC (1983) In vivo metabolic activity of the suprachiasmatic nuclei: a comparative study. Brain Res 274:184–187
33. Kumar Jha P, Challet E, Kalsbeek A (2015) Circadian rhythms in glucose and lipid metabolism in nocturnal and diurnal mammals. Mol Cell Endocrinol 418(Pt 1):74–88
34. Furuya S, Yugari Y (1974) Daily rhythmic change of L-histidine and glucose absorptions in rat small intestine in vivo. Biochim Biophys Acta 343:558–564
35. Furuya S, Yugari Y (1971) Daily rhythmic change in the transport of histidine by everted sacs of rat small intestine. Biochim Biophys Acta 241:245–248
36. Fisher RB, Gardner ML (1976) A diurnal rhythm in the absorption of glucose and water by isolated rat small intestine. J Physiol 254:821–825
37. Corpe CP, Burant CF (1996) Hexose transporter expression in rat small intestine: effect of diet on diurnal variations. Am J Phys 271:G211–G216
38. Pan X, Terada T, Irie M, Saito H, Inui K (2002) Diurnal rhythm of H+-peptide cotransporter in rat small intestine. Am J Physiol Gastrointest Liver Physiol 283:G57–G64
39. Pan X, Hussain MM (2007) Diurnal regulation of microsomal triglyceride transfer protein and plasma lipid levels. J Biol Chem 282:24707–24719
40. Iwashina I, Mochizuki K, Inamochi Y, Goda T (2011) Clock genes regulate the feeding schedule-dependent diurnal rhythm changes in hexose transporter gene expressions through the binding of BMAL1 to the promoter/enhancer and transcribed regions. J Nutr Biochem 22:334–343
41. Pan X, Zhang Y, Wang L, Hussain MM (2010) Diurnal regulation of MTP and plasma triglyceride by CLOCK is mediated by SHP. Cell Metab 12:174–186
42. Saito H, Terada T, Shimakura J, Katsura T, Inui K (2008) Regulatory mechanism governing the diurnal rhythm of intestinal H+/peptide cotransporter 1 (PEPT1). Am J Physiol Gastrointest Liver Physiol 295:G395–G402
43. Froy O (2010) Metabolism and circadian rhythms—implications for obesity endocrine reviews. Endocr Rev 31:1–24
44. Comperatore CA, Stephan FK (1987) Entrainment of duodenal activity to periodic feeding. J Biol Rhythm 2:227–242
45. Pendergast JS, Yamazaki S (2018) The mysterious food-entrainable oscillator: insights from mutant

and engineered mouse models. J Biol Rhythm 33:458–474
46. Stokkan K-A, Yamazaki S, Tei H, Sakaki Y, Menaker M (2001) Entrainment of the circadian clock in the liver by feeding. Science 291:490–493
47. Hara R, Wan K, Wakamatsu H, Aida R, Moriya T, Akiyama M, Shibata S (2001) Restricted feeding entrains liver clock without participation of the suprachiasmatic nucleus. Genes Cells 6:269–278
48. Crosby P, Hamnett R, Putker M, Hoyle NP, Reed M, Karam CJ, Maywood ES, Stangherlin A, Chesham JE, Hayter EA, Rosenbrier-Ribeiro L, Newham P, Clevers H, Bechtold DA, O'Neill JS (2019) Insulin/IGF-1 drives PERIOD synthesis to entrain circadian rhythms with feeding time. Cell 177:896
49. Pickel L, Sung H-K (2020) Feeding rhythms and the circadian regulation of metabolism. Front Nutr 7:39
50. Sender R, Milo R (2021) The distribution of cellular turnover in the human body. Nat Med 27:45–48
51. Barker N, van Es JH, Kuipers J, Kujala P, van den Born M, Cozijnsen M, Haegebarth A, Korving J, Begthel H, Peters PJ, Clevers H (2007) Identification of stem cells in small intestine and colon by marker gene Lgr5. Nature 449:1003–1007
52. Sigdestad CP, Bauman J, Lesher SW (1969) Diurnal fluctuations in the number of cells in mitosis and DNA synthesis in the jejunum of the mouse. Exp Cell Res 58:159–162
53. Ijiri K, Potten CS (1986) Radiation-hypersensitive cells in small intestinal crypts; their relationships to clonogenic cells. Br J Cancer Suppl 7:20–22
54. Ijiri K, Potten CS (1988) Circadian rhythms in the incidence of apoptotic cells and number of clonogenic cells in intestinal crypts after radiation using normal and reversed light conditions. Int J Radiat Biol 53:717–727
55. Leblond CP, Stevens CE (1948) The constant renewal of the intestinal epithelium in the albino rat. Anat Rec 100:357–377
56. Bertalanffy FD (1960) Mitotic rates and renewal times of the digestive tract epithelia in the rat. Acta Anat (Basel) 40:130–148
57. Parasram K, Bernardon N, Hammoud M, Chang H, He L, Perrimon N, Karpowicz P (2018) Intestinal stem cells exhibit conditional circadian clock function. Stem Cell Rep 11:1287–1301
58. Stokes K, Cooke A, Chang H, Weaver DR, Breault DT, Karpowicz P (2017) The circadian clock gene BMAL1 coordinates intestinal regeneration. Cell Mol Gastroenterol Hepatol 4:95–114
59. Karpowicz P, Zhang Y, Hogenesch JB, Emery P, Perrimon N (2013) The circadian clock gates the intestinal stem cell regenerative state. Cell Rep 3:996–1004
60. Matsu-Ura T, Dovzhenok A, Aihara E, Rood J, Le H, Ren Y, Rosselot AE, Zhang T, Lee C, Obrietan K, Montrose MH, Lim S, Moore SR, Hong CI (2016) Intercellular coupling of the cell cycle and circadian clock in adult stem cell culture. Mol Cell 64:900–912
61. Masri S, Sassone-Corsi P (2018) The emerging link between cancer, metabolism, and circadian rhythms. Nat Med 24:1795–1803
62. Fu L, Lee CC (2003) The circadian clock: pacemaker and tumour suppressor. Nat Rev Cancer 3:350–361
63. Benitah SA, Welz PS (2020) Circadian regulation of adult stem cell homeostasis and aging. Cell Stem Cell 26:817–831
64. Kinouchi K, Sassone-Corsi P (2020) Metabolic rivalry: circadian homeostasis and tumorigenesis. Nat Rev Cancer 20:645–661
65. Shilts J, Chen G, Hughey JJ (2018) Evidence for widespread dysregulation of circadian clock progression in human cancer. PeerJ 6:e4327
66. Papantoniou K, Devore EE, Massa J, Strohmaier S, Vetter C, Yang L, Shi Y, Giovannucci E, Speizer F, Schernhammer ES (2018) Rotating night shift work and colorectal cancer risk in the nurses' health studies. Int J Cancer 143:2709–2717
67. Wichert K, Rabstein S, Stang A, Erbel R, Eisele L, Arendt M, Keimer A, Dragano N, Hoffmann W, Lerch MM, Roskoden FC, Schmidt CO, Völzke H, Jöckel KH, Brüning T, Behrens T (2020) Associations between shift work and risk of colorectal cancer in two German cohort studies. Chronobiol Int 37:1235–1243
68. Bishehsari F, Engen PA, Voigt RM, Swanson G, Shaikh M, Wilber S, Naqib A, Green SJ, Shetuni B, Forsyth CB, Saadalla A, Osman A, Hamaker BR, Keshavarzian A, Khazaie K (2020) Abnormal eating patterns cause circadian disruption and promote alcohol-associated colon carcinogenesis. Cell Mol Gastroenterol Hepatol 9:219–237
69. Bishehsari F, Saadalla A, Khazaie K, Engen PA, Voigt RM, Shetuni BB, Forsyth C, Shaikh M, Vitaterna MH, Turek F, Keshavarzian A (2016) Light/dark shifting promotes alcohol-induced colon carcinogenesis: possible role of intestinal inflammatory milieu and microbiota. Int J Mol Sci 17(12):2017
70. Zheng D, Ratiner K, Elinav E (2020) Circadian influences of diet on the microbiome and immunity. Trends Immunol 41:512–530
71. Voigt RM, Forsyth CB, Green SJ, Engen PA, Keshavarzian A (2016) Circadian rhythm and the gut microbiome. Int Rev Neurobiol 131:193–205
72. Bishehsari F, Voigt RM, Keshavarzian A (2020) Circadian rhythms and the gut microbiota: from the metabolic syndrome to cancer. Nat Rev Endocrinol 16:731–739
73. Thaiss Christoph A, Zeevi D, Levy M, Zilberman-Schapira G, Suez J, Tengeler Anouk C, Abramson L, Katz Meirav N, Korem T, Zmora N, Kuperman Y, Biton I, Gilad S, Harmelin A, Shapiro H, Halpern Z, Segal E, Elinav E (2014) Transkingdom control of microbiota diurnal oscillations promotes metabolic homeostasis. Cell 159:514–529

74. Voigt RM, Forsyth CB, Green SJ, Mutlu E, Engen P, Vitaterna MH, Turek FW, Keshavarzian A (2014) Circadian disorganization alters intestinal microbiota. PLoS One 9:17
75. Zarrinpar A, Chaix A, Yooseph S, Panda S (2014) Diet and feeding pattern affect the diurnal dynamics of the gut microbiome. Cell Metab 20:1006–1017
76. Thaiss CA, Levy M, Korem T, Dohnalová L, Shapiro H, Jaitin DA, David E, Winter DR, Gury-BenAri M, Tatirovsky E, Tuganbaev T, Federici S, Zmora N, Zeevi D, Dori-Bachash M, Pevsner-Fischer M, Kartvelishvily E, Brandis A, Harmelin A, Shibolet O, Halpern Z, Honda K, Amit I, Segal E, Elinav E (2016) Microbiota diurnal rhythmicity programs host transcriptome oscillations. Cell 167:1495–510.e12
77. Lopez DEG, Lashinger LM, Weinstock GM, Bray MS (2021) Review circadian rhythms and the gut microbiome synchronize the host? Metabolic response to diet. Cell Metab 33:873–887
78. Choi H, Rao MC, Chang EB (2021) Gut microbiota as a transducer of dietary cues to regulate host circadian rhythms and metabolism. Nat Rev Gastroenterol Hepatol 18(10):679–689
79. Scheiermann C, Kunisaki Y, Frenette PS (2013) Circadian control of the immune system. Nat Rev Immunol 13:190–198
80. Logan RW, Sarkar DK (2012) Circadian nature of immune function. Mol Cell Endocrinol 349:82–90
81. Prior KF, Rijo-Ferreira F, Assis PA, Hirako IC, Weaver DR, Gazzinelli RT, Reece SE (2020) Periodic parasites and daily host rhythms. Cell Host Microbe 27:176–187
82. Hotta CT, Gazarini ML, Beraldo FH, Varotti FP, Lopes C, Markus RP, Pozzan T, Garcia CR (2000) Calcium-dependent modulation by melatonin of the circadian rhythm in malarial parasites. Nat Cell Biol 2:466–468
83. Hirako IC, Assis PA, Hojo-Souza NS, Reed G, Nakaya H, Golenbock DT, Coimbra RS, Gazzinelli RT (2018) Daily rhythms of TNFα expression and food intake regulate synchrony of plasmodium stages with the host circadian cycle. Cell Host Microbe 23:796–808, e6
84. Westwood ML, O'Donnell AJ, de Bekker C, Lively CM, Zuk M, Reece SE (2019) The evolutionary ecology of circadian rhythms in infection. Nat Ecol Evol 3:552–560
85. Codoner-Franch P, Gombert M (2018) Circadian rhythms in the pathogenesis of gastrointestinal diseases. World J Gastroenterol 24:4297–4303
86. Cederroth CR, Albrecht U, Bass J, Brown SA, Dyhrfjeld-Johnsen J, Gachon F, Green CB, Hastings MH, Helfrich-Förster C, Hogenesch JB, Lévi F, Loudon A, Lundkvist GB, Meijer JH, Rosbash M, Takahashi JS, Young M, Canlon B (2019) Medicine in the fourth dimension. Cell Metab 30:238–250
87. Godinho-Silva C, Cardoso F, Veiga-Fernandes H (2019) Neuro-immune cell units: a new paradigm in physiology. Annu Rev Immunol 37:19–46
88. Muller PA, Koscso B, Rajani GM, Stevanovic K, Berres M-L, Hashimoto D, Mortha A, Leboeuf M, Li X-M, Mucida D, Stanley ER, Dahan S, Margolis KG, Gershon MD, Merad M, Bogunovic M (2014) Crosstalk between muscularis macrophages and enteric neurons regulates gastrointestinal motility. Cell 158:300–313
89. Bayliss WM, Starling EH (1899) The movements and innervation of the small intestine. J Physiol 24:99–143
90. Fung C, Vanden Berghe P (2020) Functional circuits and signal processing in the enteric nervous system. Cell Mol Life Sci 77(22):4505–4522
91. Spencer NJ, Hu H (2020) Enteric nervous system: sensory transduction, neural circuits and gastrointestinal motility. Nat Rev Gastroenterol Hepatol 17:338–351
92. Furness JB (2012) The enteric nervous system and neurogastroenterology. Nat Rev Gastroenterol Hepatol 9:286–294
93. Gwynne RM, Bornstein JC (2007) Synaptic transmission at functionally identified synapses in the enteric nervous system: roles for both ionotropic and metabotropic receptors. Curr Neuropharmacol 5(1):1–17
94. Bornstein JC, Costa M, Grider JR (2004) Enteric motor and interneuronal circuits controlling motility. Neurogastroenterology and Motility 16:34–38
95. Gwynne RM, Thomas EA, Goh SM, Sjövall H, Bornstein JC (2004) Segmentation induced by intraluminal fatty acid in isolated guinea-pig duodenum and jejunum. J Physiol 556:557–569
96. Cannon WB, Washburn A (1912) An explanation of hunger. Am J Physiol 29:441–454
97. Bornstein J, Furness J, Kunze W, Bertrand P (2002) Innervation of the gastrointestinal tract. CRC Press, pp 21–76
98. Furness JB, Callaghan BP, Rivera LR, Cho HJ (2014) The enteric nervous system and gastrointestinal innervation: integrated local and central control. Adv Exp Med Biol 817:39–71
99. Tack J, Demedts I, Meulemans A, Schuurkes J, Janssens J (2002) Role of nitric oxide in the gastric accommodation reflex and in meal induced satiety in humans. Gut 51:219–224
100. Ramkumar D, Schulze KS (2005) The pylorus. Neurogastroenterol Motil 17(Suppl 1):22–30
101. Huizinga JD, Lammers WJ (2009) Gut peristalsis is governed by a multitude of cooperating mechanisms. Am J Physiol Gastrointest Liver Physiol 296:G1–G8
102. Sanders KM, Koh SD, Ward SM (2006) Interstitial cells of cajal as pacemakers in the gastrointestinal tract. Annu Rev Physiol 68:307–343
103. Kunze WAA, Furness JB (1999) The enteric nervous system and regulation of intestinal motility. Annu Rev Physiol 61:117–142
104. Swaminathan M, Hill-Yardin E, Ellis M, Zygorodimos M, Johnston LA, Gwynne RM, Bornstein JC (2016) Video imaging and spatiotemporal maps to analyze gastrointestinal motility in mice. J Vis Exp (108):53828

105. Costa M, Keightley LJ, Hibberd TJ, Wiklendt L, Smolilo DJ, Dinning PG, Brookes SJ, Spencer NJ (2021) Characterization of alternating neurogenic motor patterns in mouse colon. Neurogastroenterol Motil 33:e14047
106. Li Z, Hao MM, Van den Haute C, Baekelandt V, Boesmans W, Vanden Berghe P (2019) Regional complexity in enteric neuron wiring reflects diversity of motility patterns in the mouse large intestine. Elife 8:e42914
107. Barnes KJ, Beckett EA, Brookes SJ, Sia TC, Spencer NJ (2014) Control of intrinsic pacemaker frequency and velocity of colonic migrating motor complexes in mouse. Front Neurosci 8:96
108. Grubišić V, Verkhratsky A, Zorec R, Parpura V (2018) Enteric glia regulate gut motility in health and disease. Brain Res Bull 136:109–117
109. Huizinga JD, Chen JH (2014) Interstitial cells of Cajal: update on basic and clinical science. Curr Gastroenterol Rep 16:363
110. Sanders KM, Kito Y, Hwang SJ, Ward SM (2016) Regulation of gastrointestinal smooth muscle function by interstitial cells. Physiology (Bethesda) 31:316–326
111. Wang H, Foong JPP, Harris NL, Bornstein JC (2021) Enteric neuroimmune interactions coordinate intestinal responses in health and disease. Mucosal Immunol 15:27–39
112. Shin A, Preidis GA, Shulman R, Kashyap PC (2019) The gut microbiome in adult and pediatric functional gastrointestinal disorders. Clin Gastroenterol Hepatol 17:256–274
113. Dimidi E, Christodoulides S, Scott SM, Whelan K (2017) Mechanisms of action of probiotics and the gastrointestinal microbiota on gut motility and constipation. Adv Nutr 8:484–494
114. Bron R, Furness JB (2009) Rhythm of digestion: keeping time in the gastrointestinal tract. Clin Exp Pharmacol Physiol 36:1041–1048
115. Kentish SJ, Frisby CL, Kennaway DJ, Wittert GA, Page AJ (2013) Circadian variation in gastric vagal afferent mechanosensitivity. J Neurosci 33:19238–19242
116. Chitkara DK, Fortunato C, Nurko S (2004) Prolonged monitoring of esophageal motor function in healthy children. J Pediatr Gastroenterol Nutr 38:192–197
117. Stein HJ, Singh S, Demeester TR (2004) 'Efficacy' of esophageal peristalsis: a manometric parameter to quantify esophageal body dysfunction. Dis Esophagus 17:297–303
118. Hagger R, Kumar D, Benson M, Grundy A (2002) Periodic colonic motor activity identified by 24-h pancolonic ambulatory manometry in humans. Neurogastroenterol Motil 14:271–278
119. Auwerda JJ, Bac DJ, Schouten WR (2001) Circadian rhythm of rectal motor complexes. Dis Colon Rectum 44:1328–1332
120. Rao SS, Sadeghi P, Beaty J, Kavlock R, Ackerson K (2001) Ambulatory 24-h colonic manometry in healthy humans. Am J Physiol Gastrointest Liver Physiol 280:G629–G639
121. Konturek PC, Brzozowski T, Konturek SJ (2011) Gut clock: implication of circadian rhythms in the gastrointestinal tract. J Physiol Pharmacol 62:139–150
122. Janssen P, Vanden Berghe P, Verschueren S, Lehmann A, Depoortere I, Tack J (2011) Review article: the role of gastric motility in the control of food intake. Aliment Pharmacol Ther 33:880–894
123. Ang D, Nicolai H, Vos R, Mimidis K, Akyuz F, Kindt S, Vanden Berghe P, Sifrim D, Depoortere I, Peeters T, Tack J (2009) Influence of ghrelin on the gastric accommodation reflex and on meal-induced satiety in man. Neurogastroenterol Motil 21:528–533, e8-9
124. Kentish SJ, Vincent AD, Kennaway DJ, Wittert GA, Page AJ (2016) High-fat diet-induced obesity ablates gastric vagal afferent circadian rhythms. J Neurosci 36:3199–3207
125. Trout DL, King SA, Bernstein PA, Halberg F, Cornelissen G (1991) Circadian variation in the gastric-emptying response to eating in rats previously fed once or twice daily. Chronobiol Int 8:14–24
126. Aviv R, Policker S, Brody F, Bitton O, Haddad W, Kliger A, Sanmiguel CP, Soffer EE (2008) Circadian patterns of gastric electrical and mechanical activity in dogs. Neurogastroenterol Motil 20:63–68
127. Suzuki A, Asahina M, Ishikawa C, Asahina KM, Honma K, Fukutake T, Hattori T (2005) Impaired circadian rhythm of gastric myoelectrical activity in patients with multiple system atrophy. Clin Auton Res 15:368–372
128. Grammaticos PC, Doumas A, Koliakos G (2015) Morning and night gastric emptying half-time differed more than 220% in two young healthy adults. Hell J Nucl Med 18:60–62
129. Jensen MM, Wegeberg AL, Jensen SL, Sørensen PS, Wigh IMN, Zaugg VS, Færch K, Quist JS, Brock C (2021) The day-night pattern of colonic contractility is not impaired in type 1 diabetes and distal symmetric polyneuropathy. Chronobiol Int 38:801–806
130. Keller J, Layer P (2002) Circadian pancreatic enzyme pattern and relationship between secretory and motor activity in fasting humans. J Appl Physiol (1985) 93:592–600
131. Keller J, Gröger G, Cherian L, Günther B, Layer P (2001) Circadian coupling between pancreatic secretion and intestinal motility in humans. Am J Physiol Gastrointest Liver Physiol 280:G273–G278
132. Narducci F, Bassotti G, Gaburri M, Morelli A (1987) Twenty four hour manometric recording of colonic motor activity in healthy man. Gut 28:17–25
133. Bishehsari F, Levi F, Turek FW, Keshavarzian A (2016) Circadian rhythms in gastrointestinal health and diseases. Gastroenterology 151:e1–e5
134. Caruso CC, Lusk SL, Gillespie BW (2004) Relationship of work schedules to gastrointestinal diagnoses, symptoms, and medication use in auto factory workers. Am J Ind Med 46:586–598
135. Vener KJ, Szabo S, Moore JG (1989) The effect of shift work on gastrointestinal (GI) function: a review. Chronobiologia 16:421–439

136. Nojkov B, Rubenstein JH, Chey WD, Hoogerwerf WA (2010) The impact of rotating shift work on the prevalence of irritable bowel syndrome in nurses. Am J Gastroenterol 105:842–847
137. Schernhammer ES, Leitzmann MF, Michaud DS, Speizer FE, Giovannucci E, Colditz GA, Fuchs CS (2003) Cholecystectomy and the risk for developing colorectal cancer and distal colorectal adenomas. Br J Cancer 88:79–83
138. Chang WP, Peng YX (2021) Differences between fixed day shift nurses and rotating and irregular shift nurses in work-related musculoskeletal disorders: a literature review and meta-analysis. J Occup Health 63:e12208
139. Sveinsdóttir H (2006) Self-assessed quality of sleep, occupational health, working environment, illness experience and job satisfaction of female nurses working different combination of shifts. Scand J Caring Sci 20:229–237
140. Koh SJ, Kim M, Oh DY, Kim BG, Lee KL, Kim JW (2014) Psychosocial stress in nurses with shift work schedule is associated with functional gastrointestinal disorders. J Neurogastroenterol Motil 20:516–522
141. Yifan T, Ying L, Chunhong G, Jing S, Rong W, Zhenyu L, Zejuan G, Peihung L (2020) Symptom cluster of ICU nurses treating COVID-19 pneumonia patients in Wuhan, China. J Pain Symptom Manage 60:e48–e53
142. Omholt ML, Tveito TH, Ihlebæk C (2017) Subjective health complaints, work-related stress and self-efficacy in Norwegian aircrew. Occup Med (Lond) 67:135–142
143. Lindgren T, Runeson R, Wahlstedt K, Wieslander G, Dammström BG, Norbäck D (2012) Digestive functional symptoms among commercial pilots in relation to diet, insomnia, and lifestyle factors. Aviat Space Environ Med 83:872–878
144. Knutsson A, Bøggild H (2010) Gastrointestinal disorders among shift workers. Scand J Work Environ Health 36:85–95
145. Enck P, Müller-Sacks E, Holtmann G, Wegmann H (1995) Gastrointestinal problems in airline crew members. Z Gastroenterol 33:513–516
146. Laermans J, Vancleef L, Tack J, Depoortere I (2015) Role of the clock gene Bmal1 and the gastric ghrelin-secreting cell in the circadian regulation of the ghrelin-GOAT system. Sci Rep 5:16748
147. Yamaguchi M, Kotani K, Tsuzaki K, Takagi A, Motokubota N, Komai N, Sakane N, Moritani T, Nagai N (2015) Circadian rhythm genes CLOCK and PER3 polymorphisms and morning gastric motility in humans. PLoS One 10:e0120009
148. Assas BM, Miyan JA, Pennock JL (2014) Cross-talk between neural and immune receptors provides a potential mechanism of homeostatic regulation in the gut mucosa. Mucosal Immunol 7:1283–1289
149. Plumel M, Dumont S, Maes P, Sandu C, Felder-Schmittbuhl MP, Challet E, Bertile F (2019) Circadian analysis of the mouse cerebellum proteome. Int J Mol Sci 20(8):1852
150. Besing RC, Rogers CO, Paul JR, Hablitz LM, Johnson RL, McMahon LL, Gamble KL (2017) GSK3 activity regulates rhythms in hippocampal clock gene expression and synaptic plasticity. Hippocampus 27:890–898
151. Bowden JB, Abraham WC, Harris KM (2012) Differential effects of strain, circadian cycle, and stimulation pattern on LTP and concurrent LTD in the dentate gyrus of freely moving rats. Hippocampus 22:1363–1370
152. Chaudhury D, Wang LM, Colwell CS (2005) Circadian regulation of hippocampal long-term potentiation. J Biol Rhythm 20:225–236
153. Raghavan AV, Horowitz JM, Fuller CA (1999) Diurnal modulation of long-term potentiation in the hamster hippocampal slice. Brain Res 833:311–314
154. Hegazi S, Lowden C, Rios Garcia J, Cheng AH, Obrietan K, Levine JD, Cheng HM (2019) A symphony of signals: intercellular and intracellular signaling mechanisms underlying circadian time-keeping in mice and flies. Int J Mol Sci 20(9):2363

19 Generation of Gut Motor Patterns Through Interactions Between Interstitial Cells of Cajal and the Intrinsic and Extrinsic Autonomic Nervous Systems

Jan D. Huizinga, Amer Hussain, and Ji-Hong Chen

Abstract

The musculature of the gastrointestinal tract is a vast network of collaborating excitable cell types. Embedded throughout are the interstitial cells of Cajal (ICC) intertwined with enteric nerves. ICC sense external stimuli such as distention, mediate nerve impulses to smooth muscle cells, and provide rhythmic excitation of the musculature. Neural circuitry involving both the intrinsic and extrinsic autonomic nervous systems, in collaboration with the ICC, orchestrate an array of motor patterns that serve to provide mixing of content to optimize digestion and absorption, microbiome homeostasis, storage, transit, and expulsion. ICC are specialized smooth muscle cells that generate rhythmic depolarization to the musculature and so provide the means for peristaltic and segmenting contractions. Some motor patterns are purely myogenic, but a neural stimulus initiates most, further depolarizing the primary pacemaker cells and the musculature and/or initiating transient pacemaker activity in stimulus-dependent secondary ICC pacemaker cells. From stomach to rectum, ICC networks rhythmically provide tracks along which contractions advance.

J. D. Huizinga (✉) · A. Hussain · J.-H. Chen
McMaster University, Farncombe Family Digestive Health Research Institute, Department of Medicine, Division of Gastroenterology, Hamilton, ON, Canada
e-mail: huizinga@mcmaster.ca

Keywords

Interstitial cells of Cajal · Gastrointestinal motility · Peristalsis

Interstitial cells of Cajal (ICC) caught the attention of Santiago Ramon y Cajal when he was searching for a simpler nervous system compared to the brain, only to find a network of "accessory nerve cells" in the myenteric plexus of the rabbit intestine that did not fit his idea of synaptic communication [28]. He correctly interpreted ICC to be a gateway of the autonomic nervous system (ANS) to the gut musculature. Later, Tinel referred to the "interstitial neurons" as the terminal cells of the parasympathetic and sympathetic nervous system [35]. We now know that interstitial cells originate from the same precursor cells as the smooth muscle cells and transform into several subtypes of intestinal pacemaker cells which will become either intrinsically active or provide stimulus-dependent pacemaker activity. In addition, ICC perform sensory functions [25] and are intermediates for innervation from the enteric nervous system (ENS) to the gastrointestinal musculature [36].

N. J. Spencer et al. (eds.), *The Enteric Nervous System II*, Advances in Experimental Medicine and Biology 1383, https://doi.org/10.1007/978-3-031-05843-1_19

19.1 ICC as Intermediary of Sensory and Motor Activities of the Vagus

A connection between the extrinsic ANS and ICC was established by Powley and co-workers when they discovered that vagal sensory nerves were intimately connected to the intramuscular ICC of the stomach [25, 26]. The intramuscular arrays monitor the physical transformation of the gastric intramuscular ICC; the ICC are ideally positioned to respond to stretch or contractions of the many smooth muscle cells with which they are connected. The sensory information is relayed to the nucleus tractus solitarius (NTS), where it is processed while influenced by other brain regions, and the decision can be made to initiate fundic relaxation and propulsive motor activity. The NTS will communicate with the dorsal motor nucleus of the vagus to send excitatory stimuli via vagal motor neurons back to the musculature [4]. Pacemaker activity in the stomach is orchestrated by both the ICC associated with the myenteric plexus (ICC-MP, the primary pacemaker) and intramuscular ICC (ICC-IM). ICC-IM help transmit the electrical slow wave of the ICC-MP into the musculature and, via vagal excitation, can change slow wave frequency [14]. ICC-IM are also directly innervated by inhibitory and excitatory enteric nerves [37]. ICC-IM are not the primary pacemaker cells; however, they can generate stimulus-dependent pacemaker activity, thereby modifying the original pacemaker activity and its function. A major target of gastric vagal excitatory neurons is the fundus, to facilitate fundic relaxation (adaptive accommodation). Another target is the myenteric plexus in the corpus where enteric nerves can be excited to stimulate the musculature into propulsive motor activity, the characteristic 3 cpm propulsive waves that travel along the stomach from corpus to pylorus, functioning to mix and grind content for emulsification and to allow transit into the small intestine.

It appears that the brain is continuously monitoring the gastric pacemaker activity generated by the ICC-MP. A study by Richter et al. [30] found significant electrophysiological communication between gastric pacemaker activity and the alpha rhythm within certain regions of the cerebral cortex, including the right anterior insula. In other words, they found a role for interstitial cells of Cajal in interoception, the sensory system responsible for detecting internal regulation responses. The gastric slow wave, as monitored by the right anterior insula, modified its alpha rhythm through phase amplitude coupling; the phase of the lower frequency gastric pacemaker modulated the amplitude of alpha waves in the anterior insula [17]. It is likely that the communication between the stomach and the anterior insula is mediated by vagal sensory nerves that monitor the network of ICC-MP and ICC-IM.

A sensory role for ICC-IM has also been proposed for the rectum. De Lorijn et al. showed that rectal distension-induced inhibition of the internal anal sphincter was markedly impaired in WWv mice that lack ICC-IM [8]. In these experiments, nitrergic innervation and functional inhibition were not impaired, just the sensory aspect of nitrergic inhibition. Since no abnormality was found in the ENS, it is likely that the absence of ICC hindered sensory information processing into the spinal cord so that sacral parasympathetic nerves would not be triggered to activate nitrergic nerves to relax the internal anal sphincter.

19.2 The Migrating Motor Complex (MMC)

Vagal excitation is responsible for the migrating motor complex (MMC), the characteristic motor pattern that occurs at regular intervals in the stomach during fasting [13]. The MMC is also called the housekeeping complex, first characterized by Szurszewski [33]. All components of the MMC in the lower oesophageal sphincter and stomach normally require vagosympathetic integrity [13]. The vagus exerts a wave of excitation over the stomach during phase III (a period of continuous propulsive activity) of the MMC. The vagal innervation of the myenteric plexus is not to excite a special set of command enteric neurons to initiate a programmed excitation [24], but rather the vagus sends numerous varicose neurons into the ganglia of the myenteric plexus, thereby likely exciting numerous cellular structures including the ICC-MP and glial cells. Hence the varicose nerve endings of the vagal motor neurons are part of a network of excitable cells within the ganglia [15] (Fig. 19.1a) that, as a network, generate motor patterns [24]. Varicosities send neurotransmitters into the extracellular space, thereby no doubt exciting ICC that occupy this space [20] (Fig. 19.1b); thus, we speculate that the vagal motor neurons directly communicate with the ICC-MP. Other cells that are likely influenced by vagal excitation are glial cells and specialized PDGFalpha$^+$ fibroblast-like cells that have been demonstrated to form gap junctions with ICC [16]. PDGFalpha$^+$ cells might even be gateways of sympathetic innervation to ICC [32].

The wave of neural activity induced by the vagus excites the ICC-MP network and the smooth musculature, and the consequence is propagating pressure waves at the frequency and propagation velocity of the slow waves generated by ICC-MP. This is illustrated with high-resolution manometry in humans (Fig. 19.2). The ICC-driven pressure waves propel content in the pyloric direction, and when the wave of neural excitation enters the small intestine, the motor pattern changes into intestinal ICC-MP-directed activity, increasing the frequency of the propulsive waves of contraction as well as their propagation velocity (Fig. 19.2). The frequency and duration of the wave of neural excitation remain in the realm of the vagus, with the ENS taking over from the vagus in the middle and distal part of the small intestine.

19.3 Duodenal Propulsive Activity

The importance of intestinal motility is illustrated by the fact that even a temporary absence due to surgical manipulation can be fatal [38]. The small intestine generates both pure myogenic motor patterns and motor activity initiated by neural activity. A pure myogenic motor pattern was documented by an in vivo study in the mouse duodenum [9] and a neurally mediated motor pattern, incorporating myogenic control systems, was illustrated by an in vitro study of the human small intestine [21].

When contrast fluid is gavaged into the stomach of a mouse, its arrival in the duodenum sets off rhythmic waves of contraction, propagating in anal direction to free the duodenum from content to allow more arrivals from the stomach. Spatiotemporal mapping shows that lumen-occluding contractions propel the contrast fluid at the frequency of the ICC-MP generated slow waves that travel into and dominate the electrical activity of the musculature (Fig. 19.3a) [9].

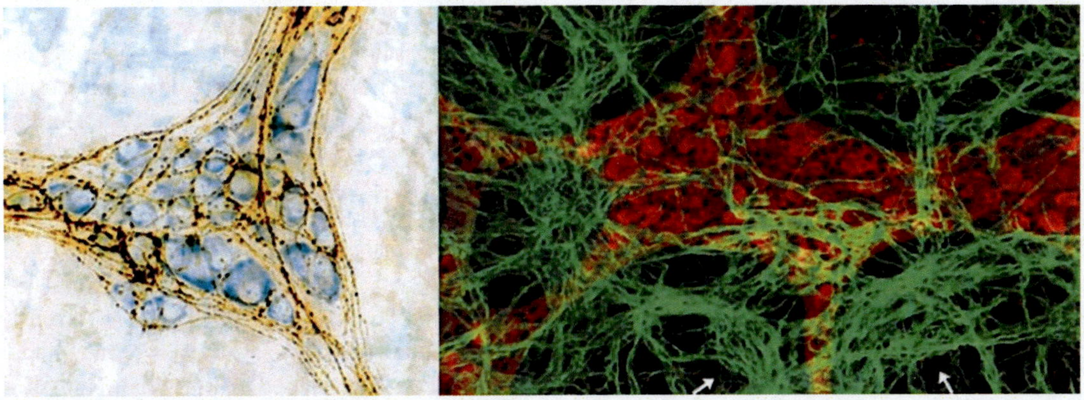

Fig. 19.1 Do varicose vagal nerve endings directly communicate with ICC-MP? (**a**) Vagal preganglionic label in ganglia of the myenteric plexus of the rat stomach showing the many varicose nerve endings [15] (**b**) Myenteric plexus of the guinea pig antrum with c-kit positive ICC in green [20]

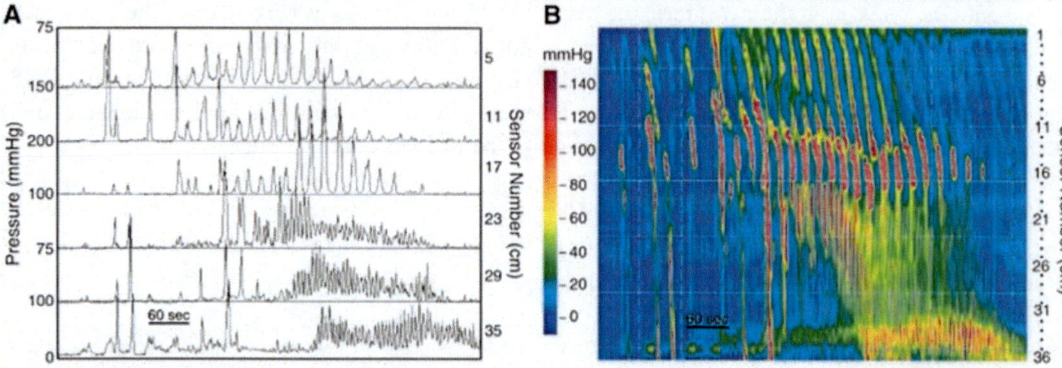

Fig. 19.2 Phase III of the migrating motor complex traversing the human stomach and small intestine. Phase III can be seen to be composed of propulsive contractions at the slow wave frequency of the stomach, and when the vagal excitation enters the small intestine, the contractions obey the laws of the ICC-MP of the small intestine. (From Conklin et al. [7])

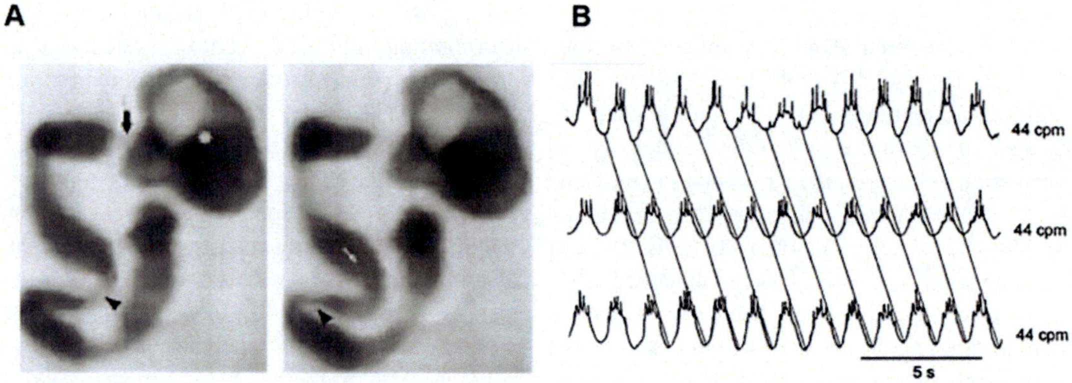

Fig. 19.3 (**a**) Lumen occluding (see arrows) contractions travels the proximal duodenum at the frequency and propagation velocity of the ICC-MP of the mouse small intestine in vivo. (**b**) Extracellular electrical activity of the mouse duodenum. (From Der-Silaphet et al. [9])

Although this myogenic motor pattern is often referred to as "ripples," this may give the wrong impression of minor effectiveness. As can be seen in Fig. 19.3a, the contractions can be lumen occluding and move the entire content of the small intestine in an anal direction [9].

19.4 The Minute Rhythm Contraction Pattern in the Human Small Intestine

Rhythmic waves of neural excitation by the ENS were seen to travel along a section of the human terminal ileum in vitro, facilitated by distension [21]. Its duration at any time point was about 50 seconds. This transient excitation depolarized the ICC of the myenteric plexus as well as the musculature leading to the generation of propulsive contractions at the frequency and propagation velocity of the ICC-MP pacemaker activity and primarily propagating in anal direction. In short segments, where no ripple frequency gradient could be detected, the propagation direction was mixed, with simultaneous pressure waves dominating. Without neural stimulation, ripples occurred at the intestinal ICC-driven pacemaker activity, with relatively low amplitude and propagating in oral or aboral direction or appearing simultaneously. After neural stimulation, the propagation direction became predominantly in anal direction, and the force of contraction increased. The ICC-driven pressure waves reached values of ~150 mmHg as shown in Fig. 19.4. When the amplitude increased, the individual contractions could fuse to generate a wide contraction [21].

19.5 The Segmentation Motor Pattern as Described by Cannon

The slow waves of the small intestine are not only responsible for rhythmic propulsive pressure waves but are also involved in segmentation, non-propulsive motor patterns that promote absorption and mixing of content with digestive enzymes and microbiota. Any local non-propulsive motor pattern can provide segmentation, but in the mouse small intestine, we observed a persistent, highly organized segmentation motor pattern that was in its architecture identical to the pattern observed by Cannon in the cat intestine [5]. At any point along the intestine during segmentation, non-propulsive contractions alternate with quiescence of similar duration at the slow wave frequency. When we analysed the frequency spectrum of the transient contractions within this motor pattern, two frequencies appeared; one was the dominant pacemaker activity, and one was much slower. The slower oscillation caused waxing and waning of the amplitude of the pacemaker-driven contractions [18]. We discovered that the ICC-DMP could generate oscillatory activity in the low frequency range. Hence, we hypothesized that the induction of the low frequency slow wave in the ICC-DMP by fatty acids, and possibly other stimuli, either directly or via enteric sensory neurons, would induce the segmentation motor pattern [18, 40]. Interestingly, a similar pattern was observed in the study by Kuizenga et al. in the human small intestine [21]. As shown in Fig. 19.5, two oscillations are seen: one due to the 6 cpm human ileal pacemaker activity and the other creating waxing and waning [21]. The periods of waxing and waning appeared semi-random in this section, suggesting that the entire pattern is likely causing the segmentation activity [18]. A reliable stimulus for the segmentation activity in the mouse small intestine is decanoic acid, and when its effect was studied with intracellular electrodes, it evoked a slow oscillation that caused waxing and waning of the pacemaker slow wave [22]. This suggests that fatty acids may induce this motor pattern in humans. Consistently, Bornstein and co-workers showed that decanoic acid, L-phenylalanine and L-tryptophan could induce segmentation motor patterns in the guinea pig small intestine [12].

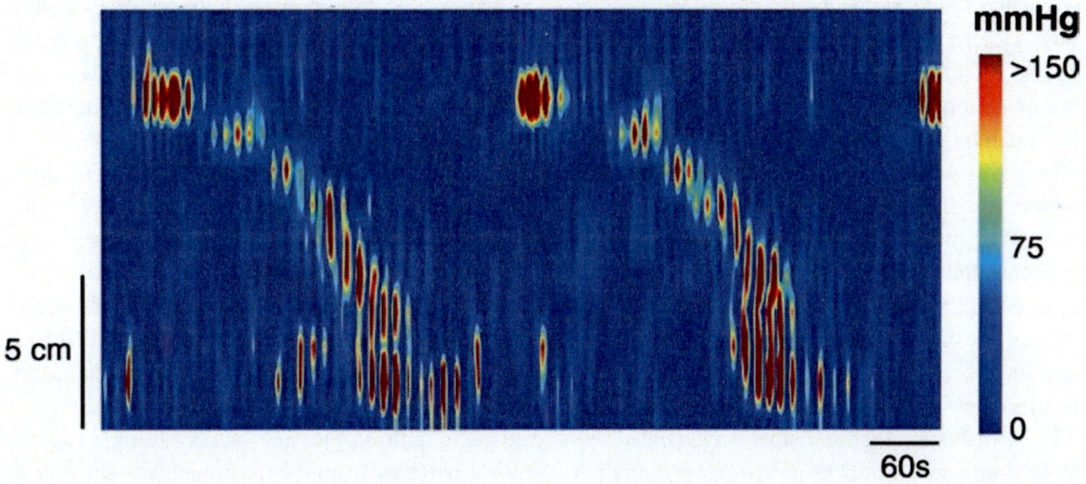

Fig. 19.4 A propagating wave of neural excitation in the human ileum in vivo evokes clustered contractions orchestrated by the ICC-MP pacemaker cells. (From Kuizenga et al. [21])

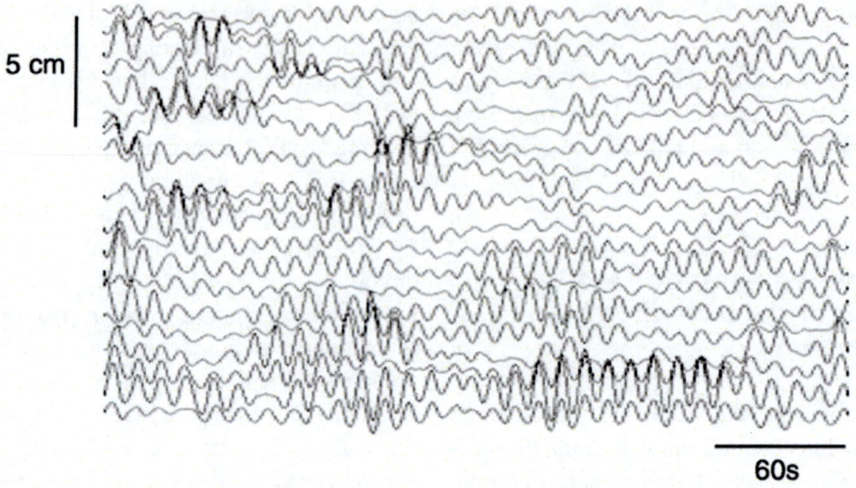

Fig. 19.5 Fibreoptic manometry pressure recordings of a section of the human ileum in vitro. Two oscillatory activities can be seen, a ~ 6 cpm oscillations directed by ICC-MP and a much slower oscillation identified by the waxing and waning of the 6 cpm pressure transients. (From Kuizenga et al. [21])

19.6 The High-Amplitude Propagating Pressure Wave (HAPW) in the Human Colon

The HAPW is a critical propulsive activity of the human colon that serves to propel content anally and is essential for the preparatory phase of defecation [1, 2, 10] and for defection itself [29]. A pancolonic HAPW initiated by rectal distension or rectal bisacodyl is a response to a wave of neural excitation that involves sacral sensory nerves, vagal and sacral parasympathetic autonomic nervous system activity, and activity of the ENS [3, 11, 39]. The HAPW is associated with a cyclic motor pattern at ~12 cpm [23], and the HAPW itself can be seen to be composed of 12 cpm pressure waves when the strength of the stimulus that generates the HAPW is diminishing, see Fig. 19.6. and Huizinga et al. [19]. We hypothesize that the motor pattern underlying both the HAPW and the subsequent 12 cpm cyclic motor pattern are orchestrated by excitation of ICC-MP that have been shown to generate this frequency in electrophysiological in vitro experiments [6, 27]. A ~ 12 cpm electrical slow wave is also seen to be dominant in human in vivo recordings [31, 34]. The neural stimulus determines the amplitude, propagation direction and velocity of the HAPW, whereas the ICC-MP determine the orchestration of the individual propulsive pressure waves within the HAPW. The 12 cpm pressure waves that underlie the HAPW are summating to generate the high pressure, and the individual pressure waves cannot be distinguished except in the tail end of the HAPW. This summation of myogenic pressure waves was also suggested to occur in the human ileum [21].

In summary, the gastrointestinal musculature is principally an expansive network of excitable cells that form close connections and together generate and command electrical slow wave patterns that evoke motor activities underlying transit and segmentation. Many stimuli make use of the ever-present and active network of ICC pacemaker cells to finalize the program of motor pattern generation. Some motor patterns are purely myogenic, but a neural stimulus initiates most, either depolarizing the primary pacemaker cells and the musculature or initiating transient pacemaker activity in stimulus-dependent secondary ICC pacemaker cells. Both the enteric and the extrinsic autonomic nervous systems program motor patterns from the oesophagus to the colon, making use of the tracks laid down by the ICC.

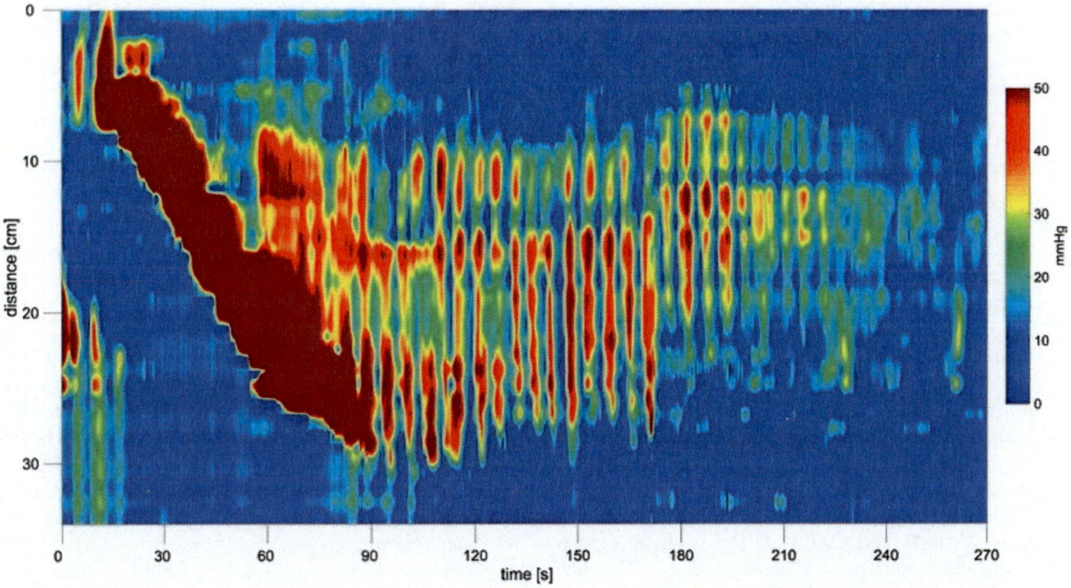

Fig. 19.6 The 12 cpm cyclic motor pattern underlies the HAPW. The tail end of the HAPW can be seen as composed of 12 cpm pressure waves. The 12 cpm pressure waves in the body of the HAPW summate into a strong propulsive motor pattern (Modified from Pervez et al. [23])

Acknowledgements The work of the Chen–Huizinga laboratory is funded by the Canadian Institutes of Health Research, the National Science and Engineering Council of Canada and the Farncombe Family Digestive Health Research Institute.

References

1. Pervez M, Ratcliffe E, Parsons SP, Chen J-H, Huizinga JD (2020) The cyclic motor patterns in the human colon. Neurogastroenterol Motil e13807:1–17 https://doi.org/10.1111/nmo.13807.
2. Bassotti G, Gaburri M (1988) Manometric investigation of high-amplitude propagated contractile activity of the human colon. Am J Phys 255:G660–G664
3. Bharucha AE, Brookes SJH (2018) Neurophysiologic mechanisms of human large intestinal motility. In: Physiology of the gastrointestinal tract. Elsevier, Amsterdam, pp 517–564
4. Browning KN, Travagli RA (2014) Central nervous system control of gastrointestinal motility and secretion and modulation of gastrointestinal functions. Compr Physiol 4:1339–1368
5. Cannon WB (1902) The movements of the intestines studied by means of the Rontgen rays. J Med Res 7:72–75
6. Choe EK, Moon JS, Moon SB, So IS, Park KJ (2010) Electromechanical characteristics of the human colon in vitro: is there any difference between the right and left colon? Int J Color Dis 25:1117–1126
7. Conklin J, Pimentel M, Soffer E (2009) Color atlas of high resolution manometry. Springer Science & Business Media, New York
8. De Lorijn F, De Jonge WJ, Wedel T, Vanderwinden JM, Benninga MA, Boeckxstaens GE (2005) Interstitial cells of Cajal are involved in the afferent limb of the rectoanal inhibitory reflex. Gut 54:1107–1113
9. Der-Silaphet T, Malysz J, Hagel S, Arsenault LA, Huizinga JD (1998) Interstitial cells of Cajal direct normal propulsive contractile activity in the mouse small intestine. Gastroenterology 114:724–736
10. Dinning PG, Zarate N, Hunt LM, Fuentealba SE, Mohammed SD, Szczesniak MM, Lubowski DZ, Preston SL, Fairclough PD, Lunniss PJ, Scott SM, Cook IJ (2010) Pancolonic spatiotemporal mapping reveals regional deficiencies in, and disorganization of colonic propagating pressure waves in severe constipation. Neurogastroenterol Motil 22:e340–e349
11. Gulbransen BD, Bains JS, Sharkey KA (2010) Enteric glia are targets of the sympathetic innervation of the myenteric plexus in the guinea pig distal colon. J Neurosci 30:6801–6809
12. Gwynne RM, Bornstein JC (2007) Mechanisms underlying nutrient-induced segmentation in isolated guinea pig small intestine. Am J Physiol Gastrointest Liver Physiol 292:G1162–G1172
13. Hall KE, El-Sharkawy TY, Diamant NE (1982) Vagal control of migrating motor complex in the dog. Am J Phys 243:G276–G284
14. Hirst GD, Dickens EJ, Edwards FR (2002) Pacemaker shift in the gastric antrum of guinea-pigs produced by

excitatory vagal stimulation involves intramuscular interstitial cells. J Physiol 541:917–928
15. Holst MC, Kelly JB, Powley TL (1997) Vagal preganglionic projections to the enteric nervous system characterized with Phaseolus vulgaris-leucoagglutinin. J Comp Neurol 381:81–100
16. Horiguchi K, Komuro T (2000) Ultrastructural observations of fibroblast-like cells forming gap junctions in the W/W(nu) mouse small intestine. J Auton Nerv Syst 80:142–147
17. Huizinga JD (2017) The role of ICC in interoception (Commentary: Phase amplitude coupling at the organism level: the amplitude of spontaneous alpha rhythm fluctuations varies with the phase of the infra-slow gastric basal rhythm). Front Auton Neurosci:11–102. https://doi.org/10.3389/fnins.2017.00102
18. Huizinga JD, Chen JH, Zhu YF, Pawelka A, McGinn RJ, Bardakjian BL, Parsons SP, Kunze WA, Wu RY, Bercik P, Khoshdel A, Chen S, Yin S, Zhang Q, Yu Y, Gao Q, Li K, Hu X, Zarate N, Collins P, Pistilli M, Ma J, Zhang R, Chen D (2014) The origin of segmentation motor activity in the intestine. Nat Commun 5:1–11. https://doi.org/10.1038/ncomms4326
19. Huizinga JD, Pervez M, Nirmalathasan S, Chen J-H (2021) Characterization of haustral activity in the human colon. Am J Physiol 320:G1067–G1080. https://doi.org/10.1152/ajpgi.00063.2021.
20. Komuro T (2012) Atlas of interstitial cells of Cajal in the Gastrointestinal tract. Springer. https://doi.org/10.1007/978-94-007-2917-9
21. Kuizenga MH, Sia TC, Dodds KN, Wiklendt L, Arkwright JW, Thomas A, Brookes SJ, Spencer NJ, Wattchow DA, Dinning PG, Costa M (2015) Neurally mediated propagating discrete clustered contractions superimposed on myogenic ripples in ex vivo segments of human ileum. Am J Physiol Gastrointest Liver Physiol 308:G1–G11
22. Pawelka AJ, Huizinga JD (2015) Induction of rhythmic transient depolarizations associated with waxing and waning of slow wave activity in intestinal smooth muscle. Am J Physiol Gastrointest Liver Physiol 308:G427–G433
23. Pervez M, Ratcliffe E, Parsons SP, Chen J-H, Huizinga JD (2020) The cyclic motor patterns in the human colon. Neurogastroenterol Motil e13807:1–17. https://doi.org/10.1111/nmo.13807
24. Powley TL (2000) Vagal input to the enteric nervous system. Gut 47 Suppl 4:iv30–iv32; discussion iv36
25. Powley TL, Wang XY, Fox EA, Phillips RJ, Liu LW, Huizinga JD (2008) Ultrastructural evidence for communication between intramuscular vagal mechanoreceptors and interstitial cells of Cajal in the rat fundus. Neurogastroenterol Motil 20:69–79
26. Powley TL, Hudson CN, McAdams JL, Baronowsky EA, Phillips RJ (2016) Vagal intramuscular arrays: the specialized mechanoreceptor arbors that innervate the smooth muscle layers of the stomach examined in the rat. J Comp Neurol 524:713–737
27. Rae MG, Fleming N, McGregor DB, Sanders KM, Keef KD (1998) Control of motility patterns in the human colonic circular muscle layer by pacemaker activity. J Physiol 510:309–320
28. Ramon y Cajal S (1911) Maloine, Paris Histologie du systéme nerveux de l' homme et des vertébrés
29. Rao SS, Sadeghi P, Beaty J, Kavlock R, Ackerson K (2001) Ambulatory 24-h colonic manometry in healthy humans. Am J Physiol Gastrointest Liver Physiol 280:G629–G639
30. Richter CG, Babo-Rebelo M, Schwartz D, Tallon-Baudry C (2017) Phase-amplitude coupling at the organism level: the amplitude of spontaneous alpha rhythm fluctuations varies with the phase of the infra-slow gastric basal rhythm. NeuroImage 146:951–958
31. Sarna SK, Waterfall WE, Bardakjian BL, Lind JF (1981) Types of human colonic electrical activities recorded postoperatively. Gastroenterology 81:61–70
32. Smith-Edwards KM, Edwards BS, Wright CM, Schneider S, Meerschaert KA, Ejoh LL, Najjar SA, Howard MJ, Albers KM, Heuckeroth RO, Davis BM (2021) Sympathetic input to multiple cell types in mouse and human colon produces region-specific responses. Gastroenterology 160:1208–1223.e4
33. Szurszewski JH (1969) A migrating electric complex of canine small intestine. Am J Phys 217:1757–1763
34. Taylor I, Duthie HL, Smallwood R, Linkens D (1975) Large bowel myoelectrical activity in man. Gut 16:808–814
35. Tinel J (1937) Le système nerveux végétatif, Masson
36. Ward SM, Sanders KM (2006) Involvement of intramuscular interstitial cells of Cajal in neuroeffector transmission in the gastrointestinal tract. J Physiol 576:675–682
37. Ward SM, Beckett EA, Wang X, Baker F, Khoyi M, Sanders KM (2000) Interstitial cells of Cajal mediate cholinergic neurotransmission from enteric motor neurons. J Neurosci 20:1393–1403
38. Yanagida H, Yanase H, Sanders KM, Ward SM (2004) Intestinal surgical resection disrupts electrical rhythmicity, neural responses, and interstitial cell networks. Gastroenterology 127:1748–1759
39. Yuan Y, Ali MK, Mathewson KJ, Sharma K, Faiyaz M, Tan W, Parsons SP, Zhang KK, Milkova N, Liu L, Chen J-H, Huizinga JD (2020) Associations between colonic motor patterns and autonomic nervous system activity assessed by high-resolution manometry and concurrent heart rate variability. Front Neurosci 13:1447. https://doi.org/10.3389/fnins.2019.01447
40. Zhu YF, Wang X-Y, Parsons SP, Huizinga JD (2016) Stimulus-induced pacemaker activity in interstitial cells of Cajal associated with the deep muscular plexus of the small intestine (ICC-DMP). Neurogastroenterol Motil. https://doi.org/10.1111/nmo.12808

20 Refining Enteric Neural Circuitry by Quantitative Morphology and Function in Mice

Marthe J. Howard

Abstract

RNA-Seq, electrophysiology and optogenetics in mouse models are used to assess function, identify disease related genes and model enteric neural circuits. Lacking a comprehensive quantitative description of the murine colonic enteric nervous system (ENS) makes it difficult to most effectively use mouse data to better understand ENS function or for development of therapeutic approaches for human motility disorders. Our goal was to provide a quantitative description of mouse colon to establish the extent to which mouse colon architecture, connectivity and function is a useful surrogate for human and other mammalian ENS. Using GCaMP imaging coupled with pharmacology and quantitative confocal and 3D image reconstruction, we present quantitative and functional data demonstrating that regional structural changes and variable distribution of neurons define neural circuit dynamics and functional connectivity responsible for colonic motor patterns and regional functional differences. Our results advance utility of multispecies and gut region-specific data.

Keywords

ENS · Neural circuits · Optogenetics · Synaptic function · Enteric morphology

M. J. Howard (✉)
University of Toledo, College of Medicine and Life Sciences, Toledo, OH, USA
e-mail: Marthe.Howard@utoledo.edu

20.1 Introduction

In the gastrointestinal tract, the enteric nervous system (ENS) controls secretion, blood flow, nutrient transport, immune interactions and peristalsis (muscle motor control). The ENS comprises two ganglionated plexuses located between the inner circular and outer longitudinal layers of smooth muscle (myenteric plexus) and in the submucosal layer of connective tissue (submucosal plexus) [16, 17, 37, 46, 51]. The integrated activity of intrinsic motor pattern-generating circuits and reflexes underpins the neural regulation of gastrointestinal function. Intrinsic neural control is integrated and modulated by signals from the epithelium [15, 26, 46, 51] as well as the parasympathetic and sympathetic branches of the autonomic nervous system [16, 38, 46, 48, 51].

Enteric neurons have been classified according to projection pattern, neurochemical code, functional activity and morphology [4, 5, 9, 14, 16, 25, 31]; significant species and gut regional differences in these criteria have been identified

N. J. Spencer et al. (eds.), *The Enteric Nervous System II*, Advances in Experimental Medicine and Biology 1383, https://doi.org/10.1007/978-3-031-05843-1_20

by a number of groups [16, 19, 22, 49, 50] suggesting that there may not be a unifying neural circuit that comprehensively defines ENS function. In the colon, there are region-specific motor patterns that likely reflect colonic architecture [6, 8, 24, 47] but details of ganglion architecture and functional connectivity have been lacking.

A roadblock to understanding functional neural circuitry in the murine ENS is the dearth of information detailing, in quantitative terms, the structural architecture, ganglion morphology, connectivity and distribution of neurons by neurochemical code. Although an enormous amount of data are available, we set out to develop a predictive neural circuit model of the mouse colonic ENS and found that comprehensive quantitative data, not available in the literature, were required. We thus undertook to describe in some detail aspects of the mouse colon required to generate such a model. In so doing, we have shown that mouse colon morphology and ENS are best described in three regions that can be delineated both by structure and function. We provide significant insight into the structural underpinnings of motor patterns and fill several important gaps in our understanding of ENS circuit connectivity.

20.2 Quantitative Morphology

The approach we took to accomplish a quantitative description of the colonic myenteric plexus was to combine tiled confocal imaging and 3D reconstruction to measure and/or describe cell-to-cell and ganglion-to-ganglion connectivity, architecture and morphology in samples obtained from mouse proximal, middle and distal colon. Ganglion size, neuron number, volume and distribution were measured. These quantitative assessments were linked with functional studies employing electrical stimulation or GCaMP6F or GCaMP6S calcium imaging and pharmacology (manuscripts detailing these studies are in revision, [27, 33]).

Tracing studies of dye-filled enteric neurons and electrophysiology has defined our understanding of ENS neuron morphology and connectivity [1, 2, 12, 18, 34]. Although incredibly important, this approach leaves gaps in our knowledge due to the limitations imposed by the size of neurons most likely to be impaled as well as their location within a ganglion. An additional limitation is that the full extent of neurite arborization as well as identification of specific neural targets is difficult to achieve. By measuring myenteric neuron area and volume in whole-mount samples or in FAC-sorted myenteric ganglion neurons, a significant number of small neurons, not likely to be impaled successfully were identified. Interestingly, although IPANS were routinely found to be the majority of the largest neurons, small IPANS were identified as well; a similar size distribution has been reported for guinea pig [23] and mouse colon [18]. A broad distribution of neuron sizes is apparent for a variety of species [20]. The distribution of neuron sizes appears to be random between neuron types defined by neurochemical code, and within ganglia; no significant differences were identified between proximal, middle and distal colon. This is potentially important as there is a significant change in ganglion size when comparing proximal, middle and distal colon in adult mouse. The myenteric ganglia in proximal colon are large with up to 300 neurons per ganglion [33]. The number of neurons per ganglion decreases gradually from proximal to middle colon and further decreases in distal colon where ganglia can be quite small (~20 neurons/ganglion). This change in ganglion size opens a number of questions related to relative cell type distribution. Recent single-cell RNA-Seq studies suggest that inferring function from neurochemical code may lead to erroneous conclusions [10, 28, 52, 53]; features such as receptor subtypes, ion channel subtypes and newly identified neuromodulators complicate our ability to classify neurons or infer function using a limited number of characteristics [24, 11].

Examination of tiled confocal images indicated that the overall architecture of the mouse colon had regional specificity which was confirmed by measuring ganglion volume, patterns of ganglion-to-ganglion connectivity and neuron-to-neuron connectivity [33]. The packing of ganglia in the mouse proximal colon is dense with ganglia much closer together in both the longitudinal and circumferential directions when compared to the middle or distal colon. The length and architecture of internodal strands in the proximal colon is much smaller than in middle or distal colon and the way that ganglia appear to be connected also differs. In the proximal colon, there is a significant amount of intraganglionic connectivity and process pass-through in the circumferential direction with axons traversing through one ganglion before projecting to a ganglion in either the oral or anal direction. Although we provide a more detailed picture of colon connectivity, especially in regard to IPANS, the pattern is similar to that found in guinea pig [18] and other species [1, 12, 22, 34]. The caliber of internodal strands in proximal colon is greater than in more distal parts of the colon, but patterns of ganglion-to-ganglia connectivity is more varied in middle and distal colon compared to proximal colon. In middle and distal colon, there is a significant amount of connectivity between one ganglion and one or more internodal strands; this pattern is not obvious in the proximal colon. Differences in the patterns of ganglion-to-ganglion connectivity are reflected in regional functional differences discussed below.

Of particular interest was our observation of intraganglionic synaptic connectivity [33]. As far as we are aware, ours is the first demonstration of intraganglionic structural or functional neural signaling in the ENS. Furness et al., [18] reported, in mouse distal colon, suggestive data indicating that IPANS within a single ganglion make functional synaptic connections in neighboring ganglia. This demonstration that IPANS in the mouse colon had functional connectivity comparable to that in guinea pig highlights that there are some unifying aspects of functional connectivity, broadening the utility of multispecies data. It has been dogma that peripheral ganglia are way-stations and that there is little to no intraganglionic neural signaling; it is significant to show this kind of neural interaction as it is necessary to understand neural circuitry and functional connectivity. Intraganglionic synaptic signaling in the colon could account for aspects of colonic motor patterns not comprehensively explained by current data; this concept is discussed in more detail below. The overall morphology and connectivity of IPANS in the mouse colon are similar to that reported in guinea pig [13, 18, 43] and also more complex. We identified a new kind of IPAN in the proximal and middle colon [32, 33, 36] that expresses CGRPA (IPANA) rather than CGRPB, has both Dogiel type I and type II morphology, and its terminals surround neurons in neighboring ganglia with basket-type endings that are more complex than those observed in the guinea pig [3, 12, 40–42]. This result is confirmed by single-cell RNA-Seq data [10]. IPANA neurons do make synapses on neurons close to their soma, indicative of intraganglionic connectivity. This could impact initiation of colonic reflexes since distension or other mechanical forces have been shown to activate some interneurons as well as IPANS [29, 30, 43, 44]. It is of interest that based on single-cell RNA-Seq data from mouse colon, somatostatin is exclusively expressed in only a single class of colonic IPAN [10]. This result is significant as it has been accepted that somatostatin is expressed in interneurons and that it impacts motility [21, 25, 35]. This could have significant functional consequences as we have shown, using GCaMP6f imaging in cholinergic neurons, that agonist binding of somatostatin significantly dampens the responses evoked by nicotinic agonists while depressing the frequency of spontaneous calcium transients [33], indicating at least two forms of synaptic regulation; one impacting presynaptic regulation and the other post-synaptic control. These results highlight the utility of single-cell RNA-Seq data in providing refined identification of functionally relevant molecules and receptors in transcriptionally defined cells.

20.3 Functional Connectivity

When thinking about functional connectivity in the mouse colon, and being cognizant of the multiple motor patterns coupled with the regional differences in myenteric ganglion architecture, morphology and connectivity we identified, we undertook to assess function using GCaMP imaging coupled with electrical stimulation and pharmacology [27, 33]. Several recent papers have focused on the mouse colon since there had not been electrophysiological recordings in proximal colon with intact neural circuitry. Studies focused on the control of smooth muscle electrical activity [8, 24, 45, 47] demonstrate that the rhythmic electrical activity in smooth muscle is neurogenic in nature [7, 45, 47] and that more complex motor patterns observed in the proximal colon, compared to the distal colon are likely due to differences in neural network connectivity [24]. Although quite significant, these investigations left unresolved issues regarding the relationships between neurogenic signaling, myenteric ganglion architecture, connectivity and neuron type distribution.

The differential distribution of ganglia, ganglion size and patterns of connectivity in the proximal compared to the distal colon has an impact on neural signaling and thus patterned motor activity [24, 27, 33, 39, 47]. We showed that summation of spontaneous activity precedes colonic motor complex formation (CMC), in both proximal and distal colon, but found that CMCs were more frequent in the proximal compared to distal colon supporting our morphological data demonstrating differences in neural network circuitry. In the colon, responses to electrical stimulation close to the image recording site, showed no differences between proximal and distal colon early response but a significant increase in summation and duration of activity in the late response, which is correlated to CMCs, in proximal colon [33]. Li et al. [24] argued that the increased neuron number and larger connective fiber density in proximal colon could not explain the larger number of neurons responding to electrical stimulation; we posit that this can be accounted for by intraganglionic connectivity coupled with differences in connectivity patterns and data showing that a significant number of neurons in the proximal colon have synaptic delays up to 65 ms indicating poly-synaptic connectivity [27, 33]. Interestingly, the delayed responses we observed in distal colon were generally of smaller amplitude and did not occur at distances greater than 20 mm from the stimulating electrode, while larger amplitude responses detectable at greater than 60 mm from the stimulation site were observed in proximal colon [33]; this pattern supports the idea that in the proximal colon, local neural circuits are arrayed in parallel, as previously suggested [24]. We suggest further, that in the proximal colon, inputs from both short and long distance are integrated resulting in self-generating circuit activity, while in the distal colon, information transfer occurs in series without the modulation or integration observed in proximal colon. Together these data support and extend previous conclusions that circuit connectivity, ganglion morphology and organization underlie regionally diverse motor function. Based on single-cell RNA sequencing (Howard unpublished data; [10, 28, 52, 53]) and our assumption that differential distribution of neurotransmitter receptors could contribute to regional functional specificity, we showed regional distribution of 5HT-3 receptors and neuronal nicotinic acetylcholine receptors in the mouse colon; the percentage of neurons that respond only to nicotinic agonist was significantly higher in the proximal colon while those that only respond to 5HT was significantly higher in the distal colon [27, 33]; this has been confirmed by single-cell transcript profiling that identified a significant increase of 5HT-3 receptors in distal colon [10]. Neurons that responded to both 5HT and nicotinic agonist were most abundant in the middle colon [33]. It has been reported that significantly more neurons depend upon cholinergic inputs in the distal colon compared to proximal colon and which responses are more apparent at a distance from the stimulation site [24] suggesting polarized mono-synaptic connectivity. Our quantification of neuron distribution indicated that the number of cholinergic neurons decreased from proximal to distal colon while the number of neurons expressing NOS increased [24, 33] indicating that the increased reliance on cholinergic input in the distal colon reflects long axons originating aboral to the recording site and not an increased number of cholinergic neurons; this increased reliance on cholinergic input could also reflect an increase in the number and/or class of neuronal nicotinic acetylcholine receptors as suggested by RNA-

Seq data [10]. In the proximal colon, based on neuron-to-neuron connectivity and functional studies, we suggest that there is an increased number of short distance and poly-synaptic connectivity resulting in reverberating neural activation. In total, our results support the idea that at least two types of neuron-to-neuron connectivity account for regionally identifiable motor patterns in the colon. The differences in the data between ours and Li et al., [24] could be based on the fact that Li et al., [24] combined electrical stimulation and nicotinic receptor antagonism with calcium imaging while Nestor-Kalinoski et al., [33] assessed responses to receptor agonists avoiding issues related to potential receptor desensitization and stimulated release of additional neurotransmitter or modulators whose synaptic effects would not be directly assessable. Currently, the most parsimonious conclusion is that in the distal colon, peristaltic activity is underpinned by mono-synaptic connectivity and the more complicated patterns in the proximal colon responsible for mixing are the result of poly-synaptic and reverberating intra-and inter-ganglionic synaptic connectivity.

20.4 Conclusion

We tested the possibility that in the colon, underlying structure of the myenteric plexus, provides the underpinning for regional functional differences. This idea is not novel (Furness; Spencer Costa Li), but earlier studies leave many unanswered questions. Our data suggest that the increased complexity in neurogenic activity and motor patterning in the proximal colon is underpinned by the significantly larger number of neurons, intraganglionic and poly-synaptic connectivity, and distribution of neurotransmitter receptors.

Acknowledgements The work presented was funded by the NIH Stimulating Peripheral Activity to Relieve Conditions (SPARC) program. OT2OD023859 (Dr. Marthe J. Howard, PI) and U18EB021790 (Dr. Marthe J. Howard, PI). The studies presented were done in collaboration with Drs. Joseph Margiotta, Andrea Nestor-Kalinoski, Brian Davis, Katheryn Albers, Kristen Smith-Edwards, Robert Heuckeroth and Joel Bornstein.

References

1. Bornstein JC, Hendricks R, Furness JB, Trussell DC (1991) Ramification of theaxons of AH-neurons injected with the intracellular marker biocytin in the myenteric plexus of the guineapig small intestine. J Comp Neurol 314:437–451
2. Brehmer A, Schrodl F, Neuhuber W (1999) Morphological classification of enteric neurons-100 years after Dogiel. Anat Embryol (Berl) 200:125–135
3. Clerc N, Furness JB, Bornstein JC, Kunze WA (1998) Correlation of electrophysiological and morphological characteristics of myenteric neurons of the duodenum in the guinea-pig. Neuroscience 82:899–914
4. Costa M, Brookes SJ, Steele PA, Gibbins I, Burcher E, Kandiah CJ (1996) Neurochemical classification of myenteric neurons in the guinea-pig ileum. Neuroscience 75(3):949–967
5. Costa M, Furness JB, Pompolo S, Brookes SJH, Bornstein JC, Bredt DS, Snyder SH (1992) Projections and chemical coding of neurons with immunoreactivity for nitric oxide synthase in the guinea-pig small intestine. Neurosci Lett 148(1–2):121–112
6. Costa M, Keightley LJ, Hibberd TJ, Wiklendt L, Smolilo DJ, Dinning PG, Brookes SJ, Spencer NJ (2020) Characterization of alternating neurogenic motor patterns in mouse colon. Neurogastroenterol Motil 33:e14047
7. Costa M, Dodds KN, Wiklendt L, Spencer NJ, Brookes SJ, Dinning PG (2013) Neurogenic and myogenic motor activity in the colon of the guinea pig, mouse, rabbit, and rat. Am J Physiol Gastrointest Liver Physiol 305:G749–G759
8. Costa M, Keightley LJ, Wiklendt L, Hibberd TJ, Arkwright JW, Omari T, Wattchow DA, Brookes SJH, Dinning PG, Spencer NJ (2019) Identification of multiple distinct neurogenic motor patterns that can occur simultaneously in the guinea pig distal colon. Am J Physiol Gastrointest Liver Physiol 316:G32–G44
9. Dogiel A (1899) ber den Bau der Ganglien in den Geflechten des Darmes und der Gallenblase des Menschen und der S.ugetiere. Arch Anat Physiol Leipzig Anat Abt Jg 1899:130–158
10. Drokhlyansky E, Smillie CS, Van Wittenberghe N, Ericsson M, Griffin G K, Dionne D, Cuoco MS, Goder-Reiser MN, Sharova T, Aguirre AJ, Boland GM, Graham D, Rozenblatt-Rosen O, Xavier RJ, Regev A (2019) The enteric nervous system of the human and mouse colon at a single-cell resolution bioRxiv, 746743
11. Fung C, Vanden Berghe P (2020) Functional circuits and signal procesing in the enteric nervous system. Cell Mol Life Sci 77(22):4505–4522
12. Furness JB, Bornstein JC, Trussell DC (1988) Shapes of nerve cells in the myenteric plexus of the guinea-pig small intestine revealed by the intracellular injection of dye. Cell Tissue Res 254(3):561–571
13. Furness JB, Kunze WA, Bertrand PP, Clerc N, Bornstein JC (1998) Intrinsic primary afferent neurons of the intestine. Prog Neurobiol 54(1):1–18

14. Furness JB (2000) Types of neurons in the enteric nervous system. J Auton Nerv Syst 81:87–96
15. Furness JB (2008) The enteric nervous system: normal functions and enteric neuropathies. Neurogastroenterol Motil 20 Suppl 1:32–38
16. Furness JB (2012) The enteric nervous system and neurogastroenterology. Nat Rev Gastroenterol Hepatol 9:286–294
17. Furness JB, Callaghan BP, Rivera LR, Cho HJ (2014) The enteric nervous system and gastrointestinal innervation: integrated local and central control. Adv Exp Med Biol 817:39–71
18. Furness JB, Robbins HL, Xiao J, Stebbing MJ, Nurgali K (2004) Projections and chemistry of Dogiel type II neurons in the mouse colon. Cell Tissue Res 317:1–12
19. Furness JB, Stebbing MJ (2018) The first brain: species comparisons and evolutionary implications for the enteric and central nervous systems. Neurogastroenterol Motil 30
20. Gabella G, Trigg P (1984) Size of neurons and glial cells in the enteric ganglia of mice, guinea-pigs, rabbits and sheep. J Neurocytol 13:49–71
21. Gonkowski S, Rytel L (2019) Somatostatin as an active substance in the mammalian enteric nervous system. Int J Mol Sci 20:4461
22. Graham KD, Lopez SH, Sengupta R, Shenoy A, Schneider S, Wright CM, Feldman M, Furth E, Valdivieso F, Lemke A, Wilkins BJ, Naji A, Doolin EJ, Howard MJ, Heuckeroth RO (2020) Robust, 3-dimensional visualization of human colon enteric nervous system without tissue sectioning. Gastroenterology 158(2221–2235):e5
23. Kugler EM, Michel K, Zeller F, Demir IE, Ceyan GO, Schemann M et al (2015) Mechanical stress activates neurites and somata of myenteric neurons. Front Cell Neurosci 9:342
24. Li Z, Hao MM, Van Den Haute C, Baekelandt V, Boesmans W, Vanden Berghe P (2019) Regional complexity in enteric neuron wiring reflects diversity of motility patterns in the mouse large intestine. elife 8:e42914
25. Lomax AE, Furness JB (2000) Neurochemical classification of enteric neurons in the guinea-pig distal colon. Cell Tissue Res 302:59–72
26. Makadia PA, Najjar SA, Saloman JL, Adelman P, Feng B, Margiotta JF, Albers KM, Davis BM (2018) Optogenetic activation of colon epithelium of the mouse produces high-frequency bursting in extrinsic colon afferents and engages visceromotor responses. J Neurosci 38:5788–5798
27. Margiotta JF, Smith-Edwards KM, Nestor-Kalinoski AL, Davis BM, Albers KM, Howard MJ (2021) Synaptic components, function and modulation characterized by GCaMP6f Ca2+ imaging in mouse cholinergic myenteric ganglion neurons. Front Neurosci. In the press
28. May-Zhang AA, Tycksen E, Southard-Smith AN, Deal KK, Benthal JT, Buehler DP, Adam M, Simmons AJ, Monaghan JR, Matlock BK, Flaherty DK, Potter SS, Lau KS, Southard-Smith EM (2021) Combinatorial transcriptional profiling of mouse and human enteric neurons identifies shared and disparate subtypes in situ. Gastroenterology 160(3):755–770.e26
29. Mazzuoli G, Schemann M (2009) Multifunctional rapidly adapting mechanosensitive enteric neurons (RAMEN) in the myenteric plexus of the guinea pig ileum. J Physiol 587(Pt 19):4681–4693
30. Mazzuoli G, Schemann M (2012) Mechanosensitive enteric neurons in the myenteric plexus of the mouse intestine. PLoS One 7:e39887
31. Messenger JP, Furness JB (1990) Projections of chemically-specified neurons in the guinea-pig colon. Arch Histol Cytol 53:467–495
32. Mulderry PK, Ghatei MA, Spokes RA, Jones PM, Pierson AM, Hamid QA, Kanse S, Amara SG, Burrin JM, Legon S et al (1988) Differential expression of alpha-CGRP and beta-CGRP by primary sensory neurons and enteric autonomic neurons of the rat. Neuroscience 25:195–205
33. Nestor-Kalinoski A, Smith-Edwards KM, Meerschaert K, Margiotta JF, Rajwa B, Davis B, Howard MJ (2021) Quantitative anatomical and functional mapping reveals unique neural circuit connectivity in mouse proximal, middle and distal colon enteric nervous system. Cell Mol Gastroenterol Hepatol. In the press
34. Nurgali K, Stebbing MJ, Furness JB (2004) Correlation of electrophysiological and morphological characteristics of enteric neurons in the mouse colon. J Comp Neurol 468:112–124
35. Portbury AL, Pompolo S, Furness JB, Stebbing MJ, Kunze WA, Bornstein JC, Hughes S (1995) Cholinergic, somatostatin-immunoreactive interneurons in the guinea pig intestine: morphology, ultrastructure connections and projections. J Anat 187(Pt2):303–321
36. Qu ZD, Thacker M, Castelucci P, Bagyanszki M, Epstein ML, Furness JB (2008) Immunohistochemical analysis of neuron types in the mouse small intestine. Cell Tissue Res 334:147–161
37. Schneider S, Wright CM, Heuckeroth RO (2019) Unexpected roles for the second brain: enteric nervous system as master regulator of bowel function. Annu Rev Physiol 81:235–259
38. Smith-Edwards KM, Najjar SA, Edwards BS, Howard MJ, Albers KM, Davis BM (2019) Extrinsic primary afferent neurons link visceral pain to colon motility through a spinal reflex in mice. Gastroenterology 157:522–536 e2
39. Smith TK, Koh SD (2017) A model of the enteric neural circuitry underlying the generation of rhythmic motor patterns in the colon: the role of serotonin. Am J Physiol Gastrointest Liver Physiol 312:G1–G14
40. Smolilo DJ, Costa M, Hibberd TJ, Wattchow DA, Spencer NJ (2018) Morphological evidence for novel enteric neuronal circuitry in guinea pig distal colon. J Comp Neurol 526:1662–1672
41. Smolilo DJ, Costa M, Hibberd TJ, Brookes SJH, Wattchow DA, Spencer NJ (2019) Distribution,

projections, and association with calbindin baskets of motor neurons, interneurons, and sensory neurons in guinea-pig distal colon. J Comp Neurol 527:1140–1158
42. Smolilo DJ, Hibberd TJ, Costa M, Wattchow DA, De Fontgalland D, Spencer NJ (2020) Intrinsic sensory neurons provide direct input to motor neurons and interneurons in mouse distal colon via varicose baskets. J Comp Neurol 528:2033–2043
43. Spencer NJ, Smith TK (2004) Mechanosensory S-neurons rather than AH-neurons appear to generate a rhythmic motor pattern in guinea-pig distal colon. J Physiol 558:577–596. https://doi.org/10.1113/jphysiol.2004.063586
44. Spencer NJ, Dickson EJ, Hennig GW, Smith TK (2006) Sensory elements within the circular muscle are essential for mechanotransduction of ongoing peristaltic reflex activity in guinea-pig distal colon. J Physiol 576:519–531
45. Spencer NJ, Hibberd TJ, Travis L, Wiklendt L, Costa M, Hu H, Brookes SJ, Wattchow DA, Dinning PG, Keating DJ, Sorensen J (2018) Identification of a rhythmic firing pattern in the enteric nervous system that generates rhythmic electrical activity in smooth muscle. J Neurosci 38(24):5507–5522
46. Spencer NJ, Hu H (2020) Enteric nervous system: sensory transduction, neural circuits and gastrointestinal motility. Nat Rev Gastroenterol Hepatol 17:338–351
47. Spencer NJ, Travis L, Wiklendt L, Hibberd TJ, Costa M, Dinning P, Hu H (2020) Diversity of neurogenic smooth muscle electrical rhythmicity in mouse proximal colon. Am J Physiol Gastrointest Liver Physiol 318:G244–G253
48. Szurszewski JH, Ermilov LG, Miller SM (2002) Prevertebral ganglia and intestinofugal afferent neurones. Gut 51 Suppl 1:i6–10
49. Timmermans JP, Scheuermann DW, Stach W, Adriaensen D, De Groodt-Lasseel MH (1992) Functional morphology of the enteric nervous system with special reference to large mammals. Eur J Morphol 30:113–122
50. Timmermans JP, Adriaensen D, Cornelissen W, Scheuermann DW (1997) Structural organization and neuropeptide distribution in the mammalian enteric nervous system, with special attention to those components involved in mucosal reflexes. Comp Biochem Physiol A Physiol 118:331–340
51. Vanner S, Greenwood-Van Meerveld B, Mawe G, Shea-Donohue T, Verdu EF, Wood J, Grundy D (2016) Fundamentals of neurogastroenterology: basic science. Gastroenterology 150(6):1280–1291
52. Wright CM, Schneider S, Smith-Edwards KM, Mafra F, Leembruggen AJL, Gonzalez MV, Kothakapa DR, Anderson JB, Maguire BA, Gao T, Missall TA, Howard MJ, Bornstein JC, Davis BM, Heuckeroth RO (2021) scRNA-sequencing reveals new enteric nervous system roles for GDNF, NRTN, and TBX3. Cell Mol Gastroenterol Hepatol 11(5):1548–1592.e1
53. Zeisel A, Hochgerner H, Lonnerberg P, Johnsson A, Memic F, Van Der Zwan J, Haring M, Braun E, Borm LE, La Manno G, Codeluppi S, Furlan A, Lee K, Skene N, Harris KD, Hjerling-Leffler J, Arenas E, Ernfors P, Marklund U, Linnarsson S (2018) Molecular architecture of the mouse nervous system. Cell 174:999–1014 e22

21 Molecular Targets to Alleviate Enteric Neuropathy and Gastrointestinal Dysfunction

Lauren Sahakian, Rachel McQuade, Rhian Stavely, Ainsley Robinson, Rhiannon T. Filippone, Majid Hassanzadeganroudsari, Raj Eri, Raquel Abalo, Joel C. Bornstein, Mark R. Kelley, and Kulmira Nurgali

Abstract

Enteric neuropathy underlies long-term gastrointestinal (GI) dysfunction associated with several pathological conditions. Our previous studies have demonstrated that structural and functional changes in the enteric nervous system (ENS) result in persistent alterations of

L. Sahakian · A. Robinson · R. T. Filippone · M. Hassanzadeganroudsari
Institute for Health and Sport, Victoria University, Western Centre for Health, Research and Education, Sunshine Hospital, Melbourne, VIC, Australia

R. McQuade
Department of Medicine Western Health, Faculty of Medicine, Dentistry and Health Sciences, The University of Melbourne, Melbourne, VIC, Australia

R. Stavely
Department of Pediatric Surgery, Pediatric Surgery Research Laboratories, Massachusetts General Hospital, Harvard Medical School, Boston, MA, USA

R. Eri
University of Tasmania, School of Health Sciences, Launceston, TAS, Australia

R. Abalo
Department of Basic Health Sciences, Faculty of Health Sciences, University Rey Juan Carlos (URJC), Alcorcón, Spain

High Performance Research Group in Physiopathology and Pharmacology of the Digestive System (NeuGut-URJC), URJC, Alcorcón, Spain

Associated I+D+i Unit to the Institute of Medicinal Chemistry (IQM), Scientific Research Superior Council (CSIC), Madrid, Spain

Working Group of Basic Sciences in Pain and Analgesia of the Spanish Pain Society (Grupo de Trabajo de Ciencias Básicas en Dolor y Analgesia de la Sociedad Española del Dolor), Madrid, Spain

J. C. Bornstein
Department of Physiology, Melbourne University, Melbourne, VIC, Australia

M. R. Kelley
Indiana University Simon Cancer Center, Departments of Pediatrics, Biochemistry & Molecular Biology and Pharmacology & Toxicology, Program in Pediatric Molecular Oncology & Experimental Therapeutics, Herman B Wells Center for Pediatric Research, Indianapolis, IN, USA

K. Nurgali (✉)
Institute for Health and Sport, Victoria University, Western Centre for Health, Research and Education, Sunshine Hospital, Melbourne, VIC, Australia

Department of Medicine Western Health, Faculty of Medicine, Dentistry and Health Sciences, The University of Melbourne, Melbourne, VIC, Australia

Regenerative Medicine and Stem Cells Program, Australian Institute of Musculoskeletal Science (AIMSS), Melbourne, VIC, Australia
e-mail: kulmira.nurgali@vu.edu.au

N. J. Spencer et al. (eds.), *The Enteric Nervous System II*, Advances in Experimental Medicine and Biology 1383, https://doi.org/10.1007/978-3-031-05843-1_21

intestinal functions long after the acute insult. These changes lead to aberrant immune response and chronic dysregulation of the epithelial barrier. Damage to the ENS is prognostic of disease progression and plays an important role in the recurrence of clinical manifestations. This suggests that the ENS is a viable therapeutic target to alleviate chronic intestinal dysfunction. Our recent studies in preclinical animal models have progressed into the development of novel therapeutic strategies for the treatment of enteric neuropathy in various chronic GI disorders. We have tested the anti-inflammatory and neuroprotective efficacy of novel compounds targeting specific molecular pathways. *Ex vivo* studies in human tissues freshly collected after resection surgeries provide an understanding of the molecular mechanisms involved in enteric neuropathy. *In vivo* treatments in animal models provide data on the efficacy and the mechanisms of actions of the novel compounds and their combinations with clinically used therapies. These novel findings provide avenues for the development of safe, cost-effective, and highly efficacious treatments of GI disorders.

Keywords

Enteric neuropathy · Inflammatory bowel disease · Chemotherapy · Apurinic/apyrimidinic endonuclease/redox factor-1 (APE1/Ref-1) · High mobility group box protein 1 (HMGB1)

Enteric neuropathy associates with many gastrointestinal (GI) disorders and underlies symptoms of GI dysfunction, including inflammatory bowel diseases (IBD) and chemotherapy-induced toxicity.

More than 7 million people suffer from IBD worldwide. Current treatments include corticosteroids and immunosuppressive therapy, which cause severe side effects including liver toxicity and immunosuppression. Biological therapies have fewer side effects but require hospitalization and intravenous infusions. Most IBD patients undergo surgery during the course of the disease. About 40% of Crohn's disease patients require repeated surgeries to remove inflamed parts of the intestine. Many IBD patients end up with colostomy after the resection of the entire colon. Moreover, IBD patients are at a high risk of developing colorectal cancer [1]. Colorectal cancer (CRC) is the second most common cancer worldwide; it is often asymptomatic at the early stages, and, therefore, in most cases, it is diagnosed at the advanced stages, when the tumour penetrates through the wall of the GI tract and blood appears in the stool. If diagnosed at the early stages, resection surgery is followed by chemotherapy; if diagnosed at the later stages, chemotherapy is the main treatment and in the case of rectal cancer radiotherapy. All chemotherapeutic drugs currently used clinically have severe neurotoxic and GI side effects [17]. Peripheral sensory neuropathy and GI side effects, such as nausea, vomiting, diarrhoea and constipation, are the main reasons for the reduction of the dose of the chemotherapy and treatment cessation, which significantly reduces the efficacy of anticancer treatment. Traditionally it was accepted that the GI side effects are due to mucosal damage (mucositis) [30]. Although mucositis plays an important role in pathophysiology of chemotherapy-induced GI adverse effects, however, mucosa regenerates very fast, but these side effects last many years after the end of chemotherapy in cancer survivors [15], suggesting that not only mucosa but also other structures are also affected. Our studies in colon tissues from IBD and CRC patients demonstrate that chronic inflammation and chemotherapy cause functional and structural damage to the enteric nervous system (ENS) innervating GI tract, making enteric neurons a viable target for effective therapies to attenuate GI dysfunction [5]. Our previous studies found that oxidative stress is a major contributor to enteric neuronal damage and death induced by intestinal inflam-

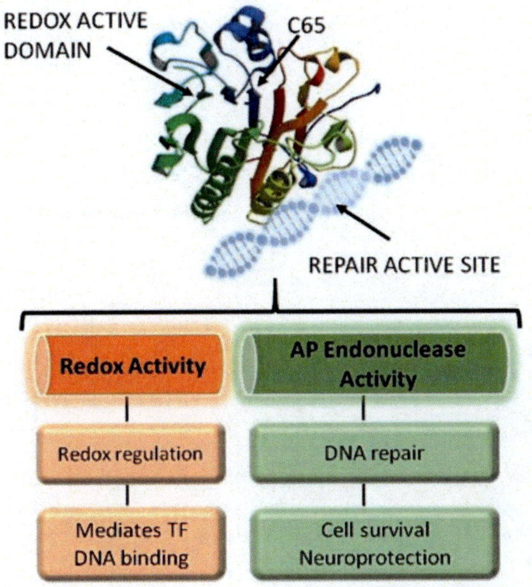

Fig. 21.1 APE1/Ref-1 is a vital dual functioning molecule containing a redox activating domain and a DNA repair domain

mation and platinum-based chemotherapeutic agent, oxaliplatin [11, 14].

One of the molecular targets identified by our studies is apurinic/apyrimidinic endonuclease/redox factor-1 (APE1/Ref-1). APE1/Ref-1 is a dual functional protein with endonuclease activity involved in the base excision repair (BER) pathway, which is critical for neuronal mitochondrial and nuclear DNA repair (Fig. 21.1). DNA repair is an important neuroprotective mechanism used for limiting neurotoxicity due to oxidative stress, direct toxicity and inflammation. APE1 also functions as a major reducing-oxidising (redox) factor that augments the binding of many transcription factors to DNA, including activator protein-1 (AP-1), nuclear factor kappa B (NF-κB), p53, the cAMP response element-binding protein (CREB), and hypoxia-inducible factor-1-α (HIF-1α), that play important roles in stress responses and various disorders [29, 32]. Professor Kelley's group is the first to identify that the repair function of APE1 contributes to the survival of non-dividing post-mitotic cells following oxidative DNA damage. Novel small-molecule compounds, APX3330 and its analogues, developed by Professor Kelley, directly and specifically inhibited APE1's redox signalling and enhanced DNA repair in dorsal root ganglia (DRG) neurons [7]. These compounds are highly effective in preventing or reversing cisplatin-induced sensory neuropathy without diminishing its anticancer efficacy [10]. Moreover, APX3330 demonstrated effective tumour-killing properties [26]. APX3330 is widely reported to be a direct, highly selective inhibitor of APE1 redox activity that does not affect the protein's endonuclease activity in tumours and enhances DNA repair in neurons. APX3330 is well absorbed orally with a bioavailability of >60%.

We hypothesised that using APX3330 in our animal models of IBD and chemotherapy, we will be able to block oxidative stress, prevent DNA damage and enhance DNA repair, which will alleviate enteric neuropathy and improve GI functions (Fig. 21.2).

We used the *Winnie* mouse model of spontaneous chronic colitis. These mice have a point mutation in the *Muc2* mucin gene (C57BL/6 background) leading to intestinal inflammation resulting from a primary intestinal epithelial defect [4, 8]. All *Winnie* mice develop mild spontaneous inflammation in the colorectum, which is developed in by 6–7 weeks of age (young adults) in pathogen-free conditions; it progresses over time and results in severe colitis by the age of 12–16 weeks. This is due to a thinner mucus layer allowing increased intestinal permeability and thus enhanced susceptibility to luminal toxins normally present within the gut. *Winnie* mice display symptoms of chronic diarrhoea, ulcerations, rectal bleeding and pain, as well as changes in microbiota composition similar to human IBD. Our RNA sequencing studies of the colons from *Winnie* mice demonstrate that this model accurately represents ulcerative colitis with 91% similarity in the expression of inflammation-associated genes, as well as 88% similarity to males with Crohn's disease. In comparison, 16.1% of genes in dextran sodium sulfate (DSS)-treated mice [6] and 12.5% in TNBS-treated rats

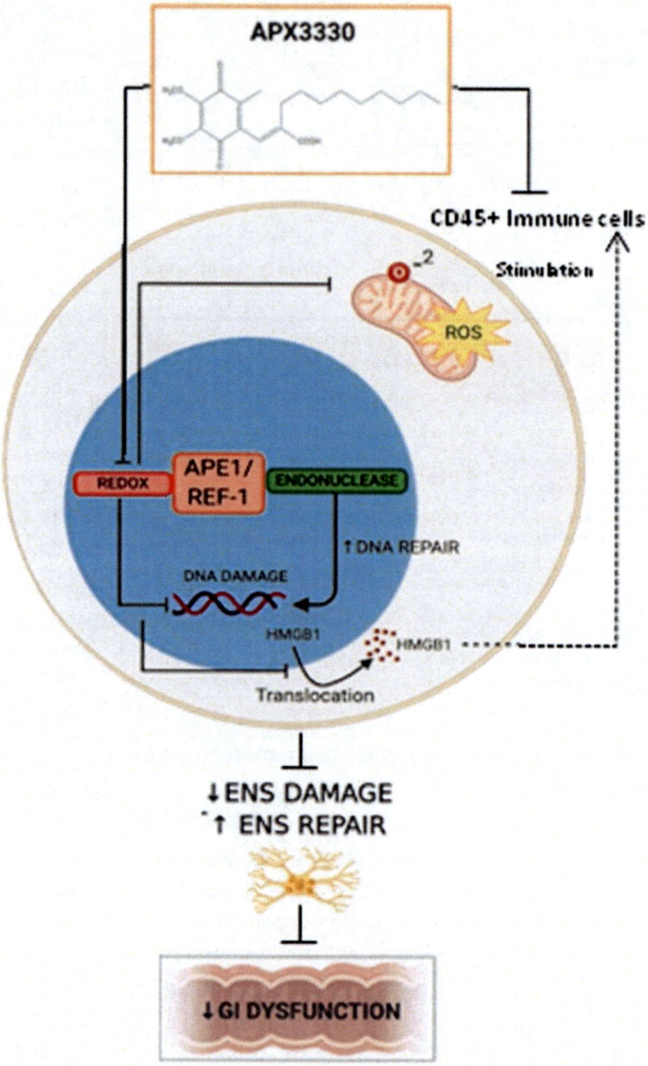

Fig. 21.2 APX3330 selectively inhibits APE1's redox signalling and DNA damage, reduces oxidative stress, alleviates enteric neuropathy, and reduces gastrointestinal dysfunction. (Created with Biorender.com)

[2] show concordance with IBD. Other models, including piroxicam-accelerated colitis in interleukin (IL)-10 knockout mice and adoptive transfer of CD4+CD25- leukocytes in immunodeficient mice, demonstrated a concordance of 77% and 64%, respectively, compared to 92 IBD-associated genes [9]. We have established the *Winnie* colony and published several papers on the characterisation of the enteric nervous system, gut functions and microbiota in this model [21–24]. Thus, the *Winnie* model of spontaneous chronic colitis closely represents human IBD and is not subjective to the variability of experimental techniques used to induce colitis.

Experiments were performed in 12 w.o (male and female) *Winnie* mice with active inflammation. APX3330 (25 mg kg^{-1} dissolved in 2% Cremophor, 2% EtOH and 96% sterile water) was administered twice a day with 8 h interval for 14 days via intraperitoneal injections. Sham-treated mice received the same volume and regimen of the vehicle solution. Age-matched C57BL/6 mice or non-*Winnie* littermates were used as naïve controls; no difference was found between them. The results demonstrate that APX3330 treatment had a significant positive effect on clinical symptoms in *Winnie* mice: it reduced rectal bleeding and rectal prolapse, alleviated diarrhoea and improved body weight in treated mice. APX3330 reduced inflammatory markers, a pan leukocyte marker, CD45, and faecal lipocalin-2 (Lcn-2), also known as neutrophil gelatinase-associated lipocalin (NGAL), which is a highly sensitive non-invasive biomarker of intestinal inflammation. APX3330 restored GI transit measured by in vivo X-ray imaging method and parameters of colonic motility measured ex vivo in organ bath studies of the excised colons from *Winnie* mice to control levels.

Winnie mice have significant damage to the enteric nervous system (ENS): reduction in numbers of nerve fibres projecting to the mucosa, loss of myenteric neurons and loss of enteric glial cells. APX3330 treatment increased density of nerve fibres, improved number of myenteric neurons and increased number of GFAP-positive glial cells in the myenteric plexus [25]. Overexpression of APE1/Ref-1 observed in the colonic cross section and the myenteric ganglia from *Winnie* mice, confirmed by immunohistochemical and western blot studies, was reduced by APX3330 treatment to the level comparable to control mice. Myenteric neurons from *Winnie* mice have a high level of oxidative stress measured by MitoSOX assay, which labels mitochondrial superoxide production. APX3330 reduced levels of MitoSOX fluorescence to control levels. DNA damage was measured by immunoreactivity to 8-Oxo-2′-deoxyguanosine (8-Oxo-dG), one of the major products of DNA oxidation widely used reliable marker of oxidative stress-induced DNA damage, and was observed only in the myenteric ganglia from *Winnie* mice, but not control mice and in APX3330-treated mice [25]. Another marker of the DNA damage used in this study is a high mobility group box protein 1 (HMGB1), a nuclear protein that acts as a chromatin-binding factor involved in the maintenance of nucleosome structure and regulation of gene transcription. HMGB1 is released by glial cells and neurons upon inflammasome activation [20]. HMGB1 is present in the nuclei of all enteric neurons; when neurons undergo cellular stress or injury, HMGB1 translocates from the nucleus into the cytoplasm and further into the extracellular space. This translocation was very prominent in the myenteric neurons of sham-treated *Winnie* mice. APX3330 inhibited this cytoplasmic translocation of HMGB1 in the myenteric neurons of *Winnie* mice.

Since HMGB1 is a downstream product of APE1/Ref-1 activation, we performed further studies testing the efficacy of HMGB1 inhibitor, a small-molecule compound glycyrrhizin, in two animal models: (1) chemotherapy treatment without colorectal cancer induction and (2) chemotherapy treatment in mice with an orthotopic model of colorectal cancer induced by implanting murine CT26 colorectal cancer cells into the caecum. Glycyrrhizin directly binds to a nuclear protein that maintains nucleosome structure and regulates gene transcription and has strong anti-inflammatory and neuroprotective properties. Inhibition of HMGB1 by glycyrrhizin modulates TLRs efficiently reducing the neuroinflammatory response resulting in neuroprotection from ischemic and traumatic brain damage [20]. It has been shown that HMGB1 inhibition reduces cisplatin-induced increases in iNOS levels and prevents ototoxicity [12], potentiates anticancer effects of platinum-based chemotherapeutics and has anticancer efficacy [31]. These properties make APX3330 and glycyrrhizin ideal candidates for the prevention and treatment of chemotherapy-induced enteric neuropathy.

Our previous studies demonstrate that treatment with first-line anti-colorectal cancer drugs, oxaliplatin, irinotecan and 5-fluorouracil induces death of enteric neurons, axonal damage, changes in their electrophysiological properties and significant morphological and functional alterations in neuronal nitric oxide synthase (nNOS)-immunoreactive neurons [3, 16, 18, 33]. Our studies provide evidence that oxidative stress is a key player in enteric neuropathy and colonic dysmotility associated with chemotherapy [14]. HMGB1 was observed within the nuclei of myenteric neurons from the control group. Our studies revealed that overexpression of HMGB1 and its translocation from the nuclei to cytoplasm were prominent in myenteric neurons from oxaliplatin-treated animals with and without

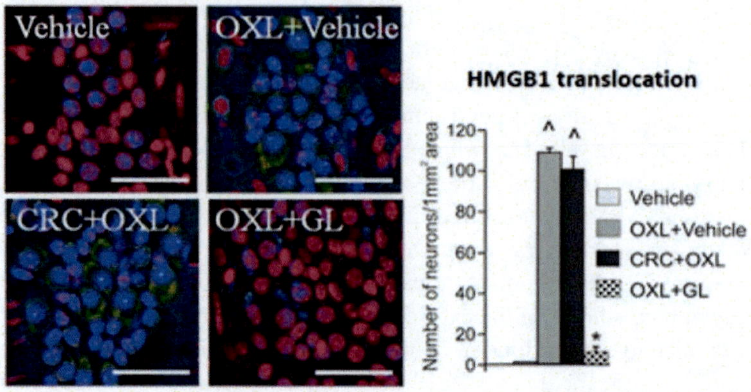

Fig. 21.3 Cytoplasmic translocation of HMGB1 (green) in myenteric neurons from mice oxaliplatin (OXL)-treated mice with and without CRC was prevented by co-treatment with glycyrrhizin (GL) (n = 6/group). ^$P < 0.0001$ compared to the control group, *$P < 0.001$ compared to OXL+Vehicle and CRC+OXL

colorectal cancer, but not in the neurons from untreated animals with colorectal cancer. HMGB1 translocated and released into the extracellular space acts as an antigen-presenting cytokine stimulating immune cells to release cytokines and chemokines, which activate toll-like receptors (TLRs), leading to changes in the microbiota [28]. These changes further exacerbate neuroinflammation and stimulate cellular damage and death of enteric neurons.

Both glycyrrhizin and APX3330 (25 mg/kg) combined with oxaliplatin were given to mice without cancer and in mice with CRC starting at day 7 post-surgery twice a day with 8 h interval for 14 days via *i.p.* injections starting on the same day as oxaliplatin treatment. The volumes for all injections were calculated to each animal's body weight with less than 200 μL per injection.

Co-treatment of glycyrrhizin (10 mg/kg) with oxaliplatin (3 mg/kg/dose; 3 times a week, 2 weeks) significantly reduced oxaliplatin-induced HMGB1 translocation in myenteric neurons (Fig. 21.3). The number of neurons in the myenteric plexus was not affected by induction of CRC and treatment with two vehicles. Treatment with oxaliplatin+vehicle in both mice with and without CRC induced loss of about 28% of myenteric neurons. The number of myenteric neurons in the CRC mice treated with oxaliplatin+glycyrrhizin was similar to the control group and significantly different to both oxaliplatin+vehicle-treated groups.

Co-treatment of glycyrrhizin and APX3330 with oxaliplatin inhibited tumour growth and vascularisation and prevented the loss of neurons induced by oxaliplatin treatment. These results provide a basis for further studies on the neuroprotective and anticancer efficacy of compounds targeting APE1/Ref-1 and HMGB1 proteins.

In summary, both treatments, APX3330 and glycyrrhizin, effectively alleviate enteric neuropathy and improve symptoms associated with intestinal inflammation and chemotherapy. APX3330 completed Phase I clinical trials for safety and toxicity in adult cancer patients who have failed all other treatments (NCT03375086) [27]. Clinical trials with glycyrrhizin have been approved by the World Health Organisation for the treatment of COVID-19 patients (ChiCTR2000029768 and ChiCTR2000030490) [13, 19]. Importantly, both of these drugs are not toxic and can be administered orally. Therefore, a fast translation of both compounds into clinical practice for the treatment of inflammatory bowel disease and the side effects of chemotherapy is plausible.

Acknowledgements This work was supported by the Institute for Health and Sport, Victoria University. R.T.F. is supported by the Australian NHMRC Project grant GNT1158952 to JCB, and KN. M.R.K. was supported by grants from the National Institute of Health CA167291, CA205166, CA231267, HL140961 and DOD grant W81XWH1910217. He is also supported by the Riley Children's Foundation and the Tom Wood Lexus Foundation.

Conflict of Interest Statement The authors declare no conflict of interest.

References

1. Axelrad JE, Lichtiger S, Yajnik V (2016) Inflammatory bowel disease and cancer: the role of inflammation, immunosuppression, and cancer treatment. World J Gastroenterol 22:4794–4801
2. Brenna Ø, Furnes MW, Drozdov I, van Beelen Granlund A, Flatberg A, Sandvik AK, Zwiggelaar RTM, Mårvik R, Nordrum IS, Kidd M, Gustafsson BI (2013) Relevance of TNBS-colitis in rats: a methodological study with endoscopic, histologic and Transcriptomic [corrected] characterization and correlation to IBD. PLoS One 8:e54543
3. Carbone SE, Jovanovska V, Brookes SJ, Nurgali K (2016) Electrophysiological and morphological changes in colonic myenteric neurons from chemotherapy-treated patients: a pilot study. Neurogastroenterol Motil 28:975–984
4. Eri RD, Adams RJ, Tran TV, Tong H, Das I, Roche DK, Oancea I, Png CW, Jeffery PL, Radford-Smith GL, Cook MC, Florin TH, McGuckin MA (2011) An intestinal epithelial defect conferring ER stress results in inflammation involving both innate and adaptive immunity. Mucosal Immunol 4:354–364
5. Escalante J, McQuade RM, Stojanovska V, Nurgali K (2017) Impact of chemotherapy on gastrointestinal functions and the enteric nervous system. Maturitas 105:23–29
6. Fang K, Bruce M, Pattillo CB, Zhang S, Stone R 2nd, Clifford J, Kevil CG (2011) Temporal genomewide expression profiling of DSS colitis reveals novel inflammatory and angiogenesis genes similar to ulcerative colitis. Physiol Genomics 43:43–56
7. Fehrenbacher JC, Guo C, Kelley MR, Vasko MR (2017) DNA damage mediates changes in neuronal sensitivity induced by the inflammatory mediators, MCP-1 and LPS, and can be reversed by enhancing the DNA repair function of APE1. Neuroscience 366:23–35
8. Heazlewood CK, Cook MC, Eri R, Price GR, Tauro SB, Taupin D, Thornton DJ, Png CW, Crockford TL, Cornall RJ, Adams R, Kato M, Nelms KA, Hong NA, Florin TH, Goodnow CC, McGuckin MA (2008) Aberrant mucin assembly in mice causes endoplasmic reticulum stress and spontaneous inflammation resembling ulcerative colitis. PLoS Med 5:e54
9. Holgersen K, Kutlu B, Fox B, Serikawa K, Lord J, Hansen AK, Holm TL (2015) High-resolution gene expression profiling using RNA sequencing in patients with inflammatory bowel disease and in mouse models of colitis. J Crohns Colitis 9:492–506
10. Kelley MR, Wikel JH, Guo C, Pollok KE, Bailey BJ, Wireman R, Fishel ML, Vasko MR (2016) Identification and characterization of new chemical entities targeting apurinic/apyrimidinic endonuclease 1 for the prevention of chemotherapy-induced peripheral neuropathy. J Pharmacol Exp Ther 359:300–309
11. Lam G, Apostolopoulos V, Zulli A, Nurgali K (2015) NADPH oxidases and inflammatory bowel disease. Curr Med Chem 22:2100–2109
12. Li G, Liu W, Frenz D (2006) Cisplatin ototoxicity to the rat inner ear: a role for HMG1 and iNOS. Neurotoxicology 27:22–30
13. Luo P, Liu D, Li J (2020) Pharmacological perspective: glycyrrhizin may be an efficacious therapeutic agent for COVID-19. Int J Antimicrob Agents 55:105995
14. McQuade RM, Carbone SE, Stojanovska V, Rahman A, Gwynne RM, Robinson AM, Goodman CA, Bornstein JC, Nurgali K (2016) Role of oxidative stress in oxaliplatin-induced enteric neuropathy and colonic dysmotility in mice. Br J Pharmacol 173:3502–3521
15. McQuade RM, Stojanovska V, Abalo R, Bornstein JC, Nurgali K (2016) Chemotherapy-induced constipation and diarrhea: pathophysiology, current and emerging treatments. Front Pharmacol 7:414
16. McQuade RM, Stojanovska V, Donald E, Abalo R, Bornstein JC, Nurgali K (2016) Gastrointestinal dysfunction and enteric neurotoxicity following treatment with anticancer chemotherapeutic agent 5-fluorouracil. Neurogastroenterol Motil 28:1861–1875
17. McQuade RM, Stojanovska V, Bornstein JC, Nurgali K (2017) Colorectal cancer chemotherapy: the evolution of treatment and new approaches. Curr Med Chem 24:1537–1557
18. McQuade RM, Stojanovska V, Donald EL, Rahman AA, Campelj DG, Abalo R, Rybalka E, Bornstein JC, Nurgali K (2017) Irinotecan-induced gastrointestinal dysfunction is associated with enteric neuropathy, but increased numbers of cholinergic myenteric neurons. Front Physiol 8:391
19. Murck H (2020) Symptomatic protective action of glycyrrhizin (licorice) in COVID-19 infection? Front Immunol 11:1239
20. Paudel YN, Shaikh MF, Chakraborti A, Kumari Y, Aledo-Serrano Á, Aleksovska K, Alvim MKM, Othman I (2018) HMGB1: a common biomarker and potential target for TBI, neuroinflammation, epilepsy, and cognitive dysfunction. Front Neurosci 12:628
21. Rahman AA, Robinson AM, Jovanovska V, Eri R, Nurgali K (2015) Alterations in the distal colon innervation in Winnie mouse model of spontaneous chronic colitis. Cell Tissue Res 362:497–512

22. Rahman AA, Robinson AM, Brookes SJH, Eri R, Nurgali K (2016) Rectal prolapse in Winnie mice with spontaneous chronic colitis: changes in intrinsic and extrinsic innervation of the rectum. Cell Tissue Res 366:285–299
23. Robinson AM, Gondalia SV, Karpe AV, Eri R, Beale DJ, Morrison PD, Palombo EA, Nurgali K (2016) Fecal microbiota and metabolome in a mouse model of spontaneous chronic colitis: relevance to human inflammatory bowel disease. Inflamm Bowel Dis 22:2767–2787
24. Robinson AM, Rahman AA, Carbone SE, Randall-Demllo S, Filippone R, Bornstein JC, Eri R, Nurgali K (2017) Alterations of colonic function in the Winnie mouse model of spontaneous chronic colitis. Am J Physiol Gastrointest Liver Physiol 312:G85–g102
25. Sahakian L, Filippone RT, Stavely R, Robinson AM, Yan XS, Abalo R, Eri R, Bornstein JC, Kelley MR, Nurgali K (2021) Inhibition of APE1/Ref-1 redox signaling alleviates intestinal dysfunction and damage to myenteric neurons in a mouse model of spontaneous chronic colitis. Inflamm Bowel Dis 27:388–406
26. Shah F, Logsdon D, Messmann RA, Fehrenbacher JC, Fishel ML, Kelley MR (2017) Exploiting the Ref-1-APE1 node in cancer signaling and other diseases: from bench to clinic. NPJ Precis Oncol 1:19
27. Shahda S, Lakhani NJ, O'Neil B, Rasco DW, Wan J, Mosley AL, Liu H, Kelley MR, Messmann RA (2019) A phase I study of the APE1 protein inhibitor APX3330 in patients with advanced solid tumors. J Clin Oncol 37:3097–3097
28. Stojanovska V, McQuade RM, Fraser S, Prakash M, Gondalia S, Stavely R, Palombo E, Apostolopoulos V, Sakkal S, Nurgali K (2018) Oxaliplatin-induced changes in microbiota, TLR4+ cells and enhanced HMGB1 expression in the murine colon. PLoS One 13:e0198359
29. Su D, Delaplane S, Luo M, Rempel DL, Vu B, Kelley MR, Gross ML, Georgiadis MM (2011) Interactions of apurinic/apyrimidinic endonuclease with a redox inhibitor: evidence for an alternate conformation of the enzyme. Biochemistry 50:82–92
30. Sylvester CL, Anderson PH, Stringer AM (2020) New therapeutic strategies for combatting gastrointestinal toxicity. Curr Opin Support Palliat Care 14:142–152
31. Ugrinova I, Pasheva E (2017) HMGB1 protein: a therapeutic target inside and outside the cell. Adv Protein Chem Struct Biol 107:37–76
32. Vasko MR, Guo C, Thompson EL, Kelley MR (2011) The repair function of the multifunctional DNA repair/redox protein APE1 is neuroprotective after ionizing radiation. DNA Repair (Amst) 10:942–952
33. Wafai L, Taher M, Jovanovska V, Bornstein JC, Dass CR, Nurgali K (2013) Effects of oxaliplatin on mouse myenteric neurons and colonic motility. Front Neurosci 7:30

22 Ca^{2+} Signaling Is the Basis for Pacemaker Activity and Neurotransduction in Interstitial Cells of the GI Tract

Kenton M. Sanders, Salah A. Baker, Bernard T. Drumm, and Masaaki Kurahashi

Abstract

Years ago gastrointestinal motility was thought to be due to interactions between enteric nerves and smooth muscle cells (SMCs) in the *tunica muscularis*. Thus, regulatory mechanisms controlling motility were either myogenic or neurogenic. Now we know that populations of interstitial cells, c-Kit$^+$ (interstitial cells of Cajal or ICC), and PDGFRα$^+$ cells (formerly "fibroblast-like" cells) are electrically coupled to SMCs, forming the SIP syncytium. Pacemaker and neurotransduction functions are provided by interstitial cells through Ca^{2+} release from the endoplasmic reticulum (ER) and activation of Ca^{2+}-activated ion channels in the plasma membrane (PM). ICC express Ca^{2+}-activated Cl$^-$ channels encoded by *Ano1*. When activated, Ano1 channels produce inward current and, therefore, depolarizing or excitatory effects in the SIP syncytium. PDGFRα$^+$ cells express Ca^{2+}-activated K$^+$ channels encoded by *Kcnn3*. These channels generate outward current when activated and hyperpolarizing or membrane-stabilizing effects in the SIP syncytium. Inputs from enteric and sympathetic neurons regulate Ca^{2+} transients in ICC and PDGFRα$^+$ cells, and currents activated in these cells conduct to SMCs and regulate contractile behaviors. ICC also serve as pacemakers, generating slow waves that are the electrophysiological basis for gastric peristalsis and intestinal segmentation. Pacemaker types of ICC express voltage-dependent Ca^{2+} conductances that organize Ca^{2+} transients, and therefore Ano1 channel openings, into clusters that define the amplitude and duration of slow waves. Ca^{2+} handling mechanisms are at the heart of interstitial cell function, yet little is known about what happens to Ca^{2+} dynamics in these cells in GI motility disorders.

Keywords

SIP syncytium · Ca^{2+} stores · Voltage-dependent Ca^{2+} channels · ANO1 channels · Enteric nervous system · GI motility

K. M. Sanders (✉) · S. A. Baker · B. T. Drumm
Department of Physiology and Cell Biology, University of Nevada, Reno, School of Medicine, Reno, NV, USA
e-mail: ksanders@med.unr.edu

M. Kurahashi
Department of Physiology and Cell Biology, University of Nevada, Reno, School of Medicine, Reno, NV, USA

Department of Internal Medicine, Division of Gastroenterology and Hepatology, University of Iowa, Iowa, Iowa City, USA

N. J. Spencer et al. (eds.), *The Enteric Nervous System II*, Advances in Experimental Medicine and Biology 1383, https://doi.org/10.1007/978-3-031-05843-1_22

22.1 Introduction

Interstitial cells of Cajal (ICC or c-Kit$^+$) and platelet-derived growth factor receptor alpha (PDGFRα$^+$) cells provide important regulatory behaviors in the generation of motility patterns in the gastrointestinal (GI) tract. ICC and PDGFRα$^+$ cells are electrically coupled to smooth muscle cells (SMCs) [32, 44, 66], so changes in membrane conductances in these cells influence the excitability and contractility of SMCs. Together these cells make up an electrical syncytium known as the SIP syncytium [64]. The functions of interstitial cells were suggested by their anatomical characteristics and associations, such as formations of networks, abundant close proximity to varicose processes of motor neurons, and gap junction connectivity with SMCs. Thus, it was proposed that ICC might be pacemaker cells and innervated by motor neurons [24, 69, 79]. The role of PDGFRα$^+$ cells, referred to as "fibroblast-like" by anatomists, was assumed to be only structural in nature. Recent work has begun to dissect the mechanisms of action of interstitial cells and reveal the many important functions of these cells in GI motility.

Major breakthroughs for understanding the functions of interstitial cells came when (i) mutants in which ICC failed to develop were used to investigate the role of ICC [36, 73, 74]: (ii) specific immunolabels for ICC and PDGFRα$^+$ cells were discovered [42, 51, 52, 56, 70]; (iii) reporter strains were developed [31, 63, 83], allowing observation of specific types of interstitial cells in intact muscles and unequivocal identification of cells after enzymatic dispersion of muscle tissues [47, 51, 52, 70, 83]; (iv) Cre-loxP technology was used to allow highly specific deletion or expression of genes in ICC and allowed cell specific expression of genetically encoded Ca^{2+} sensors to monitor Ca^{2+} dynamics [3, 7, 39, 48, 68]. Evaluations of genes expressed in cellular components of the SIP syncytium showed highly specific expression of genes important for the functions of these cells such as *Ano1* (originally called *Tmem16a*) in ICC [12] and *Kcnn3* in PDGFRα$^+$ cells [52]. Electrophysiological experiments on isolated cells using the patch clamp technique confirmed the importance of the highly expressed membrane conductances in pacemaker activity and responses to neurotransmitters [52, 83]. Activation of Ano1 channels results in an inward current, and activation of SK3 channels activates outward current. Thus, ICC and PDGFRα$^+$ cells provide opposing regulatory inputs in the SIP syncytium.

The two major ionic currents of ICC and PDGFRα$^+$ cells are activated by intracellular Ca^{2+}, suggesting that Ca^{2+} dynamics regulate the electrophysiological behaviors of these cells. This chapter summarizes some of the current understanding about Ca^{2+} handling mechanisms in ICC and PDGFRα$^+$ cells. Realization of the importance of Ca^{2+} dynamics has coincided with the ability to create transgenic mice with expression of genetically encoded Ca^{2+} sensors in ICC. Ca^{2+} dynamics in ICC can be monitored in these mice in situ, making it possible to follow the activation of cells, whether spontaneous or in response to neurotransmitters and other bioagonists. Cell-specific expression of genetically encoded sensors has not been accomplished for PDGFRα$^+$ cells due to the lack of highly selective Cre recombinase strains for these cells. However, PDGFRα$^+$ cells can be distinguished in intact muscles due to nuclear expression of eGFP in a reporter strain [52] and the ability to monitor Ca^{2+} transients in cell cytoplasms after loading with membrane-permeable Ca^{2+} sensors [5, 6].

22.2 Basal Ca^{2+} Transients in Interstitial Cells

There are two major classes of ICC, intramuscular cells and networks of interconnected cells. Intramuscular ICC (ICC-IM and ICC in the region of the deep muscular plexus of the small intestine, ICC-DMP) lie in close association with enteric motor nerve processes in most smooth muscle regions of the GI tract. In mice expressing GCaMPs exclusively in ICC, we have found that all intramuscular ICC studied generate spontaneous Ca^{2+} transients that originate from multiple sites within individual cells [3, 7, 16, 17]. These

Ca^{2+} transients occur in a stochastic and localized manner and show no coupling between firing sites only a few microns from each other in the same cell or in cells nearby. The concept of coupled oscillators, used to describe the behaviors of ICC in the past [35, 72], is not apparent at this level of organization.

Ca^{2+} transients in ICC result from brief release events from intracellular Ca^{2+} stores. Drugs that block the uptake of Ca^{2+} into stores via SERCA pumps (CPA or thapsigargin) inhibit the generation of spontaneous Ca^{2+} transients within a few minutes [3, 16, 17]. Drugs that inhibit Ca^{2+} release through IP_3 receptor (IP_3R) channels or ryanodine (RyR) receptor channels also inhibit Ca^{2+} transients; however, the dependence upon IP_3Rs and RyRs varies between different types of ICC [3, 16, 17]. Ca^{2+} release appears to occur in microdomains (now known as endoplasmic reticulum–plasma membrane junctions; ER/PM junctions) formed by close associations between the ER and the PM [65]. In the excluded volumes of ER/PM junctions, Ca^{2+} released from the ER reaches high concentrations and activates Ca^{2+}-dependent conductances, such as Ano1, in the plasma membrane. Indeed, ICC-DMP generate spontaneous transient inward currents (STICs) due to transient activation of Ano1 channels by Ca^{2+} release [85]. In this manner, the stochastic release of Ca^{2+} from many sites within individual cells and from thousands of ICC within tissues can generate a net inward current that is conducted to SMCs via gap junctions. Thus, ICC exert a net excitatory influence in the SIP syncytium (Fig. 22.1a).

When Ca^{2+} is released from stores into an ER/PM junction, some of the Ca^{2+} can be recovered via the SERCA pump, but some is lost through general diffusion to the bulk cytoplasm or to the extracellular space via plasma membrane Ca^{2+} ATPase (PMCA) or Na^+/Ca^{2+} exchange, both of which are expressed by ICC [3, 16, 81]. Therefore, mechanisms to recover Ca^{2+} must exist to maintain the large ER-to-cytoplasm gradient and sustain the ability of the ER to release Ca^{2+}. A major contributor to Ca^{2+} recovery in ICC occurs by the process of store-operated Ca^{2+} entry (SOCE) [60–62, 67, 71]. The apparatus for SOCE consists of a protein, stromal interaction protein (STIM) that spans the ER membrane. The portion of this protein within the ER lumen contains a Ca^{2+} sensor. When ER Ca^{2+} is depleted by release events, STIM molecules oligomerize, and the cytoplasmic portion of STIM binds to and activates ORAI, a highly selective Ca^{2+} channel in the PM. Thus, STIM and ORAI form a complex that senses a reduction in ER Ca^{2+} and activates Ca^{2+} entry to facilitate store refilling. Drugs that block ORAI can reduce or stop spontaneous Ca^{2+} transients in ICC [80].

While much less is known about Ca^{2+} dynamics in PDGFRα^+ cells, these cells also have the ability to generate spontaneous Ca^{2+} transients [5, 6]. These events, however, couple to SK3 channels, that are also Ca^{2+} activated but cause the development of outward currents [52]. As a result of Ca^{2+} transients and activation of SK3 channels, isolated PDGFRα^+ cells can generate spontaneous transient outward currents (STOCs). Generation of STOCs by many PDGFRα^+ cells within GI muscles exerts a net hyperpolarizing or membrane-stabilizing effect, thereby reducing the excitability and contractility of muscles (Fig. 22.1b).

22.3 Neurotransduction by Interstitial Cells

Both types of interstitial cells in GI muscles contribute to neural regulation of motility. Although this topic has been somewhat controversial, some investigators have clung to the notion that SMCs are the cells innervated and responsible for post-junctional responses. Monitoring of Ca^{2+} signaling in ICC and PDGFRα^+ cells clearly shows that these cells are innervated and display appropriate responses and temporal characteristics, which suggests that they mediate significant components of post-junctional responses. Blocking the ionic conductances specific to these cells, Ano1 or SK3, can reduce or block post-junctional electrical and mechanical responses to motor nerve stimulation [2, 4, 6, 19].

ICC form very close associations with the varicosities of motor neurons. This is not to say

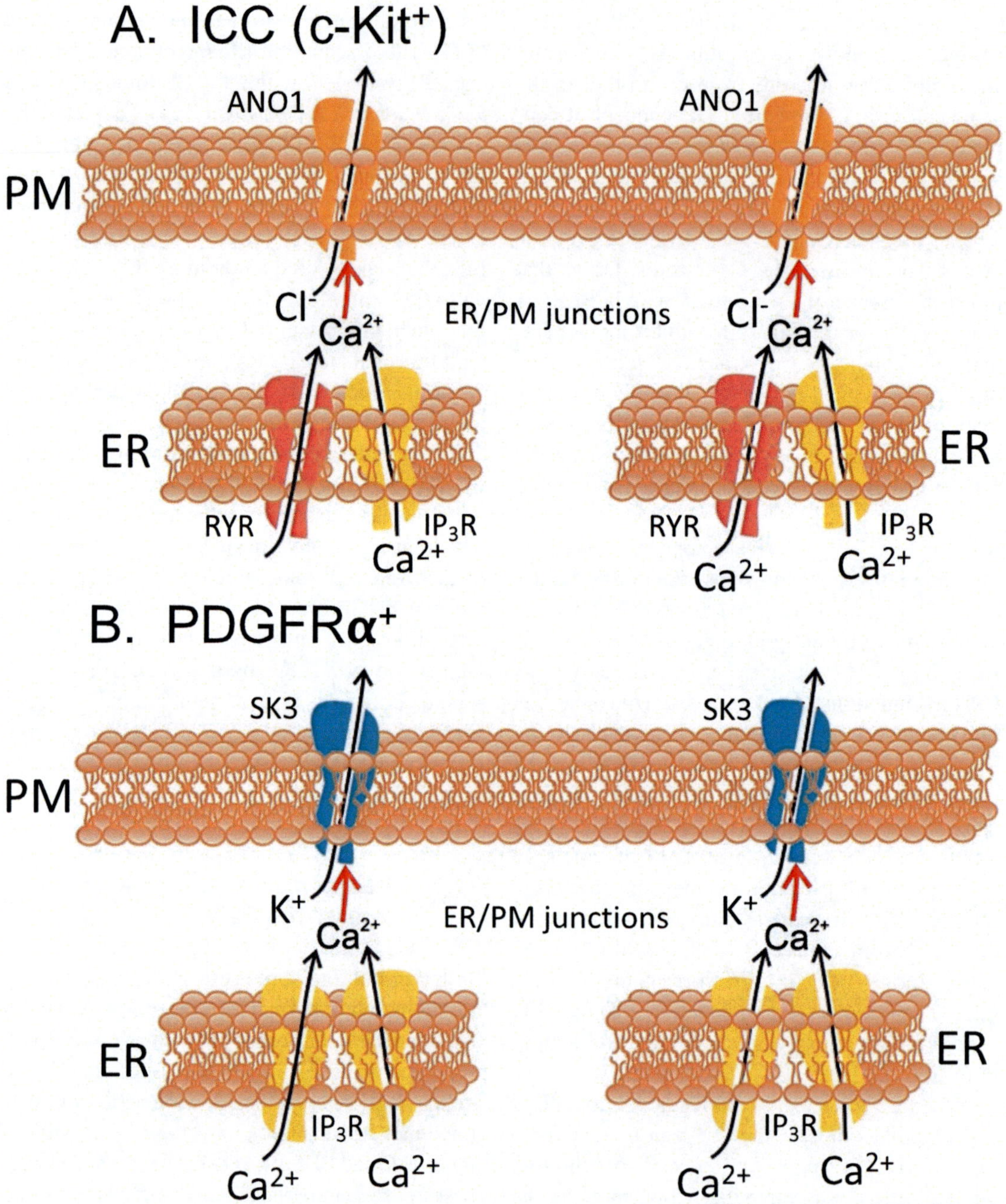

Fig. 22.1 Schematics showing fundamental Ca^{2+} dynamics in ICC (**a**) and PDGFRα⁺ cells (**b**). These cells display localized Ca^{2+} transients due to release of Ca^{2+} from the endoplasmic reticulum (ER). The ER forms close associations with the plasma membrane (PM). These junctions are referred to as ER/PM junctions. High concentrations of $[Ca^{2+}]_i$ are achieved during Ca^{2+} transients in the excluded volumes of the ER/PM junctions. (**a**) The rise in $[Ca^{2+}]_i$ activates Ca^{2+}-activated Cl^- channels in the PM (encoded by *Ano1*) of ICC, producing spontaneous transient inward currents (STICs) and depolarization. STICs activated by Ca^{2+} transients in ICC conduct to adjoining SMCs electrically coupled via gap junctions (not shown). Generation of Ca^{2+} transients is highly localized within cells and independent of Ca^{2+} transients even at nearby ER/PM junctions or events occurring in nearby cells (i.e., purpose of showing two discrete ER/PM junctions in the schematic). Coupling between active Ca^{2+} release sites has not been observed, even after addition of excitatory neurotransmitters. (**b**) The rise in $[Ca^{2+}]_i$ activates SK3 in the PM (encoded by *Kcnn3*) of PDGFRα⁺ cells, producing spontaneous transient outward currents (STOCs) and hyperpolarization or stabilization of membrane potential. STOCs activated by Ca^{2+} transients in PDGFRα⁺ cells conduct to adjoining SMCs that are electrically coupled via gap junctions (not shown)

that such junctions are never found between varicosities and SMCs [57], but a morphometric study of esophageal muscles revealed a high propensity of these synaptic-like junctions between motor neurons and ICC [15]. These junctions may facilitate the availability of high concentrations of neurotransmitters near receptors of ICC. Expression of the appropriate receptors by ICC is another indication that these cells are involved in transduction of neural inputs. In the case of excitatory neurotransmission, ICC express type 3 muscarinic (M3) receptors and neurokinin type 1 (NK1) receptors [4, 12, 19, 43]. These receptors dominate responses to neurotransmitters released from enteric excitatory motor neurons (Fig. 22.2). The metabolic enzyme for ACh (acetylcholine esterase (AChE)) is expressed by enteric motor neurons [78], so it is likely that ACh is broken down rapidly in junctional spaces and sufficiently high concentrations of the transmitter may not reach muscarinic receptors expressed by SMCs. Our experiments have shown this to be the case in the murine gastric fundus, but after inhibition of AChE or when gastric ICC-IM fail to develop in the fundus, as in W/W^V mutants, thus removing the synaptic-like junctions, new post-junctional mechanisms are recruited, suggesting that in these conditions ACh reaches SMC receptors [10]. Enteric inhibitory responses due to release of nitric oxide (NO) are compromised in parts of the GI tract where ICC-IM are depleted in W/W^V mutants. ICC express soluble guanylate cyclase (sGC), the receptor for NO, suggesting these cells mediate at least a portion of post-junctional nitrergic responses [40, 41]. However, diffusion of NO may not be so heavily confined, and it may spread to other components of the SIP syncytium. Cell-specific knockdown of sGC suggests that nitrergic, post-junctional responses are mediated by ICC and SMCs [9, 28, 29, 54, 55]. Pathways for nitrergic inhibitory regulation are shown in Fig. 22.2.

Effector pathways in ICC, including Ca^{2+} handling mechanisms and dominant ion channels, transduce neural inputs. Stimulation of intrinsic motor neurons with electrical field stimulation (EFS) under conditions favoring excitatory neurotransmission (i.e., by blocking inhibitory pathways) causes significant enhancement in the frequency and amplitude of Ca^{2+} transients in ICC-IM in the colon and ICC-DMP in the small intestine. Stimulation of ICC-DMP appears to be dominated through tachykinin release and binding of NK1 receptors [4], while these receptors are hardly functional and responses are dominated by M3 receptors in the colon [19]. M3 and NK1 receptors couple through G proteins ($G_{q/11}$) and activation of phospholipase Cβ, causing generation of IP_3 [1, 76]. Enhanced production of IP_3 causes dramatic increases in Ca^{2+} release from ER via IP_3Rs. These events, like spontaneous Ca^{2+} transients, activate Ano1 channels in the PM and elicit a depolarizing trend in the SIP syncytium, enhancing SMC excitability and contraction (Fig. 22.2). Although Ca^{2+} transient frequency increases in ICC, entrainment of Ca^{2+} release or coupled oscillations between discrete Ca^{2+} release sites have not been observed. Thus, no evidence for coupled oscillators is substantiated at this level of organization.

Opposite effects on Ca^{2+} transients are observed in ICC-IM and ICC-DMP in response to enteric inhibitory neural stimulation (i.e., when excitatory pathways are blocked). The inhibitory effects are mediated by NO and transduced through activation of sGC and generation of cGMP. Downstream from production of cGMP, the mechanisms for the inhibitory effects of NO on Ca^{2+} release in ICC are less well understood, but may occur through cGMP-dependent protein kinase I and phosphorylation of IP_3R-associated cGMP kinase substrate (IRAG) [26, 75]. Initiation of enteric inhibitory responses blocks Ca^{2+} transients for the duration of stimulation (Fig. 22.2). Inhibition of Ca^{2+} release causes deactivation of Ano1 channels and cessation of the basal inward current in the SIP syncytium. In colonic muscles, this resulted in a 9 mV hyperpolarization of muscles [17]. Hyperpolarization reduces the likelihood that the threshold for Ca^{2+} action potentials is reached in SMCs, so this is a mechanism through which nitrergic neurotransmission reduces SMC excitability and inhibits contractions.

A long-acknowledged phenomenon in GI muscles is tonic inhibition [77]. In some muscles,

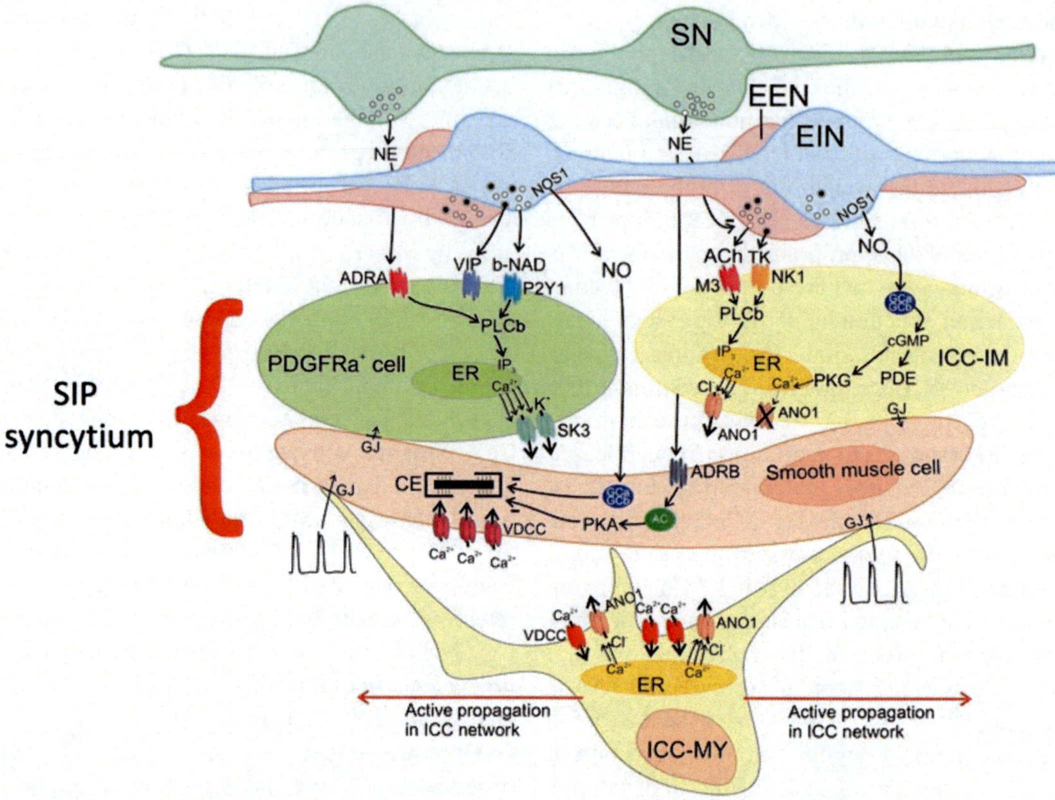

Fig. 22.2 Elements and mechanisms of the SIP syncytium and how enteric and sympathetic motor neurons regulate the output of the SIP syncytium. The SIP syncytium is composed of *S*MCs, *I*CC (intramuscular and pacemaker types of ICC), and *P*DGFRα+ cells. Ca^{2+} release regulates the open probabilities of Ca^{2+}-activated conductances in ICC and PDGFRα+ cells (see Fig. 22.1). Ca^{2+} release in ICC is regulated by excitatory and inhibitory neurotransmission by release of ACh and neurokinins from enteric excitatory enteric motor neurons (EEN) and NO from enteric inhibitory motor neurons (EIN). Pathways activated either increase (excitatory neurotransmission) or inhibit (nitrergic neurotransmission) the release of Ca^{2+} and therefore the activation of Ano1 channels in the PM. Ca^{2+} release in PDGFRα+ cells is regulated by release of purines from enteric inhibitory motor neurons and norepinephrine (NE) from sympathetic neurons. These neurotransmitters increase release of Ca^{2+} from the ER and activate SK3 channels, producing an outward current. Currents generated by ICC and PDGFRα+ cells conduct to SMCs and regulate the excitability of the SMC component of the SIP syncytium. SMCs also receive input from pacemaker ICC, such as ICC-MY. The pacemaker cells generate slow waves that cause periodic depolarizations of SMCs, activation of L-type Ca^{2+} channels (VDCC in SMCs), and phasic contractions of these cells. Active propagation of slow waves in the network of ICC-MY coordinates contractions of the muscles, producing behaviors such as gastric peristalsis and intestinal segmentation. Active propagation of slow waves depends upon T-type Ca^{2+} channels in ICC-MY (shown as VDCC in ICC). SMCs also receive direct stimulation by neurotransmission from, at a minimum, NO and NE. These neurotransmitters regulate smooth muscle contractions through electrophysiological responses (not shown) and by altering the Ca^{2+} sensitivity of contractile elements (CE). Based on the expression of receptors in SIP cells, stimulation of PDGFRα+ cells by inhibitory neuropeptides (e.g., VIP) is likely to occur

excitability and contractions are suppressed by ongoing release of inhibitory neurotransmitter, and SMC excitability and contractions of the muscles are greatly enhanced by TTX or by inhibition of nNOS. We found that tonic inhibition is linked to ongoing partial suppression of Ca^{2+} transients in ICC. TTX and antagonists of nNOS or sGC greatly increased the occurrence and amplitude of Ca^{2+} transients in ICC concomitantly with the increase in contractions [18]. The increased contractions resulting from suppression of tonic inhibition were reversed by an antagonist of Ano1 channels, supporting the idea that tonic inhibition in the proximal colon is mediated by nitrergic effects on ICC.

Enteric inhibitory neurons exhibit co-transmission whereby single neurons release multiple neurotransmitters. The synthetic enzyme for NO (nNOS) and the peptide VIP co-localize in enteric motor neurons [14, 45]. It is likely that these neurons also release purine neurotransmitters; however, this has never been shown definitively. A somewhat controversial story exists around the identity of the purine neurotransmitter in the GI tract. Classically, the transmitter was believed to be ATP [11], but more recent studies demonstrate that β-NAD⁺ fulfills the criteria for a neurotransmitter much better than ATP in mouse, monkey, and human GI muscles [38, 58]. Actually, enteric inhibitory neurons may release a cocktail of purines that include β-NAD⁺, ADPR, and Up4A [21, 22]. This story is detailed and not among the main topics of this review.

Another gene highly expressed by mouse and human PDGFRα⁺ cells is *P2ry1* [5, 51, 52]. Activation of $P2Y_1$ receptors by several purines, including ATP, ADP, β-NAD⁺, and the highly selective agonist MRS2365, greatly increases the frequency and amplitude of Ca^{2+} transients in PDGFRα⁺ cells in gastric fundus and colon [5, 6]. $P2Y_1$ receptors are coupled through G proteins ($G_{q/11}$) [23], so like M3 and NK1 receptors in ICC, $P2Y_1$ agonists increase the activity of PLCβ, production of IP_3, and release of Ca^{2+} through IP_3Rs (Fig. 22.2). Interesting to note is that MRS2500, a selective antagonist of $P2Y_1$ receptors, blocked all of these responses except the responses of some PDGFRα⁺ cells to ATP. Genetic deactivation of *P2ry1* ablates purinergic enteric inhibitory responses in mice [25, 37] and the increase in Ca^{2+} transients in PDGFRα⁺ cells caused by ADP, β-NAD⁺, and MRS2365. Responses to ATP, however, persist in some PDGFRα⁺ cells in GI muscles of *P2ry1*$^{-/-}$ animals [5, 6]. Similar to the Ca^{2+} transients enhanced by $P2Y_1$ agonists, these compounds elicited STOCs in PDGFRα⁺ cells [52]. Confidence that purinergic inhibitory responses are mediated by PDGFRα⁺ cells increased dramatically when it was shown that large amplitude hyperpolarization responses are elicited in PDGFRα⁺ cells that are equivalent to purinergic inhibitory junction potentials (IJPs) elicited in whole muscles [50]. These responses, like IJPs, are blocked by apamin or MRS2365. Of importance, $P2Y_1$ agonists failed to cause hyperpolarization of SMCs isolated from the same muscles.

Evaluation of the transcriptomes of SIP cells has revealed other receptors in these cells that might mediate regulatory effects (Fig. 22.2). An example is the expression of α adrenergic receptors (*Adra1a* and *Adra1b*) in PDGFRα⁺ cells [30]. Expression of these receptors was verified by real-time PCR [48]. Processes of sympathetic neurons, as labeled with antibodies to tyrosine hydroxylase, are distributed in the plane of the myenteric plexus and within circular muscles in proximity to PDGFRα⁺ cells. As with $P2Y_1$ agonists, NE elicited STOCs in voltage-clamped PDGFRα⁺ cells that were blocked by both RS100329, a specific adrenergic α1a receptor antagonist, and apamin. In current-clamped PDGFRα⁺ cells, significant hyperpolarization responses were caused by NE. NE also initiated or increased Ca^{2+} transients in PDGFRα⁺ cells in situ, and these responses were also blocked by RS100329. Contractions of colonic muscles were inhibited by phenylephrine, and these responses were blocked by RS100329 and apamin. Inhibitory effects of phenylephrine did not occur in *Adra1a*$^{-/-}$ mice. A preparation with the inferior mesenteric ganglion attached to the colon via the lumbar colonic nerve was used to isolate electrical stimulation of sympathetic neurons. Sympathetic nerve stimulation (SNS) caused hyperpolarization of colonic muscles that were partially blocked by prazosin and apamin [48]. SNS also inhibited colonic migrating motor complexes (CMMCs) through the mid and distal colon. This dramatic sympathetic effect on CMMCs did not occur in colons of *Adra1a*$^{-/-}$ mice. Similar sympathetic neurotransduction in PDGFRα⁺ cells and regulation of contractions also occur in human colons [49].

22.4 Pacemaker Activity in Interstitial Cells

ICC generate pacemaker activity in the GI tract that is responsible for electrical slow waves [53, 83]. Networks of pacemaker cells occur along the boundaries of the muscle layers, between the circular and longitudinal muscle layers in the stomach, small bowel, and colon (ICC-MY) and at the submucosal surface of the circular muscle in the colon (ICC-SM). Recent studies have also shown that a type of pacemaker activity is also generated by the ICC along the serosal surface of the proximal colon (ICC-SS), and a type of ICC within muscle bundles (ICC-IM type II) appears to generate the high-frequency pacemaker activity responsible for tone in the internal anal sphincter [33]. A major difference between pacemaker types of ICC (i.e., ICC-MY and ICC-SM) and intramuscular types of ICC (i.e., ICC-IM and ICC-DMP) is the expression of voltage-dependent Ca^{2+} conductances in pacemaker cells [7, 12, 16, 27, 33, 82].

ICC-MY and ICC-SM generate spontaneous Ca^{2+} transients that activate Ano1 channels in the PMs of these cells (Fig. 22.2). In the small intestine, ICC-MY express T-type Ca^{2+} channels ($Ca_V3.2$) [12, 16, 27, 82]. These channels are activated by the small depolarizations caused by the activation of Ano1 channels (STICs). Increasing the open probability of T-type Ca^{2+} channels can result in development of a Ca^{2+} action potential that constitutes the upstroke phase of slow waves. The dV/dt of the slow wave upstroke, when recorded directly from ICC-MY, is 2 V/s in the mouse intestine and 11 V/s in the rabbit intestine [46]. As in other excitable cells connected by gap junctions, the upstroke phase of slow waves depolarizes neighboring ICC-MY, activates the T-type conductance in these cells, and regenerates the upstroke potential, facilitating active propagation of slow waves in the ICC-MY network. Propagation is seen optically as a coherent intracellular Ca^{2+} wave front that proceeds at about 2 mm/sec in ICC-MY networks in mouse small intestine [16, 59]. SMCs do not express the same ion channels as ICC-MY, and in spite of electrical coupling between SMCs and ICC-MY, SMCs cannot regenerate slow waves and sustain active propagation. Therefore, slow waves conduct passively and decay with distance in the SMC compartment of the SIP syncytium.

Active propagation of slow waves is an important factor in the generation of normal motility behaviors. For example, in the stomach of larger mammals, slow waves propagate without decrement for many centimeter from the dominant pacemaker in the orad corpus to the pylorus [8]. Slow wave propagation was studied in canine gastric muscles using a dual chamber apparatus where slow waves could be initiated by pacing in one chamber, and recordings could be made in the second, electrically isolated chamber at various distances from the site of initiation or after addition of antagonists of specific conductances [8]. Nicardipine had no effect on the dV/dt of the upstroke depolarization nor on the rate of propagation. Antagonists of T-type channels, however, dramatically reduced both the upstroke velocity and the propagation rate. At higher concentrations, these antagonists blocked active slow wave propagation, as did reducing extracellular Ca^{2+} to 0.5 mM. These experiments also showed that reducing the availability of T-type Ca^{2+} channels by depolarization, which causes voltage-dependent inactivation, also greatly reduced upstroke and propagation velocity. These experiments clearly showed that propagation of slow waves is dependent upon voltage-dependent activation of a Ca^{2+} conductance with characteristics of T-type channels.

The Ca^{2+} waves that sweep across a network of ICC-MY are also dependent upon T-type Ca^{2+} channels in the murine small intestine (Fig. 22.2) [16]. Clusters of Ca^{2+} transients occur regularly in jejunal ICC-MY networks at a frequency of about 30 cycles per min in the mouse. The coherent spread of waves and the clustering of Ca^{2+} transients are blocked by NNC 55-0396 and TTA-A2, two specific T-type Ca^{2+} channel antagonists. However, these compounds do not block all Ca^{2+} transients, and in fact the occurrence of Ca^{2+} transients in the presence of T-type Ca^{2+} channel antagonists reverts to a stochastic pattern, as seen in ICC-IM. These experiments suggest that stochastic Ca^{2+} release is a basic behavior of ICC, and addition of a voltage-dependent Ca^{2+} conductance organizes Ca^{2+} transients into periodic clusters of events. While T-type channels are of primary importance in the stomach and small intestine, L-type channels are also important in some ICC, such as those along the submucosal surface of the circular muscle (ICC-SM) in the murine colon [7]. Muscles of the murine proximal colon are relatively more depolarized than cells in the small bowel and stomach. It would be impractical for T-type channels to be the dominant conductance providing active propagation in these muscles because T-type channels are inactivated at the depolarized potentials of colonic muscles. Instead, ICC-SM in murine colon utilize L-type Ca^{2+} as the dominant conductance. However, when ICC-SM are hyperpolarized, T-type channels become available and contribute to slow wave propagation. Presence of both T- and L-type channels provides a safety factor such that slow waves can persist over a broad range of membrane potentials.

Ca^{2+} entry due to the activation of the voltage-dependent Ca^{2+} conductance elicits not only depolarization but also Ca^{2+}-induced Ca^{2+} release (CICR) from the ER. Multiple firing sites generate these events both in the soma of ICC and in their processes. As above, Ca^{2+} entry organizes the otherwise stochastic Ca^{2+} transients into clusters of events. Ca^{2+} transients and corresponding activation of Ano1 channels through the network of ICC cause sustained net activation of the Cl^- conductance and clamp membrane potentials of ICC close to the equilibrium potential for the Cl^- gradient (about −10 mV; [84]). Ca^{2+} release events persist for more than a second or until Ca^{2+} stores are depleted to an extent where they cannot continue to release Ca^{2+}. The durations of the Ca^{2+} transient clusters define the durations of the plateau phase of slow waves. Here again, the concept of coupled oscillators is fallacious. Ca^{2+} release events are independent of each other and are organized into clusters by propagation of Ca^{2+} action potentials (i.e., the upstroke of the slow wave event), Ca^{2+} entry, and CICR.

A question to be considered is why the plateau phase is sustained for more than a second. Ca^{2+} entry through T-type Ca^{2+} channels is brief because the ensuing depolarization rapidly inactivates these channels. Therefore, it is unlikely that entry of Ca^{2+} through T-type channels is capable of sustaining CICR. The increased open probability of Ano1 channels is maintained by elevated intracellular Ca^{2+}. Thus, a source of Ca^{2+} is needed to maintain openings of Ano1 channels. Several potential sources are possible. The plateau phase of slow waves recorded from ICC is in the range of potentials that generate "window currents" from L-type Ca^{2+} channels [13]. Therefore, openings of these Ca^{2+} channels could provide persistent Ca^{2+} entry that could sustain CICR. This is likely a mechanism for sustaining the plateau in canine gastric antrum because nicardipine dramatically decreases and shortens the plateau phase of slow waves [8]. In contrast, dihydropyridines have little to no effect on slow waves in the small intestine. In murine, small intestine reverse mode Na^+/Ca^{2+} exchange appears to be responsible for sustained Ca^{2+} entry during the plateau phase [81]. During the period of Ca^{2+} release from stores, Ca^{2+} entry can also occur through ORAI [80]. When ER Ca^{2+} stores are depleted, Ca^{2+} release terminates, the open probability for Ano1 channels drops to low levels, and membrane potential repolarizes. After repolarization, membrane potential rests at negative levels, Ca^{2+} entry is minimal, and the stores refill via SOCE [80].

There is another ICC behavior in which membrane potential of the SIP syncytium is conditioned or tuned into a range where SMCs become rhythmic. In colonic longitudinal muscle, rhythmic intercellular Ca^{2+} waves occur that are due to the periodic firing of Ca^{2+} action potentials in longitudinal muscle cells [34]. These events are inhibited by an Ano1 antagonist [20]. Ano1 is expressed in ICC but not in SMCs, so it is likely that the inward current due to Ano1 develops in the ICC along the serosal surface, ICC-SS. This hypothesis was investigated in muscles of murine proximal colon [20]. ICC-SS form network-like structures and fire stochastic, localized Ca^{2+} transients. These events did not spread cell to cell as observed in ICC-MY or ICC-SM. Thus, they appear to be a hybrid type of ICC with stellate morphologies reminiscent of pacemaker ICC, but with Ca^{2+} dynamics similar to ICC-IM and ICC-DMP. As in all other ICC, however, Ca^{2+} transients in ICC-SS activate Ano1 channels in the PM. The inward currents produced summate to provide a depolarizing influence on longitudinal muscle cells, bringing membrane potential into a range where the SMCs fire Ca^{2+} action potentials. Thus, this example of electrical and mechanical rhythmicity represents an emergent property due to the electrophysiological characteristics of ICC-SS and SMCs and electrical coupling between these cells.

22.5 Conclusions

Ca^{2+} imaging has provided revelations about the mechanisms and functions of interstitial cells in GI muscles. These cells have dynamic Ca^{2+} handling mechanisms that utilize activation of Ano1 or SK3 channels in the PM and conduction of electrical responses to SMCs to distribute slow waves and responses to neural inputs from enteric and sympathetic motor neurons. Much has been learned about the organization and functions of ICC and PDGFRα$^+$ cells in normal GI muscles. It is now clear that disease or genetic mutations causing loss of function in interstitial cells can result in GI motor disorders. It is also possible that responses to immune mediators and phenotypic changes in interstitial cells could be a cause of fibrosis. However, we still lack knowledge about what causes defects or alters the phenotypes of interstitial cells and have no therapeutic means of restoring their functions if damaged in disease or aging. It is possible that defects in the SIP syncytium are the primary cause of motor disorders, and additional studies are needed to learn how to manipulate and potentially repair interstitial cell networks and connectivity with motor neurons. In-depth study of the pathophysiology of interstitial cells may provide new opportunities for therapeutics.

Acknowledgments Writing of this paper and much of the data reviewed was supported by R01 DK120759 to KMS and SAB and R01 DK-091336 to KMS and MK.

References

1. Almeida TA, Rojo J, Nieto PM, Pinto FM, Hernandez M, Martín JD, Candenas ML (2004) Tachykinins and tachykinin receptors: structure and activity relationships. Curr Med Chem 11:2045–2081
2. Baker SA, Drumm BT, Cobine CA, Keef KD, Sanders KM (2018a) Inhibitory neural regulation of the Ca(2+) transients in intramuscular interstitial cells of Cajal in the small intestine. Front Physiol 9:328
3. Baker SA, Drumm BT, Saur D, Hennig GW, Ward SM, Sanders KM (2016) Spontaneous Ca(2+) transients in interstitial cells of Cajal located within the deep muscular plexus of the murine small intestine. J Physiol 594:3317–3338
4. Baker SA, Drumm BT, Skowronek KE, Rembetski BE, Peri LE, Hennig GW, Perrino BA, Sanders KM (2018b) Excitatory neuronal responses of Ca(2+) transients in interstitial cells of Cajal in the small intestine. eNeuro 5(2):ENEURO.0080-18.2018. https://doi.org/10.1523/ENEURO.0080-18.2018
5. Baker SA, Hennig GW, Salter AK, Kurahashi M, Ward SM, Sanders KM (2013) Distribution and Ca(2+) signalling of fibroblast-like (PDGFR(+)) cells in the murine gastric fundus. J Physiol 591:6193–6208
6. Baker SA, Hennig GW, Ward SM, Sanders KM (2015) Temporal sequence of activation of cells involved in purinergic neurotransmission in the colon. J Physiol 593:1945–1963
7. Baker SA, Leigh WA, Del Valle G, De Yturriaga IF, Ward SM, Cobine CA, Drumm BT, Sanders KM (2021) Ca2+ signaling driving pacemaker activity in submucosal interstitial cells of Cajal in the murine colon. elife 10:e64099
8. Bayguinov O, Ward SM, Kenyon JL, Sanders KM (2007) Voltage-gated Ca2+ currents are necessary for slow-wave propagation in the canine gastric antrum. Am J Physiol Cell Physiol 293:C1645–C1659
9. Beck K, Friebe A, Voussen B (2018) Nitrergic signaling via interstitial cells of Cajal and smooth muscle cells influences circular smooth muscle contractility in murine colon. Neurogastroenterol Motil 30(6):e13300
10. Bhetwal BP, Sanders KM, An C, Trappanese DM, Moreland RS, Perrino BA (2013) Ca2+ sensitization pathways accessed by cholinergic neurotransmission in the murine gastric fundus. J Physiol 591:2971–2986
11. Burnstock G, Satchell DG, Smythe A (1972) A comparison of the excitatory and inhibitory effects of non-adrenergic, non-cholinergic nerve stimulation and exogenously applied ATP on a variety of smooth muscle preparations from different vertebrate species. Br J Pharmacol 46:234–242
12. Chen H, Ordog T, Chen J, Young DL, Bardsley MR, Redelman D, Ward SM, Sanders KM (2007) Differential gene expression in functional classes of interstitial cells of Cajal in murine small intestine. Physiol Genomics 31:492–509

13. Cohen NM, Lederer WJ (1987) Calcium current in isolated neonatal rat ventricular myocytes. J Physiol 391:169–191
14. Costa M, Furness JB, Pompolo S, Brookes SJ, Bornstein JC, Bredt DS, Snyder SH (1992) Projections and chemical coding of neurons with immunoreactivity for nitric oxide synthase in the guinea-pig small intestine. Neurosci Lett 148:121–125
15. Daniel EE, Posey-Daniel V (1984) Neuromuscular structures in opossum esophagus: role of interstitial cells of {Cajal}. Am J Phys 246:G305–G315
16. Drumm BT, Hennig GW, Battersby MJ, Cunningham EK, Sung TS, Ward SM, Sanders KM, Baker SA (2017) Clustering of Ca(2+) transients in interstitial cells of Cajal defines slow wave duration. J Gen Physiol 149:703–725
17. Drumm BT, Hwang SJ, Baker SA, Ward SM, Sanders KM (2019a) Ca2+ signalling behaviours of intramuscular interstitial cells of Cajal in the murine colon. J Physiol 597:3587–3617
18. Drumm BT, Rembetski BE, Baker SA, Sanders KM (2019b) Tonic inhibition of murine proximal colon is due to nitrergic suppression of Ca. Sci Rep 9:4402
19. Drumm BT, Rembetski BE, Huynh K, Nizar A, Baker SA, Sanders KM (2020a) Excitatory cholinergic responses in mouse colon intramuscular interstitial cells of Cajal are due to enhanced Ca2+ release via M3 receptor activation. FASEB J 34(8):10073–10095
20. Drumm BT, Rembetski BE, Messersmith K, Manierka MS, Baker SA, Sanders KM (2020b) Pacemaker function and neural responsiveness of subserosal interstitial cells of Cajal in the mouse colon. J Physiol 598:651–681
21. Durnin L, Hwang SJ, Kurahashi M, Drumm BT, Ward SM, Sasse KC, Sanders KM, Mutafova-Yambolieva VN (2014) Uridine adenosine tetraphosphate is a novel neurogenic P2Y1 receptor activator in the gut. Proc Natl Acad Sci U S A 111:15821–15826
22. Durnin L, Hwang SJ, Ward SM, Sanders KM, Mutafova-Yambolieva VN (2012) Adenosine 5-diphosphate-ribose is a neural regulator in primate and murine large intestine along with beta-NAD(+). J Physiol 590:1921–1941
23. Erb L, Weisman GA (2012) Coupling of P2Y receptors to G proteins and other signaling pathways. Wiley Interdiscip Rev Membr Transp Signal 1:789–803
24. Faussone Pellegrini MS, Cortesini C, Romagnoli P (1977) Ultrastructure of the tunica muscularis of the cardial portion of the human esophagus and stomach, with special reference to the so-called Cajal's interstitial cells. Arch Ital Anat Embriol 82:157–177
25. Gallego D, Gil V, Martinez-Cutillas M, Mane N, Martin MT, Jimenez M (2012) Purinergic neuromuscular transmission is absent in the colon of P2Y(1) knocked out mice. J Physiol 590:1943–1956
26. Geiselhoringer A, Werner M, Sigl K, Smital P, Worner R, Acheo L, Stieber J, Weinmeister P, Feil R, Feil S, Wegener J, Hofmann F, Schlossmann J (2004) IRAG is essential for relaxation of receptor-triggered smooth muscle contraction by cGMP kinase. EMBO J 23:4222–4231
27. Gibbons SJ, Strege PR, Lei S, Roeder JL, Mazzone A, Ou Y, Rich A, Farrugia G (2009) The alpha1H Ca2+ channel subunit is expressed in mouse jejunal interstitial cells of Cajal and myocytes. J Cell Mol Med 13:4422–4431
28. Groneberg D, Konig P, Lies B, Jager R, Seidler B, Klein S, Saur D, Friebe A (2013) Cell-specific deletion of nitric oxide-sensitive guanylyl cyclase reveals a dual pathway for nitrergic neuromuscular transmission in the murine fundus. Gastroenterology 145(1):188–196
29. Groneberg D, Zizer E, Lies B, Seidler B, Saur D, Wagner M, Friebe A (2015) Dominant role of interstitial cells of Cajal in nitrergic relaxation of murine lower oesophageal sphincter. J Physiol 593:403–414
30. Ha SE, Lee MY, Kurahashi M, Wei L, Jorgensen BG, Park C, Park PJ, Redelman D, Sasse KC, Becker LS, Sanders KM, Ro S (2017) Transcriptome analysis of PDGFRalpha+ cells identifies T-type Ca2+ channel CACNA1G as a new pathological marker for PDGFRalpha+ cell hyperplasia. PLoS One 12:e0182265
31. Hamilton TG, Klinghoffer RA, Corrin PD, Soriano P (2003) Evolutionary divergence of platelet-derived growth factor alpha receptor signaling mechanisms. Mol Cell Biol 23:4013–4025
32. Hanani M, Farrugia G, Komuro T (2005) Intercellular coupling of interstitial cells of cajal in the digestive tract. Int Rev Cytol 242:249–282
33. Hannigan KI, Bossey AP, Foulkes HJL, Drumm BT, Baker SA, Ward SM, Sanders KM, Keef KD, Cobine CA (2020) A novel intramuscular Interstitial Cell of Cajal is a candidate for generating pacemaker activity in the mouse internal anal sphincter. Sci Rep 10:10378
34. Hennig GW, Smith CB, O'Shea DM, Smith TK (2002) Patterns of intracellular and intercellular Ca2+ waves in the longitudinal muscle layer of the murine large intestine in vitro. J Physiol 543:233–253
35. Huizinga JD, Chen JH, Zhu YF, Pawelka A, Mcginn RJ, Bardakjian BL, Parsons SP, Kunze WA, Wu RY, Bercik P, Khoshdel A, Chen S, Yin S, Zhang Q, Yu Y, Gao Q, Li K, Hu X, Zarate N, Collins P, Pistilli M, Ma J, Zhang R, Chen D (2014) The origin of segmentation motor activity in the intestine. Nat Commun 5:3326
36. Huizinga JD, Thuneberg L, Kluppel M, Malysz J, Mikkelsen HB, Bernstein A (1995) W/kit gene required for interstitial cells of Cajal and for intestinal pacemaker activity. Nature 373:347–349
37. Hwang SJ, Blair PJ, Durnin L, Mutafova-Yambolieva V, Sanders KM, Ward SM (2012) P2Y1 purinoreceptors are fundamental to inhibitory motor control of murine colonic excitability and transit. J Physiol 590:1957–1972
38. Hwang SJ, Durnin L, Dwyer L, Rhee PL, Ward SM, Koh SD, Sanders KM, Mutafova-Yambolieva VN (2011) Beta-nicotinamide adenine dinucleotide is an enteric inhibitory neurotransmitter in human

and nonhuman primate colons. Gastroenterology 140(608–617):e6

39. Hwang SJ, Pardo DM, Zheng H, Bayguinov Y, Blair PJ, Fortune-Grant R, Cook RS, Hennig GW, Shonnard MC, Grainger N, Peri LE, Verma SD, Rock J, Sanders KM, Ward SM (2019) Differential sensitivity of gastric and small intestinal muscles to inducible knockdown of anoctamin 1 and the effects on gastrointestinal motility. J Physiol 597(9):2337–2360
40. Iino S, Horiguchi K, Nojyo Y (2008) Interstitial cells of Cajal are innervated by nitrergic nerves and express nitric oxide-sensitive guanylate cyclase in the guinea-pig gastrointestinal tract. Neuroscience 152:437–448
41. Iino S, Horiguchi K, Nojyo Y, Ward SM, Sanders KM (2009) Interstitial cells of Cajal contain signalling molecules for transduction of nitrergic stimulation in guinea pig caecum. Neurogastroenterol Motil 21(542–50):e12–e13
42. Iino S, Nojyo Y (2009) Immunohistochemical demonstration of c-Kit-negative fibroblast-like cells in murine gastrointestinal musculature. Arch Histol Cytol 72:107–115
43. Iino S, Ward SM, Sanders KM (2004) Interstitial cells of Cajal are functionally innervated by excitatory motor neurones in the murine intestine. J Physiol 556:521–530
44. Ishikawa K, Komuro T, Hirota S, Kitamura Y (1997) Ultrastructural identification of the c-kit-expressing interstitial cells in the rat stomach: a comparison of control and Ws/Ws mutant rats. Cell Tissue Res 289:137–143
45. Keef KD, Shuttleworth CW, Xue C, Bayguinov O, Publicover NG, Sanders KM (1994) Relationship between nitric oxide and vasoactive intestinal polypeptide in enteric inhibitory neurotransmission. Neuropharmacology 33:1303–1314
46. Kito Y, Mitsui R, Ward SM, Sanders KM (2015) Characterization of slow waves generated by myenteric interstitial cells of Cajal of the rabbit small intestine. Am J Physiol Gastrointest Liver Physiol 308:G378–G388
47. Komuro T (2012) Atlas of interstitail cells of Cajal in the gastrointestinal tract. Springer, Dordrecht
48. Kurahashi M, Kito Y, Baker SA, Jennings LK, Dowers JGR, Koh SD, Sanders KM (2020a) A novel postsynaptic signal pathway of sympathetic neural regulation of murine colonic motility. FASEB J 34(4):5563–5577
49. Kurahashi M, Kito Y, Hara M, Takeyama H, Sanders KM, Hashitani H (2020b) Norepinephrine has dual effects on human colonic contractions through distinct subtypes of alpha 1 adrenoceptors. Cell Mol Gastroenterol Hepatol 10:658–671, e1
50. Kurahashi M, Mutafova-Yambolieva V, Koh SD, Sanders KM (2014) Platelet-derived growth factor receptor-alpha-positive cells and not smooth muscle cells mediate purinergic hyperpolarization in murine colonic muscles. Am J Physiol Cell Physiol 307:C561–C570
51. Kurahashi M, Nakano Y, Hennig GW, Ward SM, Sanders KM (2012) Platelet-derived growth factor receptor alpha-positive cells in the tunica muscularis of human colon. J Cell Mol Med 16:1397–1404
52. Kurahashi M, Zheng H, Dwyer L, Ward SM, Koh SD, Sanders KM (2011) A functional role for the 'fibroblast-like cells' in gastrointestinal smooth muscles. J Physiol 589:697–710
53. Langton P, Ward SM, Carl A, Norell MA, Sanders KM (1989) Spontaneous electrical activity of interstitial cells of Cajal isolated from canine proximal colon. Proc Natl Acad Sci U S A 86:7280–7284
54. Lies B, Beck K, Keppler J, Saur D, Groneberg D, Friebe A (2015) Nitrergic signalling via interstitial cells of Cajal regulates motor activity in murine colon. J Physiol 593:4589–4601
55. Lies B, Gil V, Groneberg D, Seidler B, Saur D, Wischmeyer E, Jimenez M, Friebe A (2014) Interstitial cells of Cajal mediate nitrergic inhibitory neurotransmission in the murine gastrointestinal tract. Am J Physiol Gastrointest Liver Physiol 307:G98–G106
56. Maeda H, Yamagata A, Nishlkawa S, Yoshinaga K, Kobayashi S, Nishi K, Nishikawa SI (1992) Requirement of c-kit for development of intestinal pacemaker system. Development 116:369–375
57. Mitsui R, Komuro T (2002) Direct and indirect innervation of smooth muscle cells of rat stomach, with special reference to the interstitial cells of Cajal. Cell Tissue Res 309:219–227
58. Mutafova-Yambolieva VN, Hwang SJ, Hao X, Chen H, Zhu MX, Wood JD, Ward SM, Sanders KM (2007) Beta-nicotinamide adenine dinucleotide is an inhibitory neurotransmitter in visceral smooth muscle. Proc Natl Acad Sci U S A 104:16359–16364
59. Park KJ, Hennig GW, Lee HT, Spencer NJ, Ward SM, Smith TK, Sanders KM (2006) Spatial and temporal mapping of pacemaker activity in interstitial cells of Cajal in mouse ileum in situ. Am J Physiol Cell Physiol 290:C1411–C1427
60. Prakriya M, Lewis RS (2015) Store-operated calcium channels. Physiol Rev 95:1383–1436
61. Putney JW (1999) "Kissin' cousins": intimate plasma membrane-ER interactions underlie capacitative calcium entry. Cell 99:5–8
62. Putney JW (1986) A model for receptor-regulated calcium entry. Cell Calcium 7:1–12
63. Ro S, Chanjae P, Jin J, Zheng H, Blair P, Redelman D, Ward SM, Yan W, Sanders KM (2010) A model to study the phenotypic changes of interstitial cells of Cajal in gastrointestinal diseases. Gastroenterology 138:1068–1078.e2
64. Sanders KM, Koh SD, Ro S, Ward SM (2012) Regulation of gastrointestinal motility--insights from smooth muscle biology. Nat Rev Gastroenterol Hepatol 9:633–645
65. Sanders KM, Ward SM, Koh SD (2014) Interstitial cells: regulators of smooth muscle function. Physiol Rev 94:859–907
66. Seki K, Komuro T (1998) Further observations on the gap-junction-rich cells in the deep muscular plexus of the rat small intestine. Anat Embryol (Berl) 197:135–141

67. Stathopulos PB, Zheng L, Li GY, Plevin MJ, Ikura M (2008) Structural and mechanistic insights into STIM1-mediated initiation of store-operated calcium entry. Cell 135:110–122
68. Sung TS, Hwang SJ, Koh SD, Bayguinov Y, Peri LE, Blair PJ, Webb TI, Pardo DM, Rock JR, Sanders KM, Ward SM (2018) The cells and conductance mediating cholinergic neurotransmission in the murine proximal stomach. J Physiol 596:1549–1574
69. Thuneberg L (1982) Interstitial cells of Cajal: intestinal pacemaker cells? Adv Anat Embryol Cell Biol 71:1–130
70. Torihashi S, Ward SM, Nishikawa S, Nishi K, Kobayashi S, Sanders KM (1995) c-kit-dependent development of interstitial cells and electrical activity in the murine gastrointestinal tract. Cell Tissue Res 280:97–111
71. Trebak M, Zhang W, Ruhle B, Henkel MM, Gonzalez-Cobos JC, Motiani RK, Stolwijk JA, Newton RL, Zhang X (2013) What role for store-operated Ca(2)(+) entry in muscle? Microcirculation 20:330–336
72. Van Helden DF, Imtiaz MS (2003) Ca2+ phase waves: a basis for cellular pacemaking and long-range synchronicity in the guinea-pig gastric pylorus. J Physiol 548:271–296
73. Ward SM, Burns AJ, Torihashi S, Harney SC, Sanders KM (1995) Impaired development of interstitial cells and intestinal electrical rhythmicity in steel mutants. Am J Phys 269:C1577–C1585
74. Ward SM, Burns AJ, Torihashi S, Sanders KM (1994) Mutation of the proto-oncogene c-kit blocks development of interstitial cells and electrical rhythmicity in murine intestine. J Physiol 480(Pt 1):91–97
75. Werder A, Mayr M, Schneider G, Oesterle D, Fritsch RM, Seidler B, Schlossmann J, Hofmann F, Schemann M, Allescher HD, Schmid RM, Saur D (2011) Truncated IRAG variants modulate cGMP-mediated inhibition of human colonic smooth muscle cell contraction. Am J Physiol Cell Physiol 301:C1445–C1457
76. Wess J (1996) Molecular biology of muscarinic acetylcholine receptors. Crit Rev Neurobiol 10:69–99
77. Wood JD (1972) Excitation of intestinal muscle by atropine, tetrodotoxin, and xylocaine. Am J Phys 222:118–125
78. Worth AA, Forrest AS, Peri LE, Ward SM, Hennig GW, Sanders KM (2015) Regulation of gastric electrical and mechanical activity by cholinesterases in mice. J Neurogastroenterol Motil 21:200–216
79. Yamamoto M (1977) Electron microscopic studies on the innervation of the smooth muscle and the interstitial cell of Cajal in the small intestine of the mouse and bat. Arch Histol Jpn 40:171–201
80. Zheng H, Drumm BT, Earley S, Sung TS, Koh SD, Sanders KM (2018) SOCE mediated by STIM and Orai is essential for pacemaker activity in the interstitial cells of Cajal in the gastrointestinal tract. Sci Signal 11(534):eaaq0918
81. Zheng H, Drumm BT, Zhu MH, Xie Y, O'Driscoll KE, Baker SA, Perrino BA, Koh SD, Sanders KM (2020) Na+/Ca2+ exchange and pacemaker activity of interstitial cells of Cajal. Front Physiol 11:230
82. Zheng H, Park KS, Koh SD, Sanders KM (2014) Expression and function of a T-type Ca2+ conductance in interstitial cells of Cajal of the murine small intestine. Am J Physiol Cell Physiol 306:C705–C713
83. Zhu MH, Kim TW, Ro S, Yan W, Ward SM, Koh SD, Sanders KM (2009) A Ca(2+)-activated Cl(−) conductance in interstitial cells of Cajal linked to slow wave currents and pacemaker activity. J Physiol 587:4905–4918
84. Zhu MH, Sung TS, Kurahashi M, O'Kane LE, O'Driscoll K, Koh SD, Sanders KM (2016) Na+-K+-Cl- cotransporter (NKCC) maintains the chloride gradient to sustain pacemaker activity in interstitial cells of Cajal. Am J Physiol Gastrointest Liver Physiol 311:G1037–G1046
85. Zhu MH, Sung TS, O'Driscoll K, Koh SD, Sanders KM (2015) Intracellular Ca(2+) release from endoplasmic reticulum regulates slow wave currents and pacemaker activity of interstitial cells of Cajal. Am J Physiol Cell Physiol 308:C608–C620

Identifying Types of Neurons in the Human Colonic Enteric Nervous System

23

Simon Brookes, Nan Chen, Adam Humenick, Marcello Costa, Phil Dinning, Paul Heitmann, Dominic Parker, David Smolilo, Nick J. Spencer, and David Wattchow

Abstract

Distinguishing and characterising the different classes of neurons that make up a neural circuit has been a long-term goal for many neuroscientists. The enteric nervous system is a large but moderately simple part of the nervous system. Enteric neurons in laboratory animals have been extensively characterised morphologically, electrophysiologically, by projections and immunohistochemically. However, studies of human enteric nervous system are less advanced despite the potential availability of tissue from elective surgery (with appropriate ethics permits). Recent studies using single cell sequencing have confirmed and extended the classification of enteric neurons in mice and human, but it is not clear whether an encompassing classification has been achieved. We present preliminary data on a means to distinguish classes of myenteric neurons in specimens of human colon combining immunohistochemical, morphological, projection and size data on single cells. A method to apply multiple layers of antisera to specimens was developed, allowing up to 12 markers to be characterised in individual neurons. Applied to multi-axonal Dogiel type II neurons, this approach demonstrated that they constitute fewer than 5% of myenteric neurons, are nearly all immunoreactive for choline acetyltransferase and tachykinins. Many express the calcium-binding proteins calbindin and calretinin and they are larger than average myenteric cells. This methodology provides a complementary approach to single-cell mRNA profiling to provide a comprehensive account of the types of myenteric neurons in the human colon.

Keywords

Enteric nervous system · Human · Colon · Myenteric plexus · Immunohistochemistry

23.1 Classifying Neurons

The neuron doctrine recognised that the nerve cell is the fundamental building block of neural circuits. Given the large numbers of neurons involved in functional circuits in vertebrates, it is essential to be able to group neurons into classes, to simplify our understanding of how circuits work. In the human brain, each of the 100 billion neurons is unique. However, all neurons share fundamental features (axons, ion channels, transmitter release mechanisms). Thus we can classify

S. Brookes (✉) · N. Chen · A. Humenick · M. Costa · P. Dinning · P. Heitmann · D. Parker · D. Smolilo · N. J. Spencer · D. Wattchow
Human Physiology, College of Medicine and Public Health, Flinders University, Adelaide, SA, USA
e-mail: simon.brookes@flinders.edu.au

N. J. Spencer et al. (eds.), *The Enteric Nervous System II*, Advances in Experimental Medicine and Biology 1383, https://doi.org/10.1007/978-3-031-05843-1_23

the human nervous system as comprising 100 billion separate classes or having just one superclass. Neither classification is useful. The goal of neuronal classification is to group together neurons with shared features and distinguish them from other neurons with different combinations of features. The ideal is to hit the 'sweet spot' between large numbers of classes with little heterogeneity and fewer classes with too much heterogeneity. The value of such a classification is clear when trying to understand how functional circuits tune behavioural responses to stimuli. These considerations apply to classifying neurons in the human enteric nervous system as much as they do to classifying interneurons in the central nervous system.

23.1.1 Goals for Classification

Our long-term goal is to understand how the enteric nervous system mediates simple behaviours such as motor activity. A comprehensive classification of nerve cells in a circuit is useful for several purposes [1]:

1. A list of components which make up functional circuits, such as those underlying intestinal motility. Connectivity between classes can be studied and a circuit diagram developed for modelling.
2. A standard description which allow cross-regional and cross-species comparisons to be made, identifying evolutionary trends and/or development of convergent solutions to common problems.
3. A way to improve reproducibility of findings between laboratories. If a certain feature is present in about two thirds of neurons, this can be hard to interpret. If it is shown to be expressed in four identified classes of cells that add up to two thirds of all cells, this has predictive and explanatory power, upon which more advanced studies can be based.
4. New (often rare) types of cells can be discovered, which may be otherwise missed due to sampling biases. In addition, current single cell RNA-seq methods have demonstrated that new cell-specific targets can be identified, including unique type-specific markers.
5. Genetic access can be achieved by identifying a unique marker gene for a subset of cells which can be used to target those neurons via their regulatory elements. Well established in mice, fruit flies and zebra fish; such approaches are becoming more widely available.
6. Changes of neuronal features over time can be studied. A robust classification allows the sequence of class-specific changes during development or in response to inflammation or disease processes to be documented systematically, rather than averaging changes indiscriminately across all classes.

23.2 The Opportunity to Study Human Enteric Nervous System

Most studies of nerve cells at the level of single cells have been undertaken in animal models, for obvious ethical reasons. In the last decade this has started to change, with opportunities for study of human stem cell-derived neurons and organoids becoming possible. However, the bulk of studies of enteric neurons have been carried out in a few laboratory species, mostly mice, rats and guinea pigs, with smaller numbers of studies in pigs, rabbits and other species.

It is also possible to obtain specimens of live human gastrointestinal tissue from patients undergoing elective surgery as a considerable amount healthy tissue is usually included at the margins of bowel resected for tumours. Thus, full-wall-thickness specimens of bowel up to 20–30-cm long may be obtained which are suitable for laboratory study, with the appropriate ethics permits. Moreover, bowel cancer is a common condition, so there is an abundant supply of tissues. In addition, bowel from patients with ulcerative colitis, Crohn's disease and diverticulitis is also available, making it possible to identify changes associated with some disorders.

This is a remarkable opportunity; for most other parts of the nervous system, human samples are restricted to post-mortem tissue, which often shows significant signs of degradation.

23.2.1 Technical Issues with Use of Human Tissue

While human tissue is available from abdominal surgery, there are significant issues that must be addressed before tissue can be studied in the laboratory. Obtaining prior written informed consent from patients who are capable of making an informed decision imposes significant additional workload on staff of the surgical unit. Other ethical issues include potential compromise of patient confidentiality and interference with pathologists' analysis (which is rarely a problem with tumour resections as care is taken not to take tumoral or peri-tumoral tissues). Other ethical considerations include the responsibilities for informing patients if genetic analysis identifies disease-causing mutations and inaccurate perception of duress or benefit to the patient during the consent process. In addition, there are significant occupational health and safety issues for laboratory staff involved in collecting or processing human tissue before fixation occurs, highlighted recently by the discovery of significant SARS-coV-2 load in human bowel content [2], in addition to many other pathogens including HIV and hepatitis C.

In addition to ethical and procedural matters, the physical handling of tissue can affect the quality of results. Delays between when the blood supply to the tissue is interrupted and the tissue is placed in cooled solution (i.e. 'warm ischaemia time') and time taken to transport the specimen to the laboratory may affect the health of the cells. The degree of stretch during fixation affects the quality of dissection for wholemount tissues, as well as quantitation of cell densities. Removal of excess tissue by sharp dissection and choice of tissue-permeabilisation protocols strongly affect the thickness of the preparation and quality of staining, especially signal and background ratio, with fluorescence staining. Many of these considerations are reviewed by Knowles et al. (2011) [3]. Recent technical advances have greatly improved the visualisation of full thickness wholemount preparations of human bowel, using a variety of clearing protocols [4–6], although these are time-consuming and are only recently becoming more widely used.

23.3 Classification of Enteric Neurons

The classification of neurons is currently undergoing a transformation due to the proliferation of studies using high-throughput single-cell RNA sequencing [7]. Nevertheless, complete reliance on analysis of transcriptomics is likely to miss important classes of cells that are revealed by other methods [8], so it is worth reprising the basis of previous taxonomies of enteric neurons.

Early methods to distinguish types of enteric neurons were based on soma-dendritic morphology revealed by silver stains [9] or methylene blue [10]. These have been reviewed and refined [11, 12], and three to six classes of neurons can be distinguished on the basis of numbers of axons, morphology of dendrites and shape and size of cell body. A taxonomy based largely on soma-dendritic morphology has recently been published [13]. Classifications by electrophysiological features revealed just a few distinct types [14, 15], although later studies suggested that these may be subdivided on the basis of responses to injected current pulses [16, 17] although this has not been comprehensively studied to date in the gastrointestinal tract. To a large extent, electrophysiological types of neurons (S cells and AH cells) were shown to correspond to the two major morphological types (type I and type II) [18]. Pharmacological and ultrastructural features also added options for classifying enteric neurons [19].

23.4 Chemical Coding

Meanwhile, the presence of neuropeptides in nerve fibres and cell bodies in the enteric nervous system had been detected using immunohistochemical methods. First, in mice, substance P [20], VIP, ENK and SOM were characterised [21]. Later, a range of neuropeptides were mapped in other species of laboratory animals. This quickly led to the conclusion that single neurons could contain and release multiple transmitters [22]. From this, the concept of 'chemical coding' of neurons was born: the idea that combinations of markers could be used to distinguish functional classes [23, 24]. Many types of enteric neurons (>10) could be distinguished using this approach. This was later supplemented by the systematic application of retrograde tracing of enteric neurons combined with immunohistochemical labelling which allowed chemical coding to be directly related to targets and polarities of different functional classes of enteric neurons [25, 26] including cholinergic and 'nitrergic' neurones, acetylcholine and nitric oxide being the major excitatory and inhibitory neurotransmitters in the gut wall.

The ability of chemical coding to distinguish multiple types of enteric neurons was soon applied to human bowel samples. VIP- SP and bombesin-like immunoreactivity was detected in nerve fibres in human bowel [27], followed by somatostatin [28], and since then many other neuropeptides, enzymes, calcium-binding proteins and cytoskeletal proteins have been characterised. Retrograde-tracing studies have also been used to extend the identification of functional classes in the human bowel [29–31], and similarities and difference to small laboratory animals became apparent [32].

23.5 Limits of Chemical Coding

Chemical coding was capable of distinguishing markers in different populations of neurons, but it was also constrained by the number of markers that could be studied simultaneously. Typically, only three to four markers could be labelled in the same piece of tissue due to availability of fluorescence filters. This meant that many dozens of combinations of staining had to be analysed to achieve comprehensive coverage of all markers. Since each stained sample had a finite number of cells (usually a few hundred), sampling errors accumulated when results were combined arithmetically. Despite this, a comprehensive, quantitative description of the enteric nervous system of the guinea pig small intestine was developed [33] and was followed by quantitative accounts in the guinea pig stomach [34], guinea pig distal colon [35] and mouse small and large intestines [36, 37]. However, in each case, descriptions of classes tended to be incomplete, with minority markers in many classes remaining unquantified (often labelled as 'plus or minus' a marker). To date, no comprehensive quantitative accounts of the human enteric nervous system have been developed although a recent report shows progress towards this goal [13].

In contrast, recent studies in the mouse using transcriptomic methods have identified nine clusters of myenteric neurons [38] or in a larger sample (of 4932 neurons) 12 classes [39] or 18 or 21 classes [40]. The mathematics of clustering is complex and different methods yield different results [41]. Strikingly, within all of these schemes, there is considerable variability in expression of transcripts by neurons within the same cluster – the issue of 'plus or minus' markers persists in single cell RNA-seq, albeit with better quantitation. This strongly suggests that even within a defined class, major markers may be constant, but other markers may show considerable variability in expression levels. However, while single-cell RNA-seq can be combined with studies of cell morphology, biophysics, projection or connectivity [42], this is a very low-throughput technology. In contrast, electrophysiological, morphological, immunohistochemical and tracing methods are readily compatible and relate relevant features of a functional taxonomy.

To summarise, transcriptome analysis is providing a richness and depth of quantitative classification of nerve cells that far surpasses other methods. However, the older approaches of multiple labelling immunohistochemistry and compatible techniques have a proven capability in producing a functional classification of neurons, which in many cases has been used to interpret putative roles for clusters identified using single-cell RNA-seq methods.

23.6 A New Approach to Chemical Coding

We have recently developed an extension of multiple labelling immunohistochemistry which is proving useful for classification of human enteric neurons. Essentially the technique allows multiple layers of multi-labelling immunohistochemistry to be applied to the same set of neurons. An example of this this approach has been used to identify the chemical coding of putative intrinsic primary afferent neurons (IPANs) in human colon for the first time.

First, all myenteric nerve cell bodies were labelled with HuC/D. We then analysed cell morphology for 2596 neurons immunohistochemically labelled for the Neurofilament 200, a marker which has been used as a means to classify human enteric neuron soma-dendritic morphology [43, 44]. It should be noted that these cells were treated with colchicine before fixation, to enhance neuropeptide immunoreactivity in cell bodies. They were also pre-treated with 5-hydroxytryptamine to load 5-HT-accumulating cell bodies. A total of 55 neurons were identified with definitive Dogiel type II morphology, with medium to large usually ovoid cell bodies with two or more axonal processes or pseudo-unipolar axonal structure. Because Neurofilament 200 immunoreactive cells are so densely packed, it was impossible to count axons arising from many cell bodies, so 55 is certainly an underestimate of the true numbers of multi-axonal cells.

We then examined the immunoreactivity of these 55 cells for 11 other markers applied in five layers of staining separated by antibody elution. In decreasing order of abundance, human colonic Dogiel type II neurons were immunoreactive for choline acetyltransferase (100%), SP (89%), calretinin (76%), calbindin (65%), somatostatin (22%), 5-hydroxytryptamine (22%), CGRP (15%), VIP (1.8%) and NPY (1.8%). NOS and enkephalin were entirely absent from this population. Next, we estimated the total proportion of Dogiel type II neurons by counting all cells with the specific combinations of markers present in the 55 positively identified Dogiel type II neurons. This identified 119 neurons out of 2596 neurons, suggesting that about 4.6% of colonic myenteric neurons have Dogiel type II coding. This is significantly fewer than in the human small intestine where these multi-axonal nerve cells constitute approximately 10% of myenteric neurons [44]. There was no significant difference in the proportions in proximal and distal colon (4.8% vs 4.3%, NS). Lastly, we compared the size of these neurons on the basis of vertical projection area of HuC/D immunoreactivity in the cell body. Dogiel type neurons were considerably larger than most other cells, averaging 1618 ± 785 μm^2 compared to 656 ± 433 μm^2 for all myenteric neurons.

23.7 Discussion

Using the new multilayer protocol, it is possible to examine quantitatively types of human enteric neurons classified by chemical coding. In addition to the presence of immunohistochemical markers, it is possible to add data on cell soma-dendritic morphology (from neurofilament 200 immunoreactivity) and cell size (standardised using HuC/D immunoreactivity). This data can be combined with retrograde tracing studies in human colon which have used a range of immunohistochemical markers [29–31, 45–48]. By comparing the chemical coding of axons in target tissues with the chemical coding of nerve cell bodies, it is possible to identify targets for many types of neurons. It is interesting that the clusters of neurons from single-cell RT-seq studies [38–40] have primarily been assigned putative functions based on this type of data.

Here, we illustrated the use of systematic immunohistochemical data to characterise putative intrinsic primary afferent neurons (IPANs) in human colonic myenteric plexus. In the guinea pig small intestine, these have multipolar Dogiel type II morphology and AH cell electrophysiological features [49, 50]. These neurons have been characterised in human small intestinal myenteric plexus where the combination of SP+/SOM+ or calretinin+/SOM+ has high predictive values for Dogiel type II neurons. The situation appears quite different in colon where calretinin+/SOM+ neurons account for just 7.6% of Dogiel type II neurons and SP+/SOM+ neurons account for 19.3%. This reflects the lower expression of SOM in human colonic nerves compared to the small intestine [28] and may relate to the differing functions of these regions of gut. Unlike the guinea pig small intestine [51], calbindin is not an exclusive marker of Dogiel type II neurons in human colon [46, 52]. Furthermore, CGRP was only found in a small minority (15%) of Dogiel type II neurons in human colon, whereas it is an abundant marker of nearly all in the mouse colon [53]. Clearly there are major differences in the chemical coding of homologous cells in different species. Developing a systematic taxonomy of the human enteric nervous system is an important goal for neurogastroenterology.

Acknowledgements This work was supported by an NIH SPARC program grant to a consortium led by Professor Yvette Tache, of UCLA #1OT2OD24899-01.

References

1. Zeng H, Sanes JR (2017) Neuronal cell-type classification: challenges, opportunities and the path forward. Nat Rev Neurosci 18(9):530–546
2. Chen Y et al (2020) The presence of SARS-CoV-2 RNA in the feces of COVID-19 patients. J Med Virol 92(7):833–840
3. Knowles CH et al (2011) Quantitation of cellular components of the enteric nervous system in the normal human gastrointestinal tract--report on behalf of the Gastro 2009 International Working Group. Neurogastroenterol Motil 23(2):115–124
4. Graham KD et al (2020) Robust, 3-dimensional visualization of human colon enteric nervous system without tissue sectioning. Gastroenterology 158(8):2221–2235.e5
5. Liu YA et al (2012) 3-D illustration of network orientations of interstitial cells of Cajal subgroups in human colon as revealed by deep-tissue imaging with optical clearing. Am J Physiol – Gastrointest Liver Physiol 302(10):G1099–G1110
6. Liu YA et al (2013) 3-D imaging, illustration, and quantitation of enteric glial network in transparent human colon mucosa. Neurogastroenterol Motil 25(5):e324–e338
7. Yuste R et al (2020) A community-based transcriptomics classification and nomenclature of neocortical cell types. Nat Neurosci 23(12):1456–1468
8. Kim EJ et al (2020) Extraction of distinct neuronal cell types from within a genetically continuous population. Neuron 107(2):274–282.e6
9. Cajal R (1893) Sur les ganglions et plexus nerveux de l'intestin. C R Soc Biol 5:217–223
10. Dogiel AS (1899) Über den Bau der Ganglien in den Geflechten des Darmes und der Gallenblase des Menschen und der Säugethiere. Arch Anat Physiol Leipzig, Anat Abt 1899:130–158
11. Brehmer A (2006) Structure of enteric neurons. In: Beck FF et al (eds) 186 advances in anatomy embryology and cell biology. Springer, Berlin/Heidelberg
12. Brehmer A, Schrödl F, Neuhuber W (1999) Morphological classifications of enteric neurons — 100 years after Dogiel. Anat Embryol 200:125–135
13. Brehmer A (2021) Classification of human enteric neurons. Histochem Cell Biol 25:25
14. Hirst GDS, Holman ME, Spence I (1974) Two types of neurones in the myenteric plexus of duodenum in the guinea-pig. J Physiol (Lond.) 236:303–326
15. Nishi S, North RA (1973) Intracellular recording from the myenteric plexus of the guinea-pig ileum. J Physiol 231(3):471–491
16. Brookes SJ et al (1997) Orally projecting interneurones in the guinea-pig small intestine. J Physiol 505(Pt 2):473–491
17. Song ZM et al (1997) Characterization of myenteric interneurons with somatostatin immunoreactivity in the guinea-pig small intestine. Neuroscience 80(3):907–923
18. Bornstein JC et al (1984) Electrophysiology and enkephalin immunoreactivity of identified myenteric plexus neurones of guinea-pig small intestine. J Physiol Lond 351:313–325
19. Furness JB, Costa M (1980) Types of nerves in the enteric nervous system. Neuroscience 5(1):1–20
20. Nilsson G et al (1975) Localization of substance P-like immunoreactivity in mouse gut. Histochemistry 43(1):97–99
21. Schemann M, Camilleri M (2013) Functions and imaging of mast cell and neural axis of the gut. Gastroenterology 144(4):698–704.e4
22. Hokfelt T, Johansson O, Goldstein M (1984) Chemical anatomy of the brain. Science 225(4668):1326–1334
23. Costa M, Furness JB, Gibbins IL (1986) Chemical coding of enteric neurons. In: Hökfelt T, Changeux P (eds) Progress in brain research, vol 68. Elsevier, Amsterdam, pp 217–240
24. Furness JB, Morris JL, Gibbins IL, Costa M (1989b) Chemical coding of neurons and plurichemical transmission. Annu Rev Pharmacol Toxicol 29:289–306
25. Brookes SJH (2001) Retrograde tracing of enteric neuronal pathways. Neurogastroenterol Motil 13(1):1–18
26. Domoto T et al (1990) An in vitro study of the projections of enteric vasoactive intestinal polypeptide-immunoreactive neurons in the human colon. Gastroenterology 98(4):819–827. issn: 0016-5085

27. Ferri GL et al (1989) Intramural distribution of immunoreactive vasoactive intestinal polypeptide (VIP), substance P, somatostatin and mammalian bombesin in the oesophago-gastro-pyloric region of the human gut. Cell-Tissue-Res 256(1):191–197. issn: 0302-766x
28. Keast JR, Furness JB, Costa M (1984) Somatostatin in human enteric nerves. Distribution and characterization. Cell Tissue Res 237(2):299–308
29. Porter AJ et al (2002) Cholinergic and nitrergic interneurones in the myenteric plexus of the human colon. Gut 51(1):70–75
30. Wattchow DA, Brookes SJ, Costa M (1995) The morphology and projections of retrogradely labeled myenteric neurons in the human intestine. Gastroenterology 109(3):866–875
31. Wattchow DA et al (1997) The polarity of neurochemically defined myenteric neurons in the human colon. Gastroenterology 113(2):497–506
32. Brookes SJH, Costa M (2002) Cellular organisation of the mammalian enteric nervous system. In: Brookes SJH, Costa M (eds) Innervation of the gastrointestinal tract. Harwood
33. Costa M et al (1996) Neurochemical classification of myenteric neurons in the guinea-pig ileum. Neuroscience 75(3):949–967
34. Schemann M, Schaaf C, Mader M (1995) Neurochemical coding of enteric neurons in the guinea pig stomach. J Comp Neurol 353(2):161–178
35. Lomax AE, Furness JB (2000) Neurochemical classification of enteric neurons in the guinea-pig distal colon. Cell Tissue Res 302(1):59–72
36. Qu ZD et al (2008) Immunohistochemical analysis of neuron types in the mouse small intestine. Cell Tissue Res 334(2):147–161
37. Sang Q, Williamson S, Young HM (1997) Projections of chemically identified myenteric neurons of the small and large intestine of the mouse. J Anat 190(Pt 2):209–222
38. Zeisel A et al (2018) Molecular architecture of the mouse nervous system. Cell 174(4):999–1014.e22
39. Morarach K et al (2021) Diversification of molecularly defined myenteric neuron classes revealed by single-cell RNA sequencing. Nat Neurosci 24(1):34–46
40. Drokhlyansky E et al (2020) The human and mouse enteric nervous system at single-cell resolution. Cell 182(6):1606–1622.e23
41. Kiselev VY, Andrews TS, Hemberg M (2019) Challenges in unsupervised clustering of single-cell RNA-seq data. Nat Rev Genet 20(5):273–282
42. Fuzik J et al (2016) Integration of electrophysiological recordings with single-cell RNA-seq data identifies neuronal subtypes. Nat Biotechnol 34(2):175–183
43. Brehmer A et al (2004) Immunohistochemical characterization of putative primary afferent (sensory) myenteric neurons in human small intestine. Auton Neuroscience-Basic Clin 112(1–2):49–59
44. Weidmann S et al (2007) Quantitative estimation of putative primary afferent neurons in the myenteric plexus of human small intestine. Histochem Cell Biol 128(5):399–407
45. Humenick A et al (2019) Characterisation of projections of longitudinal muscle motor neurons in human colon. Neurogastroenterol Motil 31(10):e13685
46. Humenick A et al (2021) Characterization of putative interneurons in the myenteric plexus of human colon. Neurogastroenterol Motil 33(1):e13964
47. Porter AJ et al (1997) The neurochemical coding and projections of circular muscle motor neurons in the human colon. Gastroenterology 113(6):1916–1923
48. Porter AJ et al (1999) Projections of nitric oxide synthase and vasoactive intestinal polypeptide-reactive submucosal neurons in the human colon. J Gastroenterol Hepatol 14(12):1180–1187
49. Kunze WA, Bornstein JC, Furness JB (1995) Identification of sensory nerve cells in a peripheral organ (the intestine) of a mammal. Neuroscience 66(1):1–4
50. Kunze WA et al (1998) Intracellular recording from myenteric neurons of the guinea-pig ileum that respond to stretch. J Physiol 506(Pt 3):827–842
51. Iyer V et al (1988) Electrophysiology of guinea-pig myenteric neurons correlated with immunoreactivity for calcium binding proteins. J Auton Nerv Syst 22(2):141–150
52. Zetzmann K et al (2018) Calbindin D28k-immunoreactivity in human enteric neurons. Int J Mol Sci 19(1):08
53. Furness JB et al (2004) Projections and chemistry of Dogiel type II neurons in the mouse colon. Cell Tissue Res 317(1):1–12

24 Neurons, Macrophages, and Glia: The Role of Intercellular Communication in the Enteric Nervous System

Simona Elisa Carbone

Abstract

Neurons of the enteric nervous system (ENS) are the primary controllers of gastrointestinal functions. Although the ENS has been the central focus of research areas such as motility, this has now expanded to include the modulatory roles that non-neuronal cells have on neuronal function. This review discusses how enteric glia (EGC) and resident muscularis macrophages (mMacs) influence ENS communication. It highlights how the understanding of neuroglia interactions has extended beyond EGCs responding to exogenously applied neurotransmitters. Proposed mechanisms for neuron-EGC and glio-glia communication are discussed. The significance of these interactions is evidenced by gut functions that rely on these processes. mMacs are commonly known for their roles as immune cells which sample and respond to changes in the tissue environment. However, a more recent theory suggests that mMacs and enteric neurons are mutually dependent for their maintenance and function. This review summarizes the supportive and contradictory evidence for this theory, including potential mechanisms for mMac-neuron interaction. The need for a more thorough classification scheme to define how the "state" of mMacs relates to neuron loss or impaired function in disease is discussed. Despite the growing literature suggesting EGCs and mMacs have supportive or modulatory roles in ENS communication and gut function, conflicting evidence from different groups suggests more investigation is required. A broader understanding of why enteric neurons may need assistance from EGCs and mMacs in neurotransmission is still missing.

Keywords

Enteric neuron · Enteric glial cells (EGCs) · Muscularis macrophages (mMacs) · Intercellular communication · Calcium (Ca^{2+}) imaging · Immunohistochemistry

24.1 Introduction

Traditionally, enteric neurons were considered to be the only active cells in the *enteric nervous system* (ENS). However, with improved sensitivity of imaging technologies and advances in the genetic tools available, this paradigm has recently shifted. The idea that non-neuronal cells can

S. E. Carbone (✉)
Drug Discovery Biology, Monash Institute of Pharmaceutical Sciences, Parkville, VIC, Australia

Monash University, Parkville, VIC, Australia
e-mail: simona.carbone@monash.edu

N. J. Spencer et al. (eds.), *The Enteric Nervous System II*, Advances in Experimental Medicine and Biology 1383, https://doi.org/10.1007/978-3-031-05843-1_24

assist or modulate the functions of enteric neurons is an exciting new aspect for neurogastroenterology. The *interstitial cells of Cajal* (ICC) and telocytes such as *fibroblast-like cells* (also known as PDGFRα+ cells) are established examples of non-neuronal cells that act as an intermediary, relaying the synaptic inputs from enteric motor neurons to the target smooth muscle. However, are there networks of cells in the gut wall that can directly interact with the ENS or can communicate with each other in a manner comparable to enteric neurons?

Interspersed among the enteric neurons are the *enteric glial cells* (EGCs) originally thought to be little more than structural support cells that hold the ENS in place. EGCs are present at similar or greater ratios than the neurons themselves (1:1), and growing evidence indicates these cells regulate enteric neural transmission. The emerging roles of tissue-resident immune cells known as *muscularis macrophages* (mMacs) in intercellular communication with the ENS have been similarly explored. While mMacs are generally dispersed throughout the external muscle layer of the gut wall, a subset is closely associated with the enteric ganglia and may communicate directly with the ENS [16]. Conceptual advances largely borrowed from studies of the *central nervous system* (CNS) have triggered an interest in considering EGCs and mMacs as more than support and immune cells [13, 26]. Astrocytes and microglia are important regulators of CNS interactions in health and disease. There are similarities in the gene expression profiles and morphologies of EGCs and mMacs with CNS astrocytes and microglia, respectively. Similar cells to EGCs and mMacs in the CNS have associations with neurons. This has led researchers to investigate whether EGC and mMacs in the ENS have analogous roles to astrocytes and microglia in the CNS. There is already evidence that both cell types can influence essential functions of the gut such as motility, secretion, and inflammation. However, mechanistic understanding of how these cells contribute to these events is limited. This review will summarize the current understanding of the functions of EGCs and mMacs in communication with the ENS network.

24.2 The Role of EGCs in ENS Communication

EGCs are considered to be electrically "silent" cells as they fail to demonstrate a response to current pulses, electrical stimulation, or pharmacological agents in electrophysiological recordings (human-Carbone unpublished data, [33]). This is in contrast to neurons which routinely generate an action potential when depolarized by current pulses. It is well documented that EGC signal via the release of intracellular Ca^{2+} stores [6, 34, 38], or through cAMP production [10]. Standard methods to demonstrate the roles of EGCs in gastrointestinal physiology have employed either Ca^{2+} indicator dyes (e.g., Fluo-4) or mice that selectively express *genetically encoded calcium indicators* (GECIs) in EGC [34]. Exogenously applied transmitters such as ATP and serotonin (5-HT) trigger an elevation in intracellular Ca^{2+} in EGCs [6]. This provides indirect evidence of the mechanisms by which enteric neurons regulate EGC functions. Specialized contact sites between enteric neurons and glial termed *neuroglial junctions* have been identified in rodent intestine by electron microscopy [22, 23]. However, *what evidence is there that enteric neurons release these neurotransmitters to directly activate EGCs?*

EGCs express several receptors for neurotransmitters released by enteric neurons including purinergic, adrenergic, and metabotropic glutamate receptors [29]. Ca^{2+} imaging studies have provided examples where pharmacological or electrical activation of enteric neurons leads to subsequent responses in EGCs. Purinergic signaling is a major mechanism by which enteric neurons can activate EGCs [25, 28]. An important consideration for neuron-glia communication is that not all responses are inhibited by the voltage-sensitive Na^{+} channel inhibitor, *tetrodotoxin* (TTX) [7, 21]. While these channels are important for initiating the pathways that lead to typical synaptic transmission, this observation suggests that other mechanisms are involved. Enteric neurons can release ATP via channels formed by the protein subunits called *pannexins*. The release of ATP enteric neurons through pannexin channels has been shown to mediate neuronal death [31]. However, a recent study from Boesmans et al. [7] provided evidence that pannexin channels provide a "*communicating junction*" between enteric neurons and EGCs. Boesmans et al. used photolytic uncaging of cytosolic calcium to activate individual enteric neurons. Their results showed calcium transients in ~2 neighboring glial cells per single neuron stimulated. Only ~0.5 EGCs responded in the presence of either purinergic P2 receptor antagonist suramin or the nonselective pannexin inhibitor probenecid. The mean amplitude of responses was also attenuated. This is not surprising as it is likely that other neurotransmitters have roles in neuron-glia communication. EGCs respond to the application of other agonists for receptors such as nicotinic acetylcholine and *neurokinin 2* (NK_2) receptors. Furthermore, calcium is only one measure of cell activation, and signaling through cAMP in response to various stimuli has rarely been investigated. There is evidence that neuron-glia interactions occur during physiological events, particularly during motility patterns such as *colonic motor complexes* (CMCs) [8, 34]. Evidence from these studies indicates that only subpopulations of EGCs are activated during this physiological event and that this activation is secondary to enteric neuron stimulation. While this suggests potential roles for EGCs in modulating GI motility, research from other groups indicates that EGCs may not have a major contribution to this process [42].

It is worth noting that EGCs can also receive input from the autonomic nervous system and extrinsic sensory neurons [15, 30]. Gulbransen et al. in 2010 demonstrated that electrical field stimulation of nerve fiber tracts in the myenteric plexus of the guinea pig colon typically elicited robust Ca^{2+} transients in EGCs. However, the amplitude of these responses was significantly reduced in tissues where extrinsic nerves were chemically or surgically removed. The authors suggested that the primary neurotransmitter involved may be ATP since EGCs failed to respond to exogenous application of norepinephrine [30]. While EGCs are closely associated with TRPV1 expressing neuronal varicosities [15], stimulation of primary afferent nerves with capsaicin does not elicit a Ca^{2+} response in these cells [30]. EGCs are also closely associated with tachykinin immunoreactive varicosities, which include both extrinsic and intrinsic nerve fibers. Given that EGCs robustly respond to exogenous application of NK_2R agonists, it is likely that this provides a mechanism for interaction with extrinsic sensory fibers. While the mechanisms for transmission between extrinsic sensory nerve fibers and EGC are not clear, they may be important for the development of visceral pain following bowel inflammation [27].

24.3 Communication Between Neighboring EGCs

Functional coupling of EGCs is an important mechanism for how this network of cells interacts. While electron microscopy studies fail to demonstrate the formation of typical gap junctions between many EGCs in the myenteric plexus of the rodent intestine [23], dye filling experiments provide evidence to support functional coupling within this network of cells [32]. Gap junctions are made up of hemichannels formed from *connexin* protein subunits. Ca^{2+} imaging experiments have demonstrated the importance of this functional coupling, as responses to exogenously applied ADP are attenuated with inhibition or deletion of these hemichannels [38]. In these experiments, hemichannels were either inhibited pharmacologically in tissues from wild-type animals or deleted by glia-specific disruption of the gene encoding connexin-43. In this same study, tissue from the modified mice failed to generate contractions in response to electrical stimulation of neurons, and GI motility in vivo was generally delayed. These results highlight that coupling between EGCs is important for functional interactions within this network of cells and the functional output of the ENS in physiological processes more broadly.

24.4 The Role of mMacs in ENS Communication

mMacs are the tissue-resident macrophages of the external muscle of the gastrointestinal tract. They have diverse and dynamic morphologies which facilitate their ability to constantly sample and respond to the local tissue environment [41]. Evidence suggests that mMacs have bidirectional relationships with the ENS that are important for the maintenance and function of both cell types. One arm of this interaction centers on the secretion of colony-stimulating factor 1 (CSF-1) by enteric neurons. CSF-1, also known as the macrophage colony-stimulating factor (M-CSF), is required for the continual maintenance and survival of mMacs [20, 40, 43]. mMacs fail to develop in the intestine of the osteopetrotic (*op/op*) transgenic mice, which express an inactivating mutation to the CSF-1 gene [39]. Furthermore, pharmacological or antibody inhibition of the receptor for CSF-1 (CSF-1R) leads to depletion of mMacs [3, 14]. Although enteric neurons are not the sole source of CSF-1 in the gut wall [1], it provides a hypothesized mechanism for ENS and mMacs association. The importance of these interactions is questioned by immunohistochemical studies in tissues from Hirschsprung Disease patients or from animal models of the disease, where the ENS fails to develop in the distal portion of the gastrointestinal tract (Ret knock-out) [1]. In both groups, mMacs continued to develop and colonize the GI tract despite the absence of neurons. The authors conclude that the enteric neurons may be necessary for the continual maintenance of mMac rather than the initial patterning of the cells. This was supported by their evidence that the colonization and development of mMacs precede that of enteric neurons.

The second arm of this bidirectional relationship is centered on the release of *bone morphogenetic proteins* (BMPs), for which there are receptors on enteric neurons (BMPR). During development, BMPs are required for the colonization of neural crest cells in the embryonic gut and the development of a functional ENS in the fetal gut [9, 20, 24]. mMacs highly express bone morphogenetic protein 2 (BMP2) [40]; therefore it is hypothesized that secretion of this soluble factor provides a mechanism for mMacs to interact with enteric neurons. In intestinal and stomach tissues from *op/op* mice [11, 12], the continual absence of mMacs is associated with the development of significantly greater numbers of enteric neurons. Immunohistochemistry data from Cipriani et al. suggests, that in the stomach of *op/op* mice, the proportion of nitrergic neurons in the myenteric plexus is more affected by the absence of mMacs than the proportion of cholinergic neurons [11, 12]. The overall increase of enteric neuron number is consistent with findings from tissues in mice that globally overexpress the BMP antagonist, noggin [9]. However, the timing of mMac depletion may have important implications for how their potential roles in ENS maintenance and function are interpreted. The use of tissues from inducible knockout models has shown that chronic mMac depletion leads to significant reductions in enteric neuron number (Cx3cr1CreERT2.Rosa26-iDTR) [14]. Muller et al. employed a different strategy to deplete macrophages by administering monoclonal antibodies directed at the CSF-1 receptor to mice [40]. They found no significant effects on enteric neuron number following acute depletion of macrophages, although functional effects were noted and will be discussed later in this review. How the changes in enteric neuron number then alter the output of the ENS is a necessary aspect to validate the impact of these changes in gastrointestinal processes.

mMacs have dynamic motile processes that are constantly changing to adapt to their cellular environment [41]. Flow cytometry and transcriptional information are available to define the various mMac subtypes in the individual layers of the gut wall [14, 36]. However, the field is lacking a more thorough classification of the "when," "where," and "how" the various subtypes contribute to ENS communication. Basic immunohistochemistry studies have attempted to quantify the broad groups of cells based on their density and location within the gastrointestinal tract. However, this sort of profiling does not take into account differences in cellular morphology and the number and length of processes, as has been demonstrated by studies such as De Schepper et al. [14]. These studies have aligned the location of the cells to a tissue layer, but the proximity of these subsets to the myenteric plexus and whether their processes directly contact enteric neurons or non-neural cells such as EGCs and ICCs has yet to be included in this classification. A detailed analysis of the anatomical locations of mMacs was very recently published by Dora et al., and their findings lead the authors to speculate whether a subset of mMac may function as a "barrier" around the myenteric ganglia much like the *blood-brain barrier* in the CNS [16]. Morphological analysis of CNS microglia is routinely used to assess their activation state [2, 44]. These cells normally have a hyper-ramified morphology during physiological conditions (anti-inflammatory state) but lose this complex branching in disease (pro-inflammatory state) [5]. There are examples of studies that have associated the activation states or morphology of mMacs, with ENS damage and changes to GI function. Kinoshita et al. compared tissues from control versus TNBS-colitis mice [35]. Using immunohistochemistry, they demonstrated that mMacs were ramified in the control intestine but were non-ramified in the colitis tissues. Damage to the ENS and ICC networks and a reduction in the contractility of tissues were also associated with these changes in mMac morphology in the inflamed intestine [35]. Immunohistochemistry and flow cytometry analyses were used by Becker et al. to demonstrate similar shifts in macrophage profiles in aging from an "anti-inflammatory" profile in intestinal samples from younger mice to a pro-inflammatory state in samples from older mice. This was associated with an overall reduction in enteric neuron density and a reduction in gastrointestinal motility measured in vivo [4]. More knowledge is needed to relate why these changes occur and the mechanisms that connect these changes with associated damage to the ENS and other non-neural cells.

24.5 How Do mMac Interactions with Enteric Neurons Relate to Changes to Gastrointestinal Function?

Several studies have investigated the potential involvement of mMac in gastrointestinal motility and secretion. In a study by De Schepper et al., depletion of mMacs inhibited the ability for ileal segments to generate neurally evoked contractions and limited neurally mediated increases in short circuit current (a measure of secretion). This was supported by in vivo analysis showing that mMac depletion reduced small intestinal transit and increased transit times in treated mice [14]. The hypothesis from this paper centered on the idea that mMacs are necessary for the continual maintenance of the enteric neurons. While Muller et al. showed that fecal pellet output was delayed in mice treated with αCSF-1 to deplete mMacs, they identified that this delay was due to "dysmotility" [40]. In contrast to findings by *De Schepper* et al., contractility increased in small colonic segments from mMac-depleted mice, and this increase was perturbed by stretching the tissue. Colonic motility returns to normal in mMac-depleted tissues treated with exogenously applied BMP2, as mMacs typically release BMP2 for which there are receptors on enteric neurons. This led the authors to postulate that BMP2 is an important signaling molecule for the direct activation of enteric neurons during physiological processes. Observations by Luo et al. challenge the relative involvement of enteric neurons in mMac-mediated contractions [37]. Using optogenetics and genetic approaches (DREADD mice, Designer Receptors Exclusively Activated by Designer Drugs), they were able to selectively activate mMacs and demonstrate a corresponding contraction in segments of the colon. These contractions were TTX-insensitive suggesting mMacs directly stimulated the smooth muscle cells. This study provided evidence that the mechanisms for this interaction involved the release of *prostaglandin* E_2 (PGE2) from mMac acting on the prostaglandin E receptors on smooth muscle cells. However, others have shown that activation of mMac with the bacterially derived endotoxin *lipopolysaccharide* (LPS) inhibits circular muscle contractility, through an iNOS-dependent mechanism [18, 19]. Unpublished evidence from our laboratory questions whether mMacs have a clear role in controlling motility under physiological conditions, and we postulate whether these roles may be more pronounced in pathophysiology. The conflicting findings across the literature certainly demonstrate that our understanding of the functions of mMacs in GI motility and physiological functions more broadly is still being defined.

24.6 Evidence for EGC and mMac Interactions in the Gut Wall

The fact that mMacs still populate and develop within the intestine, despite the absence of enteric neurons, highlights that there must be other sources of CSF-1 in the gut wall. Grubisic et al. recently demonstrated using immunohistochemistry that around 10% of myenteric neurons in the mouse colon express the membrane-bound form of CSF-1, versus 60% of enteric glia. Using 3D analysis, they also showed that mMacs within the ganglia have processes that physically interact with EGCs [27]. They provided evidence that colonic inflammation stimulated CSF-1 release by EGCs, which in turn was associated with activation of mMacs toward a pro-inflammatory phenotype. This signaling may drive visceral pain in colitis.

24.7 Conclusion

Results from various studies have shown that the absence of EGCs or mMacs disrupts the interactions between enteric neurons and the smooth muscle cells. If this outcome is true, then it suggests that EGCs and mMacs have continual roles in regulating the output of enteric neurons, and it raises several questions for the field. What benefit is there for EGCs or mMacs to regulate neuromuscular transmission and why do enteric neurons require an intermediary cell to regulate the process? At present, enteric neurons are divided into several, much more diverse subtypes based on transmitter complement and expression than EGCs and mMacs [17]. While there are subsets of EGCs and mMac, they appear to be far more similar to each other than enteric neuron subsets. EGCs and mMacs form functional networks and perhaps the more limited diversity compared to enteric neurons assists in the synchrony of the system. By firing in synchrony EGCs/mMac may amplify the outcome of neural firing. Understanding the roles of EGCs and mMacs in ENS communication is a growing area, with many questions remain to be answered. Potential mechanisms for how these cells communicate with the ENS have been identified. However, future directions should aim to seek clarification as to when these interactions are important.

Acknowledgments Many thanks to Daniel Poole for reading and revising this manuscript and for mentoring me to become a better scientist.

Funding SEC is an Australian Research Council DECRA Fellow (DE200100825).

References

1. Avetisyan M, Rood JE, Huerta Lopez S, Sengupta R, Wright-Jin E, Dougherty JD, Behrens EM, Heuckeroth RO (2018) Muscularis macrophage development in the absence of an enteric nervous system. Proc Natl Acad Sci U S A 115:4696–4701
2. Bachiller S, Jimenez-Ferrer I, Paulus A, Yang Y, Swanberg M, Deierborg T, Boza-Serrano A (2018) Microglia in neurological diseases: a road map to brain-disease dependent-inflammatory response. Front Cell Neurosci 12:488
3. Bain CC, Bravo-Blas A, Scott CL, Perdiguero EG, Geissmann F, Henri S, Malissen B, Osborne LC, Artis D, Mowat AM (2014) Constant replenishment from circulating monocytes maintains the macrophage pool in the intestine of adult mice. Nat Immunol 15:929–937
4. Becker L, Nguyen L, Gill J, Kulkarni S, Pasricha PJ, Habtezion A (2018) Age-dependent shift in macrophage polarisation causes inflammation-mediated degeneration of enteric nervous system. Gut 67:827–836
5. Beynon SB, Walker FR (2012) Microglial activation in the injured and healthy brain: what are we really talking about? Practical and theoretical issues associ-

ated with the measurement of changes in microglial morphology. Neuroscience 225:162–171
6. Boesmans W, Cirillo C, Van den Abbeel V, Van den Haute C, Depoortere I, Tack J, Vanden Berghe P (2013) Neurotransmitters involved in fast excitatory neurotransmission directly activate enteric glial cells. Neurogastroenterol Motil 25:e151–e160
7. Boesmans W, Hao MM, Fung C, Li Z, Van den Haute C, Tack J, Pachnis V, Vanden Berghe P (2019) Structurally defined signaling in neuro-glia units in the enteric nervous system. Glia 67:1167–1178
8. Broadhead MJ, Bayguinov PO, Okamoto T, Heredia DJ, Smith TK (2012) Ca2+ transients in myenteric glial cells during the colonic migrating motor complex in the isolated murine large intestine. J Physiol 590:335–350
9. Chalazonitis A, Pham TD, Li Z, Roman D, Guha U, Gomes W, Kan L, Kessler JA, Gershon MD (2008) Bone morphogenetic protein regulation of enteric neuronal phenotypic diversity: relationship to timing of cell cycle exit. J Comp Neurol 509:474–492
10. Christofi FL, Hanani M, Maudlej N, Wood JD (1993) Enteric glial cells are major contributors to formation of cyclic Amp in myenteric plexus cultures from adult guinea-pig small intestine. Neurosci Lett 159:107–110
11. Cipriani G, Gibbons SJ, Miller KE, Yang DS, Terhaar ML, Eisenman ST, Ordog T, Linden DR, Gajdos GB, Szurszewski JH, Farrugia G (2018) Change in populations of macrophages promotes development of delayed gastric emptying in mice. Gastroenterology 154:2122–2136 e12
12. Cipriani G, Terhaar ML, Eisenman ST, Ji S, Linden DR, Wright AM, Sha L, Ordog T, Szurszewski JH, Gibbons SJ, Farrugia G (2019) Muscularis propria macrophages alter the proportion of nitrergic but not cholinergic gastric myenteric neurons. Cell Mol Gastroenterol Hepatol 7:689–691 e4
13. De Schepper S, Stakenborg N, Matteoli G, Verheijden S, Boeckxstaens GE (2018) Muscularis macrophages: key players in intestinal homeostasis and disease. Cell Immunol 330:142–150
14. De Schepper S, Verheijden S, Aguilera-Lizarraga J, Viola MF, Boesmans W, Stakenborg N, Voytyuk I, Schmidt I, Boeckx B, Dierckx de Casterle I, Baekelandt V, Gonzalez Dominguez E, Mack M, Depoortere I, de Strooper B, Sprangers B, Himmelreich U, Soenen S, Guilliams M, Vanden Berghe P, Jones E, Lambrechts D, Boeckxstaens G (2018) Self-maintaining gut macrophages are essential for intestinal homeostasis. Cell 175:400–415 e13
15. Delvalle NM, Dharshika C, Morales-Soto W, Fried DE, Gaudette L, Gulbransen BD (2018) Communication between enteric neurons, glia, and nociceptors underlies the effects of tachykinins on neuroinflammation. Cell Mol Gastroenterol Hepatol 6:321–344
16. Dora D, Ferenczi S, Stavely R, Toth VE, Varga ZV, Kovacs T, Bodi I, Hotta R, Kovacs KJ, Goldstein AM, Nagy N (2021) Evidence of a myenteric plexus barrier and its macrophage-dependent degradation during murine colitis: implications in enteric neuroinflammation. Cell Mol Gastroenterol Hepatol 12:1617
17. Drokhlyansky E, Smillie CS, Van Wittenberghe N, Ericsson M, Griffin GK, Eraslan G, Dionne D, Cuoco MS, Goder-Reiser MN, Sharova T, Kuksenko O, Aguirre AJ, Boland GM, Graham D, Rozenblatt-Rosen O, Xavier RJ, Regev A (2020) The human and mouse enteric nervous system at single-cell resolution. Cell 182:1606–1622 e23
18. Eskandari MK, Kalff JC, Billiar TR, Lee KK, Bauer AJ (1997) Lipopolysaccharide activates the muscularis macrophage network and suppresses circular smooth muscle activity. Am J Phys 273:G727–G734
19. Eskandari MK, Kalff JC, Billiar TR, Lee KK, Bauer AJ (1999) Lps-induced muscularis macrophage nitric oxide suppresses rat jejunal circular muscle activity. Am J Phys 277:G478–G486
20. Fu M, Vohra BP, Wind D, Heuckeroth RO (2006) Bmp signaling regulates murine enteric nervous system precursor migration, neurite fasciculation, and patterning via altered Ncam1 polysialic acid addition. Dev Biol 299:137–150
21. Fung C, Boesmans W, Cirillo C, Foong JPP, Bornstein JC, Vanden Berghe P (2017) Vpac receptor subtypes tune purinergic neuron-to-glia communication in the murine submucosal plexus. Front Cell Neurosci 11:118
22. Gabella G (1972) Fine structure of the myenteric plexus in the guinea-pig ileum. J Anat 111:69–97
23. Gabella G (1981) Ultrastructure of the nerve plexuses of the mammalian intestine: the enteric glial cells. Neuroscience 6:425–436
24. Goldstein AM, Brewer KC, Doyle AM, Nagy N, Roberts DJ (2005) Bmp signaling is necessary for neural crest cell migration and ganglion formation in the enteric nervous system. Mech Dev 122:821–833
25. Gomes P, Chevalier J, Boesmans W, Roosen L, van den Abbeel V, Neunlist M, Tack J, Vanden Berghe P (2009) ATP-dependent paracrine communication between enteric neurons and glia in a primary cell culture derived from embryonic mice. Neurogastroenterol Motil 21:870–e62
26. Grubisic V, Gulbransen BD (2017) Enteric glia: the most alimentary of all glia. J Physiol 595:557–570
27. Grubisic V, McClain JL, Fried DE, Grants I, Rajasekhar P, Csizmadia E, Ajijola OA, Watson RE, Poole DP, Robson SC, Christofi FL, Gulbransen BD (2020) Enteric glia modulate macrophage phenotype and visceral sensitivity following inflammation. Cell Rep 32:108100
28. Gulbransen BD, Sharkey KA (2009) Purinergic neuron-to-glia signaling in the enteric nervous system. Gastroenterology 136:1349–1358
29. Gulbransen BD, Sharkey KA (2012) Novel functional roles for enteric glia in the gastrointestinal tract. Nat Rev Gastroenterol Hepatol 9:625–632
30. Gulbransen BD, Bains JS, Sharkey KA (2010) Enteric glia are targets of the sympathetic innervation of the

myenteric plexus in the guinea pig distal colon. J Neurosci 30:6801–6809
31. Gulbransen BD, Bashashati M, Hirota SA, Gui X, Roberts JA, MacDonald JA, Muruve DA, McKay DM, Beck PL, Mawe GM, Thompson RJ, Sharkey KA (2012) Activation of neuronal P2X7 receptor-pannexin-1 mediates death of enteric neurons during colitis. Nat Med 18:600–604
32. Hanani M, Zamir O, Baluk P (1989) Glial cells in the guinea pig myenteric plexus are dye coupled. Brain Res 497:245–249
33. Hanani M, Francke M, Hartig W, Grosche J, Reichenbach A, Pannicke T (2000) Patch-clamp study of neurons and glial cells in isolated myenteric ganglia. Am J Physiol Gastrointest Liver Physiol 278:G644–G651
34. Hennig GW, Gould TW, Koh SD, Corrigan RD, Heredia DJ, Shonnard MC, Smith TK (2015) Use of genetically encoded calcium indicators (Gecis) combined with advanced motion tracking techniques to examine the behavior of neurons and glia in the enteric nervous system of the intact murine colon. Front Cell Neurosci 9:436
35. Kinoshita K, Horiguchi K, Fujisawa M, Kobirumaki F, Yamato S, Hori M, Ozaki H (2007) Possible involvement of muscularis resident macrophages in impairment of interstitial cells of Cajal and myenteric nerve systems in rat models of TNBS-induced colitis. Histochem Cell Biol 127:41–53
36. Koscso B, Gowda K, Schell TD, Bogunovic M (2015) Purification of dendritic cell and macrophage subsets from the normal mouse small intestine. J Immunol Methods 421:1–13
37. Luo J, Qian A, Oetjen LK, Yu W, Yang P, Feng J, Xie Z, Liu S, Yin S, Dryn D, Cheng J, Riehl TE, Zholos AV, Stenson WF, Kim BS, Hu H (2018) Trpv4 channel signaling in macrophages promotes gastrointestinal motility via direct effects on smooth muscle cells. Immunity 49(107–119):e4
38. Mcclain J, Grubisic V, Fried D, Gomez-Suarez RA, Leinninger GM, Sevigny J, Parpura V, Gulbransen BD (2014) Ca2+ responses in enteric glia are mediated by connexin-43 hemichannels and modulate colonic transit in mice. Gastroenterology 146:497–507 e1
39. Mikkelsen HB, Thuneberg L (1999) Op/op mice defective in production of functional colony-stimulating factor-1 lack macrophages in muscularis externa of the small intestine. Cell Tissue Res 295:485–493
40. Muller PA, Koscso B, Rajani GM, Stevanovic K, Berres ML, Hashimoto D, Mortha A, Leboeuf M, Li XM, Mucida D, Stanley ER, Dahan S, Margolis KG, Gershon MD, Merad M, Bogunovic M (2014) Crosstalk between muscularis macrophages and enteric neurons regulates gastrointestinal motility. Cell 158:300–313
41. Phillips RJ, Powley TL (2012) Macrophages associated with the intrinsic and extrinsic autonomic innervation of the rat gastrointestinal tract. Auton Neurosci 169:12–27
42. Rao M, Rastelli D, Dong L, Chiu S, Setlik W, Gershon MD, Corfas G (2017) Enteric glia regulate gastrointestinal motility but are not required for maintenance of the epithelium in mice. Gastroenterology 153:1068–1081 e7
43. Schneider S, Wright CM, Heuckeroth RO (2019) Unexpected roles for the second brain: enteric nervous system as master regulator of bowel function. Annu Rev Physiol 81:235–259
44. York EM, Ledue JM, Bernier LP, MacVicar BA (2018) 3DMorph automatic analysis of microglial morphology in three dimensions from ex vivo and in vivo imaging. eNeuro 5:ENEURO.0266-18.2018

Mas-Related G Protein-Coupled Receptors (Mrgprs) as Mediators of Gut Neuro-Immune Signaling

25

Samuel Van Remoortel
and Jean-Pierre Timmermans

Abstract

Over the past 15 years, the research field on Mas-related G protein-coupled receptors (Mrgprs), a relatively new family of rhodopsin A-like G protein-coupled receptors, has expanded enormously, and a plethora of recent studies have provided evidence that several of these Mrgpr family members play an important role in the underlying mechanisms of itch and pain, as well as in the initiation and modulation of inflammatory/allergic responses. Initial studies mainly focused on the skin, but more recently also visceral organs such as the respiratory and gastrointestinal (GI) tracts emerged as sites for Mrgpr involvement. It has become clear that the gastrointestinal tract and its innervation in close association with the immune system represent a novel expression site for Mrgprs where they contribute to the interoceptive mechanisms maintaining homeostasis and might constitute promising targets in chronic abdominal pain disorders. In this short review, we provide an update of our current knowledge on the expression, distribution, and function of members of this Mrgpr family in intrinsic and extrinsic neuro-immune pathways related to the gastrointestinal tract, their mediatory role(s) in gut neuro-immune signaling, and their involvement in visceral afferent (nociceptive) pathways.

Keywords

Mas-related G protein-coupled receptors · Enteric nervous system · Gut · Neuro-immune signaling · Mast cell

S. Van Remoortel · J.-P. Timmermans (✉)
Laboratory of Cell Biology and Histology, Department of Veterinary Sciences, Faculty of Pharmaceutical, Biomedical and Veterinary Sciences, University of Antwerp, Antwerpen, Belgium
e-mail: jean-pierre.timmermans@uantwerpen.be

25.1 The Family of Mas-Related G Protein-Coupled Receptors

About two decades ago, a novel family of rhodopsin A-like G protein-coupled receptors (GPCRs) was discovered in the mouse and human genome [1, 2]. The members of this family showed 30–41% sequence homology with the Mas oncogene and was accordingly termed Mas-related G protein-coupled receptors (Mrgprs). Mammalian Mrgprs are subdivided into nine separate subfamilies (A-H and X). Subfamilies D to G are evolutionarily conserved between species and thus considered direct orthologues [1–3], whereas subfamilies A, B, C, and H are only found in rodents, and subfamily X is specific to primates, including humans, macaques, and rhesus monkeys [1–5]. Even though these rodent-specific and human-specific receptors are

N. J. Spencer et al. (eds.), *The Enteric Nervous System II*, Advances in Experimental Medicine and Biology 1383, https://doi.org/10.1007/978-3-031-05843-1_25

divergent in their gene and protein sequences, they do share common expression patterns and ligands and are therefore considered functional orthologues. Since their initial discovery, it has become clear that Mrgprs might be involved in a wide array of human disorders, and our insights in their signaling pathways and polymorphisms are increasing, but there is still a long way to go [6].

25.2 Mrgprs in the Skin Sensory Innervation

In the skin sensory innervation (Fig. 25.1), Mrgprs are primarily expressed in small- to medium-diameter DRG neurons [7–12]. Although most of these Mrgpr-expressing neurons selectively bind isolectin B4 (IB4) and express the glial cell-derived neurotrophic factor (GDNF) receptor Ret, which marks a subset of non-peptidergic, nociceptive DRG neurons, distinct Mrgprs have also been reported to be expressed in subsets of peptidergic, calcitonin gene-related peptide (CGRP)-expressing nociceptive neurons [1, 4, 7]. In line with their expression in DRG neurons, a large body of evidence indicates that Mrgprs are involved in somatosensation, more specifically in nociception and pruriception, also known as itch, which are both complex sensory modalities intended to protect us from and respond to (potentially) harmful stimuli.

25.2.1 Mrgpra3 and MRGPRX1

Murine Mrgpra3 is expressed in a subset of small- to medium-diameter DRG neurons that innervate the epidermal layers of the skin [11]. Upon activation by its respective ligand, that is, the antimalarial drug chloroquine, Mrgpra3 signaling induces nocifensive scratching behavior in wild-type mice indicative of itch, a behavior that is abolished in animals lacking Mrgpra3 [10, 11]. Since Mrgpra3 activation specifically induces itch behavior and not pain behavior, it was generally accepted that Mrgpra3-expressing C-fibers represented a subpopulation specifically involved in itch detection and not in pain, but a recent report showed that Mrgpra3-expressing C-fibers transmit both itch and pain, where fast ionotropic receptor signaling of neurons triggers pain responses, whereas metabotropic signaling through Mrgpra3 in the same population of fibers encodes for itch responses, which can be alleviated by blocking TRP channels [13].

In chronic itch models, Mrgpra3-expressing DRG neurons display marked expressional changes, and Mrgpra3-expressing nerve fibers display increased spontaneous activity and excitability [14, 15]. Moreover, deletion of several Mrgpr members including Mrgpra3 markedly reduces spontaneous scratching behaviors under chronic itch conditions, indicating that Mrgpra3 represents an important underlying driver in mouse models of chronic itch conditions [15, 16].

Mrgpra3 has no direct orthologue in humans, but the human MRGPRX1 member is considered the functional orthologue of Mrgpra3 based on similar expression patterns and ligands [10]. Cutaneous activation of MRGPRX1 induces itch responses in MRGPRX1-humanized mice as well as in human subjects, strongly supporting the translational potential of the findings in mice [17].

25.2.2 Mrgpra1/MRGPRX4

For both murine Mrgpra1 and human MRGPRX4, which are expressed in small-diameter DRG neurons and trigeminal neurons, it was recently shown that these receptors are activated by bilirubin and several other bile acid metabolites [18, 19]. In mice, cutaneous application of bilirubin induces scratching behavior in an Mrgpra1-dependent manner, and similar findings were observed in humanized MRGPRX4 mice [19]. Furthermore, human DRG neurons are responsive to bile acids in an MRGPRX4-dependent manner, and pharmacological activation of MRGPRX4 induces itch sensations in human subjects [18]. In cholestatic conditions, elevated bile acids are correlated with the occurrence of

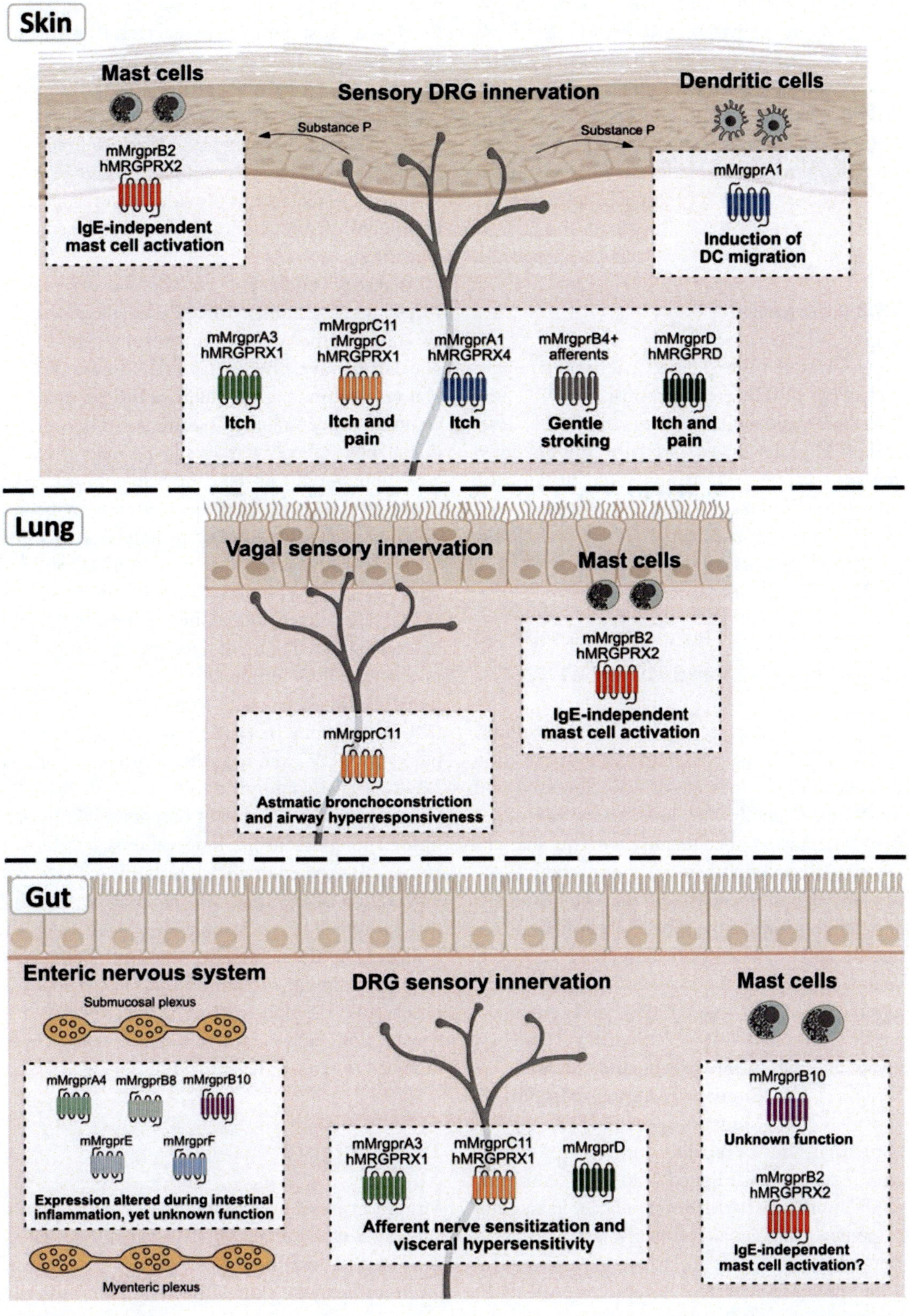

Fig. 25.1 Schematic overview of the expression and cellular location of Mrgprs in skin, in respiratory airways and in the gastrointestinal tract

itch among patients with liver disease and are sufficient to activate MRGPRX4. Similarly, humanized MRGPRX4 mice scratched more frequently compared to control mice lacking MRGPRX4 under cholestatic model conditions, and this scratching behavior was associated with increased serum bile acid levels, further supporting a role for MRGPRX4 in cholestatic itch [19, 20].

25.2.3 Mrgprb4

Mouse Mrgprb4 is expressed in a limited population of small-diameter, IB4-positive DRG neurons that terminate in the hairy skin [9, 21]. Even though Mrgprb4 ligands are currently unknown, genetic labeling and manipulation of Mrgprb4-expressing sensory neurons showed that Mrgprb4-exspressing nerve fibers are tuned toward detection of gentle, massage-like stroking [9, 21].

25.2.4 Mrgprc11 and MRGPRX1

Mrgprc11 is also expressed in small- to medium-diameter DRG neurons innervating the skin, where the receptor is largely co-expressed with the Mrgpra3 family members. Using Mrgprc11-CreERT2 mice, a recent study by Xing and colleagues showed that Mrgprc11-expressing DRG neurons exhibit free nerve endings with extensive axonal branching in the superficial epidermis and large receptive fields [12]. Mrgprc11 is specifically activated by the synthetic peptide bovine adrenal medulla (BAM)8-22, and cutaneous application of BAM(8-22) induces scratching behavior that is abolished after deletion of Mrgprc11 [10]. Similar to Mrgpra3, Mrgprc11 is considered the functional orthologue of the human MRGPRX1 receptor due to shared expression patterns and ligands. Indeed, cutaneous application of BAM(8-22) elicits scratching responses in human subjects as well [17]. Remarkably, Mrgprc11 shares several ligands with protease-activated receptor 2 (PAR-2), including the PAR-2 tether ligand sequence SLIGRL, the cysteine protease cathepsin S, and the house dust mite protease Der P1, thereby linking Mrgprc11 to protease-mediated signaling [22–24]. Specifically, in the skin, these ligands induce pain behaviors through their actions on PAR-2 receptors, whereas their actions on Mrgprc11 receptors specifically induce itch responses, indicating a previously unrecognized functional heterogeneity in protease-mediated signaling.

Whereas cutaneous activation of mouse Mrgprc11 and human MRGPRX1 induces itch, central (intrathecal) activation results in profound anti-nociceptive effects. Central Mrgprc11 activation inhibits high voltage-activated calcium channels on central DRG terminals in the spinal dorsal horn, thereby reducing peripheral excitatory inputs onto postsynaptic dorsal horn neurons [25, 26]. This has been shown to be an effective analgesic strategy in animal models of neurogenic and inflammatory pain [27]. Similar effects were observed in a humanized transgenic MRGPRX1 mouse model, indicative of the translational potential of these findings [28]. Mrgprc11 at central DRG terminals physically and functionally interacts with mu-opioid receptors (MORs), thereby increasing the potency of morphine on MORs and reducing morphine-induced MOR desensitization and tolerance [29, 30].

Interestingly, the role of Mrgprs might not be limited to the sensory innervation of the skin. Han et al. showed that the murine vagal afferent innervation of the lung also is an important site for Mrgprc11 expression [31]. Activation of Mrgprc11 in the airways resulted in nocifensive responses similar in nature to the itch responses observed in the skin, such as asthmatic bronchoconstriction and airway hyperresponsiveness intended to remove harmful stimuli (Fig. 25.1).

25.2.5 Mrgprd

Mrgprd expression is primarily encountered in a subset of small-diameter, IB4-expressing DRG neurons that terminate in the superficial *stratum granulosum* of the skin [7, 8]. Similar to Mrgpra3 and Mrgprc11, cutaneous application of the

Mrgprd ligand β-alanine induces itch behaviors that are absent in mice lacking Mrgprd and also elicits itch in human subjects [17, 32]. Apart from having a role in behavioral responses such as itch, Mrgprd controls the (mechano- and heat-) sensitivity of nociceptive somatosensory neurons. Genetic deletion of Mrgprd expression significantly reduces mechanical and heat stimulation responses of skin polymodal nociceptors under normal conditions, pointing to a role in regulating their sensitivity [33]. Mrgprd likely regulates this sensitivity by controlling neuronal excitability, as demonstrated by the observation that Mrgprd activation couples to $G_{\alpha q}$- and $G_{\alpha i/o}$-mediated signaling and inhibits non-inactivating KCNQ2/3 potassium channel currents (M-currents), thereby increasing neuronal excitability [34]. In addition to neuronal excitability under control conditions, aberrant Mrgprd signaling underlies LPS-triggered inflammatory hyperalgesia through NF-κB signaling, highlighting a possible role for Mrgprd in inflammatory pain [35].

25.3 Mrgprs in Skin Immune Cells

Apart from their expression in sensory DRG neurons, certain Mrgprs are also expressed in immune cell subsets (Fig. 25.1). Mouse Mrgprb2 and its human orthologue MRGPRX2 are two Mrgprs members that are not expressed in sensory DRG neurons but are rather specifically expressed in connective tissue mast cells (CTMCs) [36]. These receptors are typically activated by so-called "basic secretagogues," such as compound 48/80, several FDA-approved drugs, substance P (SP), and the antimicrobial beta-defensin peptides [36–40]. Strikingly, activation of Mrgprb2/MRGPRX2 results in mast cell degranulation that is both spatially and temporally distinct from classic IgE-mediated degranulation and has emerged as a novel IgE-independent mast cell activation pathway [41, 42]. Mrgprb2/MRGPRX2 signaling has emerged as a key player in immediate drug hypersensitivity [36, 43, 44] and as a novel paradigm in the pathophysiology of pseudo-allergic (drug) reactions and chronic itch, which were previously considered to be mainly driven by canonical, IgE-dependent mast cell activation [42]. Apart from these detrimental actions, Mrgprb2/MRGPRX2 in mast cells also plays a role as drivers of first-line protective immunity. For example, Mrgprb2/MRGPRX2 directly sense bacterial quorum-sensing molecules and play a central role in the innate antibacterial response [45, 46]. In addition, skin allergen exposure or acute tissue damage induces substance P (SP) release from sensory neurons, which on its turn mediates a protective neurogenic inflammatory response through Mrgprb2 signaling in mast cells [47, 48].

Very recently, a study by Perner and colleagues discovered Mrgpra1 expression in skin dendritic cells [49]. In this study, the authors found that allergen-induced activation of skin sensory neurons induces a neurogenic inflammatory response through the local release of substance P, which on its turn stimulated a nearby located population of $CD301b^+$ dendritic cells (DCs) through Mrgpra1 [49]. This Mrgpra1-induced signaling resulted in DC migration to the draining lymph node where they initiated T helper-2 cell differentiation, identifying a novel role for Mrgpra1 in allergen-induced neurogenic inflammation in the skin.

25.4 Mrgprs Expression in the Gastrointestinal (GI) Tract

Based on the above-described role for Mrgprs in the skin, it is clear that these receptors are central players in different aspects of skin neuro-immune communication. The GI tract, just like the skin, is continuously exposed to potential threats or harms from our environment and represents an organ where a delicate neuro-immune balance takes places in order to maintain its homeostasis. Therefore, it seems plausible that Mrgprs might play a relevant role in the gut as well. Indeed, there is a growing body of evidence that the gut is a relevant expression site for Mrgprs, where these receptors might exert similar functions as they do in the skin (Fig. 25.1).

25.4.1 Mrgprs in the Enteric Nervous System

Early studies focused on the expression of murine Mrgpr members at the level of the terminal ileum. Previous studies in our lab investigated the expression of several murine Mrgpra subfamily members (Mrgpra1, a2, a4, a5, and a7), multiple Mrgprb subfamily members (Mrgprb1, b2, b4, b5, b8, b10), as well as the subfamily e and f members in the murine terminal ileum [50, 51]. The results of these studies revealed low mRNA expression for most of these members in the healthy intestine but significantly increased mRNA expression for several Mrgpra and b subfamily members in intestinal inflammatory conditions, induced by parasitic infection with *Schistosoma mansoni* (*S. mansoni*) or chemically induced by trinitrobenzenesulfonic acid (TNBS). Interestingly, Mrgpre and Mrgprf expression was significantly reduced in the TNBS-inflamed ileum. Further immunohistochemical studies using custom-developed antibodies showed that Mrgpra4, b2, b8, b10, e, and f are expressed, be it in low numbers, by several enteric neuronal subtypes, such as sensory, secretomotor, and vasodilator neurons, as well as in nerve fibers in the *lamina propria* and *tunica muscularis* [50, 51]. Unfortunately, since many of these Mrgpr members currently remain orphan receptors without available ligand, further functional evaluation of their role in the ENS is hampered.

More recently, an immunohistochemical study characterized the expression and localization of Mrgprd in the mouse small and large intestine. Using a commercial antibody, the authors observed Mrgprd expression in smooth muscle cells of the *tunica muscularis*, as well as in lamina propria macrophages and T cells, which led them to suggest a novel role for Mrgprd in gut motility and immunity [52]. However, we recently demonstrated that the commercial antibody used in the latter study displayed similar staining patterns in the gut of wild-type and Mrgprd knockout mice and hence does not specifically stain for Mrgprd [53]. Furthermore, Mrgprd in situ hybridization did not reveal the presence of Mrgprd mRNA in the gut wall, further indicating that Mrgprd is not expressed by cells residing in the gut wall proper (Fig. 25.1).

25.4.2 Mrgprs in the Gut Spinal Afferent Innervation

25.4.2.1 Mrgpra3/c11 and Its Human Counterpart MRGPRX1

More recently, it has become clear that Mrgpr expression is not limited to the somatosensory DRG innervation and that the viscerosensory DRG innervation also represents an important expression site for the Mrgpr family. Using in situ hybridization and immunohistochemistry, for the first time our group reported Mrgprc11 expression in up to 20% of colon-projecting peptidergic DRG neurons [54]. Interestingly, intracolonic administration of an Mrgprc11 ligand, that is, BAM(8-22), resulted in the sensitization of the splanchnic afferent innervation and induced visceral hypersensitivity in healthy mice, an effect that was not observed in mice lacking functional Mrgprc11 expression [54]. Our results on Mrgprc11 were confirmed by an independent study by the Brierley lab, who in addition reported that Mrgpra3 plays a very similar role in causing visceral hypersensitivity [55]. Furthermore, these authors showed functional upregulation of Mrgpra3 and Mrgprc11 signaling pathways in a mouse model for chronic visceral hypersensitivity, highlighting the potential role for these Mrgprs as drivers of chronic abdominal pain. Both studies also confirmed the presence of the human functional orthologue of mouse Mrgpra3 and Mrgprc11, that is, MRGPRX1, in human thoracolumbar DRG neurons. In the skin, Mrgpra3 and Mrgprc11 are known to mediate their effects through direct coupling with TRP channels [56]. Strikingly, this downstream coupling of Mrgpra3/Mrgprc11 with TRP channels was equally demonstrated in the gut afferent innervation, where Mrgrpr-mediated effects on gut pain sensitivity were abolished in mice lacking TRPA1, indicating its relevance as downstream target for Mrgpr-mediated signaling in the gut afferent innervation [55].

25.4.2.2 Mrgprd

Other studies indicated that also Mrgprd is expressed in the gut extrinsic DRG innervation. Hockley and colleagues, in their single-cell RNAseq-based classification of colon-projecting DRG neurons, identified a subset of neurons that were clustered based on Mrgprd expression [57]. Additionally, it was recently found that Mrgprd activation in these neurons by the polyunsaturated fatty acid metabolite 5-oxo-ETE directly causes visceral hypersensitivity, thus pointing to a role for Mrgprd in the gut that is similar to that of Mrgpra3 and Mrgprc11 [58].

25.4.3 Mrgprs: Novel Targets in Chronic Abdominal Pain Disorders?

The newly established role for Mrgpr-mediated signaling in gut viscerosensory pathways shows that these receptors have a similar pro-nociceptive role to their earlier described function in the skin and hence can be considered to represent the visceral counterpart of itch-related pathways found in the skin. In this respect, Mrgpr-mediated signaling might act as a novel irritant-sensing mechanism in the bowel which, upon activation, impacts on gut sensory pathways and is aimed at expelling (potentially) harmful stimuli from the body, and as such could contribute to sensory disturbances commonly found in GI disorders. A further increase in gut afferent nerve signaling was observed in a post-inflammatory chronic abdominal pain mouse model, characterized by gut afferent nerve sensitization and visceral hypersensitivity in the absence of active inflammation, after intracolonic administration of chloroquine and BAM(8-22), the respective ligands for Mrgpra3 and Mrgprc11, suggestive of a functional upregulation of these Mrgpr-mediated pathways under chronic abdominal pain conditions.

As far as Mrgprc11 is concerned, several lines of evidence indicate that its ligands are expressed in the gut and that several of these ligands are also linked to pathological gut conditions. BAM(8-22), a peptide ligand that specifically acts on Mrgprc11 and MRGPRX1, has been described as a potential cleavage product of the opioid precursor proenkephalin A (PENK) [2]. Even though the expression of BAM(8-22) itself has not been directly detected endogenously so far, the expression of its precursor PENK and some PENK-derived opioid peptides has been reported in the GI tract, more specifically in the myenteric plexus and in colitogenic T cells of the colonic mucosa [59–62]. As such, the presence of precursors of BAM(8-22) indirectly suggests that this ligand could be present in the gut as well, although definite proof for this assumption still needs to be provided. Furthermore, Mrgprc11 is known to be activated by two FMRF-amide peptides, neuropeptide AF (NPAF) and neuropeptide FF (NPFF). Interestingly, intracolonic administration of NPAF in healthy mice induces afferent nerve sensitization and visceral hypersensitivity in an Mrgpr-dependent manner [55]. NPAF and NPFF, whose common precursor gene is expressed by myenteric neurons and mucosal-type mast cells, are associated with pain signaling and inflammation and thus might act in association with Mrgprc11, as a possible neuro-immune axis on gut afferents [63–65]. In addition to being activated by NPFF and NPAF, Mrgprc11 has also been linked to protease-mediated signaling. Established protease-activated receptor 2 (PAR-2) agonists, such as the PAR-2 tethered ligand peptide SLIGRL and distinct cysteine proteases, including cathepsin S, papain, and the house dust mice DerP1 protease, also act as Mrgprc11 agonists, thereby inducing effects that were previously considered to be solely PAR2-mediated [22–24]. In the gut, several of these PAR-2/Mrgprc11 ligands can be linked to visceral hypersensitivity. For example, intracolonic administration of SLIGRL induces an inflammatory response and drives visceral hypersensitivity, an effect considered to be mediated through PAR-2 [66]. Moreover, increased cathepsin S activity and release from macrophages have been observed in animal models of intestinal inflammation, in which this ligand was found to contribute to visceral hypersensitivity through PAR-2 [67]. Also, direct intracolonic papain administration induces visceral hypersensitivity in mice [68]. Another

PAR-2/Mrgprc11 ligand, the house dust mite protease Der P1, is also found throughout the GI tract, where it has been suggested as a possible environmental trigger for gut dysfunction in GI disorders such as IBS patients [69]. At this point, these proteases are mainly thought to exert their effects in the gut through PAR-2-mediated signaling in the gut afferent innervation, but the emerging role of Mrgprs in the gut afferent innervation adds a new dimension to the current concepts of protease-mediated signaling in the bowel, implying that protease-mediated signaling, which was previously considered to be mediated via PAR-2 only, might also be mediated—at least in part—through Mrgprs. Given the importance of proteases and their signaling pathways in chronic abdominal pain disorders, further studies exploring the role of protease-mediated signaling through Mrgprs in the gut are warranted.

25.4.4 Mrgprs in Gut Mast Cells

There is also some evidence available to suggest the expression of specific Mrgprs in gut mast cells. In our mouse studies, Mrgprb10 was found to be de novo expressed in mucosal mast cells of *Schistosoma mansoni*-infected ileum [50]. In an in vitro study, mucosal-type bone marrow-derived mast cells (BMDMCs) were found to express Mrgpra1 and Mrgpra4 mRNA, but these findings have not been confirmed in vivo [65]. With regard to the human MRGPRX2 member, a study by Plum and colleagues showed MRGPRX2 expression in gut mast cells [70]. Furthermore, it was recently shown that adrenomedullin, an important precursor of the MRGPRX2 ligand PAMP20, is significantly upregulated in ulcerative colitis patients, indicating a possible role for this receptor in driving aberrant mast cell activity [71]. Increased MRGPRX2 mRNA expression was also found in mucosal biopsies of a subset of immune-activated IBS patients [71]. With regard to the mouse counterpart of MRGPRX2, that is, Mrgprb2, using transgenic mouse lines, we are currently looking into the expression and role of Mrgprb2 as a novel IgE-independent mast cell activation pathway under healthy and pathological conditions (Fig. 25.2).

25.5 Conclusion

Over the past years, the Mrgprs research field has expanded enormously, and it has become clear that the GI tract and its innervation represent a

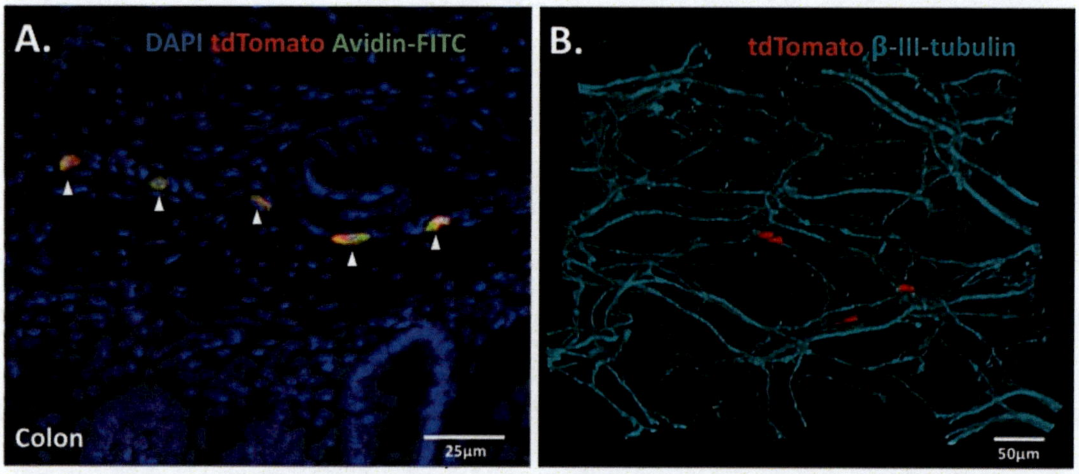

Fig. 25.2 Mrgprb2 expressing mast cells in the mouse colon. (**a**) The colon of Mrgprb2-cre:tdTomato mice, kindly provided by Prof. Dr. X. Dong, shows the presence of tdTomato-positive cells (red) that co-stain for Avidin-FITC (mast cell granule marker, green). (**b**) 3D rendering of an optically cleared colonic submucosal whole-mount of Mrgprb2-cre:tdTomato mice that was stained for tdTomato (Mrgprb2-expressing cells, red) and beta-III-tubulin (neuronal marker, magenta)

novel expression site for Mrgprs. Our lab was the first to document the expression of several murine Mrgpr family members in the gut wall, more specifically in the ENS, and has shown that several of these members undergo marked expressional changes under intestinal inflammatory conditions. More recently, we and others have shown that several Mrgprs are also expressed in the gut extrinsic splanchnic innervation and play a role in gut pain (hyper)sensitivity. Considering that the family of Mrgprs still contains many other uncharacterized members, we are probably merely looking at the tip of the iceberg. Therefore, further studies are warranted, and these future efforts need to be focused on elucidating the upstream drivers of aberrant Mrgpr-mediated signaling in gut sensory pathways under healthy and diseased conditions. Moreover, current efforts in our lab are focused on certain Mrgprs in gut mast cells and their role as novel mast cell activation pathways. This new knowledge should in turn provide us with a better understanding of the role these receptors and their signaling mechanisms play in gut neuro-immune disorders, such as IBD and IBS, and will eventually allow us to validate their therapeutic potential.

References

1. Dong X et al (2001) Diverse family of GPCRs expressed in specific subsets of nociceptive sensory neurons. Cell 106:619–632
2. Lembo PMC et al (2002) Proenkephalin A gene products activate a new family of sensory neuron–specific GPCRs. Nat Neurosci 5:201–209
3. Zylka MJ et al (2003) Atypical expansion in mice of the sensory neuron-specific Mrg G protein-coupled receptor family. Proc Natl Acad Sci 100:10043–10048
4. Zhang L et al (2005) Cloning and expression of MRG receptors in macaque, mouse, and human. Brain Res Mol Brain Res 133:187–197
5. Burstein ES et al (2006) Characterization of the Mas-related gene family: structural and functional conservation of human and rhesus MrgX receptors. Br J Pharmacol 147:73–82
6. Steele HR, Han L (2021) The signaling pathway and polymorphisms of Mrgprs. Neurosci Lett 744:135562. https://doi.org/10.1016/j.neulet.2020.135562
7. Zylka MJ et al (2005) Topographically distinct epidermal nociceptive circuits revealed by axonal tracers targeted to Mrgprd. Neuron 45:17–25
8. Wang H, Zylka MJ (2009) Mrgprd-expressing polymodal nociceptive neurons innervate most known classes of substantia gelatinosa neurons. J Neurosci 29:13202–13209
9. Liu Q et al (2007) Molecular genetic visualization of a rare subset of unmyelinated sensory neurons that may detect gentle touch. Nat Neurosci 10:946–948
10. Liu Q et al (2009) Sensory neuron-specific GPCR Mrgprs are itch receptors mediating chloroquine-induced pruritus. Cell 139:1353–1365
11. Han L et al (2013) A subpopulation of nociceptors specifically linked to itch. Nat Neurosci 16:174–182
12. Xing Y et al (2020) Visualizing the itch-sensing skin arbors. J Investig Dermatol 141(5):1308–1316. https://doi.org/10.1016/j.jid.2020.08.030
13. Sharif B et al (2020) Differential coding of itch and pain by a subpopulation of primary afferent neurons. Neuron 106:1–12
14. Qu L et al (2014) Enhanced excitability of MRGPRA3- and MRGPRD-positive nociceptors in a model of inflammatory itch and pain. Brain 137:1039–1050
15. Shi H et al (2017) MrgprA3 shows sensitization to chloroquine in an acetone–ether–water mice model. Neuroreport 28:1127–1133
16. Zhu Y et al (2017) Mrgprs activation is required for chronic itch conditions in mice. Itch (Phila) 2(3):e9
17. Sikand P et al (2011) BAM8-22 peptide produces itch and nociceptive sensations in humans independent of histamine release. J Neurosci 31:7563–7567
18. Yu H et al (2019) MRGPRX4 is a bile acid receptor for human cholestatic itch. elife 8:e48431
19. Meixiong J et al (2019) Identification of a bilirubin receptor that may mediate a component of cholestatic itch. elife 8:e44116
20. Meixiong J et al (2019) MRGPRX4 is a G protein-coupled receptor activated by bile acids that may contribute to cholestatic pruritus. Proc Natl Acad Sci 116:10525–10530
21. Vrontou S et al (2013) Genetic identification of C fibres that detect massage-like stroking of hairy skin in vivo. Nature 493:669–673
22. Liu Q et al (2011) The distinct roles of two GPCRs, MrgprC11 and PAR2, in itch and hyperalgesia. Sci Signal 4:ra45
23. Reddy VB, Lerner EA (2017) Activation of mas-related G-protein–coupled receptors by the house dust mite cysteine protease Der p1 provides a new mechanism linking allergy and inflammation. J Biol Chem 292:17399–17406
24. Reddy VB et al (2015) Redefining the concept of protease-activated receptors: cathepsin S evokes itch via activation of Mrgprs. Nat Commun 6:7864
25. He S-Q et al (2014) MrgC agonism at central terminals of primary sensory neurons inhibits neuropathic pain. Pain 155:534–544
26. Li Z et al (2015) The inhibition of high-voltage-activated calcium current by activation of MrgC11 involves phospholipase C-dependent mechanisms. Neuroscience 300:393–403

27. Jiang J et al (2013) Effect of Mas-related gene (Mrg) receptors on hyperalgesia in rats with CFA-induced inflammation via direct and indirect mechanisms. Br J Pharmacol 170:1027–1040
28. Li Z et al (2017) Targeting human Mas-related G protein-coupled receptor X1 to inhibit persistent pain. Proc Natl Acad Sci U S A 114:E1996–E2005
29. Wang D et al (2013) Activation of Mas oncogene-related gene (Mrg) C receptors enhances morphine-induced analgesia through modulation of coupling of mu-opioid receptor to Gi-protein in rat spinal dorsal horn. Neuroscience 253:455–464
30. He S-Q et al (2018) Oligomerization of MrgC11 and μ-opioid receptors in sensory neurons enhances morphine analgesia. Sci Signal 11:eaao3134
31. Han L et al (2018) Mrgprs on vagal sensory neurons contribute to bronchoconstriction and airway hyper-responsiveness. Nat Neurosci 21:324–328
32. Liu Q et al (2012) Mechanisms of itch evoked by β-alanine. J Neurosci 32:14532–14537
33. Rau KK et al (2009) Mrgprd enhances excitability in specific populations of cutaneous murine polymodal nociceptors. J Neurosci 29:8612–8619
34. Crozier RA et al (2007) MrgD activation inhibits KCNQ/M-currents and contributes to enhanced neuronal excitability. J Neurosci 27:4492–4496
35. Lan L et al (2020) Mas-related G protein-coupled receptor D participates in inflammatory pain by promoting NF-κB activation through interaction with TAK1 and IKK complex. Cell Signal 76:109813
36. McNeil BD et al (2015) Identification of a mast-cell-specific receptor crucial for pseudo-allergic drug reactions. Nature 519:237–241
37. Kashem SW et al (2011) G protein coupled receptor specificity for C3a and compound 48/80-induced degranulation in human mast cells: roles of Mas-related genes MrgX1 and MrgX2. Eur J Pharmacol 668:299–304
38. Subramanian H et al (2013) beta-Defensins activate human mast cells via Mas-related gene X2. J Immunol 191:345–352
39. Subramanian H et al (2011) PMX-53 as a dual CD88 antagonist and an agonist for Mas-related gene 2 (MrgX2) in human mast cells. Mol Pharmacol 79:1005–1013
40. Alkanfari I et al (2019) Small-molecule host-defense peptide mimetic antibacterial and antifungal agents activate human and mouse mast cells via Mas-related GPCRs. Cell 8:311
41. Gaudenzio N et al (2016) Different activation signals induce distinct mast cell degranulation strategies. J Clin Investig 126:3981–3998
42. Meixiong J et al (2019) Activation of mast-cell-expressed Mas-related G Protein-coupled receptors drives non-histaminergic itch. Immunity 50(5):1163–1171
43. Ebo DG et al (2021) Immunoglobulin E cross-linking or MRGPRX2 activation: clinical insights from rocuronium hypersensitivity. Br J Anaesth 126:e27–e29
44. Elst J et al (2020) MRGPRX2 and immediate drug hypersensitivity: insights from cultured human mast cells. J Investig Allergol Clin Immunol 31(6):489–499. https://doi.org/10.18176/jiaci.0557
45. Pundir P et al (2019) A connective tissue mast-cell-specific receptor detects bacterial quorum-sensing molecules and mediates antibacterial immunity. Cell Host Microbe 26:114–122, e8
46. Arifuzzaman M et al (2019) MRGPR-mediated activation of local mast cells clears cutaneous bacterial infection and protects against reinfection. Sci Adv 5(1):eaav0216
47. Serhan N et al (2019) House dust mites activate nociceptor–mast cell clusters to drive type 2 skin inflammation. Nat Immunol 20(11):1435–1443. https://doi.org/10.1038/s41590-019-0493-z
48. Green DP et al (2019) A mast-cell-specific receptor mediates neurogenic inflammation and pain. Neuron 101:412–420, e3
49. Perner C et al (2020) Substance P release by sensory neurons triggers dendritic cell migration and initiates the type-2 immune response to allergens. Immunity 53:1063–1077, e7
50. Avula LR et al (2013) Expression and distribution patterns of Mas-related gene receptor subtypes A-H in the mouse intestine: inflammation-induced changes. Histochem Cell Biol 139:639–658
51. Avula LR et al (2011) The effect of inflammation on the expression and distribution of the MAS-related gene receptors MrgE and MrgF in the murine ileum. Histochem Cell Biol 136:569–585
52. Zhou C et al (2019) Expression and localization of MrgprD in mouse intestinal tract. Cell Tissue Res 377:259–268
53. Van Remoortel S, Timmermans J-P (2019) Presence of MrgprD within the gastrointestinal wall: reality or fake? Cell Tissue Res 378:555–558
54. Van Remoortel S et al (2019) Mas-related G protein-coupled receptor C11 (Mrgprc11) induces visceral hypersensitivity in the mouse colon: a novel target in gut nociception? Neurogastroenterol Motil 31(8):e13623. https://doi.org/10.1111/nmo.13623
55. Castro J et al (2019) Activation of pruritogenic TGR5, MRGPRA3, and MRGPRC11 on colon-innervating afferents induces visceral hypersensitivity. JCI Insight 4(20):e131712. https://doi.org/10.1172/jci.insight.131712
56. Wilson SR et al (2011) TRPA1 is required for histamine-independent, Mas-related G protein-coupled receptor-mediated itch. Nat Neurosci 14:595–602
57. Hockley JRF et al (2019) Single-cell RNAseq reveals seven classes of colonic sensory neuron. Gut 68:633–644
58. Bautzova T et al (2018) 5-oxoETE triggers nociception in constipation-predominant irritable bowel syndrome through MAS-related G protein–coupled receptor D. Sci Signal 11:eaal2171

59. Ferri GL et al (1988) Proenkephalin A-derived peptides in the human gut. Gastroenterology 95:1011–1017
60. Porcher C et al (1999) Opioid peptide mRNA expression in the colon of the rat. Neurosci Lett 272:111–114
61. Denning GM et al (2008) Proenkephalin expression and enkephalin release are widely observed in non-neuronal tissues. Peptides 29:83–92
62. Boué J et al (2014) Endogenous regulation of visceral pain via production of opioids by colitogenic CD4(+) T cells in mice. Gastroenterology 146:166–175
63. Han S-K et al (2002) Orphan G protein-coupled receptors MrgA1 and MrgC11 are distinctively activated by RF-amide-related peptides through the Galpha q/11 pathway. Proc Natl Acad Sci U S A 99:14740–14745
64. Yang H-YT et al (2008) Modulatory role of neuropeptide FF system in nociception and opiate analgesia. Neuropeptides 42:1–18
65. Abdellah N et al (2018) Neuropeptide AF induces piecemeal degranulation in murine mucosal mast cells: a new mediator in neuro-immune communication in the intestinal lamina propria? Anat Rec 301:1103–1114
66. Coelho A-M et al (2002) Proteinases and proteinase-activated receptor 2: a possible role to promote visceral hyperalgesia in rats. Gastroenterology 122:1035–1047
67. Cattaruzza F et al (2011) Cathepsin S is activated during colitis and causes visceral hyperalgesia by a PAR2-dependent mechanism in mice. Gastroenterology 141:1864–1874, e3
68. Annaházi A et al (2013) Luminal cysteine-proteases degrade colonic tight junction structure and are responsible for abdominal pain in constipation-predominant IBS. Am J Gastroenterol 108:1322–1331
69. Tulic MK et al (2016) Presence of commensal house dust mite allergen in human gastrointestinal tract: a potential contributor to intestinal barrier dysfunction. Gut 65:757–766
70. Plum T et al (2020) Human mast cell proteome reveals unique lineage, putative functions, and structural basis for cell ablation. Immunity 52:404–416, e5
71. Aguilera-Lizarraga J et al (2019) Expression of immune-related genes in rectum and colon *descendens* of Irritable Bowel Syndrome patients is unrelated to clinical symptoms. Neurogastroenterol Motil 31:e13579

26 Analysis of Intestinal Movements with Spatiotemporal Maps: Beyond Anatomy and Physiology

Marcello Costa, Luke Wiklendt, Tim Hibberd, Phil Dinning, Nick J. Spencer, and Simon Brookes

Abstract

Over 150 years ago, methods for quantitative analysis of gastrointestinal motor patterns first appeared. Graphic representations of physiological variables were recorded with the kymograph after the mid-1800s. Changes in force or length of intestinal muscles could be quantified, however most recordings were limited to a single point along the digestive tract.

In parallel, photography and cinematography with X-Rays visualised changes in intestinal shape, but were hard to quantify. More recently, the ability to record physiological events at many sites along the gut in combination with computer processing allowed construction of spatiotemporal maps. These included diameter maps (DMaps), constructed from video recordings of intestinal movements and pressure maps (PMaps), constructed using data from high-resolution manometry catheters. Combining different kinds of spatiotemporal maps revealed additional details about gut wall status, including compliance, which relates forces to changes in length. Plotting compliance values along the intestine enabled combined DPMaps to be constructed, which can distinguish active contractions and relaxations from passive changes. From combinations of spatiotemporal maps, it is possible to deduce the role of enteric circuits and pacemaker cells in the generation of complex motor patterns. Development and application of spatiotemporal methods to normal and abnormal motor patterns in animals and humans is ongoing, with further technical improvements arising from their combination with impedance manometry, magnetic resonance imaging, electrophysiology, and ultrasonography.

Keywords

Physiological traces · Gut motility · Mechanical recording · Manometry · Cinefluoroscopy · Cineradiography · Myoelectrical activity · Calcium imaging · Transit times · Intraluminal flow · Propulsion

26.1 Introduction

The digestive tract is a long organ, comprising multiple anatomically distinct regions, each with distinctive motility patterns. Variations in diet (herbivore, omnivore, carnivore) correlate with anatomical and functional variation in gut motil-

M. Costa (✉) · T. Hibberd · N. J. Spencer · S. Brookes
College of Medicine and Public Health, Department of Human Physiology, Flinders University, Bedford Park, SA, Australia
e-mail: marcello.costa@flinders.edu.au

L. Wiklendt · P. Dinning
Department of Gastroenterology and Surgery, Flinders Medical Centre, Bedford Park, SA, Australia

N. J. Spencer et al. (eds.), *The Enteric Nervous System II*, Advances in Experimental Medicine and Biology 1383, https://doi.org/10.1007/978-3-031-05843-1_26

ity between species. There remains much uncertainty about the types of behaviour of the digestive tract. Understanding them requires insight into the nature and origin of forces (kinetics) in the gut and their relations with actual movements (kinematics or 'geometry of motion') that cause propulsion of digestive content. However, these parameters are seldom recorded/viewed simultaneously.

This brief review summarises the development of methods for recording gastrointestinal movements and the forces that drive them. In early studies, these factors were recorded separately, either as images, which are highly accurate in space but frozen in time (spatial records), or as a graphical record of localised intestinal events (e.g. pressure), which are accurate in time but have restricted spatial resolution. The advent of the graphical physiological recording technique led to the development of modern physiology and its clear separation from anatomy.

Over last century new methods have increased the temporal resolution of changes in the shape of the gut; photography and cinematography (kinematics) culminated in the construction of diameter maps (DMaps). In parallel, the number and spatial resolution of force/pressure sensors incorporated into catheters have improved significantly (i.e. kinetics). Interpolation between multiple recording sites enables the construction of intraluminal pressure maps (PMaps).

Recently DMap and PMaps have been combined (DPMaps). These composite maps provide a fuller spatiotemporal representation of intestinal events, unifying the historical disciplines of anatomy and physiology into four-dimensional space. In recent years, algorithms have been developed to represent changes in three-dimensional features whose shape changes over time [8, 183], including examples from gastrointestinal motility [79]. The current ability to differentiate muscle-mechanical states from compliance measurements in DPMaps has provided a window into the activity status of enteric motor circuits and pacemaker networks. Adding recording of intraluminal flow to DPMaps would achieve a long-awaited unification of movement, force and propulsion in gastrointestinal motility studies.

26.2 The Origin of Graphic Representation in Physiology

Historically the studies of organ structures (anatomy) preceded studies of their functions (physiology). It can be argued that anatomy developed in parallel to the realistic representation of space by artists in the Renaissance. In contrast, physiological phenomena reflect physical events changing over time; this appeared in conjunction with a cognitive revolution of 'visual thinking', in the form of maps, scatter plots and eventually graphic traces of physical events [71]. The physiological revolution revealed physical phenomena by making them visualisable. This led to graphic representation of events over time. Perhaps this development began in 1665 with the invention of the barometer. In 1684, Robert Plot used barometers to display daily changes in barometric pressure in a graph he called a 'History of the Weather'.

Friendly and Wainer [71] argue that visual thinking has had a profound effect on the way we approach problems, by revealing elements that would otherwise remain invisible. The first field of knowledge to be advantaged by graphic representation was geographical maps. In the 1620s, the Dutch cartographer Michael Florent van Langren devised what may be the first statistical graph in history. This took place in the 'Age of Discovery', when Europeans were concerned with the measurement of time, distance and location [77]. Time-related changes were plotted graphically only more recently. For example, William Playfair [147] can rightly be called the father of modern graphical methods. He described historical periods with showing the relative strength of each civilisation changing over time. Such graphics were called 'ridgeline' plots because they resembled a set of mountain ridges [71]. However, in the fields of experimental science, up until the mid-1800s, most temporal findings were based on verbal descriptions or were recorded as tables of values at various time points, as in a modern spreadsheet.

26.3 The Kymograph and the Birth of Modern Physiology

At the beginning of the nineteen century, there was an intellectual rebellion against the 'philosophy of nature' (Naturphilosophie) which assumed a 'vital force' to explain biological phenomena [13]. This ideological change was paralleled by increasing emphasis on objective, measurable events that could be shared between observers [22, 87, 167].

As far back as 1798, the French had been occupied with problems of dynamometry; examples include the study of human muscular strength, the force required to pull carriages and other loads, the tensile strength of threads, etc. In 1828, Didion invented a dynamometer, patterned on Prony's brake dynamometer. This instrument made it possible to measure the force exerted by a horse in motion. In the early 1830s, Morin added a disk or roll of paper which was driven either by the movement of the vehicle, or by a clockwork mechanism, and thus recorded distance or time. This paper was written on by a stylus mounted on the dynamometer, which traced excursions varying with time or distance. Morin is therefore both the immediate progenitor of Pouillet in the development of the chronometric disk and the great exponent of graphic representation in physics and industry. Nevertheless, it was the work of Pouillet that seems to have been the immediate stimulus for the defining work of Ludwig [87]. The Italian physicist Matteucci reported the generation of an electrical current during muscle contraction in a monograph published in 1842 [130]. He developed a method of graphic recording with a rotating drum, but his impact in the history of visualising physiological phenomena remains minor [87].

A breakthrough came with Carl Ludwig who in 1847 invented the Kymograph, that is, the rotating smoked drum ([124]; Fig. 26.1). This was the first method to transform visual observations and numerical measurements into graphical records. Kymographic images represent the temporal domain in a single spatial dimension, freezing and storing the event as a continuous trace that allows closer, thoughtful examination of the 'character and extent' of the event. Ludwig was probably the first influential scientist to clearly separate physiology from anatomy [13].

Other physiologists quickly recognised the tremendous potential of Ludwig's method. For example, du Bois-Reymond, a close colleague of Ludwig, commented that '*the dependence of the effects upon each condition is now presented in the form of a curve, whose exact law, to be sure, remains unknown, but whose general character one will most often be able to trace. It will almost always be possible to determine whether the function grows or diminishes with the variable investigated. In other cases one may be able to discover distinctive points of the curves, the sense of its bending with respect of the abscissa. Whether approaches asymptotically a constant value…*' [62]. Volkmann added: '*After Ludwig invented an instrument that permitted one to represent the variable forces of the heart through curves, it immediately suggested making other motive forces also visualisable and measurable…*' [177].

This new method marked the beginning of accurate recording of physiological parameters as physical events. It also enabled observations to be shared and facilitated quantitative, objective analysis of the traces. This approach allowed for the visual observation of phenomena occurring over time, as geometrical traces, rather than relying on descriptive observations. In the 1850s, Helmholtz was one of the first to recognise the importance of this method, when he established for the first time the speed of propagation of action potentials. He achieved this by recording the latency of isolated striated muscle contractions in response to electrical stimulation [141, 178, 179]. A few decades later Etienne-Jules Marey [128, 129], a medical student in Paris, modified a device to record the external blood pulses. He continued to develop devices that used the 'graphic method' or 'wave writer' including

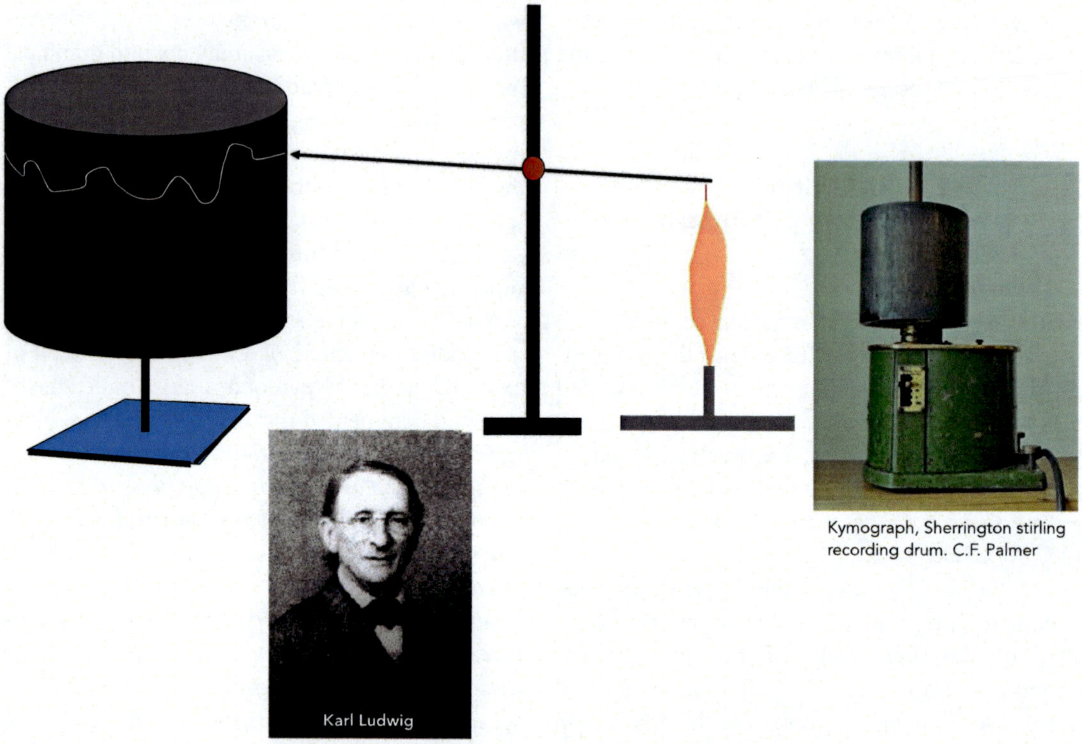

Fig. 26.1 Diagram of the smoked rotating drum (kymograph) invented by Karl Ludwig to record heart muscle and striated muscle contractions. On the right one of the earliest commercial Kymograph machines

cardiograph, polygraph, pneumography, myograph, thermograph, electrometer and plethysmograph. He developed the first force transducers and a horizontal rotating tambour to study the heartbeat and its electrical activity, blood pressure, breathing movements, micturition and several types of animal and human locomotion [23, 128, 129]. Langendorff [112] further developed these ideas, formalising the traces recorded by kymographs.

The graphic representation of physical events, particularly in the discipline of meteorology, extended the perception of observers into previously unobservable phenomena. These included thermometry, barometry, hygrometry and the measurement of rainfall and wind [23]. In 1869, the London physiologist William Rutherford championed the pedagogical progress afforded by these 'instruments of precision', explaining to his introductory physiology class that '*we no longer estimate the force of the heart's action by merely feeling the pulse, or by observing the distance to which blood is projected from a divided artery [...] movements are recorded on revolving cylinders or flat surfaces, so that a tracing, or writing, indicating the character and extent of the motion, may be preserved*'.

Sir Charles Scott Sherrington and Ernest Henry Starling developed further versions of the smoked drum. William Porter, a physiology instructor at Harvard, was the founder of the company that manufactures kymograph: Harvard Apparatus. Charles Fielding Palmer, an English mechanical engineer and bicycle maker in 1891 in London, was an independent manufacturer of scientific instruments from 1932, mostly in the field of physiology including modern kymographs. Since 1987 the Palmer Company has been a subsidiary of Harvard Apparatus.

26.4 First Recording of Intestinal Mechanical Events Using the Kymograph

Since the mid-1800s mechanical physiological events were recorded with kymographs, either as changes in length of muscles or as forces associated with their contractions. However, in the second half of the nineteenth century, the movements of the intestine were still being described verbally, albeit with great observational insight [125, 159]. In France, Legros and Onimus [115] made the first graphical recordings of intestinal motility with a kymograph. A fluid-filled balloon was connected to a magnifying tambour, from which a lever wrote on a rotating drum, to record intraluminal pressure from the rabbit small intestine. These recordings showed rhythmic pressure peaks and, in-parallel, independent respiratory movements (Fig. 26.2). The authors commented appropriately that the graphic traces '*also have the advantage of making the results of the observation indisputable and of putting the sincerity or the illusions of the observer out of question*'.

After these initial recordings, there was surprising gap before the kymograph was used again to measure activity of the digestive tract. In 1883, the German physiologist, Karl Hugo Kronecker, and his student, Samuel James Meltzer [134], used the technique to record swallowing movements. The next important step occurred at the turn of the century when Bayliss, a young medical student and Starling, his professor of physiology [14] developed a recording method to measure forces of the longitudinal muscle, called the enterography (Fig. 26.3). This method was used in the dog small intestine in parallel with an inflatable balloon to record intraluminal intestinal pressure, leading to their classic paper describing the forces of both longitudinal and circle muscle layers simultaneously.

By the beginning of the twentieth century, the kymograph had become a critical tool for most of studies in gastrointestinal motility. However, its graphic measurements were generally limited to a single recording point, thus failing to capture the spatial extent of motor patterns. It slowly became apparent that it is difficult to build a 'picture' of complex movements distributed over a length, from a single physiological trace. The twentieth century saw the beginning of recording from more than one point; the effect of this is discussed below.

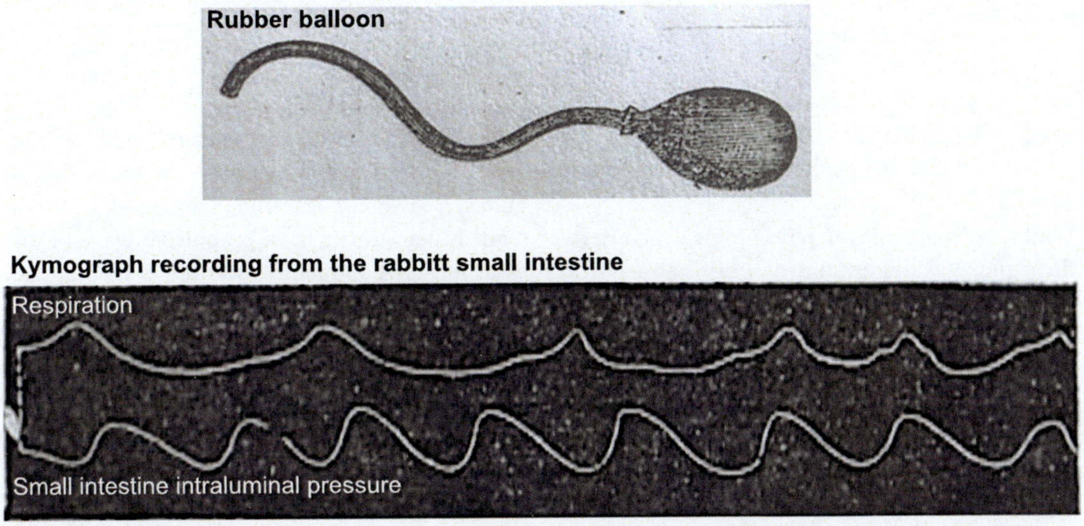

Fig. 26.2 First recording of intestinal motility with the kymograph performed by Legros and Onimus [115]. A rubber balloon was inserted into the rabbit small intestine. The traces show both the intraluminal pressure (bottom trace) and respiration (top trace)

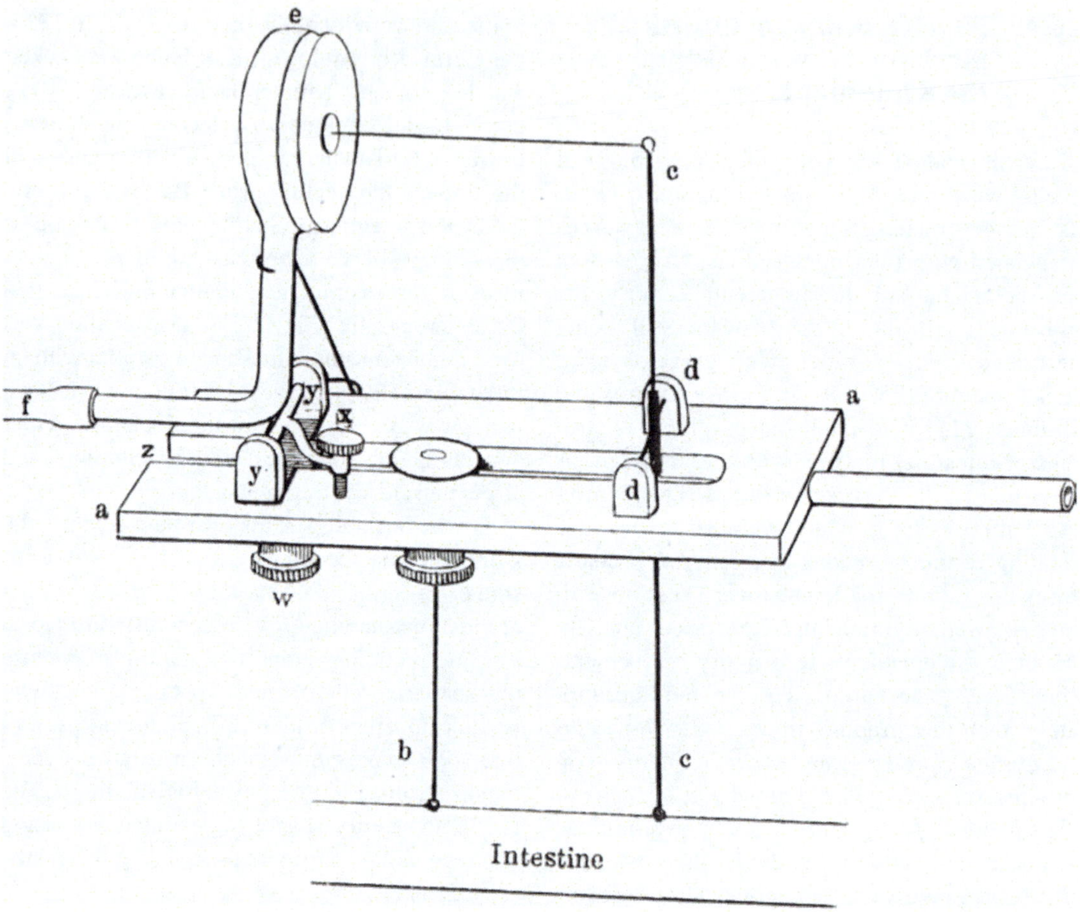

Fig. 26.3 The enterograph developed by Bayliss and Starling [14] to record mechanical forces of the intestinal longitudinal muscle. The moving lever c pivoting around d acted on a Marey's tambour, measuring isometric force exerted on the longitudinal muscle between b and c

26.5 Isolated Preparations of the Intestine

While the main aim of physiological investigations into gastrointestinal motility has been to ascertain the nature of motor patterns in the live humans or animals, an important development was the study of excised segments of intestine kept alive in organ baths. The initial attempt to keep organs alive ex vivo was developed by Carl Ludwig in the mid-1800s [13]. The first study of isolated segments of intestine was by Rudolf Magnus in 1904 [126, 127]. However, Trendelenburg's [176] development of a method to study isolated intestinal preparations was very influential. He set up segments of intestine from several experimental animals, in particular of guinea pig small intestine, and gradually distended them to trigger a propulsive wave of circular muscle contractions starting at its oral end, which emptied the content aborally [90, 176]. This method became a standard tool to investigate the neural and mechanical events underlying intestinal propulsion. Over 200 publications on guinea pig peristalsis are based on this method. In parallel, small longitudinal strips of guinea pig small intestine have been widely used to study neuromuscular transmission from enteric motor neurons and to test pharmacological agents. There have been more than 400 publications since Paton [144] developed this simple method to study longitudinal muscle contractions elicited by electrical stimulation of enteric nerves.

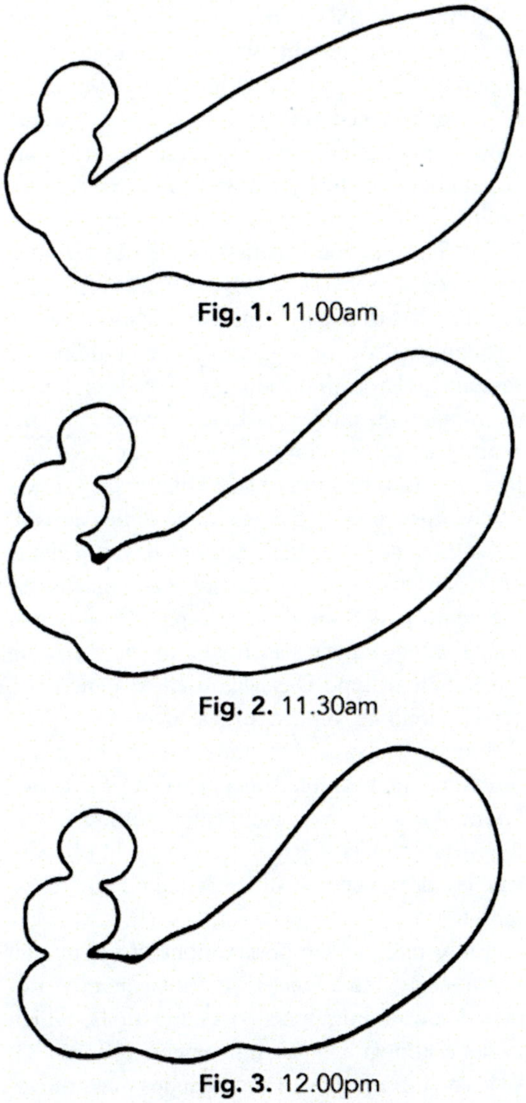

Fig. 26.4 Sequential outlines of the stomach of a dog filled with barium and visualised by X-rays by Cannon in 1898, revealing indentations of the antral muscle resulting in antral peristaltic movements

26.6 X-Rays in Gastroenterology at the Turn of the Twentieth Century

Another momentous development to view and record movements of the digestive tract in vivo was driven by a first-year student of medicine at Harvard named Walter B. Cannon. Just a year after the discovery of X-rays by Roentgen in December 1895, Cannon and a fellow student Albert Moser used them to study the movements of the digestive tract. He started by giving '*a frog a gelatin capsule filled with this bismuth salt, and after seeing clearly the shadow of the substance in the frog's stomach, repeated, the next day the use of the salt in a capsule to show deglutition in the dog*'. Then he tested the radio opaque medium in a goose to visualise oesophageal peristalsis. They communicated their findings to '*the meeting of the American Physiological Society in Boston, Dec. 29, 1896, the phenomena of deglutition as exhibited by the goose when swallowing capsules containing bismuth subnitrate was informally demonstrated to the members by means of the Roentgen rays. This was, I think, the first public demonstration of movements of the alimentary tract by use of the new method*' [34]. Cannon and Moser also pioneered the use of radio opaque material to study gastrointestinal movements in humans. They started with a girl of 7 and '*saw clearly bread and milk mush mixed with the salt, and also a suspension of the salt in water, passing down the esophagus*' [35].

The movements of the cat stomach were verbally described [35] and in addition '*...outlines were made by tracing the form of the stomach on tissue-paper laid on the fluorescent screen*'. From these silhouettes, it was possible to identify the sequential progression of muscle contractions (Fig. 26.4). In his characteristic humility, Cannon in 1914 concluded that '*... there is little warrant for any man to claim for himself or for any other the credit of inventing the method now used everywhere in Roentgen examinations of the alimentary canal*'. The advantages of this new method of visualising movements of the digestive tract were significant [20] and despite difficulty quantifying geometrical outlines of the digestive tube allowed Cannon in his monograph of 1911 [33] to describe in detail the major motor patterns of the digestive tract based on X-rays.

In the late 1800s, the advent of permanent recording of gut motility with photography, chronophotography, cinematography and cineradiography established the complexity of intestinal motor behaviour. However, these observations could not be readily quantified. X-rays were also used to produce what was called X-ray kymograph. This method was first applied by a Polish physiologist, Sabat, in 1911. An image of a moving object formed by emitted radiation passes through a narrow transparent slit in a radiopaque screen. Underneath the slit, a surface sensitive to the radiation is moved, or in some cases, the slit-bearing screen is moved over the sensitive surface. The resulting image (the kymograph) shows a wave form of the changes in the dimensions of the silhouette. By 1928, Stumpf developed the multi-slit method, which allowed beams of X-rays to pass through several parallel slits in a lead grating, forming an image of the organ on the sensitive film. This was used to generate a graphic outline of the organ over time. The method was first used to study movements of the heart and lung (quoted in [149]). As far as we know, the only work using this method in the digestive tract is a paper by Popov [148] who studied movements of stomach and rectum of the rabbit.

Early radiological observations of the human colon showed that faeces were infrequently propelled towards the anus by contractions, which became known as mass movements [10, 36, 83, 91]. The infrequency of spontaneous mass movements and increasing knowledge of the dangers of radiation (e.g. [131]) meant that these detailed studies have never been repeated. As an alternative, time-lapse cinefluorography was developed, with single images being taken at >1 min intervals [153, 154]. While this provided clear images of the gut, the long intervals between images meant that significant components of the motility events were missed. In addition, standard radiological contrast media were later shown to affect the speed of transit [3]. X-ray studies have been used in animals in vivo, and because prolonged X-ray exposure is feasible, these studies have provided a wealth of information. Tasaka and Farrar [174] used tantalum markers over opposite sites on the gut and plotted graphically the distances both in the circumferential and longitudinal directions. More recently in the early 2000s, two papers were published, detailing the relationships between flow recorded by x-ray and muscle contractions recorded by serosal strain gauges in pigs [85] and sheep [15].

26.7 Direct Visual Recording In Vivo

Zondek [188] studied colonic motility visually after implanting a transparent celluloid window in the abdominal wall of rabbits. A few years later, Fleisch and Wyss [69] filmed intestinal movements in the open abdomen of anaesthetised guinea pigs and responses to pinching. From black marks distributed along the intestine, they manually plotted the shortening of the longitudinal muscle between each mark, creating the first graphic representation of the movements of intestinal longitudinal muscle. Subsequently, Hukuhara adapted Zondek's [188] method to make movies of the guinea pig and rabbit intestine through a celluloid window, thus revealing sequential segmental movements ([92]; Fig. 26.5). As found by Cannon, the sequential progression of contractions could be identified, but the images were very difficult to quantify.

26.8 Spatiotemporal Maps of Changes in Diameter (DMaps) and Length from Videos

A next step in the development of graphical spatiotemporal mapping of motor patterns traced time-lapse silhouettes extracted from video images of intestinal movements [182]. This study identified some underlying neural mechanisms of peristalsis, but the manual analysis was excessively time-consuming and poorly quantifiable. In the late 1990s, three laboratories independently developed comparable methods to graphically record the changes in diameter or wall density along segments of intestine over time.

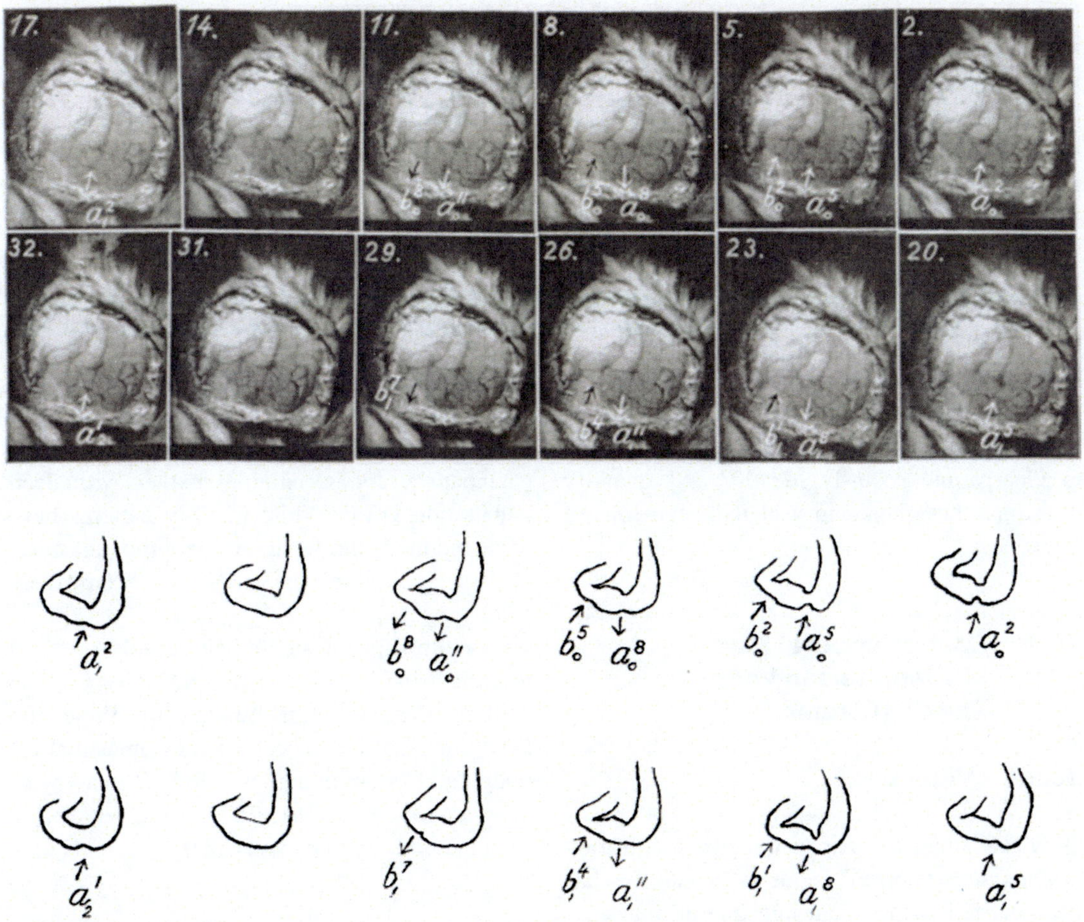

Fig. 26.5 Top: Sequence of cinema frames taken through a transparent window inserted in the abdominal wall of a guinea pig; bottom manually drawn outlines of the small intestine showing travelling of circular muscle contractions. (From [92])

Isolated specimens of gut were video recorded in an organ bath and processed to convert each frame to a spatiotemporal diameter map (DMaps). These maps represented all gut moments over a given period of time [16, 17, 80]. A sequence of photographs of isolated segments of opossum duodenum, with surface markers, enabled quantitation of longitudinal motions [132].

In the laboratory of Marcello Costa at Flinders University, Grant Hennig developed software to construct DMaps, with several substantial updates after he moved to the United States. At Flinders, Lukasz Wiklendt later developed real-time DMaps, allowing motility patterns to be observed during experimental investigation. Other groups around the world also developed versions of this software; John Furness in Melbourne [68] and Joel Bornstein also in Melbourne [171] and Roger Lentle in New Zealand [97, 117] developed similar methods to construct DMaps. Commercial software (gastrointestinal motility monitor (GIM)) and hardware were developed by Catamount, in Vermont, USA [88], and a model for interpreting DMaps was developed by Lammers and Cheng [105].

Spatiotemporal maps allowed investigators to calculate contraction frequency, origin, speed, and direction of propagation, allowing for the first detailed quantification of gut motility patterns. The technique is now used by researchers around the world to describe gastrointestinal motor patterns in a variety of different regions of gut in several species, ex vivo and in vivo. These include the *guinea stomach* [19], *rat stomach* [119], *guinea pig, mouse and rat small intestine* [1, 75, 76, 80, 120, 155, 156], *guinea pig colon* [51, 100, 135], *rabbit colon* [58, 118] and *mouse colon* [135, 155, 156] and *chicken large intestine* [98]. DMaps have been used to directly compare gut motility of different species including rabbit, guinea pig, rat and mouse [43, 44]. The method has also been used to study fish intestinal motility [152]. These studies indicate that DMaps have now become an accepted scientific methodology to clearly and robustly identify and quantify intestinal movements in isolated segments of intestine.

26.9 Spatiotemporal Maps of Diameters In Vivo: The Challenges

26.9.1 Videos

Only few attempts have been made to construct DMaps in exteriorised segments of gut; this has been carried out in small intestine of anaesthetised rat [21, 68] In anaesthetised mouse Hennig et al. [82] constructed spatiotemporal maps of changes in diameter and longitudinal muscle lengths in exteriorised loops of ileum and related them to calcium imaging.

26.9.2 Fluoroscopy

DMaps of mouse and rat small intestine were successfully recorded from fluoroscopy [18, 53]. Wang et al. [180] used videofluoroscopy and extracted DMaps to visualise myogenic motor patterns across the pyloric junction in rats and mice in vivo. More recently, spatiotemporal maps constructed from fluoroscopy have been used to investigate intestinal motility in conscious rats [123, 150]. In humans, short bouts of fluoroscopy recorded oesophageal motility during swallowing of a radiopaque bolus and were used to construct spatiotemporal maps [116].

26.9.3 Ultrasonography: Photoacoustic Imaging (PA)

Non-invasive anatomical and functional ultrasonography also called photoacoustic imaging is a recent addition to record intestinal motility [54, 56]. Imaging of the intestine is based on 20 μm micelles labelled with naphthalocyanine dyes ('nanoaps'), a non-ionizing marker with deep penetration [186]. While motility patterns have been reported, the technique requires considerable specialised training and has not been widely taken up yet.

A more sophisticated albeit complex method to extract the outlines of the gut in vivo is the method of Trans-illumination Intestine Projection (TIP) imaging. It has been used for intestinal 2D imaging of free-moving mice and 3D imaging of anesthetised mice [181, 187]. Time will tell if such methods will be used widely.

26.9.4 Magnetic Resonance Imaging (MRI)

Dynamic MRI has been used to record changes in the outline of the gastrointestinal tract [54]. Using the balanced steady-state free precession (bSSFP) gradient echo sequence, Cine MRI was developed in the cardiac field but has also been used to study the gut [86]. An excellent use of MRI to construct diameter maps of stomach was provided by Menys et al. [133]. Small bowel motility was successfully recorded using single slice cine MRI [101]. In clinical practice, defecography (evacuation proctography) can be conducted with fluoroscopy or MRI and is used to diagnose defecatory and pelvic floor dysfunction [102]. MRI has also been used to record intestinal motility in small laboratory animals [2].

26.10 Spatiotemporal Maps of Changes in Forces (PMaps)

26.10.1 Myoelectrical Activity and Its Mechanical Equivalence

By mid-way through the twentieth century, studies were under way to establish the relationship between electrical activity and contractions of the intestine [4, 25, 26, 32, 52]. Unlike ex vivo recordings, recording electrical activity was easier than recording mechanical activity in vivo. With advances in technology, it became possible to record electrical activity from multiple closely spaced electrodes. This increasingly revealed the spatiotemporal complexity of motor patterns, including the interdigestive myoelectric migrating motor complex in the small intestine of many species [40, 63, 74, 172]. Not surprisingly, propagating electrical activity, in particular spikes that generate contractions, correlated positively with flow of contents [31].

High spatiotemporal resolution recording of slow waves and their association to spike activity of the intestine was achieved by Lammers and collaborators. They used multi-electrode plates applied to the serosal surface of the intestine ex vivo and in exposed loops of intestine under anaesthesia. These recordings generated ‘activation maps’ which displayed the origin and spread of slow waves ([106, 139]; Fig. 26.6). This method has been used to map similar electrical events in a variety of animal models including mouse, rat, rabbit and cat [106–108, 110, 111]. Multisite electromyography has distinguished the ‘ripples’ generated by myogenic slow waves, from neurally mediated peristaltic contractions [109]. A group in New Zealand has further developed multi-electrode plates, with highly flexible sheets that can be placed directly upon an organ [138]. With this technique the group recorded slow wave activity directly from the stomach of anaesthetised human patients during surgical procedures [138].

26.10.2 Calcium Waves in Smooth Muscle and Pacemaker Cells

Recording electrical firing activity from the intestinal muscle generally reflects the contractility of the muscle. Smooth muscle action potentials and pacemaker cell currents feature calcium ion flux across their cell membranes and from intracellular stores. Accordingly, transient changes in intracellular calcium are also used as a proxy for muscle contractility [173]. Indeed, calcium activity has been used to describe the spatiotemporal features of contractility in mouse and guinea pig intestine [81, 82, 170]. Spatiotemporal maps based on calcium imaging were used to visualise patterns of activity of the pacemaker cells in isolated segments of mouse small intestine [143].

26.10.3 Recording Forces of Contractions

26.10.3.1 Intraluminal Balloons

Since the first recordings of intraluminal pressure by Legros and Onimus [115], Luderitz and Hess (see [159]), Bayliss and Starling [14] and Langley and Magnus [113], balloons have been extensively used to apply local distension, to record polarised responses, to measure intraluminal pressure and to observe their propulsion. Studies using fluid or air-filled balloons to record pressure have typically been limited to one or two locations and, therefore, detailed spatiotemporal maps of intestinal motor patterns have not been constructed. Sets of tandem balloons (six balloons) were used in dog colon to generate a spatiotemporal graphic representations of motor patterns plotted as active and quiescent periods [24]. This method allowed the effect of enemas on colonic activity to be assessed.

26.10.3.2 Force Transducers

Costa and Furness applied two clips connected to force transducers to an isolated guinea pig colon to record the isometric contractions of the circular muscle during propulsion of artificial pellets [42]. Since then, multiple force transducers have

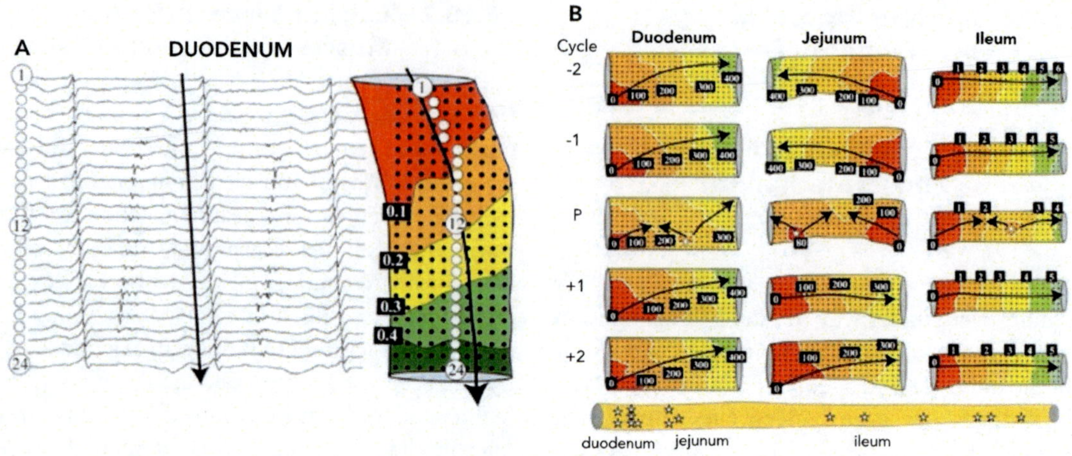

Fig. 26.6 Spatiotemporal activation map of slow wave isochrones in the dog small intestine recorded with a 340-electrode array showing changing patterns of propagation with prevalent directions. (From [111])

been applied to the mouse colon to reveal motor complexes [168, 169]. As for balloons, spatiotemporal maps cannot be constructed from this method of recording. In contrast, high-resolution fibre optic manometry catheters have multiple recording sensors which can be used as simple force transducers to generate a spatiotemporal maps. These have, for example, revealed neurally mediated clustered contractions in isolated ex vivo segments of human ileum [104].

26.10.3.3 Strain Gauges

Mechanical activity of the muscle of the intestine has also been recorded by using strain gauges attached to the serosal surface at multiple sites. Mostly used in dogs, this approach generated the first spatiotemporal maps of muscle force associated with motor patterns ([11, 95, 96, 151, 157, 165, 185]; Bass et al. [11]; also see studies by [15, 85] in section 5).

26.10.3.4 Multiple Manometry: The Birth of PMaps

Since the early use of one or two intraluminal balloons to record intraluminal pressure or forces, improvements in catheters have enabled intraluminal pressures to be recorded at multiple sites. Tasaka and Farrar [175] in fasted dogs developed spatiotemporal representations of intraluminal pressure, recorded at several points as propagated and simultaneous pressure waves.

Bassotti and Gaburri [12] used a commercially available eight-lumen PVC manometric catheter with an external diameter of 4.5 mm and an internal diameter of 0.8 mm for each lumen. The lumens terminated distally with side openings spaced 12 cm apart. More advanced intestinal manometry was developed by John Dent, based on perfused tubes with endings only 3 cm apart (Dentsleeve). Spatiotemporal pressure maps were constructed to visualise both oesophageal peristalsis [38] and antro-pyloro-duodenal propagating pressure waves ([5]; Fig. 26.7). With a 16-channel catheter, featuring side holes at 3 cm intervals, more accurate spatiotemporal depiction of propagating pressure peaks could be achieved. With similar methods, the spatiotemporal features of interdigestive myoelectric motor complexes were analysed [37]. Prolonged, multipoint perfusion manometry (with up to 17 sensors) of the ileocecal junction and unprepared colon provided improved spatial resolution of colonic motor patterns and better descriptions of colonic motility patterns [9, 55].

A comparison between water-perfused manometry catheters and solid-state catheters suggested that the latter better represent intralu-

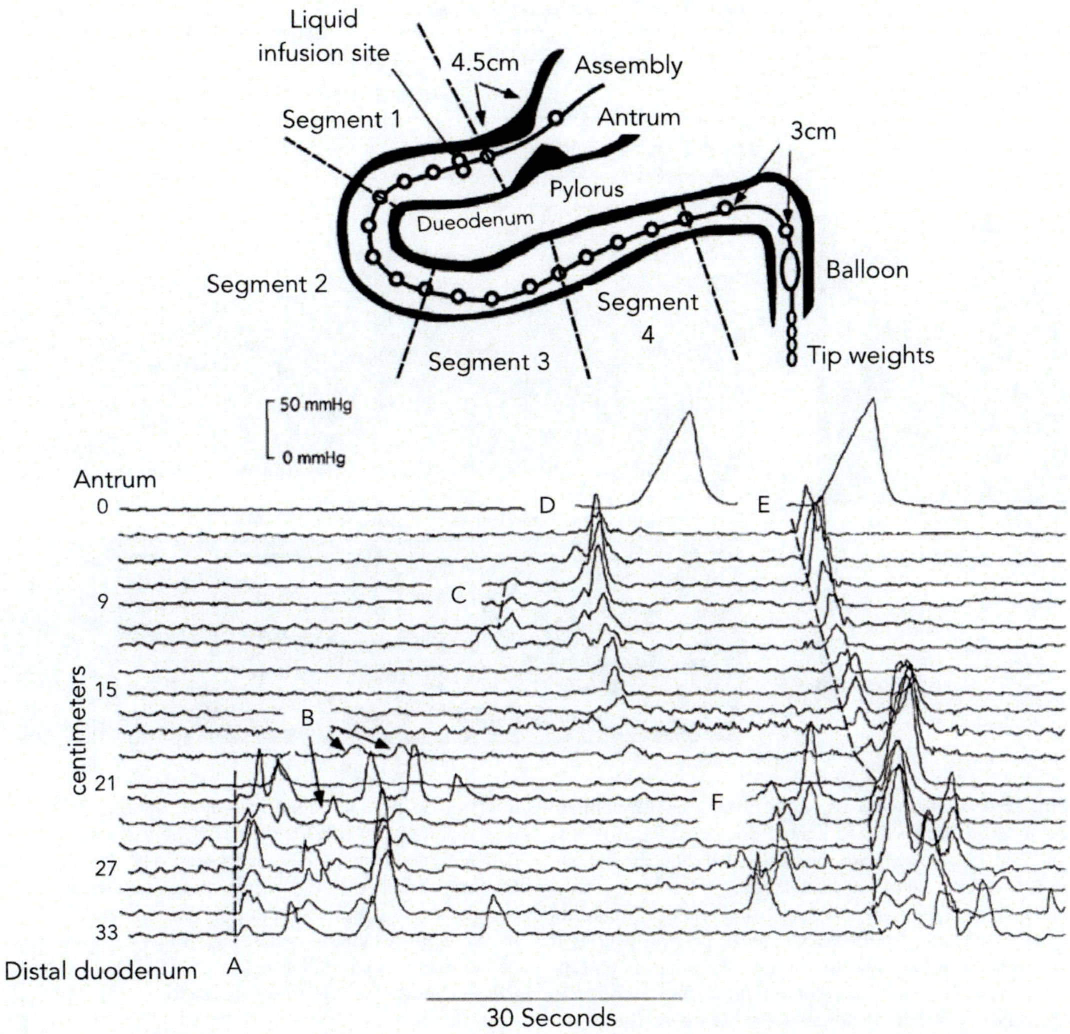

Fig. 26.7 Array of multiple intraluminal pressure with sensors in a human subject to record intraluminal pressures from the antro-pyloro-duodenal region. Right: traces of intraluminal pressures arranged to show the spatiotemporal organisation of propagating contractions. (From [5])

minal pressures [122]. However, early manometry probes based on perfused tubes or solid-state transducers often had distances between sensors of many centimetre, limiting their spatial resolution.

The advent of high-resolution fibre-optic manometry marked a substantial improvement in spatial discrimination because sensors were located 10 mm apart. In the human colon, up to 70 sensors could be used [6, 7, 54, 59], providing high-resolution motility recordings. High-resolution intraluminal pressure recording led to the identification of cyclic motor patterns as the most abundant physiological colonic motor patterns in humans (PMaps) ([60, 61]; Fig. 26.8). Using a miniature fibre-optic catheter enabled the extent of motor complexes to be recorded in high resolution along the entire small and large intestine of the mouse, in vitro. This revealed the functional

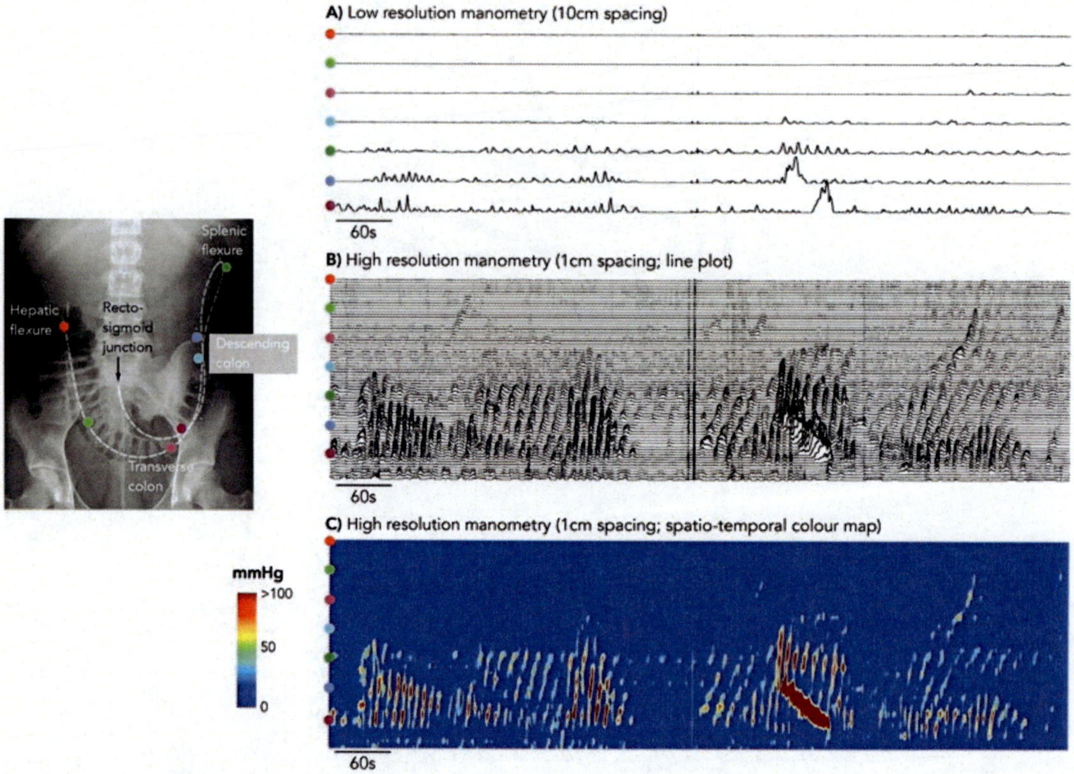

Fig. 26.8 Manometric recording from a healthy human colon. The X-ray image shows the positioning of the manometry catheter with the tip located in at the hepatic flexure. A section of the manometric recording is shown to the right of the X-ray. (**a**) shows the pressure recording from every tenth sensor effectively displaying data as it would be seen in a 'low-resolution' recording. Each sensor is associated with a coloured circle, and each circle corresponds to the location of that senor in the X-ray image. (**b**) shows the 'high-resolution' image of the same data displayed in (**a**); here the recording shows pressure data recorded from all 72 sensors spaced at 1 cm intervals. Unlike the image in (**a**), propagating contractions, their speed, direction and extent of propagation can all be seen in the high-resolution recording. (**c**) A spatiotemporal colour map of the intraluminal pressures (PMap) shown in (**b**). The motor patterns in (**c**) are readily recognizable and quantifiable

continuity of the underlying enteric circuits along the entire intestine, from the pylorus to the rectum [47–49].

26.11 Examples of Kinetic and Kinematic Recordings Combined

26.11.1 Earlier Combinations of Recording Methods

The process of combining kinetics and kinematics has been gradual in both in vivo and ex vivo experiments. By combining ultrasonography and water-perfused manometry and assuming that the oesophageal muscle itself is incompressible, the changes in dimensions of the muscle layers of the oesophageal wall could be ascertained. Thus, cross-sectional area of the muscle layer(s) could be related quantitatively to local shortening (or lengthening) of the same muscle layer(s) inferred from displacement between two clips [136]. The relationships between gastric wall motion and intraluminal pressure are believed to be major determinants of flow within the stomach and through the pylorus. Gastric antral wall motion and intraluminal pressures were monitored in five healthy subjects by concurrent antropyloroduodenal manometry and transabdominal ultrasound [93].

Schulze-Delrieu [162] studied how visual parameters change over the course of individual contractions in the isolated guinea pig ileum and how they relate to simultaneous recordings of luminal pressure and outflow. Simultaneous 'cineradiogram and kymogram' recordings of the stomach allowed the changes in shape to be correlated to intragastric pressure during antral peristalsis [167]. Combination of cinefluorographic, myoelectric and pressure recordings gave more accurate description of the mechanical events during interdigestive myoelectric motor complexes [41].

A few investigations using X-rays in humans occurred after 1960, including a correlation of X-ray imaging with manometry [70]. Oigaard and Dorph [140] measured spike potential frequency in combination with frequency, duration and amplitude of pressure waves and related these events to visible motor activity by synchronised cinefluorography. Ehrlein developed a powerful combination of cineradiography or cinefluoroscopy, with localised mechanical recordings [64, 65]. He and his colleagues applied this approach to describe in considerable detail the movements of the small and large intestine in several animal species. Colonic motility was also measured in unanaesthetised rabbits using strain-gauge transducers with simultaneous radiography. Representative still images of characteristic motor patterns were prepared manually, derived either from photographs or from video recordings [66]. Using closely spaced serosal strain gauge transducers and simultaneous fluoroscopy in dog small intestine, Schemann and Ehrlein [158] and Ehrlein et al. [67] demonstrated the spatial and temporal relationships of intestinal contractions responsible for the transport of luminal contents; stationary contractions underlay mixing activity while propagated contractions were associated with aboral transport. These studies provided major new insights.

Common cavity antroduodenal pressure difference ('pressure pump') was compared with propagating high-pressure waves in the human distal antrum ('peristaltic pump') by high-resolution manometry. This was recorded concurrently with time-resolved three-dimensional magnetic resonance imaging during intraduodenal nutrient infusion. Spatiotemporal maps of changes in internal diameter were correlated with pressure maps [94]. A catheter assembly including a pH electrode, pressure transducers and an ultrasound transducer was used to record events across the lower oesophageal sphincter. This showed that sustained oesophageal contractions during heartburn episodes were associated with acid reflux [145]. Imaging outlines of the gut wall were combined with electrical recordings in mouse-isolated intestine to show that distension elicited propagating slow waves with superimposed spikes rather that peristaltic contractions [163].

These studies provided detailed descriptions of the different patterns of motor activity which are incrementally refined with the ongoing development of combined recording methodologies.

26.11.2 Full Combination of Kinematics and Kinetics: DPMaps

Paradoxically, the classic monograph of Cannon entitled *the mechanical factors of digestion* [33] contains little quantitation of mechanics of either kinematics or kinetics. Weems outlined clearly the principles behind an appropriate mechanical physics of the intestine [184], stating that '*when only kinematics information is available, it is impossible to predict how certain previously unobserved changes in the environment surrounding the moving object will affect the motion*'. Conversely, '*the central problem of classic mechanics is to determine what will be the subsequent motion of an object having known characteristics after the object is placed in an environment that is completely described*'.

The length-tension and force-velocity relationship of smooth muscles are in principle similar to those extensively studied in striated muscles. A good start was the study of the length-tension and force-velocity relationship of the rabbit taenia coli by Gordon and Siegman [72]. The latter relationship followed Hill [84] relating the constant load, *P*, and under that load the speed of contraction, *V*, of a muscle in a tetanised state, with the equation

$$(P+a)(V+b)=(\text{Pmax}+a)b.$$

where constants *a*, *b* and Pmax were calculated experimentally.

The advent of high-resolution fibre-optic manometry made it feasible for the first time to compare PMaps with DMaps in isolated segments of rabbit small intestine, using video analysis in combination with high-resolution manometry [57]. Compliance is a measure of the change in cross-sectional area of the intestine (kinematics) with respect to change in internal pressure (kinetics) [73]. Changes in compliance reflect the effects of the neural and myogenic activity that generate motor behaviour. Temporally correlating PMaps and DMaps generated spatiotemporal maps that revealed both active and passive excitatory and inhibitory mechanical states of the muscle along an isolated segment of intestine [43, 44].

The application of DPMaps in isolated preparations of intestine in experimental animals gave an experimental demonstration of the idea of Bayliss and Starling [14] that propulsion of a bolus is due to sequential activation of enteric polarised neural pathways [60, 61]. Excitatory and inhibitory active states of the intestinal muscle during bolus propulsion were extracted using DPMaps and confirmed that the bolus is propelled by an active muscle contraction orally and preceded by an active muscle relaxation. These changes in muscle state are dynamically activated in a self-sustained neuromechanical loop [60, 61]. Later, it was shown that luminal impedance reflects internal diameter when the content is conductively homogeneous [43, 44]. This enabled the combination of pressure and diameter spatiotemporal maps to be extended to studies in conscious human subjects both for oesophageal [39, 142] and colonic motility [78]. The addition of electrical recording of muscle activity is giving further insights into the nature of propulsive and non-propulsive colonic motor patterns in both guinea pig and mouse colon ([46, 48, 50], Fig. 26.9).

26.11.3 Combining Kinetics, Kinematics Forces and Flow, Propulsion and Transit Times

A major aim of gastrointestinal motility studies is to relate forces, muscle movements and the resulting propulsion of intraluminal contents. Early studies with this aim recorded the force of circular muscle contractions during propulsion of solid boluses [14, 42, 113]. However, the nature of the content can range from fluid through viscous to solid; this has profound effects on propulsion and flow [45]. Few studies using spatiotemporal maps have directly studied the association of flow of contents and intestinal movements. Video, luminal pressures, volume outflow and wall movements of isolated segments of intestine have been recorded during peristaltic contractions in several experimental animals. Contractions generated characteristic luminal pressure transients that peaked as the contraction approached its maximum (see [159], section 4 and [27, 42, 176]). It should be noted that simple measurements of transit time over longer segments of the gastrointestinal tract in vivo do not reveal which motor patterns are responsible. Propulsion of contents in the human colon has been measured with radiopaque markers (shapes), radio-telemetric capsules, perspex capsules containing 51Cr, free 51Cr sodium chromate without radiological contrast media [89] or by magnetic pills [29].

Bueno et al. [31] investigated how myoelectrical activity relates to transit time and flow of intestinal contents, estimated by appearance of phenol red and by dilution of a marker continuously infused. The flow of contents was measured directly at a cannula in the proximal part of the jejunum and by an electromagnetic flow meter. This work established for the first time that most of propulsion in the small intestine occurs during 'irregular' activity in phase II of the interdigestive myoelectric motor complex. A similar study in conscious dogs used electromyography combined with recording outflow from intestinal loops

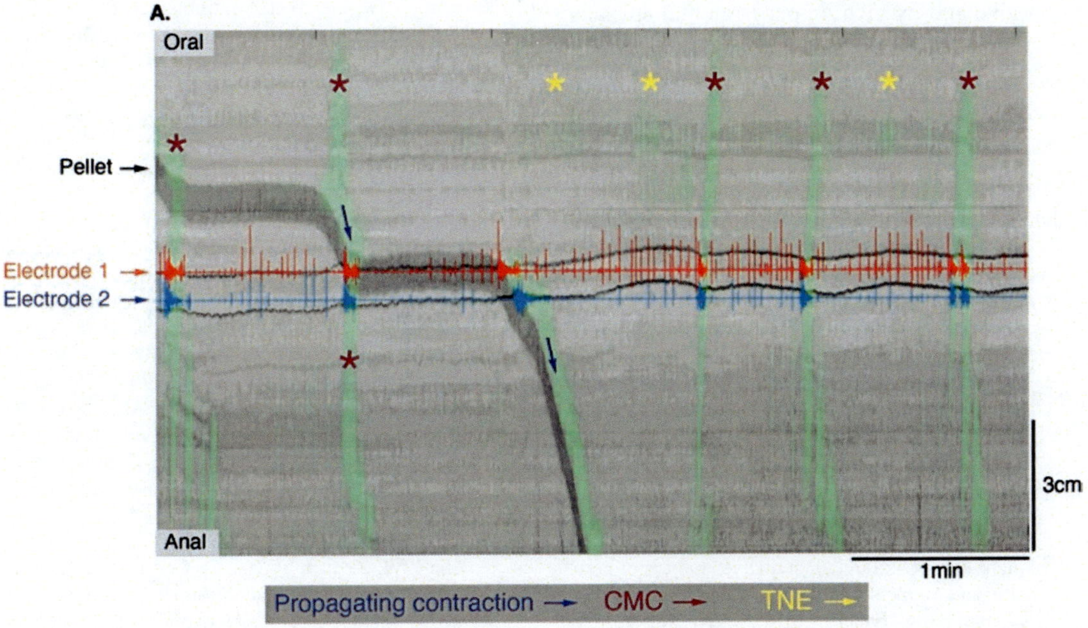

Fig. 26.9 Combined DPMap of an isolated segment of guinea pig distal colon with spatiotemporal distribution of pressures recorded with a fibre-optic catheter (green) over a grey scale DMap, during intermittent propulsion of an artificial pellet. In addition, two extracellular electrodes show bursts of action potentials associated with propulsive and non-propulsive motor complexes. (From [48])

[103]. A promising approach involved video recording isolated segment of guinea pig small intestine with intraluminal ink to identify flow [160]. In theory, this method could be combined with spatiotemporal mapping methods to establish the relationship between movements of the intestinal wall and movements of the intraluminal contents, during different motor patterns. Another interesting approach was developed by Seidl et al. [164] who used impedance manometry to measure bolus movement, combined with intraluminal pressure recordings in conscious human subjects. Several attempts have been made to model intestinal flow during propulsive activity [99, 114, 161]. However, the hydrodynamic analysis is fraught with difficulties [28, 121, 146]. One promising but little-explored method uses smoothed particle hydrodynamics modelling to estimate fluid movements [166]. Advanced models of the heart, which is a much simpler organ than the gastrointestinal tract, have established the numerous factors which need to be detailed, including the initial and boundary conditions and the need to compromise between complexity and computability [30, 137].

26.12 Concluding Remarks

This chapter reviews the long history of the studies that aim to comprehensively describe intestinal movements. The process started with improvements in visual representations of changes in the shape of the moving intestine (kinematics). In parallel, there were advances in the visual representation of forces, giving rise to a physiology of the gut (kinetics). The stepwise combinations of these two approaches eventually led to the current digital spatiotemporal graphic representation of gastrointestinal motor events. Composite spatiotemporal maps, involving measurement of 2 or more variables, can be supplemented with electrophysiological and flow measurements. These represent a hugely promising methodological framework to investigate both normal and abnormal gastrointestinal motility, in vivo and ex vivo, in experimental animals and in humans.

Acknowledgements This work was funded by ARC Project Grant DP190103628 to N Spencer and support from the National Health & Medical Research Council project grant no. 1156416 to N Spencer and the National Health & Medical Research Council project grant APP1162223 to P Dinning.

We thank Michael Schemann for his help in retrieving some of the papers published in German.

References

1. Abdu F, Hicks GA, Hennig G, Allen JP, Grundy D (2002) Somatostatin sst(2) receptors inhibit peristalsis in the rat and mouse jejunum. Am J Phys 282:G624–G633
2. Ailiani AC et al (2009) Quantitative analysis of peristaltic and segmental motion in vivo in the rat small intestine using dynamic MRI. Magn Reson Med 62:116–126
3. Alvarez WC, Freedlander BL (1924) The rate of progress of food residue through the bowel. J Am Med Assoc 83:576–580
4. Alvarez WC, Mahoney LJ (1924) Peristaltic rush and electroenterogram. Am J Phys 69(2):226–228
5. Andrews JM, Doran SM, Hebbard GS, Malbert CH, Horowitz M, Dent J (2001) Nutrient-induced spatial patterning of human duodenal motor function. Am J Phys 280:G501–G509
6. Arkwright JW, Underhill ID, Maunder SA, Blenman NG, Szczesniak MM, Wiklendt L, Lubowski DZ, Cook IJ, Dinning PG (2009) Design of a high-sensor count fibre optic manometry catheter for in-vivo colonic diagnostics. Opt Express 17:22423–22431
7. Arkwright J, Blenman N, Underhill I, Maunder S, Szczesniak M, Dinning P, Cook I (2009) In-vivo demonstration of a high resolution optical fiber manometry catheter for diagnosis of gastrointestinal motility disorders. Opt Express 17:4500–4508
8. Aydin B, Angryk RA (2018) Spatiotemporal frequent pattern mining from evolving region trajectories. Springer International Publishing, Cham: Imprint: Springer
9. Bampton P, Dinning P, Kennedy M, Lubowski D, Cook I (2001) Prolonged multi-point recording of colonic manometry in the unprepared human colon: providing insight into potentially relevant pressure wave parameters. Am J Gastroenterol 96:1838–1848
10. Barclay AE (1912) A note on the movements of the large intestine. Arch Roentg Ray 16:422–424
11. Bass P, Wiley JN (1965) Electrical and extraluminal contractile-force activity of the duodenum of the dog. Digest Dis Sci 10:183–200. https://doi.org/10.1007/BF02233747
12. Bassotti G, Gaburri M (1988) Manometric investigation of high amplitude propagated contractile activity of the human colon. Am J Phys 255:660–664
13. Bauereisen E (1962) Carl Ludwig as the founder of modern physiology. Physiologist 5(4):293–299
14. Bayliss WM, Starling EH (1899) The movements and innervation of the small intestine. J Physiol 24:99–143
15. Bedrich M, Ehrlein H (2001) Motor function of the large intestine and flow of digesta in sheep. Small Rumin Res 42:141–155
16. Benard T, Bouchoucha M, Dupres M, Cugnenc PH (1997) In vitro analysis of rat Intestinal wall movements at rest and during propagated contraction - a new method. Am J Physiol 36(4):G 776–G 784
17. Bercik P, Bouley L, Dutoit P, Blum AL, Kucera P (2000) Quantitative analysis of intestinal motor patterns: spatiotemporal organization of non-neural pacemaker sites in the rat ileum. Gastroenterology 119(2):386–394
18. Bercík P, Wang L, Verdú EF, Mao YK, Blennerhassett P, Khan WI, Kean I, Tougas G, Collins SM (2004) Visceral hyperalgesia and intestinal dysmotility in a mouse model of postinfective gut dysfunction. Gastroenterology 127:179–187
19. Berthoud H-R, Hennig G, Campbell M, Volaufova J, Costa M (2002) Video-based spatio-temporal maps for analysis of gastric motility in vitro: effects of vagal stimulation in guinea-pigs. Neurogastroenterol Motil 14(6):677–688
20. Bertrand PP, Bertrand RL (2007) Teaching basic gastrointestinal physiology using classic papers by Dr. Walter B. Cannon. Adv Physiol Educ 31:136–139. https://doi.org/10.1152/advan.00112.2006
21. Bogeski G, Shafton AD, Kitchener PD, Ferens DM, Furness JB (2005) A quantitative approach to recording peristaltic activity from segments of rat small intestine in vivo. Neurogastroenterol Motil 17:262–272. https://doi.org/10.1111/j.1365-2982.2004.00605.x
22. Borell M (1986) Extending the senses: the graphic method. Med Herit 2:114–112
23. Borell M (1987) Instrumentation and the rise of modern physiology. Sci Technol Stud 5(2):53–62
24. Borkon EL, Templeton RD (1937) The influence of oil enemas on colon motility in the dog. Am J Phys 118(4):775–779
25. Bozler E (1945) The action potentials of the stomach. Am J Phys 144(5):693–700
26. Bozler E (1946) The relation of the action potential to mechanical activity in intestinal muscle. Am J Physiol 146:496
27. Brann L, Wood JD (1976) Motility of the large intestine of piebald-lethal mice. Am J Dig Dis 21:633–640. https://doi.org/10.1007/BF01071956
28. Brasseur JG, Nicosia MA, Pal A, Miller LS (2007) Function of longitudinal vs circular muscle fibers in esophageal peristalsis, deduced with mathematical modeling. World J Gastroenterol 13:1335–1346

29. Brinck CE, Mark EB, Klinge MW, Ejerskov C, Sutter N, Schlageter V, Scott SM, Drewes AM, Krogh K (2020) Magnetic tracking of gastrointestinal motility. Physiol Meas 41:12TR0
30. Bryden J, Noble J (2006) Computational modelling, explicit mathematical treatments, and scientific explanation. In: Rocha LM, Yaeger LS, Bedau MA, Floreano D, Goldstone RL, Vespignani A (eds) Artificial life X: proceedings of the tenth international conference on artificial life. MIT Press, pp 520–526
31. Bueno L, Fioramonti J, Ruckebusch Y (1975) Rate of flow of digesta and electrical activity of the small intestine in dogs and sheep. J Physiol Lond 249:69–85
32. Bulbring E (1955) Correlation between membrane potential spike discharge and tension in smooth muscle. J Physiol 128:220–221
33. Cannon WB (1911) The mechanical factors of digestion. International Medical Monographs
34. Cannon WB (1914) Early use of the Roentgen ray in the study of the alimentary canal. JAMA 62(1):1–3
35. Cannon WB, Mosher A (1898) The movements of the food in the esophagus. Am J Phys 1:435–444
36. Case JT (1915) The x-ray investigation of the colon. Arch Roentgen Ray 19(375–387):187–188
37. Castedal M, Abrahamsson H (2001) High-resolution analysis of the duodenal interdigestive phase III in humans. Neurogastroenterol Motil 13:473–481
38. Clouse RE, Staiano A, Alrakawi A (2000) Topographic analysis of esophageal double-peaked waves. Gastroenterology 118:469–476
39. Cock C, Leibbrandt RE, Dinning PG, Costa MC, Wiklendt L, Omari TI (2020) Changes in specific esophageal neuro-mechanical wall states are associated with conscious awareness of a solid swallowed bolus in healthy subjects. AJP manuscript number GI-00235-2019. https://doi.org/10.1152/ajpgi.00235.2019
40. Code CF, Marlett JA (1975) The interdigestive myo-electric complex of the stomach and small bowel in dogs. J Physiol Lond 246:289–230
41. Code CF, Schlegel JF (1974) The gastrointestinal housekeeper. Motor correlates of the interdigestive myoelectric complex ofthe dog. In: Proc. Fourth Int. Symp. on gastro-intestinal motility. Mitchell Press Ltd, Vancouver, pp 631–634
42. Costa M, Furness JB (1976) The peristaltic reflex: an analysis of the nerve pathways and their pharmacology. Naunyn Schmiedeberg's Arch Pharmacol 294:47–60
43. Costa M, Dodds KN, Wiklendt L, Spencer NJ, Brookes SJH, Dinning PG (2013) Neurogenic and myogenic motor activity in the colon of the guinea-pig, mouse, rabbit and rat. Am J Physiol Gastrointest Liver Physiol 305:G749–G759
44. Costa M, Wiklendt L, Arkwright J, Spencer N, Omari T, Brookes S, Dinning PG (2013) An experimental method to identify neurogenic and myogenic active mechanical states of intestinal motility. Front Syst Neurosci. Hypothesis and Theory 7:1–18
45. Costa M, Wiklendt L, Simpson P, Spencer NJ, Brookes B, Dinning PG (2015) Neuromechanical factors involved in the formation and propulsion of fecal pellets in the guinea-pig colon. Neurogastroenterol Motil. https://doi.org/10.1111/nmo.12646
46. Costa M, Wiklendt L, Keightley L, Brookes SJH, Dinning PG, Spencer NJ (2017) New insights into neurogenic cyclic motor activity in the isolated guinea-pig colon. Neurogastroenterol Motil e13092:1–13. https://doi.org/10.1111/nmo.13092
47. Costa M, Hibberd TJ, Keightley LJ, Wiklendt L, Arkwright JW, Dinning PG, Brookes SJH, Spencer NJ (2019) Neural motor complexes propagate continuously along the full length of mouse small intestine and colon. Am J Phys. https://doi.org/10.1152/ajpgi.00185.2019
48. Costa M, Keightley LJ, Wiklendt L, Hibberd TJ, Arkwright JW, Omari T, Wattchow DA, Brookes SJ, Dinning PG, Spencer NJ (2019) Identification of multiple distinct neurogenic motor patterns that can occur simultaneously in the guinea pig distal colon. Am J Physiol Gastrointest Liver Physiol 316:G32–G44. First published October 18, 2018. https://doi.org/10.1152/ajpgi.00256.2019
49. Costa M, Keightley LJ, Wiklendt L, Hibberd TJ, Arkwright JW, Omari T, Wattchow DA, Zagorodnyuk V, Brookes SJH, Dinning PG, Spencer NJ (2019) Roles of three distinct neurogenic motor patterns during pellet propulsion in guinea pig distal colon. J Physiol 597(20):5125–5140. https://doi.org/10.1113/JP278284
50. Costa M, Keightley LJ, Hibberd TJ, Wiklendt L, Dinning PG, Brookes SJ, Spencer NJ (2021) Motor patterns in the proximal and distal mouse colon which underlie formation and propulsion of faeces. Neurogastroenterol Motil 33(7):e1409
51. D'Antona G, Hennig GW, Costa M, Humphreys CM, Brookes JH (2001) Analysis of motor patterns in the isolated guinea-pig large intestine by spatio-temporal maps. Neurogastroenterol Motil 13:483–492
52. Daniel EE, Wachter BT, Honour AJ, Bogoch A (1960) The relationship between electrical and mechanical activity of the small intestine of dog and man. Can J Biochem Physiol 38:777
53. Der T, Bercik P, Donnelly G, Jackson T, Berezin I, Collins SM, Huizinga JD (2000) Interstitial cells of cajal and inflammation-induced motor dysfunction in the mouse small intestine. Gastroenterology 119:1590–1599
54. Dinning PG, Scott SM (2011) Novel diagnostics and therapy of colonic motor disorders. Curr Opin Pharmacol 11:624–629
55. Dinning PG, Bampton PA, Kennedy ML, Cook IJ (1999) Relationship between terminal ileal pressure waves and propagating proximal colonic pressure waves. Am J Phys 277(5):G983–G982
56. Dinning PG, Arkwright JW, Gregersen H, O'Grady G, Scott SM (2010) Technical advances in monitoring human motility patterns. Neurogastroenterol Motil 22:366–380

57. Dinning PG, Arkwright JW, Costa M, Wiklendt L, Hennig G, Brookes SJH, Spencer NJ (2011) Temporal relationships between wall motion, intraluminal pressure, and flow in the isolated rabbit small intestine. Am J Physiol Gastrointest Liver Physiol 300:G577–G585
58. Dinning PG, Costa M, Brookes SJ, Spencer NJ (2012) Neurogenic and myogenic motor patterns of rabbit proximal, mid and distal colon. Am J Physiol Gastrointest Liver Physiol 303:G83–G92. https://doi.org/10.1152/ajpgi.00429.2011
59. Dinning PG, Wiklendt L, Gibbins I, Patton V, Bampton P, Lubowski D, Cook I, Arkwright J (2013) Low-resolution colonic manometry leads to a gross misinterpretation of the frequency and polarity of propagating sequences: initial results from fiber-optic high-resolution manometry studies. Neurogastroenterol Motil 25:e640–e649
60. Dinning PG, Wiklendt L, Maslen L, Gibbons I, Patton V, Arkwright JW, Lubowski DZ, O'Grady G, Bampton PA, Brookes SJ, Costa M (2014) Quantification of in vivo colonic motor patterns in healthy humans before and after a meal revealed by high-resolution fiber-optic manometry. Neurogastroenterol Motil. https://doi.org/10.1111/nmo.12408
61. Dinning PG, Wiklendt L, Omari T, Arkwright JW, Spencer NJ, Brookes SJ, Costa M (2014) Neural mechanisms of peristalsis in the isolated rabbit distal colon: a neuromechanical loop hypothesis. Front Neurosci 8:75
62. Du Bois Reymond E (1848) Untersuchungen uber thierische Elektricitat. G. Reimer, Berlin, pp 26–38
63. Dusdieker NS, Summers RW (1980) Longitudinal and circumferential spread of spike bursts in canine jejunum in vivo. Am J Phys 239(Gastrointest. Liver Physiol. 2):G311–G318
64. Ehrlein HJ (1980) A new technique for simultaneous radiography recording of gastrointestinal motility in unanesthetized dogs. Lab Anim Sci 30:879–884
65. Ehrlein H-J, Scheman JM (2006) Gastrointestinal motility; Technische Universität München: Munich. Retrieved April 20, 2014, from http://humanbiology.wzw.tum.de/fileadmin/Bilder/tutorials/tutorial.pdf
66. Ehrlein H-J, Reich H, Schwinger M (1983) Colonic motility and transit of digesta during hard and soft faeces formation in rabbits. J Physiol 338:75–86
67. Ehrlein H-J, Schemann M, Siegle M (1987) Motor patterns of small intestine determined by closely spaced extraluminal transducers and videofluoroscopy. Am J Phys 253(Gastrointest. Liver Physiol. 16):G259–G267
68. Ferens DM, Chang EC, Bogeski G, Shafton AD, Kitchener PD, Furness JB (2005) Motor patterns and propulsion in the rat intestine in vivo recorded by spatio-temporal maps. Neurogastroenterol Motil 17:714–720
69. Fleisch A, Wyss WHv (1923) Zur Kenntnis der visceralen Tiefensensibilität. Pflügers Arch 200:290–312; https://doi.org.ezproxy.flinders.edu.au/10.1007/BF01723340
70. Friedman G, Wolf BS, Waye JD, Janowitz HD (1965) Correlation of cineradiographic and intraluminal pressure changes in the human duodenum: an analysis of the functional significance of monophasic waves. Gastroenterology 49:37–49
71. Friendly M, Wainer H (2021) A history of data visualization and graphic communication. Harvard University Press, Cambridge, Massachusetts and London; England
72. Gordon AR, Siegman MJ (1971) Mechanical properties of smooth muscle. I. Length-tension and force-velocity relations. Am J Physiol 221: 1243–1249
73. Gregersen H, Kassab G (1996) Biomechanics of the gastrointestinal tract. Neurogastroenterol Motil 8(4):277–297
74. Grivel ML, Ruckebusch Y (1972) The propagation of segmental contractions along the small intestine. J Physiol (Lond) 227:611–625
75. Gwynne RM, Bornstein JC (2007) Mechanisms underlying nutrient-induced segmentation in isolated guinea pig small intestine. Am J Physiol Gastrointest Liver Physiol 292:1162–1172
76. Gwynne RM, Thomas EA, Goh SM, Sjovall H, Bornstein JC (2004) Segmentation induced by intraluminal fatty acid in isolated guinea-pig duodenum and jejunum. J Physiol Lond 556:557–569
77. Hannah Fry (2021) When graphs are a matter of life and death; pie charts and scatter plots seem like ordinary tools, but they revolutionized the way we solve problems. *The New Yorker*. 21 June issue
78. Heitmann PT, Mohd Rosli R, Maslen L, Wiklendt L, Kumar R, Omari TI, Wattchow D, Costa M, Brookes SJ, Dinning PG (2022) High-resolution impedance manometry characterizes the functional role of distal colonic motility in gas transit. Neurogastroenterol Motil 34(1):e14178. To NGM
79. Hennig GW (2016) Chapter 4: Spatio-temporal mapping and the enteric nervous system. In: Brierley S, Costa M (eds) The enteric nervous system 30 years later, Advances in experimental medicine and biology, vol 891. Springer
80. Hennig GW, Costa M, Chen BN, Brookes SJH (1999) Quantitative analysis of peristalsis in the guinea-pig small intestine using spatio-temporal maps. J Physiol 517(2):575–590
81. Hennig GW, Smith CB, O'Shea DM, Smith TK (2002) Patterns of intracellular and intercellular Ca2+ waves in the longitudinal muscle layer of the murine large intestine in vitro. J Physiol 543(1):233–253
82. Hennig GW, Spencer NJ, Jokela-Willis S, Bayguinov PO, Lee HT, Ritchie LA, Ward SM, Smith TK, Sanders KM (2010) ICC-MY coordinate smooth muscle electrical and mechanical activity in the murine small intestine. Neurogastroenterol Motil 22(5):e138–e151

83. Hertz AF, Newton A (1913) The normal movements of the colon in man. J Physiol Lond 41:57–65
84. Hill AV (1938) The heat of shortening and the dynamic constants of muscle. Proc Roy Sot, Ser B 126:136–119
85. Hipper K, Ehrlein HJ (2001) Motility of the large intestine and flow of digesta in pigs. Res Vet Sci 71(2):93–100
86. Hoad C, Clarke C, Marciani L, Graves MJ, Corsetti M (2018) Will MRI of gastrointestinal function parallel the clinical success of cine cardiac MRI? Br J Radiol 91:201804; Marciani L, Young P, Wright J, Moore R, Coleman N, Gowland PA, Spiller RC (2001) Antral motility measurements by magnetic resonance imaging. Neurogastroenterol Motil 13:511–518
87. Hoff HE, Geddes LA (1959) Graphic registration before Ludwig; the antecedents of the kymograph. Isis 50(1):5–21
88. Hoffman JM, Brooks EM, Mawe GM (2010) Gastrointestinal motility monitor JOVE
89. Holdstock DJ, Misiewicz JJ, Smith T, Rowlands EN (1970) Propulsion (mass movements) in the human colon and its relationship to meals and somatic activity. Gut 11:91–99
90. Holmstedt B, Liljestrand G (1963) Readings in pharmacology. Pergamon Press, pp 234–238
91. Holzknecht G (1909) Die normale peristaltik des kolon. Munch.med. Wschr 56:2401–2403
92. Hukuhara T (1931) Die normale Diinndarmbewegung. (Mit Hilfe der Bauchfenstermethode und Kinematographiel). Pflugers Arch 226:535
93. Hveem K, Sun WM, Hebbard G, Horowitz M, Doran S, Dent J (2001) Relationship between ultrasonically detected phasic antral contractions and antral pressure. Am J Physiol Gastrointest Liver Physiol 281:G95–G101
94. Indireshkumar K, Brasseur JG, Faas H, Hebbard GS, Kunz P, Dent J, Feinle C, Li M, Boesiger P, Fried M, Schwizer W (2000) Relative contributions of "pressure pump" and "peristaltic pump" to gastric emptying. Am J Physiol Gastrointest Liver Physiol 278:G604–G616
95. Itoh Z, Honda R, Takeuchi S, Aizawa I, Takayanagi R (1977) An extraluminal force transducer for recording contractile activity of the gastrointestinal smooth muscle in the conscious dogs: its construction and implantation i. Gastroenterol Jpn 12(4):275–283
96. Jacoby H, Bass P, Bennett DR (1985) In vivo extraluminal contractile force transducer for gastrointestinal muscle. J Appl Physiol (1985) 18(3):658–664
97. Janssen PW, Lentle RG (2013) Spatiotemporal mapping techniques for quantifying gut motility. In: Cheng LK et al (eds) New advances in gastrointestinal motility research, Lecture notes in computational vision and biomechanics, vol 10. Springer Science-Business Media Dordrecht. https://doi.org/10.1007/978-94-007-6561-0_12
98. Janssen PW, Lentle RG, Hulls C, Ravindran V, Amerah AM (2009) Spatiotemporal mapping of the motility of the isolated chicken caecum. J Comp Physiol B 179:593–604. https://doi.org/10.1007/s00360-009-0342-8
99. Jeffrey B, Udaykumar HS, Schulze KS (2003) Flow fields generated by peristaltic reflex in isolated guinea pig ileum: impact of contraction depth and shoulders. Am J Physiol Gastrointest Liver Physiol 285:G907–G918. https://doi.org/10.1152/ajpgi.00062.2003
100. Keating DJ, Spencer NJ (2009) Release of 5-hydroxytryptamine from the mucosa is not required for the generation or propagation of colonic migrating motor complexes. Gastroenterology 138:659–670, e652
101. Khalaf A, Nowak A, Menys A, Marciani L, Taylor SA, Spiller RC, Gowland PA, Moran GW, Hoad CL (2019) Cine MRI assessment of motility in the unprepared small bowel in the fasting and fed state: beyond the breath-hold. Neurogastroenterol Motil 31:e13466
102. Kim NY, Kim DH, Pickhardt PJ, Carchman EH, Wald A, Robbins JB (2018) Defecography; an overview of technique, interpretation, and impact on patient care. Gastroenterol Clin 47(3):553–568
103. Kruis W, Azpiroz F, Phillips SF (1985) Contractile patterns and transit of fluid in canine terminal ileum. Am J Phys 249(Gastrointest. Liver Physiol. 12):G264–G270
104. Kuizenga MH, Sia TC, Dodds KN, Wiklendt L, Arkwright JW, Thomas A, Brookes SJ, Spencer NJ, Wattchow DA, Dinning PG, Costa M (2015) Neurally mediated propagating discrete clustered contractions superimposed on myogenic ripples in ex vivo segments of human ileum. Am J Physiol Gastrointest Liver Physiol 30:G1–G1
105. Lammers WJEP, Cheng LK (2008) Simulation and analysis of spatio-temporal maps of gastrointestinal motility. Biomed Eng Online 7:2. https://doi.org/10.1186/1475-925X-7-2
106. Lammers WJ, al-Kais A, Singh S, Arafat K, el-Sharkawy TY (1993) Multielectrode mapping of slow-wave activity in the isolated rabbit duodenum. J Appl Physiol 74(3):1454–1461
107. Lammers WJ, Slack JR, Stephen B, Pozzan O (2000) The spatial behaviour of spike patches in the feline gastroduodenal junction in vitro. Neurogastroenterol Motil 12(5):467–473
108. Lammers WJEP, Dhanasekaran S, Slack JR, Stephen B (2001) Two-dimensional high-resolution motility mapping in the isolated feline duodenum: methodology and initial results. Neurogastroenterol Motil 13:309–323
109. Lammers WJ, Stephen B, Slack JR (2002) Similarities and differences in the propagation of slow waves and peristaltic waves. Am J Physiol Gastrointest Liver Physiol 283:G778–G786. https://doi.org/10.1152/ajpgi.00390.2001
110. Lammers WJ, Faes C, Stephen B, Bijnens L, Ver Donck L, Schuurkes JA (2004) Spatial determination of successive spikes in the isolated cat duodenum. Neurogastroenterol Motil 16(6):775–783

111. Lammers WJEP, Donck LV, Schuurkes JAJ, Stephen B (2005) Peripheral pacemakers and patterns of slow wave propagation in the canine small intestine in vivo. Can J Physiol Pharmacol 83:11
112. Langendorff Oskar (1891) Physiologische Graphik: Ein Leitfaden der in der Physiologie gebr uchlichen Registrirmethoden. Deuticke, Leipzig, Wien
113. Langley JN, Magnus R (1905) Some observations of the movements of the intestine before and after degenerative section of the mesenteric nerves. J Physiol 33(1):34–51
114. Larson M, Schulze K (2002) Appearance of peristaltic reflex in isolated guinea pig ileum in response to boluses of air, water, oil, and cellulose. Dig Dis Sci 47:2644–2650
115. Legros C, Onimus E (1869) Recherches experimentales sur les mouvements de l'intestin. J Anat Physiol (Paris) 6:37–65
116. Leibbrandt RE, Dinning PG, Costa M, Cock C, Wiklendt L, Wang G, Tack J, van Beckevoort D, Rommel N, Omari TI (2016) Characterization of esophageal physiology using mechanical state analysis. Front Syst Neurosci 10:10. https://doi.org/10.3389/fnsys.2016.00010
117. Lentle RG, Hulls CM (2018) Quantifying patterns of smooth muscle motility in the gut and other organs with new techniques of video spatiotemporal mapping. Front Physiol 9:338. https://doi.org/10.3389/fphys.2018.00338
118. Lentle RG, Janssen PWM, Asvarujanon P, Hemar Y (2008) High-definition spatiotemporal mapping of contractile activity in the isolated proximal colon of the rabbit. J Comp Physiol B 178:257–268. https://doi.org/10.1007/s00360-007-0217-9
119. Lentle RG, Janssen PW, Goh K, Chambers P, Hulls C (2010) Quantification of the effects of the volume and viscosity of gastric contents on antral and fundic activity in the rat stomach maintained ex vivo. Dig Dis Sci. https://doi.org/10.1007/s10620-010-1164-y
120. Lentle RG, De Loubens C, Hulls C, Janssen PWM, Golding MD, Chambers JP (2012) A comparison of the organization of longitudinal and circular contractions during pendular and segmental activity in the duodenum of the rat and guinea pig. Neurogastroenterol Motil. https://doi.org/10.1111/j.1365-2982.2012.01923.x
121. Liao DH, Zhao JB, Gregersen H (2009) Gastrointestinal tract modelling in health and disease. World J Gastroenterol 15(2):169–176
122. Liem O, Burgers RE, Connor FL, Benninga MA, Reddy SN, Mousa HM, Di Lorenzo C (2012) Solid-state vs water-perfused catheters to measure colonic high-amplitude propagating contractions. Neurogastroenterol Motil 24:345–e167
123. López-Gómez L, Díaz-Ruano S, Girón R, López-Pérez AE, Vera G, Herradón Pliego E, López-Miranda V, Nurgali K, Martín-Fontelles MI, Uranga JA, Abalo R (2018) Preclinical evaluation of the effects on the gastrointestinal tract of the antineoplastic drug vincristine repeatedly administered to rats. Neurogastroenterol Motil 30(11):e13399
124. Ludwig K (1847) Arch Anat Physiol Wissenschaftliche Med, 244
125. Ludwig K (1861) Lehrbuch der Physiologie des Menschen. Leipzig und Heidelberg 2:615
126. Magnus R (1904) Versuche am überlebenden Dünndarm von Säugethieren. Archiv für die gesamte Physiologie des Menschen und der Tiere 102:123–151
127. Magnus R (1904) Versuche am uberlebenden Dunndarm von Säugethieren. IV. Mitt. Rhythmizitat und refraktare Periode. Arch fur ges Physiol 103:525–540
128. Marey EJ (1876) La méthode graphique dans les sciences expérimentales. Physiologie Expérimentale. Travaux du laboratoire de M Marey 2:133–219
129. Marey EJ (1894) Le mouvement. G. Masson. https://play.google.com/books/reader?id=uoqQkfS2ge4C&hl=en_GB&pg=GBS.PA1
130. Matteucci C (1842) Sur un phenomene physiologique produit par les muscles en contraction. Ann de Chim et Phys 3S(6):339–343
131. Medical Research Council (1960) The hazards to man of nuclear and allied radiations. 2nd report. CMND. 1225. H.M.S.O., London
132. Melville J, Macagno E, Christensen J (1975) Longitudinal contractions in the duodenum: their fluid-mechanical function. Am J Phys 228:1887–1892
133. Menys A, Hoad C, Spiller R, Scott SM, Atkinson D, Marciani L, Taylor SA (2019) Spatio-temporal motility MRI analysis of the stomach and colon. Neurogastroenterol Motil 2019(31):e13557
134. Miller FR, Sherrington CS (1915) Some observations on the bucco-pharyngeal stage of reflex deglutition in the cat. Exp Physiol 9(2):147–186
135. Neal KB, Parry LJ, Bornstein JC (2009) Strain-specific genetics, anatomy and function of enteric neural serotonergic pathways in inbred mice. J Physiol 587(3):567–586
136. Nicosia MA, Brasseur JG, Liu JB, Miller LS (2001) Local longitudinal muscle shortening of the human esophagus from high-frequency ultrasonography. Am J Physiol Gastrointest Liver Physiol 281:G1022–G1033
137. Noble D (2002) Modelling the heart: insights, failures and progress. BioEssays 24:1155–1163
138. O'Grady G, Du P, Egbuji JU, Lammers WJ, Wahab A, Pullan AJ, Cheng LK, Windsor JA (2009) A novel laparoscopic device for measuring gastrointestinal slow-wave activity. Surg Endosc 23:2842–2848
139. O'Grady G, Angeli TR, Lammers WJEP (2013) The principles and practice of gastrointestinal high-resolution electrical mapping. In: New advances in gastrointestinal motility research. Springer Netherlands, Dordrecht, pp 51–69
140. Oigaard A, Dorph S (1974) Quantitative analysis of motility recordings in the human small intestine. Am J Dig Dis 19:804–810
141. Olesko M (1995) The meaning of precision: the exact sensibility in early nineteenth-century Germany.

In: Norton Wise M (ed) The values of precision. Princeton University Press, Princeton, pp 105–134
142. Omari T, Jones CA, Hammer MJ, Cock C, Dinning PG, Wiklendt L, Costa M, McCulloch TM (2016) Predicting the activation states of the muscles governing upper esophageal sphincter relaxation and opening. Am J Physiol Gastrointest Liver Physiol. https://doi.org/10.1152/ajpgi.00388.2015
143. Park KJ, Hennig GW, Lee HT, Spencer NJ, Ward SM, Smith TK, Sanders KM (2006) Spatial and temporal mapping of pacemaker activity in interstitial cells of Cajal in mouse ileum in situ. Am J Phys 290:1411–1427
144. Paton WDM (1957) The action of morphine and related substances on contraction and on acetylcholine output of coaxially stimulated guinea-pig ileum. Br J Pharmacol Chemother 11:119–127
145. Pehlivanov N, Liu J, Mittal RK (2001) Sustained esophageal contraction: a motor correlate of heartburn symptom. Am J Physiol Gastrointest Liver Physiol 281:G743–G751
146. Pescatori M, Grassetti F, Ronzoni G, Mancinelli R, Bertuzzi A, Salinari S (1979) Peristasis in distal colon of the rabbit: an analysis of mechanical events. Am J Phys 236(4):E464–E472
147. Playfair W (1805) Chart of universal commercial history
148. Popov MM (1963) An x-ray kymograph as a method for studying the physiology of the gastro-intestinal tract. Bull Exp Biol Med 53(3):352–355
149. Popov MM, Klymov KP (1963) An x-ray kymograph as a method for studying the physiology of the gastro-intestinal tract. Bull Exp Biol Med 53:352–355
150. Ramírez I, Pantrigo JJ, Montemayor AS, López-Pérez AE, Martín-Fontelles MI, Brookes SJH, Abalo R (2017) Computer vision-based diameter maps to study fluoroscopic recordings of small intestinal motility from conscious experimental animals. Neurogastroenterol Motil 29(8):e13052. https://doi.org/10.1111/nmo.1305
151. Reinke DA, Rosenbaum AH, Bennett DR (1967) Patterns of dog gastrointestinal contractile activity monitored in vivo with extraluminal force transducers. Am J Dig Dis 12(2):113–141
152. Rich A (2018) Zebra fish improved imaging of zebrafish motility. Neurogastroenterol Motil 30:e13435
153. Ritchie JA (1968) Colonic motor activity and bowel function. Part I. Normal movement of contents. Gut 9:442–456
154. Ritchie JA (1968) Colonic motor activity and bowel function. Part II Distribution and incidence of motor activity at rest and after food and carbachol. Gut 9:502–511
155. Roberts RR, Bornstein JC, Bergner AJ, Young HM (2008) Disturbances of colonic motility in mouse models of Hirschsprung's disease. Am J Physiol Gastrointest Liver Physiol 294(G996):G1008
156. Roberts RR, Ellis M, Gwynne RM, Bergner AJ, Lewis MD, Beckett EA, Bornstein JC, Young HM (2010) The first intestinal motility patterns in fetal mice are not mediated by neurons or interstitial cells of Cajal. J Physiol 588(7):1169 1153
157. Sarna SK, Condon R, Cowles V (1984) Colonic migrating and nonmigrating motor complexes in dogs. Am J Phys 246(Gastrointest. Liver Physiol. 9):G355–G360
158. Schemann M, Ehrlein HJ (1986) Postprandial patterns of canine jejunal motility and transit of luminal content. Gastroenterology 90:991–100
159. Schemann M, Mai G, Costa M, Enck P (2020) Translating the seminal findings of Carl Lüderitz: a description in English of his extraordinary studies of gastrointestinal motility accompanied by a historical view of peristalsis. Neurogastroenterol Motil:e13995. 1–19. https://doi.org/10.1111/nmo.13995
160. Schulze KS, Clark E (2008) Ink dispersion by sequential contractions in isolated segments of guinea pig ileum and duodenum. Neurogastroenterol Motil 20:1317–1327
161. Schulze-Delrieu K (1992) Clearance patterns of the isolated guinea-pig duodenum. Gastroenterology 102:84–56
162. Schulze-Delrieu K (1999) Visual parameters define the phase and the load of contractions in isolated guinea pig ileum. Am J Physiol Gastrointest Liver Physiol 276:G1417–G1424
163. Seerden TC, Lammers WJEP, De Winter BY, De Man JG, Pelckmans PA (2005) Spatiotemporal electrical and motility mapping of distension-induced propagating oscillations in the murine small intestine. Am J Physiol Gastrointest Liver Physiol 289:G1043–G1051
164. Seidl H, Gundling F, Pehl C, Kissler S, Schepp W, Schmidt TY (2010) Multiple intraluminal impedance measurement and manometry in the jejunum of healthy subjects: a new tool for diagnosis of small bowel motility. Digestion 81:172–180
165. Sethi AK, Sarna SK (1995) Contractile mechanisms of canine colonic propulsion. Am J Phys 268(Gastrointest. Liver Physiol. 31):G530–G538
166. Sinnott MD, Cleary PW, Arkwright JW, Dinning PG (2012) Investigating the relationships between peristaltic contraction and fluid transport in the human colon using Smoothed Particle Hydrodynamics. Comput Biol Med 42(4):492–503
167. Smith AW, Code CF, Schlegel JF (1975) Simultaneous cineradiographic and kymographic studies of human gastric antral motility. J Appl Physiol 11:12–16
168. Spencer NJ (2001) Control of migrating motor activity in the colon. Curr Opin Pharmacol 1:604–610
169. Spencer NJ, Costa M, Hibberd T, Wood JD (2021) Advances in colonic motor complexes in mice. Am J Physiol Gastrointest Liver Physiol 320:G12–GG2. https://doi.org/10.1152/ajpgi.00317.2020
170. Stevens RJ, Publicover NG, Smith TK (2000) Propagation and neural regulation of calcium waves in longitudinal and circular muscle layers of guinea pig small intestine. Gastroenterology 118:892–904

171. Swaminathan M, Hill-Yardin E, Ellis M, Zygorodimos M, Johnston LA, Gwynne RM, Bornstein JC (2016) Video imaging and spatio-temporal maps to analyze gastrointestinal motility in mice. J Vis Exp 108:e53828. https://doi.org/10.3791/53828
172. Szurszewski JH (1969) A migrating electric complex of the canine small intestine. Am J Phys 217:1757–1763
173. Tack J, Smith TK (2004) Calcium imaging of gut activity. Neurogastroenterol Motil 16(Suppl. 1):86–95
174. Tasaka K, Farrar JT (1969) Mechanics of small intestinal muscle function in the dog. Am J Phys 217:1224–1229
175. Tasaka K, Farrar JT (1976) Intraluminal pressure of the small intestine of the unanaesthetized dog. Pflugers Arch 364:35–44
176. Trendelenburg P (1917) Physiologische und pharmakologische Versuche uber die D unndarm Peristaltick. Naunyn-Schmiedeberg's Arch Exp Pathol Pharmakol 81:55–129
177. Volkmann AW (1851) Ueber das Zustandekommen der Muskelcontractionen im Verlaufe der Zeit; BVKSGW, 1
178. von Helmholtz HLF (1850) Note sur la vitesse de propagation de l'agent nerveux dans les nerfs rachidiens. C R Acad des Sci 30:204–206
179. von Helmholtz HLF (1850) Vorl ufiger bericht über die fortpflanzungs- geschwindigkeit der nervenreizung. Arch Anat Physiol wiss Med 17:71–73g
180. Wang XY, Lammers WJ, Bercik P, Huizinga JD (2005) Lack of pyloric interstitial cells of Cajal explains distinct peristaltic motor patterns in stomach and small intestine. Am J Physiol Gastrointest Liver Physiol 289(3):G539–G549
181. Wang D et al (2018) Ingestible roasted barley for contrast-enhanced photoacoustic imaging in animal and human subjects. Biomaterials 175:72–81
182. Waterman SA, Tonini M, Costa M (1994) The role of ascending excitatory and descending inhibitory pathways in peristalsis in the isolated guinea-pig small intestine. J Physiol Lond 481(1):223–232
183. Webb CW (1971) Spatiotemporal objects. J Philos 68:879–890
184. Weems WA, Weisbrodt NW (1986) Ileal and colonic propulsive behavior: contribution of enteric neural circuits. Am J Phys 250(Gastrointest. Liver Physiol. 13):G653–G659
185. Weisbrodt NW, Wiley JN, Overholt BF, Bass P (1969) A relation between gastroduodenal muscle contractions and gastric emptying. Gut 10:543–548
186. Zhang Y, Jeon M, Rich LJ, Hong H, Geng J, Zhang Y, Shi S, Barnhart TE, Alexandridis P, Huizinga JD, Seshadri M, Cai W, Kim C, Lovell JF (2014) Non-invasive, multimodal functional imaging of the intestine with frozen micellar naphthalocyanines. Nat Nanotechnol. https://doi.org/10.1038/NNANO.2014.130
187. Zhang Y et al (2016) Surfactant-stripped frozen pheophytin micelles for multimodal gut imaging. Adv Mater 28:8524–8530
188. Zondek B (1920) Uberdickdarm peristaltik. Beobachtungen am experimentellen Bauch fenster. Arch Verdauungskr 27:18–23

27 Rhythmicity in the Enteric Nervous System of Mice

Nick J. Spencer and Marcello Costa

Abstract

The enteric nervous system (ENS) is required for many cyclical patterns of motor activity along different regions of the gastrointestinal (GI) tract. What has remained mysterious is precisely how many thousands of neurons within the ENS are temporally activated to generate cyclical neurogenic contractions of GI-smooth muscle layers. This has been an especially puzzling conundrum, since the ENS consists of an extensive network of small ganglia, with each ganglion consisting of a heterogeneous population of neurons, with diverse cell soma morphologies, neurochemical and biophysical characteristics, and neural connectivity. Neuronal imaging studies of the mouse large intestine have provided major new insights into how the different classes of myenteric neurons are activated during cyclical neurogenic motor patterns, such as the colonic motor complex (CMC). It has been revealed that during CMCs (in the isolated mouse whole colon), large populations of myenteric neurons, across large spatial fields, coordinate their firing, via bursts of fast synaptic inputs at ~2 Hz. This coordinated firing of many thousands of myenteric neurons synchronously over many rows of interconnected ganglia occurs irrespective of the functional class of neuron. Aborally directed propulsion of content along the mouse colon is due, in large part, to polarity of the enteric circuits including the projections of the intrinsic excitatory and inhibitory motor neurons but still involves the fundamental ~2 Hz rhythmic activity of specific classes of enteric neurons. What remains to be determined are the mechanisms that initiate and terminate the patterned firing of large ensembles of enteric neurons during cyclic activity. This remains an exciting challenge for future studies.

Keywords

Colonic motor complex · Peristalsis · Colon · Enteric nervous system · Sensory neurons · Propulsion · Enteric nervous system

N. J. Spencer (✉)
Visceral Neurophysiology Laboratory, Department of Physiology, College of Medicine and Public Health & Centre for Neuroscience, Flinders University, Bedford Park, SA, Australia
e-mail: nicholas.spencer@flinders.edu.au

M. Costa
Visceral Neurophysiology Laboratory, Department of Physiology, College of Medicine and Public Health, Flinders University, Bedford Park, SA, Australia

N. J. Spencer et al. (eds.), *The Enteric Nervous System II*, Advances in Experimental Medicine and Biology 1383, https://doi.org/10.1007/978-3-031-05843-1_27

Abbreviations

CGRP	calcitonin gene-related peptide
CMC	colonic motor complex
CMMC	colonic migrating motor complex
TPH1	tryptophan hydroxylase 1
TPH2	tryptophan hydroxylase 2

27.1 Introduction

Unlike skeletal muscle, smooth muscle cells in many organs display electrical rhythmicity [29]. Examples of rhythmicity in smooth muscle cells are found in hollow-walled organs, like arterioles [26], veins [49], lymphatic vessels [4], ureters [36], and the urethra [44], each of which generates rhythmic depolarizations and or propagating contractions that facilitate the propulsion of fluid content. The mechanisms that generate these propagating contractions are myogenic in origin and do not require activity from nerves. Indeed, in most regions of the gastrointestinal (GI) tract, the smooth muscles are driven by a net of pacemaker cells (the ICCs) to generate "myogenic" waves of contraction that can also propagate short distances along the gastrointestinal tract [17, 32].

Unlike other hollow smooth muscle organs, the GI tract is the only internal organ to have evolved with its own independent nervous system, known as the enteric nervous system (ENS) [16, 21, 52]. A unique feature of the ENS is that it does not require any inputs from the spinal cord or brain to generate complex patterns of motor activity [16, 21, 22, 31, 48, 52, 62, 63]. The myenteric plexus consists of the major populations of motor neurons (both excitatory and inhibitory motor neurons) to the smooth muscle cells, in addition to multiple classes of ascending and descending interneurons [9, 23]. A unique feature of the ENS is that it also contains its own population of sensory neurons, also called intrinsic primary afferent neurons (IPANs), that are distinct and independent from sensory neurons outside the gut wall in the nodose, trigeminal or dorsal root ganglia [21, 22, 35, 52, 56]. The importance of the ENS for coordinated propulsion of content along the intestine is exemplified in mammals that lack large segments of ENS, such as those with the congenital condition, known as Hirschsprung's disease. Mammals with long segment aganglionosis often cannot survive without surgical resection of the aganglionic segment [42, 52, 61, 64, 65].

A large part of the neurochemical coding of the different classes of neurons in the ENS has now been established, including the pathways, projections, and synaptic inputs to distinct classes of enteric neurons [9, 21, 24, 31, 39, 40, 52, 62]. One of the most challenging questions that has eluded the field is how are the different classes of neurons that are located in the small independent ganglia of the ENS network temporally activated to generate the complex neurogenic motor patterns along the GI-tract. The mouse has become an excellent model to understand the neuronal basis of intestinal motor activity [47]. The fundamental focus of this review is to relate a fundamental rhythmic activity of enteric neurons recently revealed, with the different motor patterns of the mouse colon.

27.2 Motor Patterns in Mouse Colon

Maintained distension applied over the full length of mouse colon (without any propulsion of luminal content) generates a neurogenic motor pattern consisting of cyclic motor complexes, occurring with intervals of about every 1–3 minutes [1, 47, 58]. These motor complexes often tend to propagate slowly and thus have been initially described as colonic migrating motor complexes (CMMCs) [6, 7, 46], since they can propagate retrogradely or antegradely, or occur simultaneously over the full length of colon, a preferred name that has emerged as colonic motor complexes (CMCs) [8]. Similar motor complexes have also been recorded in the mouse small intestine [6] (and when the entire intact mouse intestine is distended by a mini fiber-optic manometry catheter), and motor complexes often propagate along the entire intestine across the small and large intestine [12]. These motor complexes are blocked by tetrodotoxin or hexamethonium, indicating that they involve interconnected enteric circuits by chains of cholinergic interneurons [12].

27.3 Sensitivity to Distension of CMCs

Major insights have been made into understanding the stimuli that modify CMC frequency in mouse colon [1, 12]. Originally, it was thought that CMCs in mice were an omnipresent phenomenon that occurred independently of the presence or absence of content in the colon, or the level of distension applied to the colonic wall [20]. However, more recent studies have revealed that in empty segments of colon that are not in any way stimulated by the recording apparatus, CMCs occur rarely, or infrequently, with incomplete propagations along the colon [1]. In these same preparations, when maintained circumferential stretch is applied to the colon wall, CMCs occur significantly more frequently and propagate over much longer lengths of colon [1]. Dramatically, CMCs have been shown to extend across the small and large intestine [12]. The degree of activation of intrinsic sensory neurons (IPANs) and the circuitry underlying CMCs is very much dependent upon the level of stretch applied to the colon. The presence of the submucosal plexus or the mucosa was not required for the stretch-sensitive neural circuitry to activate CMCs [66].

27.4 Potential Role of Endogenous 5HT from the Mucosa in Cyclical Neurogenic Motor Patterns

The identity of the pattern generator that underlies the cyclical generation of neurogenic motor patterns in the GI tract has remained mysterious. Different hypotheses have been proposed. One early theory was that serotonin played a major role, since antagonists of 5-HT3 receptors abolished CMCs [27]. Then, this hypothesis was developed further—as it was proposed the release of serotonin from enterochromaffin (EC) cells, the mucosa was critical for generation of cyclical CMCs, in the mouse colon [27], despite no recordings of serotonin release being made. This seemed an exciting and tenable possibility, at the time. However, support for this notion rapidly lost traction, when it was discovered that CMCs still occurred when the mucosa was removed from the colon and in preparations in which all endogenous release of 5-HT from the mucosa had been abolished [33]. Furthermore, CMCs also still robustly occurred in mutant mice lacking the enzyme to make mucosal 5-HT [28]. It is now accepted that endogenous 5-HT from the mucosa or from the ~1% of enteric neurons that express TPH2 is not required for the cyclical genesis of CMC motor activity [34]. However, it appears that endogenous 5-HT may modulate CMC characteristics [54], since most recent studies have acutely ablated the TPH1 gene, using diphtheria toxin-Cre, and found slight modifications to CMC activity [60].

27.5 Coordinated Rhythmic Firing of Myenteric Neurons During CMCs Revealed by Neuronal Imaging of the ENS

Neuronal imaging studies have provided major advances into how the ENS generates cyclical neurogenic contractions along the colon using neuronal imaging of the different classes of neurons in the ENS (Fig. 27.1). These studies from large populations of myenteric neurons during cyclical neurogenic motor activity in the mouse colon have revealed that large populations of neurons fire in coordinated bursts at 2 Hz (Fig. 27.1). When studies were undertaken to identify the neurochemical coding of these neurons (Fig. 27.2), it was discovered that both cholinergic and nitrergic neurons all discharged at ~2 Hz. It was particularly interesting that the putative intrinsic sensory neurons that expressed CGRP also fired at ~2 Hz, during the emergence of coordinated ENS firing (Fig. 27.3). When the ENS fired in coordinated bursts at ~2 Hz, these were associated with the generation of EJPs in

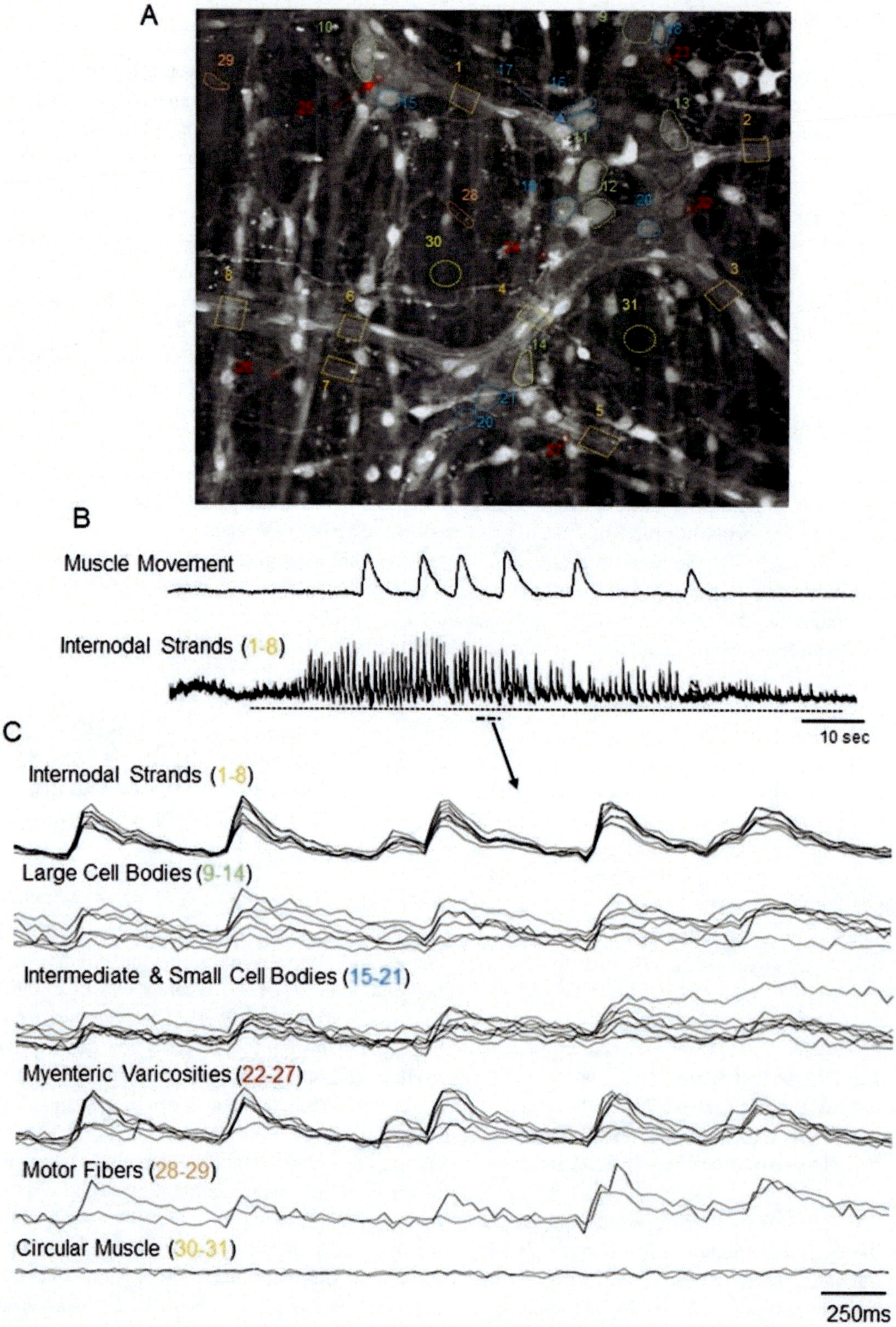

Fig. 27.1 Calcium imaging of the myenteric plexus during a CMC in isolated whole mouse colon. The recording shows coordinated neuronal firing across multiple classes of myenteric neurons in isolated mouse colon.

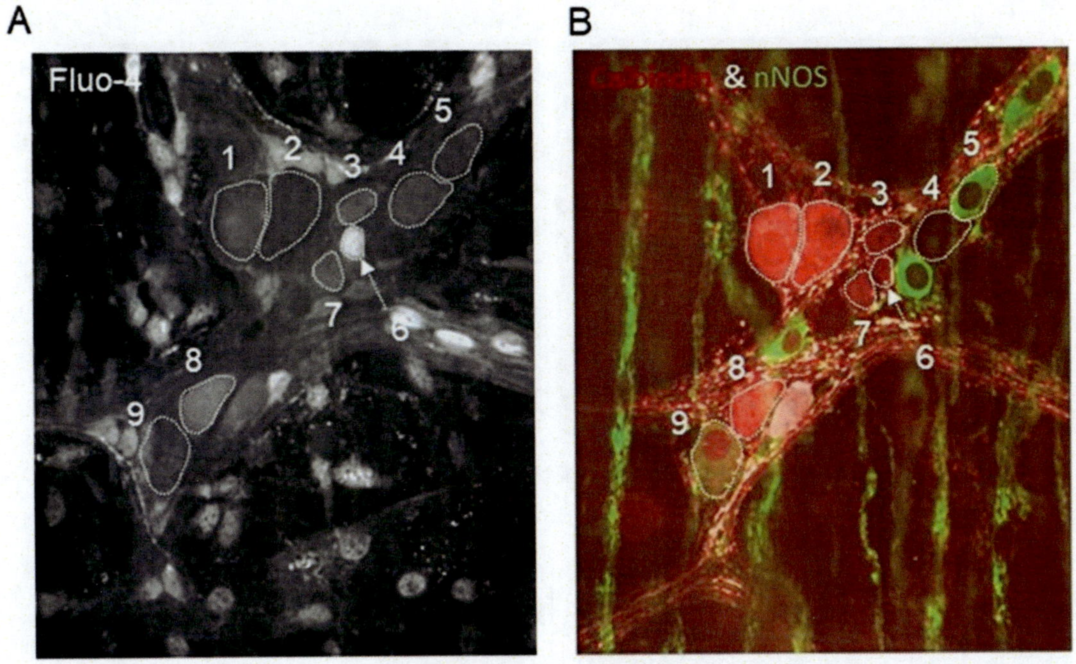

Fig. 27.2 Technique to identify the neurochemical coding of myenteric neurons in mouse colon from which calcium imaging was performed in control C57BL/6 mice. (**a**) Camera image showing a variety of different sized myenteric neurons (of unknown neurochemical coding), labeled as regions of interest from 1 to 9. (**b**) After imaging, the tissue is fixed and stained for neuronal nitric oxide synthase (NOS) in green and the largely excitatory neuronal marker calbindin (in red)

the smooth muscle at a similar frequency of ~2 Hz (Fig. 27.4) [51]. This was a major important advance, because it meant that the discharge pattern of EJPs in the smooth muscle provides a "readout" of the firing rate of ENS activity [51]. When atropine blocked the EJPs, IJPs still occurred at a similar rate indicating that the same premotor ENS circuit drives in parallel both the excitatory and the inhibitory neurons [50]. This recent work has uncovered that large assemblies of ascending and descending interneurons fire in a coordinated and repetitive fashion at ~2 Hz to drive large populations of excitatory and inhibitory motor neurons at ~2 Hz that lie both orally and aborally to the neurogenic contraction of the MCs [51]. This is based on our understanding that individual motor neurons have short projections less than 3 mm long and thus must be driven by ascending and descending interneuron circuits, which have projections up to 13 mm [50]. Exactly how the interneuronal circuits fire in a repetitive fashion is unclear, but it is apparent that they are interconnected to entrain bursting behavior at approximately 1–2 Hz. Estimates suggest

Fig. 27.1 (continued) (**a**) Camera image taken of a number of myenteric ganglia in paralyzed isolated whole mouse colon, in nifedipine. (**b**) Despite nifedipine, there is some muscle movement during a colonic motor complex, which can be detected using our spatiotemporal D-mapping. The underlying residual muscle contractions that occur in nifedipine are associated with a discharge of coordinated firing of large populations of myenteric neurons that lasts for about 1 minute. (**c**) It shows that from multiple regions of interest including internodal strands (see regions of interest 1–8), large cell bodies (9–14), intermediate and small cell bodies (15–21), myenteric varicosities (22–27), and motor axons (fibers) in the circular muscle, there is coordinated discharge of calcium transients in all regions of interest. As the circular muscle (30–31) is paralyzed with nifedipine, so does not show action potential driven calcium transients

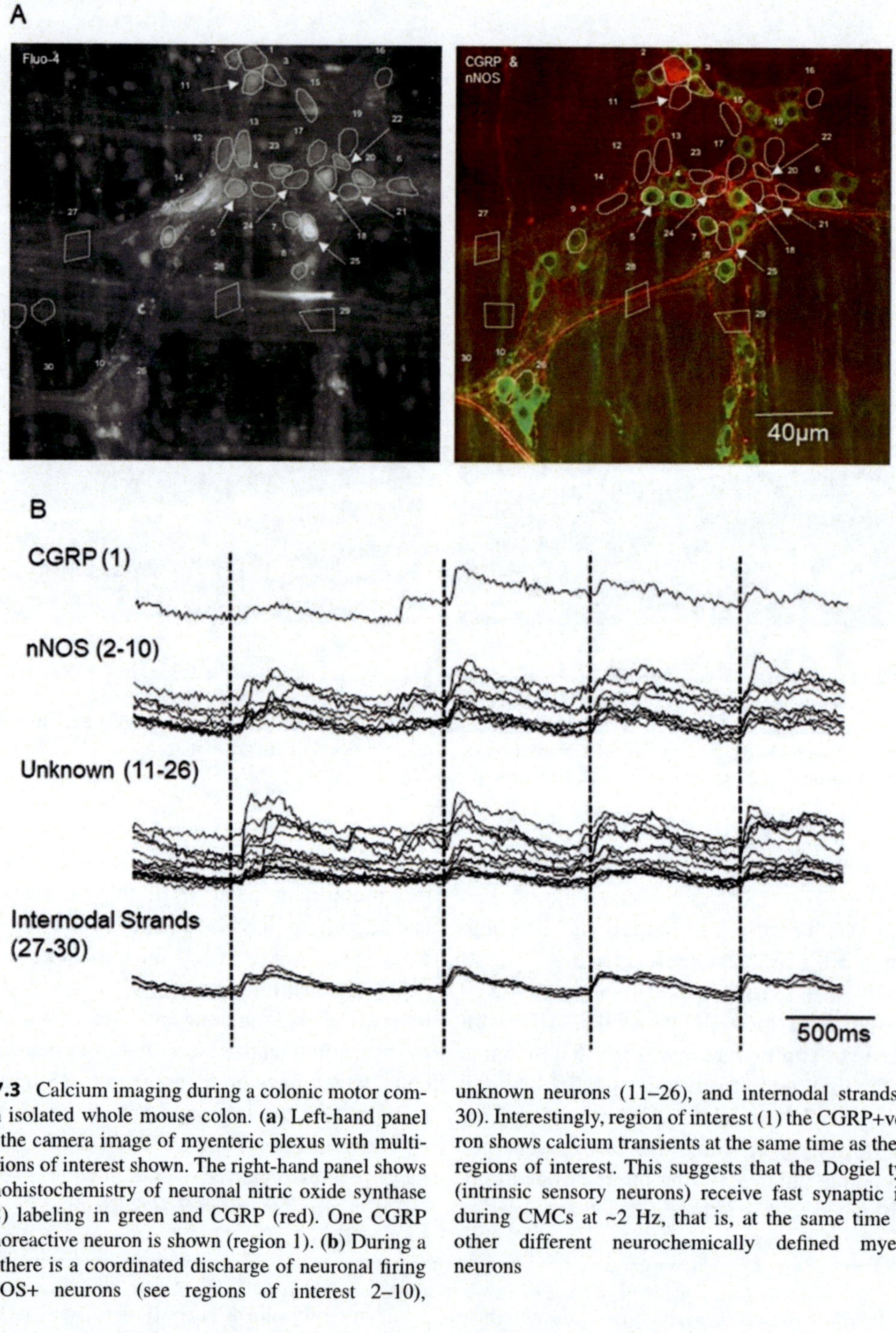

Fig. 27.3 Calcium imaging during a colonic motor complex in isolated whole mouse colon. (**a**) Left-hand panel shows the camera image of myenteric plexus with multiple regions of interest shown. The right-hand panel shows immunohistochemistry of neuronal nitric oxide synthase (nNOS) labeling in green and CGRP (red). One CGRP immunoreactive neuron is shown (region 1). (**b**) During a CMC, there is a coordinated discharge of neuronal firing of nNOS+ neurons (see regions of interest 2–10), unknown neurons (11–26), and internodal strands (27–30). Interestingly, region of interest (1) the CGRP+ve neuron shows calcium transients at the same time as the other regions of interest. This suggests that the Dogiel type II (intrinsic sensory neurons) receive fast synaptic inputs during CMCs at ~2 Hz, that is, at the same time as all other different neurochemically defined myenteric neurons

there are approximately 30,000 myenteric neurons in in the mouse colon of which approximately 8000 are inhibitory neurons and a comparable number are excitatory motor neurons [45]. The observation that during constant distension along the whole length of colon (Fig. 27.4) cholinergic motor neurons can temporally synchronize their firing at ~2 Hz across the full length of colon shows that large assemblies of interneurons from the proximal to mid-distal

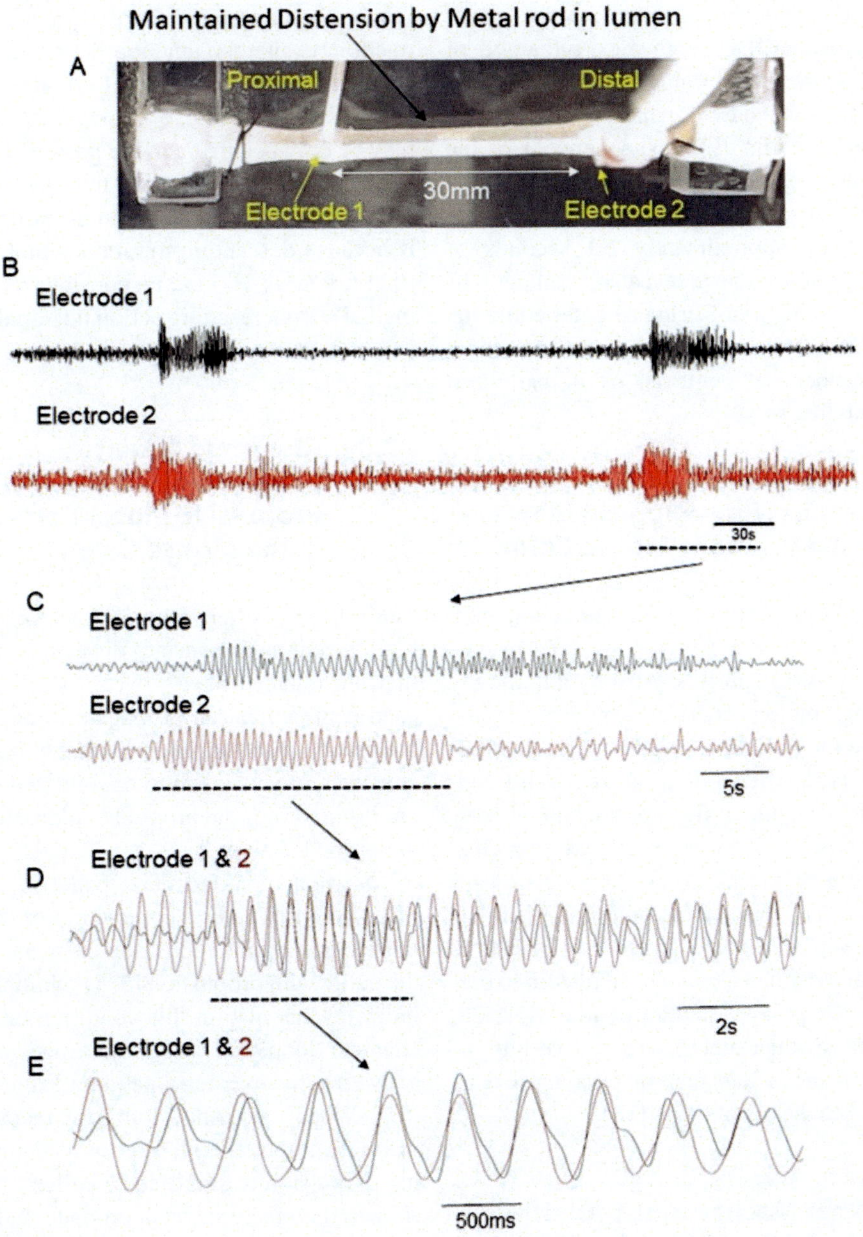

Fig. 27.4 Maintained luminal distension of mouse isolated whole colon induces cyclical colonic motor complexes (CMCs). (**a**) Photomicrograph of the whole mouse colon with a metal rod implanted in the lumen to impose luminal distension. Two extracellular recording electrodes are positioned on the serosal surface to record electrical activities from the muscularis externa. (**b**) In the presence of nicardipine (1 μM), two distinct bursts of electrical activity are recorded that occur at both proximal and distal colon. The region shown in the dotted line is shown in expanded scale in panel **c**. (**c**) The ~2 Hz discharge of EJPs in both extracellular recording electrodes. The period shown by the black dotted line is shown in expanded scale in **d**. (**d**) The two traces in C superimposed. (**e**) The region indicated by the black dotted line in panel **d** on expanded time scale. The ~2 Hz discharge of EJPs can be seen to temporally synchronize during the middle of the recording, but is phase shifted by about 20 ms at the start of the recording

region of colon can temporally synchronize their firing to generate EJPs over the near full length of colon. These recent findings uncover a fundamental circuity in the ENS, which when activated by a variety of stimuli and independent of the mechanical states of the musculature (i.e., paralyzed or not) triggers a hardwired neural net that discharges for approximately 30 seconds at ~2 Hz. These studies have revealed a unique fundamental rhythmicity of firing of enteric circuits in the mouse colon which occurs even when there is no movement of contents, or in paralyzed smooth muscle.

27.6 Distinct Physiological Motor Patterns in the Mouse Colon

In recent investigations in the mouse colon, there are at least two motor patterns involved in propulsion of natural contents. An ongoing slowly propagating motor complex appears to be involved in the formation of pellets in the proximal colon [13]. Then, when the semisolid contents reach a functional flexure, they are already formed into pellet-like boluses and are propelled by activation of ascending excitatory pathways and facilitated by descending inhibitory pathway [13]. Distinct neurogenic patterns of rhythmic electrical activity had already been identified previously in the proximal colon of mice [12, 38, 57], where distinct neurochemical coding of myenteric neurons and enteric circuitry exists, different from the distal colon [38].

27.7 What Mechanisms Underlie Aboral Propulsion?

Local stimulation of the gut is well known to activate local intrinsic neural reflexes consisting of ascending excitatory and descending inhibitory nerve pathways [2, 3, 11, 15, 18, 22, 23, 52, 55, 63]. In propulsive movements what emerges is that there is temporally coordinated activation of ascending excitatory nerve pathways orally and descending inhibitory nerve pathways aborally. Recent work in mouse colon has shown that during the aboral propagation of smooth muscle contraction, excitatory junction potentials discharge at ~2 Hz along the length of colon (from proximal to distal regions), even when the contraction may be only occurring in the proximal colon [47, 58]. It has been uncovered that the reason why the contraction commences in the proximal colon is because descending inhibitory motor neurons (that fire at ~2 Hz) are responsible for suppressing EJPs from reaching action potential threshold [47, 58].

27.8 Rhythmic Activity of Myenteric Neurons During Propulsive Motor Patterns in the Mouse Colon

Interestingly, when preparations of colon are maintained under constant distension, the CMCs that are induced have a weak polarity. That is, each contraction can propagate in any direction but predominantly from proximal to distal colon. However, as soon as fluid or solid pellet contents are inserted into the proximal colon, they always are propelled aborally.

Insertion of an artificial pellet in the mouse distal colon initiates a contraction of the circular muscle at the oral end of the segment, or oral to the pellet with propulsion [5, 51]. Initial evidence indicates that also in this condition of neuromechanical loop activation, there is coordinated activation of myenteric neurons firing at ~2 Hz [51]. When electrophysiological recordings are made from the smooth muscle of the colon during propulsion of single fecal pellets, it has been shown that there is synchronous EJPs in the smooth muscle over increasing spatial field behind the fecal pellet. It might be argued that if the ENS is simultaneously activated along the length of colon, then why does the whole colon not contract simultaneously? As mentioned above, the most recent data has shown that during the propulsion of fluid, inhibitory motor neurons are active downstream (ahead of the advancing fecal pellet) and act to suppress EJPs from reaching action potential threshold [47, 58]. The same hardwired circuit that was identified in 2018 by our laboratory is also active in these tubular preparations during fluid propulsion [47, 58]. This was revealed when inhibitory motor neurons and excitatory motor neurons are activated at the same rate by common interneurons [51]. This means that during a propulsive contraction, descending inhibitory pathways are strongly activated from the distended region, leading to aboral inhibition. It is important to note that the level of excitation in the ENS and the degree of spatial coordination of colonic neurogenic contractions are dependent upon the length of colon that is distended [12].

Fluid propulsion in the colon does not commonly occur physiologically since the colon (particularly the distal colon) usually contains solid content. However fluid infusion has been used to compare colonic propulsion in several species including mouse [10]. In a recent publication from our laboratory during the aboral propulsion of fluid, there is temporal coordination of the firing of large populations of excitatory and inhibitory motor neurons, both orally and aborally to a propagating contraction of colonic circular smooth muscle. It was shown that during propulsion of fluid, there is a repetitive discharge of cholinergic EJPs over large lengths of colon downstream (aboral) of the propagating contraction that become temporally coordinated with EJPs in the proximal colon, at a time prior to the onset of the contraction in the proximal colon. Hence, there is ongoing synchronized neural activity in the ENS over a long spatial range, even before any localized contraction commences.

A logical question that arises is if large populations of excitatory and inhibitory motor neurons temporally coordinate their firing across large populations of myenteric ganglia, then how can a polarized, aborally propagating contraction occur? This mechanism is not fully understood, but it is clear, as discussed above, that the polarity in the projections of excitatory and inhibitory motor neurons plays a major role. Evidence shows that descending inhibitory pathways to the smooth muscle suppress concurrently activated descending (and ascending) excitatory pathways to the same regions of smooth muscle. This explains why EJPs occur along the full length of colon at the same time at ~2 Hz (see [47, 58]).

27.9 Are There ENS Pacemaker Neurons That Could Underlie the Rhythmic Generation of CMCs?

In the heart, distinct populations of rhythmically excitable neurons have been identified in the cardiac plexus [19]. However, similar studies have not found rhythmically excitable neurons in the ENS. The neurogenic origin of cyclical CMCs, at least in rodents, involves the emergence of coordinated firing of many thousands of ascending and descending interneurons that synaptically converge to activate large assemblies of excitatory and inhibitory motor neurons [47, 51, 58]. The mechanism appears to be stochastic. But, during CMCs, the mechanism that underlies the onset of this coordinated neural firing in ENS circuits is elusive. Intracellular electrophysiological recordings from myenteric neurons in mouse colon have shown that following blockade of fast synaptic transmission, the vast majority of neurons fall silent and do not show intrinsic rhythmicity [25, 41]. In the study from Dr. Bywater's laboratory, some myenteric neurons showed slow rhythmic depolarizations, which they called slow waves [25].

27.10 Comparison with Other Species

The principles underlying this review could easily extend to other species. One of the most commonly recorded patterns of neurogenic motor activity in the colon of mammals constantly distended is the colonic motor complex (CMC) [8]. CMCs have been characterized in vivo and in vitro in a variety of species, including mouse [20, 30, 37, 43, 59] and humans [53]. CMCs occur in a variety of isolated preparations of colon of different species, including the whole human colon [53] and guinea pig [14].

27.11 Conclusion

Despite the different patterns of motor activity that have been identified to be involved in the formation of fecal pellets in the mouse colon, there is a common underlying rhythmicity of the enteric circuits firing at ~2 Hz. All enteric neurons appear to be involved in this rhythmicity including excitatory and inhibitory motor neurons and most interestingly the CGRP immunoreactive Dogiel Type II (sensory) neurons. A modulation of such rhythmic activity by luminal stimuli acting via intrinsic sensory inputs determines the adaptive nature of the enteric circuits to colonic contents. A challenge for future studies is identifying the origin of the neurons that initiate the emergence of coordinated firing of large ensembles of ENS activity that underlie cyclical neurogenic motor patterns in the GI-tract.

References

1. Barnes KJ, Beckett EA, Brookes SJ, Sia TC, Spencer NJ (2014) Control of intrinsic pacemaker frequency and velocity of colonic migrating motor complexes in mouse. Front Neurosci 8:96
2. Bayliss WM, Starling EH (1899) The movements and innervation of the small intestine. J Physiol 24:99–143
3. Bayliss WM, Starling EH (1900) The movement and innervation of the large intestine. J Physiol 26:107–118
4. Beckett EA, Hollywood MA, Thornbury KD, Mchale NG (2007) Spontaneous electrical activity in sheep mesenteric lymphatics. Lymphat Res Biol 5:29–43
5. Brann L, Wood JD (1976) Motility of the large intestine of piebald-lethal mice. Am J Dig Dis 21:633–640
6. Bush TG, Spencer NJ, Watters N, Sanders KM, Smith TK (2000) Spontaneous migrating motor complexes occur in both the terminal ileum and colon of the C57BL/6 mouse in vitro. Auton Neurosci 84:162–168
7. Bush TG, Spencer NJ, Watters N, Sanders KM, Smith TK (2001) Effects of alosetron on spontaneous migrating motor complexes in murine small and large bowel in vitro. Am J Physiol Gastrointest Liver Physiol 281:G974–G983
8. Corsetti M, Costa M, Bassotti G, Bharucha AE, Borrelli O, Dinning P, Di Lorenzo C, Huizinga JD, Jimenez M, Rao S, Spiller R, Spencer NJ, Lentle R, Pannemans J, Thys A, Benninga M, Tack J (2019) First translational consensus on terminology and definitions of colonic motility in animals and humans studied by manometric and other techniques. Nat Rev Gastroenterol Hepatol 16:559–579
9. Costa M, Brookes SJ, Steele PA, Gibbins I, Burcher E, Kandiah CJ (1996) Neurochemical classification of myenteric neurons in the Guinea-pig ileum. Neuroscience 75:949–967
10. Costa M, Dodds KN, Wiklendt L, Spencer NJ, Brookes SJ, Dinning PG (2013) Neurogenic and myogenic motor activity in the colon of the Guinea pig, mouse, rabbit, and rat. Am J Physiol Gastrointest Liver Physiol 305:G749–G759
11. Costa M, Furness JB (1976) The peristaltic reflex: an analysis of the nerve pathways and their pharmacology. Naunyn Schmiedeberg's Arch Pharmacol 294:47–60
12. Costa M, Hibberd TJ, Keightley LJ, Wiklendt L, Arkwright JW, Dinning PG, Brookes SJH, Spencer NJ (2020) Neural motor complexes propagate continuously along the full length of mouse small intestine and colon. Am J Physiol Gastrointest Liver Physiol 318:G99–G108
13. Costa M, Keightley LJ, Hibberd TJ, Wiklendt L, Dinning PG, Brookes SJ, Spencer NJ (2021) Motor patterns in the proximal and distal mouse colon which underlie formation and propulsion of feces. Neurogastroenterol Motil 33(7):e14098
14. Costa M, Keightley LJ, Wiklendt L, Hibberd TJ, Arkwright JW, Omari T, Wattchow DA, Zagorodnyuk V, Brookes SJH, Dinning PG, Spencer NJ (2019) Roles of three distinct neurogenic motor patterns during pellet propulsion in Guinea pig distal colon. J Physiol 597(20):5125–5140. https://doi.org/10.1113/JP278284
15. Costa M, Wiklendt L, Keightley L, Brookes SJH, Dinning PG, Spencer NJ (2017) New insights into neurogenic cyclic motor activity in the isolated Guinea-pig colon. Neurogastroenterol Motil 29:1–13
16. Costa M, Furness JB (1987) The enteric nervous system. Churchill Livingston, Edinburgh
17. Dickens EJ, Hirst GD, Tomita T (1999) Identification of rhythmically active cells in Guinea-pig stomach. J Physiol 514:515–531
18. Dinning PG, Wiklendt L, Omari T, Arkwright JW, Spencer NJ, Brookes SJ, Costa M (2014) Neural mechanisms of peristalsis in the isolated rabbit distal colon: a neuromechanical loop hypothesis. Front Neurosci 8:75
19. Edwards FR, Hirst GD, Klemm MF, Steele PA (1995) Different types of ganglion cell in the cardiac plexus of Guinea-pigs. J Physiol 486(Pt 2):453–471
20. Fida R, Lyster DJ, Bywater RA, Taylor GS (1997) Colonic migrating motor complexes (CMMCs) in the isolated mouse colon. Neurogastroenterol Motil 9:99–107
21. Furness JB (2006) The enteric nervous system. Blackwell Publishing, Oxford, U.K.
22. Furness JB (2012) The enteric nervous system and neurogastroenterology. Nat Rev Gastroenterol Hepatol 9:286–294

23. Furness JB, Jones C, Nurgali K, Clerc N (2004) Intrinsic primary afferent neurons and nerve circuits within the intestine. Prog Neurobiol 72:143–164
24. Furness JB, Kunze WA, Bertrand PP, Clerc N, Bornstein JC (1998) Intrinsic primary afferent neurons of the intestine. Prog Neurobiol 54:1–18
25. Furukawa K, Taylor GS, Bywater RA (1986) An intracellular study of myenteric neurons in the mouse colon. J Neurophysiol 55:1395–1406
26. Haddock RE, Hill CE (2005) Rhythmicity in arterial smooth muscle. J Physiol 566:645–656
27. Heredia DJ, Dickson EJ, Bayguinov PO, Hennig GW, Smith TK (2009) Localized release of serotonin (5-Hydroxytryptamine) by a fecal pellet regulates migrating motor complexes in murine colon. Gastroenterology 136:1328–1338
28. Heredia DJ, Gershon MD, Koh SD, Corrigan RD, Okamoto T, Smith TK (2013) Important role of mucosal serotonin in colonic propulsion and peristaltic reflexes: in vitro analyses in mice lacking tryptophan hydroxylase 1. J Physiol 591:5939–5957
29. Hirst GD, Edwards FR (2004) Role of interstitial cells of Cajal in the control of gastric motility. J Pharmacol Sci 96:1–10
30. Hosie S, Ellis M, Swaminathan M, Ramalhosa F, Seger GO, Balasuriya GK, Gillberg C, Rastam M, Churilov L, Mckeown SJ, Yalcinkaya N, Urvil P, Savidge T, Bell CA, Bodin O, Wood J, Franks AE, Bornstein JC, Hill-Yardin EL (2019) Gastrointestinal dysfunction in patients and mice expressing the autism-associated R451C mutation in neuroligin-3. Autism Res 12:1043–1056
31. Hu H, Spencer NJ (2018) Enteric nervous system structure and neurochemistry related to function and neuropathology. In: Said HM (ed) Physiology of the gastrointestinal tract, 6th edn. Elsevier/Academic Press, Cambridge
32. Huizinga JD, Thuneberg L, Kluppel M, Malysz J, Mikkelsen HB, Bernstein A (1995) W/kit gene required for interstitial cells of Cajal and for intestinal pacemaker activity. Nature 373:347–349
33. Keating DJ, Spencer NJ (2010) Release of 5-hydroxytryptamine from the mucosa is not required for the generation or propagation of colonic migrating motor complexes. Gastroenterology 138:659-70 670 e1-2
34. Keating DJ, Spencer NJ (2019) What is the role of endogenous gut serotonin in the control of gastrointestinal motility? Pharmacol Res 140:50–55
35. Kunze WA, Bornstein JC, Furness JB (1995) Identification of sensory nerve cells in a peripheral organ (the intestine) of a mammal. Neuroscience 66:1–4
36. Lang RJ, Hashitani H (2019) Pacemaker mechanisms driving Pyeloureteric peristalsis: modulatory role of interstitial cells. Adv Exp Med Biol 1124:77–101
37. Leembruggen AJL, Balasuriya GK, Zhang J, Schokman S, Swiderski K, Bornstein JC, Nithianantharajah J, Hill-Yardin EL (2020) Colonic dilation and altered ex vivo gastrointestinal motility in the neuroligin-3 knockout mouse. Autism Res 13:691–701
38. Li Z, Hao MM, Van Den Haute C, Baekelandt V, Boesmans W, Vanden Berghe P (2019) Regional complexity in enteric neuron wiring reflects diversity of motility patterns in the mouse large intestine. elife 8:e42914. https://doi.org/10.7554/eLife.42914
39. Mazzuoli-Weber G, Schemann M (2015) Mechanosensitive enteric neurons in the Guinea pig gastric corpus. Front Cell Neurosci 9:430
40. Mazzuoli G, Schemann M (2012) Mechanosensitive enteric neurons in the myenteric plexus of the mouse intestine. PLoS One 7:e39887
41. Nurgali K, Stebbing MJ, Furness JB (2004) Correlation of electrophysiological and morphological characteristics of enteric neurons in the mouse colon. J Comp Neurol 468:112–124
42. Roberts RR, Bornstein JC, Bergner AJ, Young HM (2008) Disturbances of colonic motility in mouse models of Hirschsprung's disease. Am J Physiol Gastrointest Liver Physiol 294:G996–G1008
43. Roberts RR, Murphy JF, Young HM, Bornstein JC (2007) Development of colonic motility in the neonatal mouse-studies using spatiotemporal maps. Am J Physiol Gastrointest Liver Physiol 292:G930–G938
44. Sergeant GP, Hollywood MA, Mccloskey KD, Thornbury KD, Mchale NG (2000) Specialised pacemaking cells in the rabbit urethra. J Physiol 526(Pt 2):359–366
45. Sibaev A, Franck H, Vanderwinden JM, Allescher HD, Storr M (2003) Structural differences in the enteric neural network in murine colon: impact on electrophysiology. Am J Physiol Gastrointest Liver Physiol 285:G1325–G1334
46. Spencer NJ, Bywater RA (2002) Enteric nerve stimulation evokes a premature colonic migrating motor complex in mouse. Neurogastroenterol Motil 14:657–665
47. Spencer NJ, Costa M, Hibberd TJ, Wood JD (2021) Advances in colonic motor complexes in mice. Am J Physiol Gastrointest Liver Physiol 320:G12–G29
48. Spencer NJ, Dinning PG, Brookes SJ, Costa M (2016) Insights into the mechanisms underlying colonic motor patterns. J Physiol 594:4099–4116
49. Spencer NJ, Greenwood IA (2003) Characterization of properties underlying rhythmicity in mouse portal vein. Auton Neurosci 104:73–82
50. Spencer NJ, Hennig GW, Dickson E, Smith TK (2005) Synchronization of enteric neuronal firing during the murine colonic MMC. J Physiol 564:829–847
51. Spencer NJ, Hibberd TJ, Travis L, Wiklendt L, Costa M, Hu H, Brookes SJ, Wattchow DA, Dinning PG, Keating DJ, Sorensen J (2018) Identification of a rhythmic firing pattern in the enteric nervous system that generates rhythmic electrical activity in smooth muscle. J Neurosci 38:5507–5522
52. Spencer NJ, Hu H (2020) Enteric nervous system: sensory transduction, neural circuits and gastrointestinal motility. Nat Rev Gastroenterol Hepatol 17(6):338–351

53. Spencer NJ, Kyloh M, Wattchow DA, Thomas A, Sia TC, Brookes SJ, Nicholas SJ (2012) Characterization of motor patterns in isolated human colon: are there differences in patients with slow-transit constipation? Am J Physiol Gastrointest Liver Physiol 302:G34–G43
54. Spencer NJ, Nicholas SJ, Sia TC, Staikopoulos V, Kyloh M, Beckett EA (2013) By what mechanism does ondansetron inhibit colonic migrating motor complexes: does it require endogenous serotonin in the gut wall? Neurogastroenterol Motil 25:677–685
55. Spencer NJ, Smith TK (2001) Simultaneous intracellular recordings from longitudinal and circular muscle during the peristaltic reflex in Guinea-pig distal colon. J Physiol 533:787–799
56. Spencer NJ, Smith TK (2004) Mechanosensory S-neurons rather than AH-neurons appear to generate a rhythmic motor pattern in Guinea-pig distal colon. J Physiol 558:577–596
57. Spencer NJ, Travis L, Wiklendt L, Hibberd TJ, Costa M, Dinning P, Hu H (2020) Diversity of neurogenic smooth muscle electrical rhythmicity in mouse proximal colon. Am J Physiol Gastrointest Liver Physiol 318:G244–G253
58. Spencer NJ, Travis L, Wiklendt L, Hibberd TJ, Costa M, Brookes SJ, Dinning PG, Hu H, Wattchow DW, Sorensen J (2021) Long range synchronization within the enteric nervous system underlies propulsion along the large intestine of mice. Commun Biol 4:955. https://doi.org/10.1038/s42003-021-02485-4
59. Vincent AD, Wang XY, Parsons SP, Khan WI, Huizinga JD (2018) Abnormal absorptive colonic motor activity in germ free mice is rectified by butyrate, an effect possibly mediated by mucosal serotonin. Am J Physiol Gastrointest Liver Physiol 315(5):G896–G907
60. Wei L, Singh R, Ha SE, Martin AM, Jones LA, Jin B, Jorgensen BG, Zogg H, Chervo T, Gottfried-Blackmore A, Nguyen L, Habtezion A, Spencer NJ, Keating DJ, Sanders KM, Ro S (2021) Serotonin deficiency is associated with delayed gastric emptying. Gastroenterology 160(2451–2466):e19
61. Wood JD (1973) Electrical activity of the intestine of mice with hereditary megacolon and absence of enteric ganglion cells. Am J Dig Dis 18:477–488
62. Wood JD (2016) Enteric neurobiology: discoveries and directions. Adv Exp Med Biol 891:175–191
63. Wood JD (2018) Enteric nervous system: brain in the gut, Chapter 15. In: Said HM (ed) Handbook of physiology, 6th edn. Academic Press, pp 361–272
64. Wood JD (2021) Motor behavior of mouse large intestine: a minireview. Neurogastroenterol Motil 33(7):e14206. https://doi.org/10.1111/nmo.14206
65. Young HM, Mckeown SJ (2016) Motility: Hirschsprung disease--laying down a suitable path. Nat Rev Gastroenterol Hepatol 13:7–8
66. Zagorodnyuk VP, Spencer NJ (2011) Localization of the sensory neurons and mechanoreceptors required for stretch-evoked colonic migrating motor complexes in mouse colon. Front Physiol 2:98

28 The Shaggy Dog Story of Enteric Signaling: Serotonin, a Molecular Megillah

Michael D. Gershon

Abstract

Historically and quantitatively, the enteric site of serotonin (5-HT) storage has primacy over those of any other organ. 5-HT, by the name of "enteramine", was first discovered in the bowel, and the gut produces most of the body's 5-HT. Not only does the bowel secrete 5-HT prodigiously but it also expresses a kaleidoscopic abundance of 5-HT receptors. The larger of two enteric 5-HT stores is mucosal, biosynthetically dependent upon tryptophan hydroxylase1 (TPH1), and located in EC cells. Mechanical stimuli, nutrients, luminal bacteria, and neurotransmitters such as acetylcholine and norepinephrine are all able to stimulate EC cells. Paracrine actions of 5-HT allow the mucosa to signal to neurons to initiate peristaltic and secretory reflexes as well as to inflammatory cells to promote intestinal inflammation. Endocrine effects of 5-HT allow EC cells to influence distant organs, including bone, liver, and endocrine pancreas. The smaller enteric 5-HT store is biosynthetically dependent upon TPH2 and is located within a small subset of myenteric neurons. 5-HT is responsible for slow excitatory neurotransmission manifested primarily in type II/AH neurons. Importantly, neuronal 5-HT also promotes enteric nervous system (ENS) neurogenesis, both pre- and postnatally, through 5-HT_{2B} and especially 5-HT_4 receptors. The early birth of serotonergic neurons allows these cells to function as sculptors of the mature ENS. The inactivation of secreted 5-HT depends on transmembrane transport mediated by a serotonin transporter (SERT; *SLC6A4*). The importance of SERT in control of 5-HT's function means that pharmacological inhibition of SERT, as well as gain- or loss-of-function mutations in *SLC6A4,* can exert profound effects on development and function of the ENS. Extra-enteric, TPH1-derived 5-HT from yolk sac and placenta promotes neurogenesis before enteric neurons synthesize 5-HT and contribute to ENS patterning. The impressive multi-functional nature of enteric 5-HT has made the precise identification of individual physiological roles difficult and sometimes controversial.

The authors' work discussed in this chapter has been supported by grant number NS15547 of the National Institutes of Health of the US Public Health Service

M. D. Gershon (✉)
Department of Pathology and Cell Biology, Columbia University Vagelos College of Physicians and Surgeons, New York, NY, USA
e-mail: mdg4@cumc.columbia.edu

N. J. Spencer et al. (eds.), *The Enteric Nervous System II*, Advances in Experimental Medicine and Biology 1383, https://doi.org/10.1007/978-3-031-05843-1_28

Keywords

Enterochromaffin (EC) cells · Enteric nervous system (ENS) · Enteric neuronal development · Peristaltic and secretory reflex · Enteric neurogenesis · Serotonin transporter (SERT)

PubMed, the bibliographic service of the National Library of Medicine (USA), provides, as of August 2021, 6265 results for the query "serotonin and intestine." That is a relatively large number of publications and indicates that scientific interest in that subject has been, if not extraordinary, at least substantial. It is thus surprising that a recent reviewer of an application to the National Institutes of Health (USA) for a research grant expressed incredulity that one might want to study a "brain chemical," like serotonin (5-HT), in the gut. The popular association of 5-HT with "happiness", or perhaps the strong focus of modern research on 5-HT in the brain, which has been the subject of ninefold as many papers as intestinal 5-HT, seems to have obscured, even for some of the scientists that study it, the many roles that 5-HT plays, not only in the gut but elsewhere outside of the CNS. In fact, enteric 5-HT is a polyfunctional molecule, acting as an endocrine hormone, a paracrine messenger, and a neurotransmitter [31]. It is also a mitogen [57] and growth factor that stimulate neurogenesis in the developing [28] and adult [48] enteric nervous system (ENS), as well as proliferation of the mucosal epithelium of the intestinal mucosa [37, 51]. The endocrine role of 5-HT links it to the control of osteocytes in bone deposition [43], the paracrine role to the initiation of peristaltic reflexes [14, 38] and inflammation [8, 23, 34, 50, 51], and the neurotransmitter role to the regulation of gastrointestinal (GI) motility [33, 41, 51]. To be sure, some of these putative functions of enteric 5-HT, such as its initiation of peristaltic reflexes, have been disputed [63], but it remains obvious that there was a good reason, or perhaps an abundance of reasons, for the biosynthesis of 5-HT in the GI tract to have evolved and thus to be studied. The enteric functions of 5-HT have been a complex tale, full of unexpected twists, turns, and reversals of fortune; thus it is very much a "shaggy dog story," and the importance of 5-HT is what makes it a molecular megillah.

28.1 The History of 5-HT

The story of 5-HT in both the brain and the gut is not very old. It began in Italy during the great depression, well after the discoveries of the enteric nervous system (ENS) [2, 46, 53] and its ability to control to the behavior of the bowel independently of input from the brain or spinal cord [3–5, 67]. At that time, despite the depression and the rise of fascism in Italy, Vittorio Erspamer observed that material extracted from the wall of the bowel with acetone caused smooth muscle preparations to contract [25]. Erspamer further noted that the active principal reacted chemically as if it were an indole and that it was concentrated in the intestinal mucosa. Histochemical staining of gut sections, done to look for that indole, revealed that enterochromaffin cells of the intestinal epithelium contained an indoleamine that he named "enteramine" [11, 27, 69]. World War II intervened; however, Erspamer was distracted, and an entirely different trail of research would first reveal the actual structure of 5-HT.

The trail of research that would ultimately lead to the discovery of the structure of 5-HT had little or nothing to do either with the gut or the brain. It had to do with the quest of Irvine Page at the Cleveland Clinic to treat hypertension [73]. Page reasoned that there might be a vasoconstrictive substance in the blood of people blood pressure was abnormally high. A big problem for Page's research was that blood clotted after it was drawn and that it had been known since 1846 that clotting released a vasoconstrictive substance into the serum. Page hired Maurice Rapport, a really good chemist who had just obtained a PhD degree, and Arda Green, a biochemist, who were given the task of getting rid of the interfering serum vasoconstrictor. Instead of doing that exactly, Rapport identified the molecule and proved the structure of 5-HT by synthesis [55]. The team of Rapport, Green, and Page chose the name "serotonin" for 5-HT because it was a serum vasoconstrictor. By 1953, Erspamer realized that "enteramine" and serotonin were the same thing [26, 27]; however, although one might argue that "enteramine" is a better name for a molecule that is most abundant in the bowel, it was too late. After Betty Twarog, working with Page, found that 5-HT was also present in the brain [68] and D. Wayne Woolley linked 5-HT to mental function [74], research aiming to determine the role of 5-HT in the brain-dominated science (as it still does) and the nomenclature had become permanent.

28.2 5-HT and the Peristaltic Reflex

The study of the functional roles that 5-HT plays in the bowel began in earnest with a pioneering series of papers from the Oxford laboratory of Edith Bülbring [14–16, 19, 20]. She was interested in the initiation of the peristaltic reflex. This reflex, or something like it, was first observed by Bayliss and Starling who showed that pinching the bowel, or increasing pressure in the lumen of the intestine, elicited a reproducible series of movements of the gut consisting of contraction oral to the point of stimulation and relaxation distal to it, which they called the "law of the intestine" [3–5]. Because Bayliss and Starling were able to block that behavior with cocaine and nicotine but evoke it after they had severed all the extrinsic nerves running to the bowel, they attributed the activity to what they called the "local nervous mechanism" of the gut. Bayliss and Starling could do this because they were aware of prior discoveries made by Auerbach [2] and Meissner [53] that a very large intrinsic enteric nervous system (ENS) is present within the bowel. After the publication of the work of Bayliss and Starling, Paul Trendelenburg demonstrated intestinal behavior, which was identical to the "law of the intestine," in preparations of gut that were isolated in vitro [67]. This observation of "the peristaltic reflex" in vitro put an exclamation point onto the conclusion of Bayliss and Starling that the behavior was due to the activity of the intrinsic ENS within the gut. Bülbring, who in her youth had lived in the same house as Trendelenburg, was anxious to determine how luminal stimulation is able to engage the ENS.

Bülbring and her associates devised a new procedure for demonstrating the peristaltic reflex. Instead of just recording the contraction of the longitudinal muscle or the pressure within the lumen of the gut, she measured both and, at the same time, the fluid the gut propelled [17]. This allowed Bülbring to conclude that she was truly looking at authentic peristaltic reflexes, which were propulsive, both in vitro and in vivo. The Bülbring group was able to use their improved preparation to follow up on the observations of Erspamer that showed that 5-HT is present in the intestinal mucosa, of Twarog and Wooley that 5-HT is present in neuronal tissue and neuroactive, and of Trendelenburg that increased intraluminal pressure evokes the peristaltic reflex. Bülbring first showed that intraluminal 5-HT enhanced peristaltic reflexes and lowered the threshold of pressure needed for their elicitation [14]. In contrast, serosal application of 5-HT inhibited or even blocked the peristaltic reflex [19]. In addition, she demonstrated that 5-HT is synthesized locally within the intestinal mucosa from 5-hydroxytryptophan and released from the mucosa when intraluminal pressure was raised and propulsive movement of the bowel was evoked [19]. These observations were important because Bülbring also showed that removal, asphyxiation, or anesthesia of the "mucous membrane" of the intestine impaired the activity that she identified as the peristaltic reflex, which was thus dependent for its initiation on the mucosa [20]. Bülbring further noted that primary afferent neurons are present in the bowel wall, she thought in the submucosal plexus, and she postulated that 5-HT is secreted from EC cells in response to mucosal stimulation and activates these intrinsic primary afferent neurons (IPANs) to initiate peristaltic reflexes [20]. Bülbring put these data together to suggest that increased intraluminal pressure releases 5-HT from EC cells, which stimulates the mucosal endings of IPANs that initiate peristaltic reflexes. The IPANs thus, in Bülbring's hypothesis, act as the gateways to circuits within the ENS that control activity of the excitatory and inhibitory motor neurons that are ultimately responsible for the oral contraction and anal relaxation that comprise the "law of the intestine" or peristaltic reflex. Bülbring's further studies confirmed that 5-HT is indeed released from EC cells in response to elevations of intraluminal pressure [16]. She subsequently went on to verify that responses of intact gut in vivo were identical to what she had demonstrated in vitro [15]. This work also established that very high concentrations of 5-HT could desensitize relevant 5-HT receptors and block its own action; however, Bülbring never was able to conclude, definitively in her mind, that mucosal 5-HT was essential for evoking peristaltic reflexes because reserpine, the agent that she employed to deplete 5-HT, was unable either to block 5-HT biosynthesis or reduce the 5-HT concentration to zero.

28.3 Did Bülbring Make a Mistake?

The deficiency that Bülbring recognized in her own work, that she could neither totally deplete the 5-HT stored in the bowel nor block its biosynthesis, has formed the basis of challenges to her hypotheses. No one, since Bülbring's original work, has challenged the ability of 5-HT to evoke the peristaltic reflex, and skepticism has all been directed at the question of whether 5-HT signaling is "essential" for its manifestation. Boullin, for example [12], used a tryptophan-free diet to deplete 5-HT in rats up to a mean of about 90%. He found that the amplitudes of peristaltic reflexes in 5-HT-depleted animals were not significantly different from those in paired rats fed a normal diet. In fact, although no deficiencies in peristaltic activity were observed due to diet-induced 5-HT depletion, the effect of 5-HT on peristaltic reflexes was amplified in rats fed diets high in tryptophan, suggesting that an elevated level of endogenous 5-HT might add to the effect of exogenous 5-HT. Boullin nevertheless concluded that 5-HT was not essential for manifestation of the peristaltic reflex. Bülbring herself accepted Boullin's conclusions and decided that mucosal 5-HT might be a modulator of the peristaltic reflex, acting to lower the pressure threshold needed to evoke the reflex although this role of 5-HT might not be an essential one [13].

More recent studies that have revisited the issue of the role of 5-HT in the peristaltic reflex have continued to focus on the question of whether 5-HT is "essential." Studies have shown, for example, that a non-absorbed inhibitor of tryptophan hydroxylase (TPH), the rate-limiting enzyme in 5-HT biosynthesis, does not affect total GI transit time or colonic motility in mice [76]. In fact, even the genetic deletion of TPH1, which is the isoform of TPH found in EC cells, fails to alter total GI transit time, the time to expel a bead inserted into the rectum, gastric emptying, or small intestinal transit [47]. In contrast, deletion of TPH2, the isoform of TPH found in neurons, greatly slows parameters of intestinal motility, although it accelerates gastric emptying [47], perhaps by eliminating a neuronal role for 5-HT in vagal compliance (relaxation) of the stomach [18].

Methods utilized to study initiation of peristaltic reflexes have not been identical. There seems to be an underlying assumption that if one elicits the intestinal behavior that fits the description of the peristaltic reflex, then that is sufficient. Variations in stimuli have not been emphasized. It is likely, however, that more than one mechanism exists for engaging the microcircuits within the ENS that underlie peristaltic reflexes. Backup systems are common in biology; elimination of one pathway is often followed by activity in another that compensates for the absence of the pathway that has been deleted. In the case of the peristaltic reflex, there is evidence that, in contrast to what Bülbring reported, behavior of the gut that appears to be a peristaltic reflex can still be evoked, even after removal of the mucosa [44, 61, 62, 78]. This evidence, however, is also contradicted by evidence that Bülbring was right after all, and that removal of the mucosa does indeed abolish the peristaltic reflex [30, 39]. The apparent contradiction in observations has led groups to suggest that the opposing experimenters did not remove the mucosa properly, either damaging delicate nerve fibers underlying the mucosal epithelium, or incompletely removing the mucosa [60, 64]. There is no question, however, that no matter how adept one might be in removing the mucosa from the bowel, the preparation that remains after the mucosa has been removed is not a physiological one. A mucosa-free mammalian gut does not exist in nature. Whatever mechanism(s) the mucosa might have had to initiate peristaltic reflexes cannot operate after the mucosa has been removed; therefore, removal of the mucosa cannot provide definitive insight into how peristaltic reflexes are evoked when the mucosa and its connections to the ENS are intact.

28.4 5-HT Is Essential for Mucosal Stimulation to Evoke the Peristaltic Reflex, But 5-HT Is Not Essential for Reflexes That Mechanosensitive Nerve Fibers Induce

Light brushing of the mucosal surface of an intact wild-type mouse gut evokes colonic migrating motor complexes (CMMCs), which are a form of peristaltic reflex; however, an identical stimulus fails to do so in mice that lack TPH1 (TPH1KO) [38]. Fecal pellets, moreover, are not propagated normally in the oral to anal direction in TPH1KO mice but may instead move irregularly or even reverse to move in the anal to oral direction. The colon of the TPH1KO mouse, furthermore, is elongated and dilated and contains grotesquely large fecal pellets. These many observations on the peristaltic reflex, taken together, suggest that more than a single mechanism exists that can engage the hardwired neuronal circuitry of the ENS that gives rise to the peristaltic reflex. In essence, at least two switches seem to exist that, when flipped, activate the machinery that drives peristaltic reflexes. One such switch or mechanism is mucosal stimulation, which operates indirectly through the secretion of 5-HT from EC cells. The secreted 5-HT excites mucosal projections of IPANs that serve as gateways to the circuitry of the ENS [38, 59]. The other mechanism is the initiation of peristaltic reflexes through the direct activation of mechanosensitive nerve fibers in the gut wall [63]. The mucosal pathway is 5-HT-dependent, while the direct activation of mechanosensitive nerve fibers is 5-HT-independent. Experimenters who stretch the bowel to evoke peristaltic reflexes thus find that the mucosa can be eliminated and that 5-HT is not essential. Experimenters who restrict their stimuli to the mucosa find that the mucosa cannot be eliminated and that 5-HT is essential. The gross normality of GI transit in TPH1KO mice probably reflects the ability of the bowel to compensate for the loss of the mucosal pathway to engage peristaltic reflexes; fecal pellets enlarge and stretch the gut, which eventually activates mechanosensors to trigger reflexes and propel luminal content. The abnormal elongated colon and massive fecal pellets of the TPH1KO mouse, no doubt, reflect the price these animals pay for not having the mucosal 5-HT-dependent pathway available.

28.5 Luminal Microbes Use EC Cells and 5-HT to Regulate the ENS

In vivo, the EC cells of the GI epithelium interact with luminal contents, including the massive microbiome of the bowel. Enteric microbiota, moreover, are not inert. Their presence enhances the ability of EC cells to synthesize and release 5-HT [56, 77]. As this happens, for example, when a germ-free mouse gut is colonized, the motility of the bowel is enhanced [24, 56, 77], an outcome that is consistent with the idea that EC cell 5-HT is linked to propulsive behavior of the gut. Strikingly, when the bowel of germ-free mice is colonized with a fecal transplant from conventionally raised mice, the neuroanatomy of the ENS changes in a way that suggests the occurrence of neurogenesis; moreover, GI transit also accelerates. These changes are associated with enhanced production of 5-HT, both in neurons and in EC cells of the intestinal mucosa [24]. The effects of depletion of endogenous 5-HT from EC cells and/or enteric neurons, as well as blockade of 5-HT_4 receptors, suggest that the action of the intestinal microbiota is to cause the release of 5-HT from EC cells and enteric neurons, which stimulates 5-HT_4 receptors that are neuroprotective and enhance maturation of the ENS, even in adult mice. The intestinal microbiome thus modifies the ENS and accelerates GI transit, not only through peristaltic reflexes, but also through trophic actions on enteric neurons [24].

28.6 5-HT Is a Growth Factor for Enteric Neurons

The idea that 5-HT is a growth factor for enteric neurons predated the observations of the effects of enteric microbiota on EC cells and 5-HT. The earliest observations were made on enteric neurogenesis in vitro [28]. 5-HT was found to enhance neurogenesis when added to cultures of enteric neuronal precursors and promoted enteric neuronal development through an action of the 5-HT_{2B} receptor [28]. That observation was followed by evidence that the 5-HT_4 receptor might be at least as important as the 5-HT_{2B} receptor in mediating the ability of 5-HT to induce enteric neuroprotection and neurogenesis [24, 48]. The postnatal development of the ENS was found to be deficient in mice that lacked 5-HT_4 receptors, and 5-HT_4 agonists, like 5-HT_{2B} agonists, were powerful promoters of the development of enteric neurons from isolated neural crest-derived precursors [48]. Studies of the development of the ENS indicated that enteric serotonergic neurons occupy a strategic position in enteric neuronal development because they are among the first neurons to be born during ENS ontogeny [54]. Neurons that are born after the birth of serotonergic neurons undergo their terminal mitoses in the presence of serotonergic axon terminals [21]. Serotonin from the early-born neurons can thus contribute both to the ultimate numbers of these follower neurons and to the choices of their phenotypes. The physiological importance of 5-HT as a growth factor has been demonstrated by studies of the ENS in mice that lack TPH2, the isoform of TPH that is critical for 5-HT biosynthesis in neurons of the CNS [72] and ENS [47]. The deletion of TPH2, but not TPH1 (the TPH isoform in EC cells), leads to hypoplasia of the ENS and specific deficiencies of late-born neurons, such as those marked by their content of GABA, CGRP, and tyrosine hydroxylase. The effects of TPH2 deletion are essentially mimicked by TPH inhibition in neurons by parachlorophenylalanine [24]. Interestingly, the effects of TPH2 depletion are not mimicked by administration of LP-920540 or LX1032, which are non-absorbed inhibitors of TPH that deplete mucosal, but not enteric neuronal 5-HT [52]. These observations indicate that enteric neuronal 5-HT is an important growth factor for the developing and mature ENS. As such, although GI transit is totally abnormal and slow in TPH2KO mice, it is impossible to know whether the defects in these animals are due to the absence of the neurotransmitter properties of 5-HT or to the absence of an appropriate number of neurons, or both. TPH2 also seems to be more important for ENS development than TPH1, although the studies of the effects of adding microbiota from conventionally raised mice to the bowel of germ-free animals [24] indicate that the mucosal as well as neuronal sources of 5-HT contribute to neuroprotection and ENS maintenance. It is possible that the effects of 5-HT from EC cells are not direct actions, but mediated indirectly through the ability, discussed above, of mucosal 5-HT to stimulate IPANs and thus engage the activity of ENS circuits.

28.7 SERT Plays a Critical Role in Enteric 5-HT Signaling

5-HT is released from the mucosa of the intestine in response to mechanical stimulation [6, 7]. Technological improvements have allowed mucosal 5-HT secretion to be followed for long periods of time in vivo and have confirmed that distension of the gut releases 5-HT in parallel with the peristaltic movements of the bowel [49]. Given the abundance of mucosal 5-HT and the observations of Bülbring that 5-HT blocks peristaltic reflexes and intestinal propulsion when it is applied to the serosal surface of the gut, it would seem to be necessary to compartmentalize mucosal 5-HT to its own layer of the gut and thus to protect the ENS from direct exposure to mucosal 5-HT. It is important to bear in mind that 5-HT is highly charged at a physiological pH; therefore, the molecule is not very membrane permeant in the absence of a specialized transporter. The enzymes that catabolize 5-HT, moreover, monoamine oxidase [11] and, in the bowel, glucuronyltransferase [32], are intracellular; thus unless 5-HT is transported into cells, it is not likely to be catabolized. In neurons, the serotonin reuptake transporter (SERT) catalyzes the uptake of 5-HT and is primarily responsible for the inactivation of 5-HT following its release at synapses and action on membrane receptors [10, 40]. The same transporter, SERT, is expressed in guinea pig brain and bowel [22] and rat gut [70], with a distribution of RNA that is consistent with its expression in enteric serotonergic neurons and mucosal epithelial cells. The enteric SERT of the guinea pig was cloned from the mucosal epithelium [22], and it is widely expressed throughout the guinea pig mucosal epithelium, although, in the rat, mucosal SERT is more heavily distributed in crypts [22]. The presence of SERT in the enteric mucosa enables the 5-HT that EC cells secrete to be inactivated rapidly, preventing excessive stimulation of mucosal receptors, their desensitization, as well as the overflow of 5-HT to swamp the ENS. In fact, the amplitude of EPSPs, recorded in cholinergic submucosal neurons (possible IPANs) following stimuli confined to the mucosa, is enhanced when SERT is inhibited with fluoxetine [22]. The number of submucosal neurons that respond to stroking of the mucosa, moreover, is also increased by the inhibition of SERT. Interestingly, the ability of mucosal stroking to excite submucosal neurons can be blocked with antagonists at 5-HT receptors, including 5-HT_3 and 5-HT_4 antagonists [29, 35, 42], as well as novel antagonists, such as anti-idiotypic antibodies that recognize 5-HT receptors [71] and a dipeptide of 5-HTP that is highly selective for actions of 5-HT on enteric neurons [65].

28.8 Inhibition of VMAT2

In assessing observations on mucosal or neuronal 5-HT signaling, it is important to consider the action of reserpine [58]. Reserpine is an old drug that, as discussed earlier, Bülbring first applied to the investigation of the peristaltic reflex [15]. Reserpine is an irreversible inhibitor of the vesicular monoamine transporter 2 (VMAT2) [9]. As such, it allows 5-HT biosynthesis to persist, although 5-HT (and other monoamine stores) is reduced to very low levels. Non-vesicular release of 5-HT, from the cytosol [45] possibly mediated by SERT acting in reverse, can still function, as Bülbring noted to her regret. To abolish the effects of endogenous 5-HT, therefore, it is necessary to delete TPH1 (for EC cells) or TPH2 (for neurons).

28.9 SERT Regulates Enteric Serotonergic Signaling

Because the inactivation of 5-HT is so SERT-dependent, alterations in SERT activity exert profound effects on the behavior of the gut and the development of the ENS [51]. The actions of SERT, moreover, are clinically relevant to autism spectrum disorder (ASD). GI disturbances are frequently seen in ASD [36]. Rare, hyperfunctional coding variants of the SERT (encoded by SLC6A4) have been identified in ASD [51]. Expression in mice of the most common of these (SERT Ala56) increases 5-HT removal and leads to behaviors that resemble ASD. Mice that express SERT Ala56 also exhibit functional GI defects that are like those observed in TPH2KO animals, which lack neuronal 5-HT, including ENS hypoplasia, slow GI transit, diminished peristaltic reflex activity, and deficient proliferation of crypt epithelial cells. These observations suggest that the SERT overactivity is the functional equivalent of lack of 5-HT. In mice that lack SERT (SERTKO) and progeny of dams treated chronically with the SERT inhibitor, fluoxetine, an opposite phenotype occurs. Inability to inactivate 5-HT leads to its overactivity. These reciprocal phenotypes thus support the concept that serotonergic signaling is a powerful and critical regulator of ENS development. Disturbances of 5-HT signaling cause long-lasting abnormalities of GI function. The critical receptor for serotonergic signaling in ENS development appears to be 5-HT_4; thus, administration of prucalopride prevents the occurrence of SERT Ala56-associated GI perturbations. That observation implies that SERT overactivity deprives the 5-HT_4 receptor of neuronal precursors of their ligand, diminishing the trophic effects of 5-HT. Prucalopride can substitute for 5-HT, despite the SERT overactivity in SERT Ala56 mice, because prucalopride is not a substrate for SERT. It seems likely that GI and behavioral features of ASD are due to deficient 5-HT signaling during development. The safety and potentially adverse effects on the gut of SERT inhibition during pregnancy to combat depression therefore must be considered.

28.10 The Importance of Extraenteric TPH1 During Early Development

The developmental importance of 5-HT in the formation of the ENS and the role of microbiota [24] are certainly consistent with the idea that TPH1 is very important. The role of SERT in the compartmentalization of the mucosa [22], however, and the severe ENS defects seen in mice that lack TPH2 [47] cast a dubious light on the relative importance of the TPH1-derived 5-HT pool in EC cells. On the other hand, serotonergic neurons themselves appear to be sensitive to the trophic effects of 5-HT, which suggests that 5-HT might influence ENS development before the appearance of serotonergic neurons or EC cells. In the CNS, the placenta and especially the yolk sac provide an early source of 5-HT in the circulation that exerts profound effects on the patterning of the forebrain [75]. This 5-HT is TPH1-dependent, like that of EC cells. We have recently confirmed these data (Margolis and Gershon, personal observation) and found that the murine placenta and yolk sac are rich in TPH1 at E12/13, a time when little or no TPH2 is expressed in the brain or gut. As a result, TPH1 deletion is potentially able to disrupt ENS development and patterning, even if the effects are subtle and not as obvious as those due to deletion of TPH2.

28.11 Tryptamine

One tool that has been able to provide insight into the role of TPH1-derived 5-HT has been to make use of tryptamine to evoke the secretion of endogenous 5-HT. Tryptamine was demonstrated years ago to act on serotonergic neurons in a manner analogous to the action of tyramine and other indirectly acting sympathomimetic compounds on catecholaminergic neurons [66]. Tryptamine enters serotonergic axon terminals, and VMAT2 transports it into synaptic vesicles. Because tryptamine is a weak base, it collapses the pH gradient across vesicular membranes, which, in turn, causes 5-HT to flow out of vesicles into the cytosol. Tryptamine further activates trace amine-associated receptors (TAAR1)[1] and phosphorylates SERT, causing it to act in reverse and pump 5-HT into the extracellular space. Once released from terminals, 5-HT gains access to its receptors and reveals the effects of endogenous 5-HT release. These effects in the ENS include the mediation of slow transmission in type II/AH neurons [66]. Chronic exposure to tryptamine, furthermore, depletes endogenous 5-HT, thereby blocking the effects of nerve stimulation on type II/AH neurons, while permitting these cells to respond normally to exogenous 5-HT and to the endogenous release of acetylcholine. More recent studies have verified the action of tryptamine on the ENS and have shown that tryptamine is without effect, when it is applied to TPH2KO bowel, which lacks 5-HT. In contrast to the TPH2KO gut, TPH1KO bowel still responds to tryptamine but does so abnormally (Gershon and Margolis, personal observations). When indices of responses to tryptamine are compared in wild-type bowel to the TPH1KO gut, neuronal uptake of FM2–10, cytochrome oxidase activity, and Fos activation in wild-type bowel are all much greater in the wild-type bowel than in that of TPH1KO mice. Strikingly, activation of glial Fos in the TPH1KO gut is stronger after administration of tryptamine than in neurons. These observations are compatible with the supposition that the deletion of TPH1 leads to abnormal patterning of enteric neurites, even though the numbers of various enteric neurons may be close to normal.

28.12 Summary and Conclusions

- The gut bristles with 5-HT and 5-HT receptors.
  - The larger of two enteric 5-HT stores is TPH1-derived and is located in EC cells.
    - EC cells respond to mechanical stimuli, nutrients, luminal bacteria, and neurotransmitters, such as acetylcholine and norepinephrine.
    - Paracrine effects allow the mucosa to signal to neurons to initiate peristaltic and secretory reflexes as well as to inflammatory cells to promote intestinal inflammation.
    - Endocrine actions of 5-HT allow EC cells to exert effects on distant organs, including bone, liver, and endocrine pancreas.
  - The smaller 5-HT store is TPH2-derived and is located within a small subset of enteric neurons.
    - 5-HT is responsible for slow excitatory neurotransmission manifested primarily in type II/AH neurons.
    - 5-HT also promotes ENS neurogenesis, postnatally mainly via 5-HT_4 receptors; moreover, the effects of 5-HT on enteric neurogenesis are manifest during development and in adult life.
- Extra-enteric, TPH1-derived 5-HT from yolk sac and placenta promote neurogenesis before enteric neurons synthesize 5-HT and contribute to ENS patterning.

References

1. Anwar MA, Ford WR, Broadley KJ, Herbert AA (2012) Vasoconstrictor and vasodilator responses to tryptamine of rat-isolated perfused mesentery: comparison with tyramine and beta-phenylethylamine. Br J Pharmacol 165:2191–2202
2. Auerbach L (1864) Fernere vorlaufige Mittielung über den Nervenapparat des Darmes. Arch Pathol Anat Physiol 30:457–460
3. Bayliss WM, Starling EH (1899) The movements and innervation of the small intestine. J Physiol Lond 24:99–143
4. Bayliss WM, Starling EH (1900) The movements and innervation of the large intestine. J Physiol Lond 26:107–118
5. Bayliss WM, Starling EH (1901) The movements and innervation of the small intestine. J Physiol Lond 26:125–138
6. Bertrand PP (2004) Real-time detection of serotonin release from enterochromaffin cells of the Guinea-pig ileum. Neurogastroenterol Motil 16:511–514
7. Bertrand PP, Bertrand RL (2010) Serotonin release and uptake in the gastrointestinal tract. Auton Neurosci 153(1–2):47–57. https://doi.org/10.1016/j.autneu.2009.08.002. Epub 2009 Sep 2. PMID: 19729349.
8. Bischoff SC, Mailer R, Pabst O, Weier G, Sedlik W, Li Z, Chen JJ, Murphy DL, Gershon MD (2009) Role of serotonin in intestinal inflammation: knockout of serotonin reuptake transporter exacerbates 2,4,6-trinitrobenzene sulfonic acid colitis in mice. Am J Physiol Gastrointest Liver Physiol 296:G685–G695
9. Black CA, Bucher ML, Bradner JM, Jonas L, Igarza K, Miller GW (2021) Assessing vesicular monoamine transport and toxicity using fluorescent false neurotransmitters. Chem Res Toxicol 34:1256–1264
10. Blakely RD, Berson HE, Fremeau RT, Caron MG, Peek MM, Prince HK, Bradley CC (1991) Cloning and expression of a functional serotonin transporter from rat brain. Nature 354:66–70
11. Blashko H, Levine WG (1966) Metabolism of indolealkylamines. In: Erspamer V (ed) Handbook of experimental pharmacology: 5-Hydroxytryptamine and related Indolealkylamines. Springer-Verlag, New York
12. Boullin DJ (1964) Observations on the significance of 5-hydroxytryptamine in relation to the peristaltic reflex of the rat. Brit J Pharmacol 23:14–33
13. Bulbring E (1961) Motility of the intestine. Proc R Soc Med 54:773–775
14. Bülbring E, Crema A (1958) Observations concerning the action of 5-hydroxytryptamine on the peristaltic reflex. Br J Pharmacol 13:444–457

15. Bülbring E, Crema A (1959a) The action of 5-hydroxytryptamine, 5-hydroxytryptophan and reserpine on intestinal peristalsis in anaesthetized Guinea-pigs. J Physiol Lond 146:29–53
16. Bülbring E, Crema A (1959b) The release of 5-hydroxytryptamine in relation to pressure exerted on the intestinal mucosa. J Physiol Lond 146:18–28
17. Bulbring E, Crema A, Saxby OB (1958) A method for recording peristalsis in isolated intestine. Br J Pharmacol Chemother 13:440–443
18. Bülbring E, Gershon MD (1967) 5-hydroxytryptamine participation in the vagal inhibitory innervation of the stomach. J Physiol Lond 192:823–846
19. Bülbring E, Lin RCY (1958) The effect of intraluminal application of 5-hydroxytryptamine and 5-hydroxytryptophan on peristalsis, the local production of 5-hydroxytryptamine and its release in relation to intraluminal pressure and propulsive activity. J Physiol Lond 140:381–407
20. Bülbring E, Lin RCY, Schofield G (1958) An investigation of the peristaltic reflex in relation to anatomical observations. Q J Exp Physiol 43:26–37
21. Chalazonitis A, Pham TD, Li Z, Roman D, Guha U, Gomes W, Kan L, Kessler JA, Gershon MD (2008) Bone morphogenetic protein regulation of enteric neuronal phenotypic diversity: relationship to timing of cell cycle exit. J Comp Neurol 509:474–492
22. Chen JX, Pan H, Rothman TP, Wade PR, Gershon MD (1998) Guinea pig 5-HT transporter: cloning, expression, distribution, and function in intestinal sensory reception. Am J Phys 275:G433–G448
23. Coates MD, Tekin I, Vrana KE, Mawe GM (2017) Review article: the many potential roles of intestinal serotonin (5-hydroxytryptamine, 5-HT) signalling in inflammatory bowel disease. Aliment Pharmacol Ther 46:569–580
24. De Vadder F, Grasset E, Manneras Holm L, Karsenty G, Macpherson AJ, Olofsson LE, Backhed F (2018) Gut microbiota regulates maturation of the adult enteric nervous system via enteric serotonin networks. Proc Natl Acad Sci U S A 115:6458–6463
25. Erspamer V (1937) Ricerche sperimentali sul significato biologico delle enuerocromaffini. Arch Fisiol 37:156–169
26. Erspamer V (1953) Über den 5-Hydroytryptamin-(Enteramin)Gehalt des Magen-Darmtraktes bei den Wirbelten. Naturwissenschaften 40:318–319
27. Erspamer V (1954) Pharmacology of indolealklyamines. Pharmacol Rev 6:425–487
28. Fiorica-Howells E, Maroteaux L, Gershon MD (2000) Serotonin and the 5-HT(2B) receptor in the development of enteric neurons. J Neurosci 20:294–305
29. Foxx-Orenstein AE, Kuemmerle JF, Grider JR (1996) Distinct 5-HT receptors mediate the peristaltic reflex induced by mucosal stimuli in human and Guinea pig intestine. Gastroenterology 111:1281–1290
30. Frigo GM, Lecchini S (1970) An improved method for studying the peristaltic reflex in the isolated colon. Br J Pharmacol 39:346–356
31. Gershon MD (2013) 5-Hydroxytryptamine (serotonin) in the gastrointestinal tract. Curr Opin Endocrinol Diabetes Obes 20:14–21
32. Gershon MD, Ross LL (1966) Radioisotopic studies of the binding, exchange, and distribution of 5-hydroxytryptamine synthesized from its radioactive precursor. J Physiol 186:451–476
33. Gershon MD, Tack J (2007) The serotonin signaling system: from basic understanding to drug development for functional GI disorders. Gastroenterology 132:397–414
34. Ghia JE, Li N, Wang H, Collins M, Deng Y, El-Sharkawy RT, Cote F, Mallet J, Khan WI (2009) Serotonin has a key role in pathogenesis of experimental colitis. Gastroenterology 137:1649–1660
35. Grider JR (2003) Neurotransmitters mediating the intestinal peristaltic reflex in the mouse. J Pharmacol Exp Ther 307:460–467
36. Grimaldi R, Gibson GR, Vulevic J, Giallourou N, Castro-Mejia JL, Hansen LH, Leigh Gibson E, Nielsen DS, Costabile A (2018) A prebiotic intervention study in children with autism spectrum disorders (ASDs). Microbiome 6:133
37. Gross ER, Gershon MD, Margolis KG, Gertsberg ZV, Cowles RA (2012) Neuronal serotonin regulates growth of the intestinal mucosa in mice. Gastroenterology 143(408–17):e2
38. Heredia DJ, Gershon MD, Koh SD, Corrigan RD, Okamoto T, Smith TK (2013) Important role of mucosal serotonin in colonic propulsion and peristaltic reflexes: in vitro analyses in mice lacking tryptophan hydroxylase 1. J Physiol 591:5939–5957
39. Heredia DJ, Grainger N, Mccann CJ, Smith TK (2012) Insights from a novel model of slow-transit constipation generated by partial outlet obstruction in the murine large intestine. Am J Physiol Gastrointest Liver Physiol 303:G1004–G1016
40. Hoffman B, Mezey E, Brownstein M (1991) Cloning of a serotonin transporter affected by antidepressants. Science 254:579–580
41. Israelyan N, Del Colle A, Li Z, Park Y, Xing A, Jacobsen JPR, Luna RA, Jensen DD, Madra M, Saurman V, Rahim R, Latorre R, Law K, Carson W, Bunnett NW, Caron MG, Margolis KG (2019) Effects of serotonin and slow-release 5-Hydroxytryptophan on gastrointestinal motility in a mouse model of depression. Gastroenterology 157(507–521):e4
42. Kadowaki M, Wade PR, Gershon MD (1996) Participation of 5-HT3, 5-HT4, and nicotinic receptors in the peristaltic reflex of the Guinea pig distal colon. Am J Phys 271:G849–G857
43. Karsenty G, Gershon MD (2011) The importance of the gastrointestinal tract in the control of bone mass accrual. Gastroenterology 141:439–442
44. Keating DJ, Spencer NJ (2010) Release of 5-hydroxytryptamine from the mucosa is not required for the generation or propagation of colonic migrating motor complexes. *Gastroenterology* 138:659-70–670 e1-2

45. Kuhn DM, Wolf WA, Youdim MB (1985) 5-Hydroxytryptamine release in vivo from a cytoplasmic pool: studies on the 5-HT behavioural syndrome in reserpinized rats. Br J Pharmacol 84:121–129
46. Langley JN (1921) The autonomic nervous system, part 1. W. Heffer, Cambridge
47. Li Z, Chalazonitis A, Huang YY, Mann JJ, Margolis KG, Yang QM, Kim DO, Cote F, Mallet J, Gershon MD (2011) Essential roles of enteric neuronal serotonin in gastrointestinal motility and the development/survival of enteric dopaminergic neurons. J Neurosci 31:8998–9009
48. Liu MT, Kuan YH, Wang J, Hen R, Gershon MD (2009) 5-HT4 receptor-mediated neuroprotection and neurogenesis in the enteric nervous system of adult mice. J Neurosci 29:9683–9699
49. Liu YL, Chen Y, Fan WT, Cao P, Yan J, Zhao XZ, Dong WG, Huang WH (2020) Mechanical distension induces serotonin release from intestine as revealed by stretchable electrochemical sensing. Angew Chem Int Ed Engl 59:4075–4081
50. Margolis KG, Gershon MD (2016) Enteric neuronal regulation of intestinal inflammation. Trends Neurosci 39:614–624
51. Margolis KG, Li ZS, Stevanovic K, Saurman V, Israelyan N, Anderson GM, Snyder I, Veenstra-Vanderweele J, Blakely RD, Gershon MD (2016) Serotonin transporter variant drives preventable gastrointestinal abnormalities in development and function. J Clin Investig 126:2221–2235
52. Margolis KG, Stevanovic K, Li Z, Yang QM, Oravecz T, Zambrowicz B, Jhaver KG, Diacou A, Gershon MD (2014) Pharmacological reduction of mucosal but not neuronal serotonin opposes inflammation in mouse intestine. Gut 63:928–937
53. Meissner G (1857) Über die Nerven der Darmwand. Z Ration Med 8:364–366
54. Pham TD, Gershon MD, Rothman TP (1991) Time of origin of neurons in the murine enteric nervous system. J Comp Neurol 314:789–798
55. Rapport MM, Green AA, Page IH (1948) Serum vasoconstrictor (serotonin). IV Isolation and characterization. J Biol Chem 176:1243–1251
56. Reigstad CS, Salmonson CE, Rainey JF 3rd, Szurszewski JH, Linden DR, Sonnenburg JL, Farrugia G, Kashyap PC (2015) Gut microbes promote colonic serotonin production through an effect of short-chain fatty acids on enterochromaffin cells. FASEB J 29:1395–1403
57. Shah, P. A., Park, C. J., Shaughnessy, M. P. & Cowles, R. A. 2021. Serotonin as a mitogen in the gastrointestinal tract: revisiting a familiar molecule in a new role. *Cell Mol Gastroenterol Hepatol*
58. Sia TC, Whiting M, Kyloh M, Nicholas S, Brookes SJ, Oliver J, Dinning P, Wattchow DA, Spencer NJ (2013) 5-HT3 and 5-HT4 antagonists inhibit peristaltic contractions in Guinea-pig distal colon by mechanisms independent of endogenous 5-HT. Front Neurosci 7:1–10
59. Smith TK, Gershon MD (2015a) CrossTalk proposal: 5-HT is necessary for peristalsis. J Physiol 593:3225–3227
60. Smith TK, Gershon MD (2015b) Rebuttal from Terence K Smith and Michael D Gershon. J Physiol 593:3233
61. Spencer NJ, Hennig GW, Smith TK (2003) Stretch-activated neuronal pathways to longitudinal and circular muscle in Guinea pig distal colon. Am J Physiol Gastrointest Liver Physiol 284:G231–G241
62. Spencer NJ, Nicholas SJ, Robinson L, Kyloh M, Flack N, Brookes SJ, Zagorodnyuk VP, Keating DJ (2011) Mechanisms underlying distension-evoked peristalsis in Guinea pig distal colon: is there a role for enterochromaffin cells? Am J Physiol Gastrointest Liver Physiol 301:G519–G527
63. Spencer NJ, Sia TC, Brookes SJ, Costa M, Keating DJ (2015a) CrossTalk opposing view: 5-HT is not necessary for peristalsis. J Physiol 593:3229–3231
64. Spencer NJ, Sia TC, Brookes SJ, Costa M, Keating DJ (2015b) Rebuttal from Nick J. Spencer, Tiong Cheng Sia, Simon J Brookes, Marcello Costa and Damien J. Keating. J Physiol 593:3235
65. Takaki M, Branchek T, Tamir H, Gershon MD (1985a) Specific antagonism of enteric neural serotonin receptors by dipeptides of 5-hydroxytryptophan: evidence that serotonin is a mediatory of slow synaptic excitation in the myenteric plexus. J Neurosci 5:1769–1780
66. Takaki M, Mawe GM, Barasch JM, Gershon MD (1985b) Physiological responses of Guinea-pig myenteric neurons secondary to the release of endogenous serotonin by tryptamine. Neuroscience 16:223–240
67. Trendelenburg P (1917) Physiologische und pharmakologische Versuche über die Dünndarm Peristaltick. Naunyn-Schmiedeberg's Arch Exp Pathol Pharmakol 81:55–129
68. Twarog BM, Page IH (1953) Serotonin content of some mammalian tissues and urine and a method for its determination. Am J Phys 175:157–161
69. Vialli M (1966) Histology of the enterochromaffin cell system. In: Erspamer V (ed) Handbook of experimental pharmacology: 5-Hydroxytryptamine and related Indolealkylamines. Springer, New York
70. Wade PR, Chen J, Jaffe B, Kassem IS, Blakely RD, Gershon MD (1996) Localization and function of a 5-HT transporter in crypt epithelia of the gastrointestinal tract. J Neurosci 16:2352–2364
71. Wade PR, Tamir H, Kirchgessner AL, Gershon MD (1994) Analysis of the role of 5-HT in the enteric nervous system using anti-idiotypic antibodies to 5-HT receptors. Am J Phys 266:G403–G416
72. Walther DJ, Peter JU, Bashammakh S, Hortnagl H, Voits M, Fink H, Bader M (2003) Synthesis of serotonin by a second tryptophan hydroxylase isoform. Science 299:76
73. Whitaker-Azmitia PM (1999) The discovery of serotonin and its role in neuroscience. Neuropsychopharmacology 21:2S–8S

74. Woolley DW, Shaw EN (1957) Evidence for the participation of serotonin in mental processes. Ann N Y Acad Sci 66:649–665; discussion, 665-7
75. Wu H-H, Choi S, Levitt P (2016) Differential patterning of genes involved in serotonin metabolism and transport in extra-embryonic tissues of the mouse. Placenta 42:74–83
76. Yadav VK, Balaji S, Suresh PS, Liu XS, Lu X, Li Z, Guo XE, Mann JJ, Balapure AK, Gershon MD, Medhamurthy R, Vidal M, Karsenty G, Ducy P (2010) Pharmacological inhibition of gut-derived serotonin synthesis is a potential bone anabolic treatment for osteoporosis. Nat Med 16:308–312
77. Yano JM, Yu K, Donaldson GP, Shastri GG, Ann P, Ma L, Nagler CR, Ismagilov RF, Mazmanian SK, Hsiao EY (2015) Indigenous bacteria from the gut microbiota regulate host serotonin biosynthesis. Cell 161:264–276
78. Zagorodnyuk VP, Spencer NJ (2011) Localization of the sensory neurons and mechanoreceptors required for stretch-evoked colonic migrating motor complexes in mouse colon. Front Physiol 2:98

29 Upper Gastrointestinal Motility, Disease and Potential of Stem Cell Therapy

Jesse Gardner-Russell, Jakob Kuriakose, Marlene M. Hao, and Lincon A. Stamp

Abstract

Many gastrointestinal motility disorders arise due to defects in the enteric nervous system. Achalasia and gastroparesis are two extremely debilitating digestive diseases of the upper gastrointestinal tract caused in part by damage or loss of the nitrergic neurons in the esophagus and stomach. Most current pharmacological and surgical interventions provide no long-term relief from symptoms, and none address the cause. Stem cell therapy, to replace the missing neurons and restore normal gut motility, is an attractive alternative therapy. However, there are a number of hurdles that must be overcome to bring this exciting research from the bench to the bedside.

Keywords

Gastrointestinal motility disorders · Stem cell therapy · Achalasia · Gastroparesis

J. Gardner-Russell · J. Kuriakose · M. M. Hao
L. A. Stamp (✉)
Department of Anatomy & Physiology,
University of Melbourne, Parkville, VIC, Australia
e-mail: lstamp@unimelb.edu.au

29.1 Introduction

Congenital and acquired deficiencies of the enteric nervous system (ENS) drive gastrointestinal (GI) motility disorders [1]. Known as enteric neuropathies, the resultant diseases lead to debilitating chronic symptoms and compromised quality of life, and include upper GI motility diseases such gastroparesis and achalasia [1–3]. Achalasia, characterized by loss in lower esophageal sphincter relaxation (LES), and gastroparesis, described as a loss in pylorus motility, are primarily caused by disease- or age-related degeneracy of nitric oxide synthase (NOS)-expressing inhibitory motor neurons [2–4]. However, in many cases, the central nervous system (CNS) and ENS circuitry of the inhibitory reflexes remain intact, with only inhibitory motor neurons and interstitial cells of Cajal (ICC) being lost [5–9]. Cell replacement therapy has the potential to restore stomach motility by replacing deficient or depleted neurons with healthy ones [1].

Until now, cell therapies for enteric neuropathies have largely focused on the developmental disorder Hirschsprung disease, which is characterized by the lack of all enteric neurons in the distal bowel. Studies have demonstrated neurogenesis and functional integration of different stem cell-derived neurons and glia into the colon of various animal models of Hirschsprung disease, showing the promise of cell therapy for GI

N. J. Spencer et al. (eds.), *The Enteric Nervous System II*, Advances in Experimental Medicine and Biology 1383, https://doi.org/10.1007/978-3-031-05843-1_29

motility disorders [10–12]. Cell replacement therapy for achalasia and gastroparesis differs in two key ways: (1) it only requires the colonization of relatively small regions of the gut compared to even short segment Hirschsprung disease and (2) it requires the replacement of only a specific subtype of cell lost in the disease, the neuronal nitric oxide synthase (nNOS)-expressing neurons. Transplanted cells still need to survive, proliferate, and integrate into an environment which may be actively depleting nNOS neurons [13, 14]. Additionally, although strong genetic links to achalasia and gastroparesis are yet to be identified, deficiency in nNOS may necessitate manipulation of stem cells prior to transplantation, in the form of corrective gene therapy [3]. Despite these challenges, the recent advances of cell replacement therapies in rodent models, integration of optogenetic technologies, and development of methods to differentiate ENS-like cells from pluripotent stem cells make developing ENS cell therapy an attractive and feasible solution to treating enteric neuropathies.

29.2 Normal Motility in the Upper GI Tract: Esophagus and Stomach

29.2.1 Esophagus

The primary function of the esophagus is to guide consumed solids and liquids (bolus) from the pharynx into the stomach. Movement of the bolus, down the esophagus, is controlled by a process known as peristalsis – a sequential and coordinated contraction wave that travels down the esophageal body [15]. The occurring peristalsis can be primary or secondary, where the former is swallow-induced and the latter is elicited by esophageal distension [16].

Primary peristalsis begins with a pharyngeal contraction that pushes the bolus into the esophagus. Here, it quickly transits the proximal third of the esophagus that is composed of striated muscle. This segment is directly innervated by vagal nerves originating in the nucleus ambiguus of the medulla oblongata [17]. The striated muscle portion of the esophagus also contains a myenteric plexus with numerous ganglia that form an ENS of conventional appearance [18]. However, the ENS has little influence on the pattern of propulsive activity, with the CNS remaining responsible for controlling primary peristalsis. This is evident by an irreversible loss of primary peristaltic function when vagal innervation is severed [19]. Continuing in the aboral direction, the bolus then passes an intermediate transition zone where striated and smooth muscles merge; traveling toward the distal two-thirds of the esophagus composed of only smooth muscle. Although this segment is innervated by the sympathetic vagal nerves from the dorsal motor nucleus in the medulla oblongata, it also synapses with postganglionic motor neurons in the myenteric plexus [20].

Continuous propulsion of the bolus down the esophagus is facilitated by the activation of both the inner circular and outer longitudinal muscular layers in the muscularis propria (externa), which encircles the myenteric plexus. Here, the circular muscle layer contracts behind the bolus to propel it forward, while maintaining a relaxed state ahead of it to accommodate the traveling mass [21, 22]. The activity of both muscular layers is modulated by excitatory (contractile) and inhibitory (relaxatory) motor neurons within the myenteric plexus, which are innervated by separate sets of preganglionic vagal fibers. Here, excitatory motor neurons are characterized by the release of acetylcholine, whereas the major inhibitory pathway is provided by release of nitric oxide (NO), vasoactive intestinal peptide (VIP), and ATP-like transmitters [23]. Ultimately, contractions of the inner circular and outer longitudinal muscular layers travel the length of the esophagus, in waves, to move the bolus toward the lower esophageal sphincter (LES).

As a physical barrier against the retrograde flow of gastric contents, the muscles of the LES differ from the rest of the esophagus in that it is contracted during its resting state. Passage of the bolus is enabled by swallow-induced LES relaxation, triggered by stimuli from the oropharyngeal region. Under the control of nerves rising from the caudal portion of the dorsal motor nucleus of the vagus nerve (DMV) [24], the activation of local myenteric inhibitory motor neurons which release nitric oxide (NO) is promoted, leading to the relaxation of the LES and allowing passage of the bolus into the stomach [25].

29.2.2 Stomach

Normal motor activity in the stomach is crucial for its function – to store food and release it to the intestines in a form and at a rate that can be processed effectively. This is shared between two parts of the organ, where the proximal stomach is involved in the accommodation process while the distal stomach carries out the mixing and emptying processes [26, 27].

The process of accommodation is enabled by the relaxation of the gastric wall in the proximal stomach and is commonly associated with the fundus. Also known as receptive relaxation, it is triggered as the bolus travels down through the esophagus and into the stomach, controlled by the vagovagal reflex pathway [28]. Activation of this pathway promotes the release of several inhibitory neurotransmitters (such as NO and VIP) from the endings of inhibitory motor neurons. Neural inhibition of this region ultimately results in decreased fundic tone and a subsequent increase in the capacity of the region to accommodate additional content [29, 30]. Remarkably, this is achieved without an increase in intragastric pressure, minimizing the risk of gastroesophageal reflux [31, 32]. The distensible ability of the proximal stomach is facilitated by structural characteristics uniquely specialized to accommodate food, namely, its comparatively thin external smooth muscle layers, lack of a pacemaker network, and substantial inhibitory innervation.

In the distal stomach, which includes the corpus (body), antrum, and pylorus, the focus transitions to mixing and emptying the bolus into the duodenum of the small intestine [26, 27]. Here, the muscular layers are observed to increase in thickness (particularly the circular muscular layer) and be less distensible in the direction of the antrum and pylorus. This reflects the ability of the distal stomach to generate powerful contractions and contain the pressures generated [33]. Contractions within the stomach are locally dictated by a network of pace-making interstitial cells of Cajal (ICC) found within the myenteric plexus (ICC-MY) [34]. Presence of these cells are noted to begin from the mid-corpus, and like the muscular layers, its density proceeds to increase distally toward the antrum and pylorus [35]. ICC-MY generate cyclical electrical "slow waves" and transmit these waves to smooth muscle cells [36]. This is believed to be facilitated by the intramuscular ICC (ICC-IM), which links the ICC-MY to both excitatory and inhibitory motor neurons in the circular and longitudinal muscular layers [37, 38].

The bolus moves through the corpus and the antrum by way of peristaltic contractions at the pace of the slow wave. It undergoes gastric mixing while progressing toward the pyloric sphincter, whereby the bolus is broken down by its exposure to progressively intensifying phasic contractions and secreted acids/enzymes. During this time, the pyloric sphincter contracts to slow gastric emptying such that the retroceding mixture is subject to further processing and homogenization by antral contractions [39]. Here, continuous trituration and mixing of solids result in the production of chyme – a semifluid mass of partially digested food. During postprandial emptying, the pylorus relaxes at intermittent intervals to allow the exit of chyme (processed to a size <0.5–2 mm) in small amounts into the duodenum of the small intestine. Relaxation of the pyloric sphincter, as in other areas of the GI tract, is enabled by the NO-releasing inhibitory myenteric motor neurons [40].

During development in the mouse, nNOS neurons are one of the first enteric neuron subpopulations to emerge at embryonic day 11.5 and rapidly increase in numbers over the subsequent 2 days to E13.5 (Fig. 29.1a) [41, 42]. In the adult, NOS neurons comprise approximately 26% of total intestinal neurons, with ~90% of NOS neurons being inhibitory motor neurons and 10% interneurons (Fig. 29.1b) [42, 43]. Produced by neuronal nitric oxide synthase (nNOS), NO relaxes smooth muscle through actin cytoskeleton reorganization [44–47]. Loss of NOS neurons, or defi-

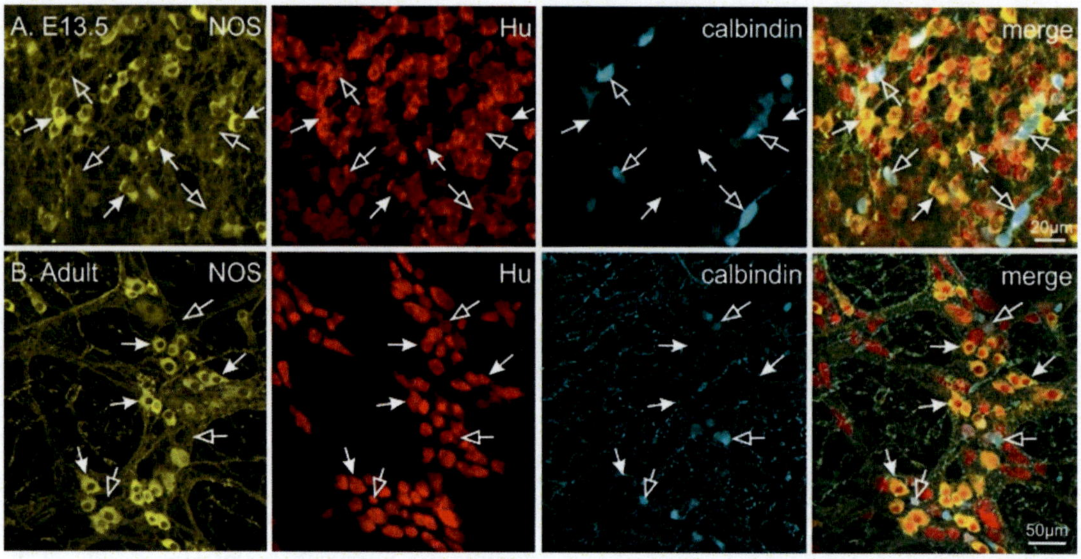

Fig. 29.1 NOS neurons in the enteric nervous system. Expression of nNOS (yellow; closed arrows) and calbindin (cyan; open arrows) in mutually exclusive populations of Hu+ neurons (red) in the pyloric region of the stomach at (**a**) embryonic day 13.5 and (**b**) the adult mouse

ciency in nNOS production of NO, can lead to a loss in swallow-induced LES relaxation, known as esophageal achalasia, and a loss in pylorus relaxation, known as gastroparesis [1, 2].

29.3 Upper GI Dysfunction at the Lower Esophageal and Pyloric Sphincters

29.3.1 Esophageal Achalasia

Esophageal achalasia is an idiopathic upper GI motility disorder characterized by the loss of LES relaxation and is estimated to affect 1–10 in every 100,000 people [2, 3, 8]. The LES acts as a one-way valve, allowing a bolus to pass into the stomach, without allowing reflux [21]. Patients suffering from achalasia experience reduced quality of life due to chronic symptoms including regurgitation, transfer dysphagia and vomiting, as well as a 10–50 times greater risk of developing esophageal cancer [48, 49]. Clinical studies in early achalasia observed maintenance in cholinergic populations and severe reduction in NOS neurons [50]. The loss of nitrergic neurons causes an imbalance between excitatory and inhibitory responses, increasing LES pressure [50]. Cytotoxic T-cell infiltration into the myenteric plexus of achalasia sufferers indicates a possible role of immune-mediated destruction on neurons [7]. This process beginning with NOS neuron and ICC degeneration progresses to cholinergic neurons, leading to complete loss of esophageal peristalsis [2, 3, 7, 48]. Infectious agents such Herpes Simplex Virus (HSV) and protozoal infection may initiate inflammation, with HSV DNA fragments being found in tissue of primary achalasia sufferers [51, 52]. This is further supported by recent detection of anti-HU neuronal antibodies circulating in the serum of achalasia sufferers [53]. Likewise, recent investigation into NOS neuron depletion from cytotoxic secretory products released by eosinophils in the esophagus is further illuminating the role of immune dysregulation in achalasia [54, 55].

29.3.2 Gastroparesis

Described as delayed or failed gastric emptying into the small intestine, gastroparesis occurs when neuro-musculature control of the pylorus is lost, leading to a loss of stomach peristalsis [2, 3]. Affecting 10–40 in 100,000 people, with a higher prevalence in women, gastroparesis presents as early satiety, gastric bloating, vomiting, and weight loss [3, 5]. Gastroparesis is classified into three groups: idiopathic, diabetic, and post-surgical [5, 56, 57]. There is a higher prevalence of gastroparesis in diabetic patients, with data suggesting that gastroparesis can develop in 20–55% of patients with type 1 diabetes and up to 30% of patients with type 2 diabetes [58]. Recent studies have investigated the role of distinct dysregulated immune profiles in idiopathic gastroparesis patients, where they observed histological abnormalities, such as increased macrophage immunoreactivity and CD45+ leukocyte infiltration of the mucosa [57, 59–61]. It is now believed that dysregulation of the immune system can drive idiopathic and diabetic gastroparesis via the loss of NOS neurons and ICC [56, 62]. Decreased nNOS expression has also been linked to diabetic gastroparesis, and in some cases idiopathic gastroparesis, and may be due to reduced dimerization of nNOS [6, 59, 60]. The loss of NOS can affect ICC levels, with nNOS-deficient mice showing a 50–70% decrease in ICC [59]. There is an emerging link between diabetic gastroparesis and PDGFα+ interstitial cells, with findings of increased glucose levels correlating to increased PDGFα+ expression of nNOS, potentially explaining why low glucose can cause a loss of nitrergic signal transduction [63].

29.4 Advances in Cell Replacement of nNOS Neurons

To investigate upper GI motility disorders, mouse models of enteric neuropathy have been generated. Using conventional gene targeting, an nNOS-knockout mouse was generated by replacing exon 1 which contains the promoter of *NOS1* with a neomycin antibiotic resistance gene [64]. The resultant mice exhibit features similar to that of humans with nNOS deficiencies, such as an absence of relaxation of the LES, delayed gastric emptying of both liquids and solids, dilation of the gastrointestinal tract, reduction in ICC, and hypertrophic pyloric stenosis [40, 59, 64]. However, this model does not always present in the same way as humans. For example, in nNOS-deficient mice, there is a global loss of nNOS, whereas in humans, NOS neurons are lost in a localized manner [40]. This means that nNOS-deficient mice exhibit non-classical pathophysiology of diseases, such as pyloric hypertrophy, which is driven by generalized gastric thickening instead of pyloric NOS neuron depletion in humans [40]. Likewise, these mice still exhibit inhibitory neuronal responses, mediated by other inhibitory neurotransmitters or peptides.

Cell therapy to restore nitrergic responses requires donor nNOS precursors to surviving, proliferate, and integrate into an environment which may be actively depleting nNOS neurons [1, 12]. In 2005, Micci et al. were the first laboratory to explore the efficacy of pyloric transplantation of CNS neural precursors on gastric emptying and pyloric function in the nNOS-deficient gastroparesis model [65]. They were able to show that 1 week after transplantation, the gastric emptying of the nNOS-deficient mice significantly improved and that grafted neurons expressed nNOS [65]. However, long-term survival and integration of grafts were not shown, with grafts only surviving 1 week, even in the presence of caspase inhibitors to increase survival [65, 66]. While this was an important proof of principal study, CNS precursors may not be a practical or ethical source for cell therapies [11]. Endogenous ENS precursors are both more easily accessible and efficient at colonizing the gut [67] and have been shown to be long-lasting following engraftment into the colon [68].

Numerous laboratories have subsequently shown survival, migration, and proliferation of donor ENS precursors in the GI tract [10, 68–71]. However, most of these studies had not investigated nNOS-deficient mice [25, 40]. More recently, McCann et al. transplanted endogenously derived ENS neurospheres, composed of multipotent ENS precursors, into the colon of nNOS-deficient mice [72]. Four weeks post-transplantation, the donor cells formed networks, colonizing on average 5.46 ± 0.5 mm from the site of transplant, forming ganglionic structures, and generating NOS neurons [72]. This indicated that donor progenitors retain their migratory ability in the adult environment, which is especially important for patients who may have adult-onset loss of neurons. Electrophysiology and electrical field stimulation (EFS) of basal contractile patterns showed that donor cells were electrically active and contributed to EFS-induced reflexes, translating to reduced total intestinal transit time [72].

Alterations in the ICC population are often observed in gastroparesis and achalasia patients. Interestingly, ENS precursor transplants restored ICC numbers to wild-type levels in nNOS-KO mouse colon, indicating that donor cells can modulate the neuro-musculature of the host environment [72]. Ultimately, these findings indicate that cell replacement therapy is capable of rescuing nitrergic responses, altering motility in the colon, and restoring ICC networks. This, combined with the work of Micci et al., could be clinically beneficial to target upper GI motility disorders such as gastroparesis and achalasia.

29.5 Analysis of Neural Function and Network Integration

One of the challenges in assessing the efficacy of cell therapy is the analysis of functional integration and contribution of donor cells to the ENS reflex circuits [11]. In ex vivo preparations of gut after donor cell transplantation, classical focal electrical stimulation will elicit responses in endogenous neurons, making the determination of effects of grafted cells difficult. Recent development of mice expressing channelrhodopsin-2 (ChR2) in enteric neural crest-derived cells allows the use of optogenetics for selective stimulation of donor neurons [71]. Isolated from the green algae *Chlamydomonas reinhardtii*, ChR2 is a light-gated, cation-selective ion channel which can be activated using 473 nm blue light to depolarize cells within tissues and activate neurons [73]. In our previous studies, we combined optogenetics, physiological recordings, and pharmacological perturbations which selectively stimulated our donor cells and demonstrated innervation of the gut by different functional classes of graft-derived enteric neurons, including interneurons and excitatory and inhibitory motor neurons [71]. Electrophysiological recordings from graft-derived cells have shown that they generate neurons which receive synaptic inputs and can fire action potentials [10]. Calcium imaging has also be used to show calcium transients in graft-derived neurons, indicative of their electrical activity [74]. In this study, the authors transduced the ENS precursors with the genetically encoded calcium indicator GCAMP3. Subsequently, they showed that their transplanted cells displayed spontaneous calcium transients, as well as transients evoked by focal electrical stimulation of the endogenous nerve fibers, indicating not only that the cells were electrically active but also that they had integrated into the existing neural network ([68]).

29.6 Protocols for Generation of ENS-Like Cells from iPSCs

Cell therapy for upper GI motility disorders requires accessible sources of ENS progenitors for transplantation [11, 12]. Stem cell sources need to be easily cultivated and safe for long-term transplant. Endogenous multipotent ENS precursors demonstrate long-term colonic integration and could be patient-derived [11, 12, 68]. However, difficulty in isolating the ENS progenitors, their limited capacity for self-renewal under existing in vitro culture conditions, and the potential difficulty in correcting disease-causing genetic defects are some of the caveats for their use in clinical application. Embryonic stem cells, while capable of unlimited expansion and capable of forming all cell lineages, are associated with some ethical hurdles which hinder their application in cell therapy. Induced pluripotent stem cells (iPSCs), generated from somatic cells by expression of *Oct4, Sox2, Klf4*, and *c-Myc*, possess in somatic, differentiated cells, possess the self-renewal and pluripotent capacities of embryonic stem cells, but avoid the use of human embryos and also have the added benefit of potentially be patient-derived [75].

Since 2005, there have been significant developments in determining how to program pluripotent stem cells into neural crest cells, which comprise much of the nervous system [76]. The majority of the ENS is derived from vagal neural crest cells, which migrate extensively at the wavefront of the developing gut [77–79]. However, early methods had challenges in reliably generating vagal neural crest cells, containing mainly HOX-negative cranial neural crest cells [70, 76, 80, 81]. Likewise, the adoption of feeder-free maintenance of iPSCs has also changed how the methods are approached [82]. Neural crest lineages through free-floating embryoid bodies and subsequent FACS sorting for P75 have been utilized to generate ENS-like neurons [70, 83]. These ENS-like cells generated from iPSCs were capable of colonizing aganglionic in vitro embryonic chick gut and differentiating into many of the key neural lineages with faithful recapitulation of expression patterns [70].

More recently, there has been the development of chemically defined protocols utilizing dual SMAD inhibition for neural induction, then subsequent vagal neural crest specification and ENS induction [82, 84, 85]. Small molecule inhibition of SMAD via TGF-β and BMP signaling pathways has been regularly used to generate neural lineages for the CNS [86]. These protocols have been more recently adopted and refined for neural crest and ENS induction [87]. Controlled Wnt, BMP4, and retinoic acid signaling direct neural crest induction, delamination of dorsal neural crest cells, and regional specification for vagal neural crest cells [82, 84, 85, 87]. Hence, the resulting cells express the key vagal neural crest markers HOXB3 and HOXB5 and key markers of enteric neural precursors such as SOX10, PHOX2B, EDNRB, and ASCL1 [82, 84, 85]. The ENS-like precursors are capable of giving rise to excitatory and inhibitory neurons, as well as enteric glial cells. Such observations indicate that these ENS-like precursors faithfully recapitulate neural commitment and subtype development. One of the key unanswered questions regarding cell therapy for upper GI motility disorders, which often involve the loss of a specific subtype of neuron, is whether a mixed population of ENS precursor cells is sufficient to restore the ENS network and normal gut function or whether a targeted approach to replace only the nNOS neurons, for example, is required.

29.7 Conclusion

There is an urgent, unmet need for effective long-term therapies that treat the cause, rather than just the symptoms of digestive diseases such as achalasia and gastroparesis. Using stem cells to replace the damaged ENS in the achalasia and gastroparesis patients has the potential to lead to a significant improvement of quality of life of the patients. These chronic disorders are extremely debilitating and have significant impact on the well-being of sufferers. Stem cell therapies have the potential to establish a new paradigm for the treatment of a range of gastrointestinal motility disorders.

References

1. Burns AJ, Thapar N (2014) Neural stem cell therapies for enteric nervous system disorders. Nat Rev Gastroenterol Hepatol 11(5):317–328
2. Di Nardo G, Blandizzi C, Volta U, Colucci R, Stanghellini V, Barbara G et al (2008) Review article: molecular, pathological and therapeutic features of human enteric neuropathies. Aliment Pharmacol Ther 28(1):25–42
3. De Giorgio R, Camilleri M (2004) Human enteric neuropathies: morphology and molecular pathology. Neurogastroenterol Motil 16(5):515–531
4. Walzer N (2008) Hirano I. Achalasia. Gastroenterol Clin N Am 37(4):807–825, viii
5. Hasler WL (2011) Gastroparesis: pathogenesis, diagnosis and management. Nat Rev Gastroenterol Hepatol 8(8):438–453
6. Rivera LR, Poole DP, Thacker M, Furness JB (2011) The involvement of nitric oxide synthase neurons in enteric neuropathies. Neurogastroenterol Motil 23(11):980–988
7. Boeckxstaens GE, Zaninotto G, Richter JE (2014) Achalasia. Lancet 383(9911):83–93
8. Pandolfino JE, Gawron AJ (2015) Achalasia: a systematic review. JAMA 313(18):1841–1852
9. Schlottmann F, Patti MG (2018) Esophageal achalasia: current diagnosis and treatment. Expert Rev Gastroenterol Hepatol 12(7):711–721

10. Hotta R, Stamp LA, Foong JP, Bergner AJ, McConnell SN, Anderson RB et al (2013) Transplanted progenitors generate functional enteric neurons in the postnatal colon. J Clin Invest 123(3):1182–1191
11. Burns AJ, Goldstein AM, Newgreen DF, Stamp L, Schafer KH, Metzger M et al (2016) White paper on guidelines concerning enteric nervous system stem cell therapy for enteric neuropathies. Dev Biol 417:229–251
12. Stamp LA (2017) Cell therapy for GI motility disorders: comparison of cell sources and proposed steps for treating Hirschsprung disease. Am J Physiol 312(4):G348–GG54
13. Delalande JM, Natarajan D, Vernay B, Finlay M, Ruhrberg C, Thapar N et al (2014) Vascularisation is not necessary for gut colonisation by enteric neural crest cells. Dev Biol 385(2):220–229
14. Stamp LA, Young HM (2017) Recent advances in regenerative medicine to treat enteric neuropathies: use of human cells. Neurogastroenterol Motil 29(1):e12993
15. Dubner R, Sessle BJ, Storey AT (1978) The neural basis of oral and facial function. Plenum Press, New York. xi, 483 p
16. Paterson WG, Hynna-Liepert TT, Selucky M (1991) Comparison of primary and secondary esophageal peristalsis in humans: effect of atropine. Am J Phys 260(1 Pt 1):G52–G57
17. Lang IM, Shaker R (1997) Anatomy and physiology of the upper esophageal sphincter. Am J Med 103(5A):50S–55S
18. Kallmunzer B, Sorensen B, Neuhuber WL, Worl J (2008) Enteric co-innervation of striated muscle fibres in human oesophagus. Neurogastroenterol Motil 20(6):597–610
19. Cannon WB (1907) Oesophageal peristalsis after bilateral vagotomy. Am J Physiol 19(3):436–444
20. Niedringhaus M, Jackson PG, Evans SR, Verbalis JG, Gillis RA, Sahibzada N (2008) Dorsal motor nucleus of the vagus: a site for evoking simultaneous changes in crural diaphragm activity, lower esophageal sphincter pressure, and fundus tone. Am J Physiol Regul Integr Comp Physiol 294(1):R121–R131
21. Vogt CD, Panoskaltsis-Mortari A (2020) Tissue engineering of the gastroesophageal junction. J Tissue Eng Regen Med 14(6):855–868
22. Brasseur JG, Nicosia MA, Pal A, Miller LS (2007) Function of longitudinal vs circular muscle fibers in esophageal peristalsis, deduced with mathematical modeling. World J Gastroenterol 13(9):1335–1346
23. Furness JB (2000) Types of neurons in the enteric nervous system. J Auton Nerv Syst 81(1–3):87–96
24. Rossiter CD, Norman WP, Jain M, Hornby PJ, Benjamin S, Gillis RA (1990) Control of lower esophageal sphincter pressure by two sites in dorsal motor nucleus of the vagus. Am J Phys 259(6 Pt 1):G899–G906
25. Sivarao DV, Mashimo HL, Thatte HS, Goyal RK (2001) Lower esophageal sphincter is achalasic in nNOS(−/−) and hypotensive in W/W(v) mutant mice. Gastroenterology 121(1):34–42
26. Collins PJ, Houghton LA, Read NW, Horowitz M, Chatterton BE, Heddle R et al (1991) Role of the proximal and distal stomach in mixed solid and liquid meal emptying. Gut 32(6):615–619
27. Kelly KA (1980) Gastric emptying of liquids and solids: roles of proximal and distal stomach. Am J Phys 239(2):G71–G76
28. Takahashi T, Owyang C (1997) Characterization of vagal pathways mediating gastric accommodation reflex in rats. J Physiol 504(Pt 2):479–488
29. Desai KM, Zembowicz A, Sessa WC, Vane JR (1991) Nitroxergic nerves mediate vagally induced relaxation in the isolated stomach of the guinea pig. Proc Natl Acad Sci U S A 88(24):11490–11494
30. Desai KM, Sessa WC, Vane JR (1991) Involvement of nitric oxide in the reflex relaxation of the stomach to accommodate food or fluid. Nature 351(6326):477–479
31. Tack J, Demedts I, Meulemans A, Schuurkes J, Janssens J (2002) Role of nitric oxide in the gastric accommodation reflex and in meal induced satiety in humans. Gut 51(2):219–224
32. Carbone F, Tack J, Hoffman I (2017) The intragastric pressure measurement: a novel method to assess gastric accommodation in functional dyspepsia children. J Pediatr Gastroenterol Nutr 64(6):918–924
33. Rau W, Hohaus C, Jessen E (2019) A differential approach to form and site of peptic ulcer. Sci Rep 9(1):8683
34. Cajal Ry (1909) Histologie du système nerveux de l'homme & des vertébrés
35. Yun HY, Sung R, Kim YC, Choi W, Kim HS, Kim H et al (2010) Regional distribution of interstitial cells of Cajal (ICC) in human stomach. Korean J Physiol Pharmacol 14(5):317–324
36. Liu LW, Thuneberg L, Huizinga JD (1998) Development of pacemaker activity and interstitial cells of Cajal in the neonatal mouse small intestine. Dev Dyn 213(3):271–282
37. Blair PJ, Bayguinov Y, Sanders KM, Ward SM (2012) Relationship between enteric neurons and interstitial cells in the primate gastrointestinal tract. Neurogastroenterol Motil 24(9):e437–e449
38. Klein S, Seidler B, Kettenberger A, Sibaev A, Rohn M, Feil R et al (2013) Interstitial cells of Cajal integrate excitatory and inhibitory neurotransmission with intestinal slow-wave activity. Nat Commun 4:1630
39. Meyer JH, Elashoff J, Porter-Fink V, Dressman J, Amidon GL (1988) Human postprandial gastric emptying of 1–3-millimeter spheres. Gastroenterology 94(6):1315–1325
40. Mashimo H, Kjellin A, Goyal RK (2000) Gastric stasis in neuronal nitric oxide synthase-deficient knockout mice. Gastroenterology 119(3):766–773
41. Hao MM, Moore RE, Roberts RR, Nguyen T, Furness JB, Anderson RB et al (2010) The role of neural activity in the migration and differentiation of enteric neuron precursors. Neurogastroenterol Motil 22(5):e127–e137
42. Bergner AJ, Stamp LA, Gonsalvez DG, Allison MB, Olson DP, Myers MG Jr et al (2014) Birthdating of

myenteric neuron subtypes in the small intestine of the mouse. J Comp Neurol 522(3):514–527
43. Qu ZD, Thacker M, Castelucci P, Bagyanszki M, Epstein ML, Furness JB (2008) Immunohistochemical analysis of neuron types in the mouse small intestine. Cell Tissue Res 334(2):147–161
44. Forstermann U, Sessa WC (2012) Nitric oxide synthases: regulation and function. Eur Heart J 33(7):829–837, 37a-37d
45. Rodriguez-Colman MJ, Schewe M, Meerlo M, Stigter E, Gerrits J, Pras-Raves M et al (2017) Interplay between metabolic identities in the intestinal crypt supports stem cell function. Nature 543(7645):424–427
46. Zhou L, Zhu DY (2009) Neuronal nitric oxide synthase: structure, subcellular localization, regulation, and clinical implications. Nitric Oxide 20(4):223–230
47. Sharma NM, Patel KP (2017) Post-translational regulation of neuronal nitric oxide synthase: implications for sympathoexcitatory states. Expert Opin Ther Targets 21(1):11–22
48. Boeckxstaens GE (2005) The lower oesophageal sphincter. Neurogastroenterol Motil 17(Suppl 1):13–21
49. Zaninotto G, Bennett C, Boeckxstaens G, Costantini M, Ferguson MK, Pandolfino JE et al (2018) The 2018 ISDE achalasia guidelines. Dis Esophagus 31(9). https://doi.org/10.1093/dote/doy071
50. Kraichely RE, Farrugia G (2006) Achalasia: physiology and etiopathogenesis. Dis Esophagus 19(4):213–223
51. Boeckxstaens GE (2008) Achalasia: virus-induced euthanasia of neurons? Am J Gastroenterol 103(7):1610–1612
52. Facco M, Brun P, Baesso I, Costantini M, Rizzetto C, Berto A et al (2008) T cells in the myenteric plexus of achalasia patients show a skewed TCR repertoire and react to HSV-1 antigens. Am J Gastroenterol 103(7):1598–1609
53. Bergeron KF, Nguyen CM, Cardinal T, Charrier B, Silversides DW, Pilon N (2016) Upregulation of Nr2f1-A830082K12Rik gene pair in murine neural crest cells results in a complex phenotype reminiscent of waardenburg syndrome type 4. Dis Model Mech 9(11):1283–1293
54. Spechler SJ (2019) Eosinophilic esophagitis: novel concepts regarding pathogenesis and clinical manifestations. J Gastroenterol 54(10):837–844
55. Rieder F, Nonevski I, Ma J, Ouyang Z, West G, Protheroe C et al (2014) T-helper 2 cytokines, transforming growth factor beta1, and eosinophil products induce fibrogenesis and alter muscle motility in patients with eosinophilic esophagitis. Gastroenterology 146(5):1266–1277, e1–9
56. Takahashi T, Nakamura K, Itoh H, Sima AA, Owyang C (1997) Impaired expression of nitric oxide synthase in the gastric myenteric plexus of spontaneously diabetic rats. Gastroenterology 113(5):1535–1544
57. Liu N, Abell T (2017) Gastroparesis updates on pathogenesis and management. Gut Liver 11(5):579–589
58. Vittal H, Farrugia G, Gomez G, Pasricha PJ (2007) Mechanisms of disease: the pathological basis of gastroparesis–a review of experimental and clinical studies. Nat Clin Pract 4(6):336–346
59. Oh JH, Pasricha PJ (2013) Recent advances in the pathophysiology and treatment of gastroparesis. J Neurogastroenterol Motil 19(1):18–24
60. Grover M, Farrugia G, Lurken MS, Bernard CE, Faussone-Pellegrini MS, Smyrk TC et al (2011) Cellular changes in diabetic and idiopathic gastroparesis. Gastroenterology 140(5):1575–85 e8
61. Gottfried-Blackmore A, Namkoong H, Adler E, Martin B, Gubatan J, Fernandez-Becker N et al (2021) Gastric mucosal immune profiling and dysregulation in idiopathic gastroparesis. Clin Transl Gastroenterol 12(5):e00349
62. Bashashati M, McCallum RW (2015) Motility: is 'ICC-opathy' present in gastroparesis-like syndrome? Nat Rev Gastroenterol Hepatol 12(7):375–376
63. Mussa BM, Khan AA, Srivastava A, Abdallah SH (2021) Differentiated PDGFRalpha-positive cells: a novel in-vitro model for functional studies of neuronal nitric oxide synthase. Int J Mol Sci 22(7):3514
64. Huang PL, Dawson TM, Bredt DS, Snyder SH, Fishman MC (1993) Targeted disruption of the neuronal nitric oxide synthase gene. Cell 75(7):1273–1286
65. Micci MA, Kahrig KM, Simmons RS, Sarna SK, Espejo-Navarro MR, Pasricha PJ (2005) Neural stem cell transplantation in the stomach rescues gastric function in neuronal nitric oxide synthase-deficient mice. Gastroenterology 129(6):1817–1824
66. Micci MA, Learish RD, Li H, Abraham BP, Pasricha PJ (2001) Neural stem cells express RET, produce nitric oxide, and survive transplantation in the gastrointestinal tract. Gastroenterology 121(4):757–766
67. Findlay Q, Yap KK, Bergner AJ, Young HM, Stamp LA (2014) Enteric neural progenitors are more efficient than brain-derived progenitors at generating neurons in the colon. Am J Physiol 307(7):G741–G748
68. Cooper JE, McCann CJ, Natarajan D, Choudhury S, Boesmans W, Delalande JM et al (2016) In vivo transplantation of enteric neural crest cells into mouse gut; engraftment, functional integration and long-term safety. PLoS One 11(1):e0147989
69. Tsai YH, Murakami N, Gariepy CE (2011) Postnatal intestinal engraftment of prospectively selected enteric neural crest stem cells in a rat model of Hirschsprung disease. Neurogastroenterol Motil 23(4):362–369
70. Li W, Huang L, Zeng J, Lin W, Li K, Sun J et al (2018) Characterization and transplantation of enteric neural crest cells from human induced pluripotent stem cells. Mol Psychiatry 23(3):499–508
71. Stamp LA, Gwynne RM, Foong JPP, Lomax AE, Hao MM, Kaplan DI et al (2017) Optogenetic demonstration of functional innervation of mouse colon by neurons derived from transplanted neural cells. Gastroenterology 152(6):1407–1418
72. McCann CJ, Cooper JE, Natarajan D, Jevans B, Burnett LE, Burns AJ et al (2017) Transplantation of enteric nervous system stem cells rescues nitric

oxide synthase deficient mouse colon. Nat Commun 8:15937

73. Nagel G, Szellas T, Huhn W, Kateriya S, Adeishvili N, Berthold P et al (2003) Channelrhodopsin-2, a directly light-gated cation-selective membrane channel. Proc Natl Acad Sci U S A 100(24):13940–13945
74. Cooper JE, Natarajan D, McCann CJ, Choudhury S, Godwin H, Burns AJ et al (2017) In vivo transplantation of fetal human gut-derived enteric neural crest cells. Neurogastroenterol Motil 29(1):e12900. https://doi.org/10.1111/nmo.12900
75. Takahashi K, Yamanaka S (2006) Induction of pluripotent stem cells from mouse embryonic and adult fibroblast cultures by defined factors. Cell 126(4):663–676
76. Pomp O, Brokhman I, Ben-Dor I, Reubinoff B, Goldstein RS (2005) Generation of peripheral sensory and sympathetic neurons and neural crest cells from human embryonic stem cells. Stem Cells 23(7):923–930
77. Obermayr F, Hotta R, Enomoto H, Young HM (2013) Development and developmental disorders of the enteric nervous system. Nat Rev Gastroenterol Hepatol 10:43–57
78. Young HM, Bergner AJ, Anderson RB, Enomoto H, Milbrandt J, Newgreen DF et al (2004) Dynamics of neural crest-derived cell migration in the embryonic mouse gut. Dev Biol 270(2):455–473
79. Uesaka T, Nagashimada M, Enomoto H (2015) Neuronal differentiation in Schwann cell lineage underlies postnatal neurogenesis in the enteric nervous system. J Neurosci 35(27):9879–9888
80. Lee G, Kim H, Elkabetz Y, Al Shamy G, Panagiotakos G, Barberi T et al (2007) Isolation and directed differentiation of neural crest stem cells derived from human embryonic stem cells. Nat Biotechnol 25(12):1468–1475
81. Hotta R, Pepdjonovic L, Anderson RB, Zhang D, Bergner AJ, Leung J et al (2009) Small-molecule induction of neural crest-like cells derived from human neural progenitors. Stem Cells 27(12):2896–2905
82. Barber K, Studer L, Fattahi F (2019) Derivation of enteric neuron lineages from human pluripotent stem cells. Nat Protoc 14(4):1261–1279
83. Workman MJ, Mahe MM, Trisno S, Poling HM, Watson CL, Sundaram N et al (2017) Engineered human pluripotent-stem-cell-derived intestinal tissues with a functional enteric nervous system. Nat Med 23(1):49–59
84. Gogolou A, Frith TJR, Tsakiridis A (2021) Generating enteric nervous system progenitors from human pluripotent stem cells. Curr Protoc 1(6):e137
85. Abu-Bonsrah KD, Viventi S, Newgreen DF, Dottori M (2019) Generation of neural crest progenitors from human pluripotent stem cells. Methods Mol Biol 1976:37–47
86. Chambers SM, Fasano CA, Papapetrou EP, Tomishima M, Sadelain M, Studer L (2009) Highly efficient neural conversion of human ES and iPS cells by dual inhibition of SMAD signaling. Nat Biotechnol 27(3):275–280
87. Fattahi F, Steinbeck JA, Kriks S, Tchieu J, Zimmer B, Kishinevsky S et al (2016) Deriving human ENS lineages for cell therapy and drug discovery in Hirschsprung disease. Nature 531:105–109

Epithelial 5-HT$_4$ Receptors as a Target for Treating Constipation and Intestinal Inflammation

30

Gary M. Mawe, Molly Hurd, Grant W. Hennig, and Brigitte Lavoie

Abstract

Because of their importance in the regulation of gut functions, several therapeutic targets involving serotonin-related proteins have been developed or repurposed to treat motility disorders, including serotonin transporter inhibitors, tryptophan hydroxylase blockers, 5-HT3 antagonists, and 5-HT4 agonists. This chapter focuses on our discovery of 5-HT4 receptors in the epithelial cells of the colon and our efforts to evaluate the effects of stimulating these receptors. 5-HT4 receptors appear to be expressed by all epithelial cells in the mouse colon, based on expression of a reporter gene driven by the 5-HT4 receptor promoter. Application of 5-HT4 agonists to the mucosal surface causes serotonin release from enterochromaffin cells, mucus secretion from goblet cells, and chloride secretion from enterocytes. Luminal administration of 5-HT4 agonists speeds up colonic motility and suppresses distention-induced nociceptive responses. Luminal administration of 5-HT4 agonists also decreases the development of, and improves recovery from, experimental colitis. Recent studies determined that the prokinetic actions of minimally absorbable 5-HT4 agonists are just as effective as absorbable compounds. Collectively, these findings indicate that targeting epithelial receptors with non-absorbable 5-HT4 agonists could offer a safe and effective strategy for treating constipation and colitis.

Keywords

Mucosal drug target · Serotonin · Motility · Colitis

Serotonin (5-hydroxytryptamine; 5-HT) is widely considered to be an important signaling molecule in the gastrointestinal (GI) tract. Because serotonin signaling influences various sensory, motor, and secretomotor functions in the gut, a variety of therapeutic strategies involving serotonin-related molecules have been tested and used clinically [9]. These include drugs that interact with the serotonin selective reuptake transporter (SERT), tryptophan hydroxylase 1,

G. M. Mawe (✉)
Department of Neurological Sciences, The University of Vermont, Burlington, VT, USA

Department of Pharmacology, The University of Vermont, Burlington, VT, USA
e-mail: gary.mawe@uvm.edu

M. Hurd · B. Lavoie
Department of Neurological Sciences, The University of Vermont, Burlington, VT, USA

G. W. Hennig
Department of Pharmacology, The University of Vermont, Burlington, VT, USA

N. J. Spencer et al. (eds.), *The Enteric Nervous System II*, Advances in Experimental Medicine and Biology 1383, https://doi.org/10.1007/978-3-031-05843-1_30

5-HT_1 receptors, 5-HT_3 receptors, and 5-HT_4 receptors. Here, we focus on 5-HT_4 receptors located in the intestinal epithelium as potential targets for the treatment of constipation and inflammation.

30.1 Expression of 5-HT_4 Receptors by Colonic Epithelial Cells and Effects of Their Stimulation on Motility and Nociception [4]

Luminal Administration of 5-HT_4 Agonists Promotes Propulsive Motility Our first indication that 5-HT_4 receptors might be expressed by epithelial cells came from propulsive motility studies using the guinea pig distal colon motility assay. At the time, 5-HT_4 agonists were thought to activate receptors located on the myenteric nerve fibers. When we added 5-HT_4 agonists to the bathing solution where the agonist would be able to access these fibers, we did not detect an increase and propulsive motility. Rather in various experiments motility either stayed the same or slowed down. On the other hand, when the agonist was infused into the lumen of the colon, propulsive motility increased by about 20%, and this effect was inhibited by 5-HT_4 antagonists. This result was consistent with findings of Grider and colleagues [5, 6], who had studied the effects of various luminally administered 5-HT_4 agonists using the same preparation and proposed that the agonists probably penetrated the epithelial barrier and activated receptors located on intrinsic afferent axons located in the lamina propria. In various experiments involving agonists and antagonists in combination, we could not find evidence of these compounds passing through the epithelium; therefore, we investigated whether 5-HT_4 receptors might be expressed by epithelial cells.

Identification of Epithelial Cells That Express the 5-HT_4 Receptor The most expeditious way of determining whether a given set of cells expresses a protein of interest is with immunohistochemistry. Unfortunately, we failed in our efforts to immunostain sections of the GI tract because the available anti-5-HT_4 receptor antibodies did not yield specific staining in our hands. We therefore resorted to evaluating intestinal samples from BAC transgenic mice in which the promotor for the 5-HT_4 receptor gene drives expression of green fluorescent protein. In tissue sections from these mice, we observed GFP in enteroendocrine (EE) cells of the duodenum and jejunum, and in the ileum, GFP was expressed by EE cells and in cells at the base of the crypts. By far, the most extensive expression was found in the colon, where it appeared that all epithelial cells were GFP positive. This pattern of expression in the mouse GI tract was consistent with the mRNA levels of 5-HT_4 receptor detected in epithelial scrapings from these regions, with a progressive increase in 5-HT_4 receptor RNA observed along the gut, highest levels in the colon, and nondetectable in the stomach. In human mucosal biopsy samples, 5-HT_4 receptor RNA levels were relatively high in the duodenum, proximal colon, and distal colon and highest in the terminal ileum. Recently, we have obtained a 5-HT_4 antiserum that provides specific immunestaining in the GI tract, and we found a similar pattern of immunoreactivity, with epithelial staining along the intestines, and the most intense immunoreactivity in the ileum [8].

The idea that most if not all cells in the mouse colonic epithelium appeared to express the 5-HT_4 receptor was somewhat surprising to us. It was clear that enterocytes were GFP positive, but we used immunohistochemistry to directly determine whether enterochromaffin (EC) cells and/or goblet cells also expressed the receptor. We found that both 5-HT immunoreactive and MUC2 immunoreactive cells were also GFP positive in tissue from the 5-HT_4-BAC mouse.

Functional studies focused on these three cell types demonstrated that secretory responses are activated in enterocytes, EC cells, and goblet cells by application of 5-HT_4 agonists to the luminal surface. An Ussing chamber was used to evaluate agonist-induced chloride currents, continuous amperometry was used to measure 5-HT release, and histological evaluation of cavitation in goblet cells was used to test for mucus secretion. In all cases, 5-HT_4 agonists evoked release events that were blocked by a selective 5-HT_4 antagonist, but not by tetrodotoxin, indicating that the agonists were acting directly on the epithelial cells as opposed to stimulating neurogenic secretion.

Luminal Administration of 5-HT_4 Agonists Dampens Nociceptive Responses Arising in the Colon Studies of animal models and data from human clinical trials provided evidence for an anti-nociceptive effect of 5-HT_4 agonists [1, 2]. Therefore, in collaboration with Dr. Beverley Greenwood-Van Meerveld at the University of Oklahoma, we tested whether intraluminal infusion of a 5-HT_4 agonist into the large intestine would dampen viscerosomatic reflex (VSR) responses in rats that were hyper-responsive to balloon distension. Infusion of the 5-HT_4 agonist into the cecum dampened the VSR response in a dose-dependent manner, and this effect was blocked by a 5- HT_4 antagonist.

Taken together, these studies demonstrated that 5-HT_4 receptors are expressed by a number of cell types in the colonic epithelium whose secretory activities could promote motility and alleviate constipation. Furthermore, luminal administration of the agonist dampened nociceptive responses originating in the colon.

We concluded at this point that targeting these receptors with nonabsorbable 5-HT_4 agonists might provide a safe and effective treatment for constipation. While newer 5-HT4 receptor agonists have had a good safety profile, the existence of 5-HT4 receptors in the brain and peripheral tissues such as the heart provides an incentive to limiting exposure to effective target tissues when possible.

30.2 Attenuation of Colitis by Luminally Administered 5-HT_4 Agonists [12]

Effects of Agonist Treatment on Experimental Colitis in Mice Disruption of the mucus barrier with mucolytic agents or deletion of the mucin 2 gene in mice results in colitis [7, 13]. The finding that 5-HT_4 receptors are expressed by goblet cells and that stimulation of these receptors causes goblet cell degranulation led us to hypothesize that 5-HT_4 agonists might be protective in colitis. We therefore tested the effects of enema administration of a 5-HT_4 agonist on the development and recovery from colitis using two established murine models of colitis: the dextran sodium sulfate (DSS) and trinitrobenzene sulfonic acid (TNBS) models. Outcome measures were disease activity index throughout the course of treatment and postmortem histological damage scores. In both models, the extent of colitis was decreased when agonist administration was initiated on the same day that colitis was induced. Furthermore, recovery from colitis was accelerated when agonist administration began after colitis was established. In both cases, the effects of the agonist were blocked by 5-HT_4 antagonist administration. The protective effects of the agonist were not observed in 5-HT_4 receptor knockout mice, or when the agonist was administered by intraperitoneal injection.

It is possible that one of the mechanisms responsible for the protective and healing actions of 5-HT_4 receptors stimulation involves epithelial healing. Enema administration of 5-HT_4 agonists increased epithelial proliferation in mice, and experiments involving Caco-2 cells demonstrated that 5-HT_4 agonists promote epithelial migration and cause a resistance to oxidative stress-induced apoptosis.

The results summarized above support the view that pharmacological stimulation of epithelial 5-HT_4 receptors could be helpful in accelerated remission in cases of colitis. To test whether 5-HT_4 receptors exert actions physiologically, we treated healthy mice with a 5-HT_4 receptor antagonist alone. In these mice, we detected a decrease in epithelial proliferation, an increase in bacterial translocation to lymph nodes, and elevated disease activity index and histological damage scores. Higher histological damage scores were also encountered in 5-HT_4 receptor knockout mice as compared to wild-type littermates. These findings indicate that in addition to having potential as a target for treating inflammatory bowel disorders, 5-HT_4 receptors likely play a role in maintaining the epithelial layer of the colon.

30.3 Prokinetic Effects of Luminally Acting 5-HT_4 Receptor Agonists [8]

The results described above strongly support the existence of epithelial 5-HT_4 receptors and strongly suggest that stimulation of these receptors could have beneficial actions in cases of constipation and colitis. The previously described studies were testing the actions of intra-luminally infused 5-HT_4 agonists using compounds that are absorbable and designed with the idea that they would be distributed via the circulation. Therefore, it is possible that the actions we were observing involved actions of agonist that was absorbed and acted on other 5-HT_4 receptors, such as those expressed by enteric neurons. To more rigorously test the hypothesis that epithelial 5-HT_4 receptors are a potential target to treat constipation and colitis, we entered a collaborative partnership with Dr. Jill Wykosky and her colleagues in the GI Drug Discovery Group at Takeda Pharmaceuticals. They generated compounds based on the 5-HT_4 agonists, prucalopride and naronapride, that were modified to prevent absorption. These compounds, which were found to be highly selective for the 5-HT_4 receptor, are called 5HT4-LA1 and 5HT4-LA2, respectively, for their luminally acting (LA) properties. When applied by gastric gavage in bioavailability studies, they were detected at concentrations 10,000-fold higher in the colon than in the serum, where the compounds were barely, and transiently, detectable.

To assess the effects of luminally acting 5-HT_4 agonists on intestinal motility, their actions were evaluated using a series of motility assays including whole GI transit, colonic motility, fecal output, and fecal water content. Both 5HT4-LA1 and 5HT4-LA2 significantly shortened whole GI transit time, and this effect was inhibited by 5-HT_4 antagonist treatment and was absent in 5-HT_4 receptor knockout mice. The decrease in transit time was more significant in response to 5HT4-LA1 ($P \leq 0.0001$) than for prucalopride ($P \leq 0.01$), the absorbable compound it is based on, supporting the view that targeting the 5-HT_4 receptor with this deliver strategy could improve effectiveness of the treatment. Both luminally acting agonist compounds also increased the speed of colonic motility, assessed by the glass bead expulsion assay, the number of fecal pellets expelled per hour, and the water content of the feces.

The prokinetic effects of the two luminally acting 5-HT_4 agonists were also evaluated in two conditions that have been demonstrated to be associated with slower motility. Chronic constipation is frequently associated with altered GI function in elderly humans, and a constipation phenotype has also been demonstrated in aged mice [10, 14]. In our hands, aged mice (18–24 months) exhibited decreased fecal output as compared to young mice at baseline, and 5HT4-LA1 reversed this age-induced constipation by causing an increase in fecal pellet output.

Multiple sclerosis is another condition that is associated with constipation [3], and we have previously demonstrated that mice with experimental autoimmune encephalomyelitis (EAE) exhibit slowed whole gut transit and colonic motility, as well as decreased fecal water content [11]. In our recent study we confirmed the constipation phenotype in EAE mice, and we demonstrated that 5HT4-LA1 completely restored whole gut transit, colonic motility, and fecal water content to baseline levels in these mice.

30.4 Concluding Remarks

While the net effects of the activation of epithelial 5-HT$_4$ receptors, namely enhanced motility, decreased nociception, and improved healing from inflammation, are becoming clear, the precise mechanisms that are responsible for these responses have not yet been resolved. In the case of the prokinetic actions, water secretion, 5-HT release, and/or mucus secretion could all potentially contribute to enhanced motility. Regarding the anti-nociceptive action that was clearly observed in response to intraluminal infusion of 5-HT$_4$ agonists, it is simply not clear at this point how activation of a receptor that activates the cAMP-PKA pathway could dampen nociceptive signals from the gut. As for the ability of 5-HT$_4$ receptor stimulation dampen the extent of, and improve recovery from, inflammation, several contributing mechanisms could be involved. As described above, recovery of epithelial integrity, involving epithelial proliferation, enhanced epithelial wound healing, and resistance to oxidative stress is likely to be a contributing factor. This is important because recovery of mucosal integrity is a key feature in reaching and maintaining remission from inflammation. Other factors to be explored in the future include barrier function, mucus secretion, and direct interactions with immune cells in the lamina propria.

In conclusion, the results of the studies outlined here provide clear evidence for a role of 5-HT$_4$ receptors in the intestinal epithelium, and these receptors deserve further investigation as targets for treating constipation and intestinal inflammation.

Funding Support Funding for the studies described here that were performed in the Mawe Laboratory was obtained from the NIH (DK113800 and DK62267) and from Takeda Pharmaceuticals.

References

1. Chey WD, Pare P, Viegas A, Ligozio G, Shetzline MA (2008) Tegaserod for female patients suffering from IBS with mixed bowel habits or constipation: a randomized controlled trial. Am J Gastroenterol 103:1217–1225
2. Greenwood-Van Meerveld B, Venkova K, Hicks G, Dennis E, Crowell MD (2006) Activation of peripheral 5-HT receptors attenuates colonic sensitivity to intraluminal distension. Neurogastroenterol Motil 18:76–86
3. Hennessey A, Robertson NP, Swingler R, Compston DA (1999) Urinary, faecal and sexual dysfunction in patients with multiple sclerosis. J Neurol 246:1027–1032
4. Hoffman JM, Tyler K, MacEachern SJ, Balemba OB, Johnson AC, Brooks EM, Zhao H, Swain GM, Moses PL, Galligan JJ (2012) Activation of colonic mucosal 5-HT4 receptors accelerates propulsive motility and inhibits visceral hypersensitivity. Gastroenterology 142:844–854. e844
5. Jin J-G, Foxx-Orenstein AE, Grider J (1999) Propulsion in guinea pig colon induced by 5-hydroxytryptamine (HT) via 5-HT4 and 5-HT3 receptors. J Pharmacol Exp Ther 288:93–97
6. Jin JG, Foxx-Orenstein AE, Grider JR (1997) Stimulation of colonic propulsion by 5-HT4 receptor agonists: synergism by delta opioid receptor antagonists. Gastroenterology 112:A754
7. Johansson ME, Gustafsson JK, Holmen-Larsson J, Jabbar KS, Xia L, Xu H, Ghishan FK, Carvalho FA, Gewirtz AT, Sjovall H, Hansson GC (2014) Bacteria penetrate the normally impenetrable inner colon mucus layer in both murine colitis models and patients with ulcerative colitis. Gut 63:281–291
8. Konen JR, Haag MM, Guseva D, Hurd M, Linton AA, Lavoie B, Kerrigan CB, Joyce E, Bischoff SC, Swann S, Griffin L, Matsukawa J, Falk MD, Gibson TS, Hennig GW, Wykosky J, Mawe GM (2021) Prokinetic actions of luminally acting 5-HT4 receptor agonists. Neurogastroenterol Motil 33:e14026
9. Mawe GM, Hoffman JM (2013) Serotonin signalling in the gut—functions, dysfunctions and therapeutic targets. Nat Rev Gastroenterol Hepatol 10:473
10. Patel BA, Patel N, Fidalgo S, Wang C, Ranson RN, Saffrey MJ, Yeoman MS (2014) Impaired colonic motility and reduction in tachykinin signalling in the aged mouse. Exp Gerontol 53:24–30
11. Spear E, Holt E, Joyce E, Haag M, Mawe S, Hennig G, Lavoie B, Applebee A, Teuscher C, Mawe G (2018) Altered gastrointestinal motility involving autoantibodies in the experimental autoimmune encephalomyelitis model of multiple sclerosis. Neurogastroenterol Motil 30:e13349
12. Spohn SN, Bianco F, Scott RB, Keenan CM, Linton AA, O'Neill CH, Bonora E, Dicay M, Lavoie B, Wilcox RL, MacNaughton WK, De Giorgio R, Sharkey KA, Mawe GM (2016) Protective actions of epithelial 5-hydroxytryptamine 4 receptors in normal and inflamed colon. Gastroenterology 151:933–944 e933
13. Van der Sluis M, De Koning BA, De Bruijn AC, Velcich A, Meijerink JP, Van Goudoever JB, Buller HA, Dekker J, Van Seuningen I, Renes IB, Einerhand

AW (2006) Muc2-deficient mice spontaneously develop colitis, indicating that MUC2 is critical for colonic protection. Gastroenterology 131:117–129
14. West CL, Amin JY, Farhin S, Stanisz AM, Mao YK, Kunze WA (2019) Colonic motility and Jejunal vagal afferent firing rates are decreased in aged adult male mice and can be restored by an aminosterol. Front Neurosci 13:955

Index

N. J. Spencer et al. (eds.), *The Enteric Nervous System II*, Advances in Experimental Medicine and Biology 1383, https://doi.org/10.1007/978-3-031-05843-1